The Simian Tongue

THE SIMIAN TONGUE
The Long Debate about Animal Language

Gregory Radick

The University of Chicago Press   Chicago and London

GREGORY RADICK is senior lecturer in history and philosophy of science
at the University of Leeds. He is coeditor (with Jonathan Hodge) of *The
Cambridge Companion to Darwin*.

The University of Chicago Press, Chicago 60637
The University of Chicago Press, Ltd., London
© 2007 by The University of Chicago
All rights reserved. Published 2007
Printed in the United States of America

16 15 14 13 12 11 10 09 08 07   1 2 3 4 5

ISBN-13: 978-0-226-70224-7 (cloth)
ISBN-10: 0-226-70224-3 (cloth)

Library of Congress Cataloging-in-Publication Data
Radick, Gregory.
    The simian tongue : the long debate about animal language / Gregory
Radick.
        p. cm.
    Includes bibliographical references and index.
    ISBN-13: 978-0-226-70224-7 (cloth : alk. paper)
    ISBN-10: 0-226-70224-3 (cloth : alk. paper) 1. Primates—Psychology.
2. Animal communication. 3. Language and languages—Origin.
4. Human evolution. I. Title.
QL737 .P9R25 2007
156'.36—dc22

                                                              2007026793

For Lindsay

Contents

List of Illustrations / ix

Preface / xi

Introduction / 1

PART ONE

15   CHAPTER ONE
      The Language Barrier

50   CHAPTER TWO
      Brains and Minds across the Barrier

84   CHAPTER THREE
      Professor Garner's Phonograph

PART TWO

123   CHAPTER FOUR
      Congo Fever

159   CHAPTER FIVE
      The Anthropologists and Animal Language

199   CHAPTER SIX
      The Psychologists and Animal Language

PART THREE

243   CHAPTER SEVEN
       Mr. Marler's Spectrograph

280   CHAPTER EIGHT
       Simian Semantics

320   CHAPTER NINE
       Playbacks in Amboseli

       Conclusion / 365

       Notes / 381

       Bibliography / 497

       Index / 555

Illustrations

1.1   Friedrich Max Müller / 17
1.2   Charles Darwin / 30
1.3   William Dwight Whitney / 42
2.1   Frederic Bateman / 55
2.2   The left hemisphere of the human brain / 59
2.3   Conwy Lloyd Morgan / 74
3.1   Richard Lynch Garner / 87
3.2   The Edison cylinder phonograph / 92
3.3   Garner with his phonograph in Central Park / 98
3.4   Garner's designs for a double-spindle phonograph / 110
4.1   Professor Johausen's empty cage / 125
4.2   The Fernan Vaz region of West Africa / 131
4.3   Henry Du Pré Labouchere / 138
4.4   Garner and his servant boy at Fort Gorilla / 145
4.5   Joachim Buléon / 150
5.1   Charles Myers on Mer with his cylinder phonograph / 164
5.2   A gathering in Garner's West Africa / 168
5.3   John P. Harrington with three Cuna Indian informants / 171
5.4   Horatio Hale / 177
5.5   The La Naulette lower jaw and the lower jaw of a chimpanzee / 178
5.6   Aleš Hrdlička / 183
5.7   Franz Boas / 192
6.1   Edward Lee Thorndike / 205
6.2   One of Thorndike's original puzzle boxes / 208
6.3   A chimpanzee solving a stacking problem / 223
6.4   Three learning or error curves / 224
6.5   Robert Yerkes's chimpanzee pupils / 227

6.6   C. Ray Carpenter in the jungles of Siam / 237

7.1   Peter Marler / 244

7.2   The Sona-Graph / 266

7.3   Spectrograms of the song and alarm call of the male chaffinch / 268

8.1   Sherwood Washburn / 297

8.2   Peter Marler with a chimpanzee at Gombe Stream / 304

8.3   Thomas Struhsaker with vervets in Amboseli / 306

8.4   David Premack's chimpanzee pupil Sarah / 317

9.1   Robert Seyfarth and Dorothy Cheney / 332

9.2   Peter Marler surrounded by the chambers used for rearing songbirds in acoustic isolation / 337

9.3   Cheney, Marler, and Seyfarth together in Amboseli / 344

9.4   Dorothy Cheney recording in Amboseli / 348

9.5   Stuffed python in front of the Seyfarths' *banda* / 352

Preface

How did language begin? Where should we look for clues? These questions—about what is true concerning the origins of language and how best to find it out—have puzzled thoughtful people time out of mind. In the middle of the nineteenth century, the rise of the evolutionary or "ape" theory of human origins gave the puzzles a more definite form and made the search for their solution more urgent. Was it the case, as some evolutionists suggested, that the human capacity for language evolved gradually from more rudimentary capacities in prehuman ancestors? If so, could evidence of this evolutionary beginning be recovered from the communication systems of living nonhuman animals, in particular monkeys and apes? At stake, in Darwin's day, was nothing less than the fate of the evolutionary theory itself. Even today, the issues raised seem to bear in a fundamental way on human self-understanding: who we are as a species, what made us the species we are, and where our species stands in relation to the rest of nature. This book is about attempts since Darwin to come to grips with animal communication as bearing (or not) on the evolutionary origins of language.

A single, interlinked set of events forms the book's narrative spine, tracing the emergence, disappearance, and re-emergence, over a span of more than a century, of an experimental study of the meanings of monkey and ape vocalizations. I have called this study the "primate playback experiment." Its inventor, Richard Garner, called what he investigated thereby "the simian tongue." My central ambition in what follows is to explain how the primate playback experiment came to be performed to wide acclaim in the early 1890s, why it nevertheless vanished from the scientific repertoire for decades, and what brought it back again in the late 1970s. It is with this aim in view, and the related one of balancing comprehensiveness and readability, that I have decided what to include in this history and how to include it. Accordingly, this is not a book that sits comfortably within a single genre of history of science. History of ideas, social history, history of books and periodicals, disciplinary history, history of material cultures and practices, history of popular science, philosophical history of science: all are here represented and, I hope, integrated.

Even a sound principle of inclusion can make for omissions that some readers might find surprising. I am mindful, for instance, that, aside from this preface, the book makes no mention of the Paris Linguistic Society's banning of language-origin papers at its founding in the mid-1860s—a fact ritually cited nowadays to illustrate the scientific disrepute into which the language-origin debate has long been sunk. Still, the book offers the fullest reconstruction of the post-Darwinian debate now available. It is also the first book to show the importance of this debate for understanding the shape and trajectory of a number of the modern biological and human sciences, including anthropology, psychology, and ethology.

At its heart this is a book about possibility, and the conditions under which certain kinds of achievement become, or cease to be, possible. So it is all the more appropriate here to thank some of the very many people and institutions who have made it possible for me to conceive and then complete this book. The project began twelve years ago at the Department of History and Philosophy of Science at the University of Cambridge, where a master's essay on the history of the animal language debate grew into a master's dissertation and eventually a doctoral dissertation. Although I owe much to a large number of people at Cambridge, I wish to thank especially Simon Schaffer, my doctoral supervisor, and Michael Bravo, Jim Secord, Peter Lipton, Marina Frasca-Spada, and Nick Jardine—superb teachers all. After Cambridge I moved to the Division of History and Philosophy of Science in the Department (then School) of Philosophy at the University of Leeds, where, taking my cue from the perceptive suggestions of my examiners, Jim Secord and Bob Richards, I began revising and expanding the dissertation. My Leeds colleagues and students have been exceptionally generous in their support of the project, and I have acquired large debts to many of them, but above all to Jonathan Hodge, a source of boundless knowledge and encouragement, and Graeme Gooday, who in the gentlest way pushed me to think harder about what technologies do and do not determine.

Outside the Cambridge HPS and Leeds HPS units, there are teams of experts that I have turned to for guidance in particular areas: Bob Richards and Jim Moore for the history of evolutionary biology; Chip Burkhardt, Cheryl Logan, Marion Thomas, Georgina Montgomery, Tania Munz, Paul Griffiths, Charlotte Sleigh, Amanda Rees, Jonathan Burt, and Donald Dewsbury for the history of the sciences of animal behavior; Gerry Fabris, Lisa Gitelman,

Peter Martland, Doug Tarr, George Tselos, and Paul Israel for the history of the phonograph; Samuel Greenblatt, Heini Hakosalo, Anthony Batty Shaw, Tom Anderson, and Stephen Jacyna for the history of neurolinguistics; Regna Darnell for the history of anthropology; and Sara Scharf and Kyle Stanford for the relationship between factual and counterfactual histories of science. Anna Mayer and Pascale Aebischer helped me with translations. Thomas Dixon commented improvingly on the whole of an early draft. Over the last few years, Annie Jamieson at Leeds has proved an indispensable ally, tracking down articles, sorting out inconsistent citation styles, and putting my illustrations into publishable form. Alex Santos at Leeds created the perfect map for chapter 4, and Lisa Hobson a superb index. My family have chipped in too, passing on references and a whole lot more. To all of these, I offer my thanks. Many other friends and colleagues have contributed individual references, and I thank them individually in the notes.

The latter part of the book would not have been possible in anything like its present form without the cooperation of several of the scientists whose lives and work I discuss: Dorothy Cheney, Irven DeVore, Marc Hauser, Robert Hinde, Phyllis Lee, Peter Marler, Robert Seyfarth, and Thomas Struhsaker. All were unstinting in giving me their time—several days' worth in the cases of Peter Marler, Dorothy Cheney, and Robert Seyfarth—as well as their papers, photographs, old correspondence and clippings, and expert comments on drafts. I am grateful for and somewhat amazed by the hospitality, intellectual and personal, that these busy men and women showed a junior scholar and disciplinary stranger. They have been a historian's dream to work with, and I hope the book proves worth their investment in it.

Crucial investment of another sort came in the form of a studentship from the Cambridge HPS Department; a research fellowship from Darwin College, Cambridge; travel grants from the Smithsonian, the British Academy, the Royal Society, and the Leeds Philosophy Department and Faculty of Arts; and funding for a year's study leave from the Leeds Philosophy Department and the Leverhulme Trust. I am most grateful to these institutions for their support, as for that of friends and family who put me up (and put up with me) on research trips to the States: Scott Powell, Rich Trott, Ken and Lisa Gordon, Jim and Wendy Zorzi, Robyn and Thierry Lints, Stuart Radick and Ginger Cooper, Esther and Lee Erman, and Caryn Radick. I am likewise grateful to the many librarians and archivists, in several countries, on whose skill a project like this one depends. Steve Johnson in the Bronx, Gerard Vieira in Paris, Jake Homiak in Washington, DC, and the Darwin Correspondence staff in Cambridge deserve special thanks, as do the indefatigable members of the Document Supply team in Leeds. The process of turning the manuscript into a book has been eased by the good people of the University of Chicago Press, in particular Catherine Rice, Christie Henry, Tisse Takagi, Pete Beatty, Kate Frentzel, Stephanie Hlywak, and Joann Hoy. To them, and to the press's referees, I offer thanks too.

Parts of this book formed the basis for talks in Bradford, Cambridge, Chicago, Durham, Hamden (CT), Leeds, London, Oaxaca, Oxford, Paris, Plzeň, Southampton, Vancouver, and Vienna. I wish to thank the participants on those occasions for much stimulating discussion. The book incorporates portions of some previously published essays, which originally appeared in the *British Journal for the History of Science*, the *Journal of the History of Biology*, *Selection*, *Studies in History and Philosophy of Biological and Biomedical Sciences*, the *Dictionary of Nineteenth-Century British Scientists*, and the collection *New Media, 1740–1915*, edited by L. Gitelman and G. B. Pingree (MIT).

As I write these words, I hear my son Ben in another part of the house, playing with his little brother, Matthew. For them, for now, this book is the thing that lures me from playing trains. The book I write for the boys is in the future. This one is for their mother Lindsay, without whom it could never have been started, much less, at long last, finished.

Introduction

"Studies in Africa Find Monkeys Using Rudimentary 'Language'": that was how the *New York Times* broke the news. The report on its front page told of novel experiments with monkeys at a national park in Kenya. For some while, scientists had known that these monkeys give one sort of alarm call on seeing leopards, a second sort on seeing eagles, and a third sort on seeing pythons. The monkeys were also known to respond to these calls appropriately: at the leopard alarm, running up into the trees; at the eagle alarm, dashing into the bushes; at the python alarm, scrutinizing the ground. What Rockefeller University ethologists Dorothy Cheney, Robert Seyfarth, and Peter Marler had hoped to discover was what the calls meant to the monkeys. Were the calls, as they seemed, informative about the nature of the threat, telling monkeys roughly what the human words "leopard!" "eagle!" and "python!" told humans? Or were the calls merely expressing different levels of arousal, leaving to listeners the task of figuring out, from experience or from looking around, what was up (or down)? The experimental results amounted, in the *Times*'s words, to "strong evidence that the monkeys respond not just to the urgency of the calls but to the semantic content as well, specifying different categories of animals or types of danger." The calls thus seemed to be "elements of a rudimentary language"—a finding the researchers expected to be controversial, because, the famous dance language of the bees excepted, the "ability to convey messages that carry such specific information content has always been considered a distinguishing feature of human speech."[1]

That was in November 1980. The well-informed monkeys were vervet monkeys, and the experimental technique used to study their communication was known as "playback": the playing of recorded animal vocalizations back to the animals. In its general form, the playback experiment was, at that time, commonplace. In its specific form, as used to determine whether monkeys have a rudimentary form of language, it was, before the Rockefeller team's work, nonexistent. When they started playing recorded vervet alarms to vervets with no predators around, to see whether the calls themselves evoked the escape behaviors, Cheney, Seyfarth, and Marler were aware of trying something new and untested. It is a measure of their success that since then there have been countless playback experiments with nonhuman primates, in laboratories and in the field. The primate playback experiment has stabilized into an off-the-shelf scientific resource, susceptible of endless variation. The facts about vervet predator calls, meanwhile, have stabilized into factoids. People who know almost nothing else about monkeys often have heard about the ones with "words" for leopard, eagle, and python. That can now be learned from the inside of soft-drink bottle caps, or from television, where footage of Cheney and Seyfarth in the field has become a staple of natural history programming. For these junior members of the team, the predator-call playbacks served as a point of entry into a decade-long inquiry into vervet social lives and social intelligence. Thanks in part to the advocacy of the philosopher Daniel Dennett, the vervet research became widely known as a shining example both of a new "cognitive ethology" and of the traditional virtues of assumptions of design in nature.[2] Cheney and Seyfarth's farewell to the vervets, *How Monkeys See the World* (1990), has the status of a classic. Over the last quarter century, serious thinkers on the evolutionary origins of language—a topic with a notoriously thin empirical base—have needed to confront the vervet alarm-call experiments and their interpretation.[3]

Yet there *was* a precedent for the vervet alarm-call research, close in conception but distant in time. Playbacks in the service of understanding simian vocalizations were inaugurated in September 1890 in Washington, DC, in the zoological garden then under construction behind the Smithsonian building. Near the monkey cage, reported a St. Louis newspaper, a "group of eminent doctors and professors," about a dozen, including the Smithsonian's secretary, had gathered to observe a test of whether monkeys talk. In charge was Richard L. Garner, "Prof. Garner," one of the Smithsonian's "honorary curators." Garner had with him a graphophone—a version of the phonograph—and had inserted its large tin horn through the cage door. As the learned company watched attentively, "Prof. Garner ground away at the hand-graphophone with its crank attachment," while "the keeper of the animals poked the monkeys up with a stick to make them talk." The cage held two monkeys, one wild and one tame. The wild monkey kept quiet, apart from the occasional enraged scream. The tame one "did nothing but chatter and gibber most unintelligibly, as it seemed to the rest of the audience; but Prof. Garner was inclined to think this was really conversation worth taking down," and proceeded to fill six wax cylinders with recorded monkey utterances. These records were the means to a novel scientific end:

> Prof. Garner was very far from imagining that he would be able to understand this monkey-talk when repeated to him by the machine. But his notion was to record the remarks of one monkey and grind them out through the horn for the benefit of the other monkey, so as to observe what sort of responses the second one would make. By comparing the original observations and the replies, he hoped to get some few clews that would eventually enable him to translate the monkey language.[4]

The history of science is replete with research unnoticed until some later, celebrated achievement made the prior work stand out from obscurity, as an "anticipation" or a "premature discovery."[5] Garner's work was not of this kind. His phonographic experiments with stateside simians between 1890 and 1892 launched a research program that made him one of the best-known scientific men of his day. The popular press could not get enough of "the monkey man," especially, from 1891, his plan to take a phonograph to the African jungle, install the machine and himself in a metal cage, and set to work on the utterances of wild gorillas and chimpanzees.[6] Years after his death in 1920, journalists could count on readers knowing who Garner was and what he was famous for.[7] Among philologists, psychologists, and anthropologists, he was acknowledged as a pioneering student of primate communication and the inventor of a promising new method for its study. The philologist E. P. Evans, for instance, reckoned the phonograph a "scientific weapon of phonetic precision," which, in Garner's hands, had eliminated the need to rely on paltry human powers to imitate and describe simian sounds. Evans foresaw the phonograph doing for philology what the microscope had done for medicine. Reviewing Garner's 1892 book *The Speech of Monkeys*, Joseph Jastrow, founder of the new psychological laboratory at Madison, congratulated Garner on "the happy idea of studying the chatterings of monkeys by recording them in a phonograph, reproducing [these sounds] before other monkeys and recording the effect produced upon them by the sounds." No less than William James and Wilhelm Wundt, the dominant psychologists of the era, took notice. James annotated his copy of his *Principles of Psychology* (1890) with a reference to Garner's first article on the "simian tongue," whose words differed only in degree from human words, thus representing, as Garner had put it, "the rudiments from which the tongues of mankind could easily develop." Some years later, when an American student in Germany, John P. Harrington, asked Wundt if apes have language, Wundt directed the young man to Garner's work. Harrington later used the phonograph extensively in legendary field studies of American Indian languages—studies, as he saw it, inspired by Garner's example.[8]

Twice, then, the playback technique became the pivot of an admired research program in the study of primate communication. What led Garner to begin his experiments? How could research so celebrated at the time have been forgotten? What, nearly ninety years later, led the members of the Rockefeller team, who were not at all conscious of following in the footsteps of Garner—he was unknown to Seyfarth and Cheney, and little more than a dim and superseded historical figure to Marler—to reinvent what Garner had invented? These are the questions that this book seeks to answer. In pursuing its goals, it reconstructs a history much wider than that of the primate playback experiment narrowly construed. It tells, for the first time, the story of the post-Darwinian debate on the animal origins of language, and how that debate shaped some of the key behavioral sciences of the twentieth century.[9]

◆　◆　◆

A high-altitude pass over the book's territory will be helpful at the outset. The three chapters making up part 1 chart the nineteenth-century developments that put animal language on public and scientific agenda, culminating in Garner's phonographic efforts. The dominant figure in this period was not Darwin but F. Max Müller, an Oxford-based Sanskrit scholar who, within two years of Darwin's publishing *On the Origin of Species* (1859), took to the Royal Institution platform in London to argue that no evolutionary theory could account for human language. According to Müller, human words are composed of irreducible roots, all of which express concepts. Because, he went on, concepts are constitutive of reason, and there is no evidence for animal reason, nor any possibility that concepts might have arisen gradually out of the sense impressions filling animal minds, the concept-expressing roots must have come into being in full conceptual flower among the first humans. For Müller, then, language and reason formed an impassable barrier between humans and all other animals, and so between the human mind and evolutionary explanation. The first chapter, "The Language Barrier," examines Müller's influential argument together with the responses it provoked among the first generation of Victorian commentators, principally Darwin and the combative American philologist W. D. Whitney, but also the many other writers who argued over the question of animal language in the general periodicals and elsewhere.

Disagreements over animal language are often disagreements over the principles that should guide objective inquiry into the human-animal border. A constant theme for Müller and his critics was whether scientific explanations could deal in causes no longer active in the world. The disagreements surveyed in the second chapter, "Brains and Minds across the Barrier," touch on the similarly broad questions of whether accounts of human mental powers and their disturbance can posit souls independent of bodies, and whether the human mind is a reliable model of the animal mind, even supposing the former to have evolved out of the latter. More concretely, the chapter looks at how the Müller-centered controversy over animal language and human evolution spilled into debates over the relationship between language and the brain and over anthropomorphism in the interpretation of animal behavior. The emissaries of Müllerian doctrine were, respectively, the physician Frederic Bateman and the psychologist Conwy Lloyd Morgan. The antievolutionist Bateman stirred up debate in the 1870s over language and cerebral localization, arguing that there was no single seat of speech in the brain, that the theory of evolution needed there to be such a seat (so that apes might have a more rudimentary version of it), and that the theory was therefore refuted. Although an evolutionist, Morgan likewise drew on Müller's work in taking a strong stance in the 1880s and 1890s against anthropomorphic interpretations of animal behavior, maintaining that animals, lacking language, probably lacked reason, and therefore that investigators needed to be skeptical about imputing human reason to animal minds, no matter how apparently reasonable the observed actions.

While the events recounted in this chapter have been largely forgotten, animal language, the cerebral localization of language, and the role of anthropomorphism in science remained entangled ever after. Separately and together, the questions they raise thread throughout the chapters that follow. Concerns about the evidence for speech centers in the brains of monkeys, apes, and extinct hominids have pressed on those wishing to interpret simian vocalizations. And under the influence of Müller's antianthropomorphic views, Morgan promoted a methodological rule that shaped the science of animal behavior in the twentieth century more profoundly than any other single development. Known as Morgan's canon, the rule is a commandment to explain animal behavior in terms of the "lowest" psychological faculty that will account for that behavior. In practice, that has meant supposing that animals do not reason about means and ends, but adapt themselves to their worlds through blind trial and error; or, in the case of animal vocalizations, that they serve not as symbols, representing ideas, but as involuntary accompaniments of emotion. Antianthropomorphism has been the badge of professionalism. Not quite coincidentally, Morgan introduced his canon in the same year, 1892, that Richard Garner traveled to Africa, promising to return with incontrovertible evidence that Müller was wrong about animal language and therefore about animal reason. The third chapter, "Professor Garner's Phonograph," aims to recapture the moment when, in the opinion of a number of observers, Morgan included, Garner's experiments with the phonograph posed a formidable challenge to anything like the canon's stricture on supposing that animals think and talk like us, only less perfectly.

Where the first two chapters cut new paths through more or less elite precincts—the Royal Institution lecture hall, Darwin's study at Down House, T. H. Huxley's rooms at the Royal School of Mines, the 1868 Norwich meeting of the British Association for the Advancement of Science (BAAS), and so on—the third chapter strikes out for altogether more heterogeneous territory. Following Garner, the narrative travels from the battlefields of the American Civil War to the imagined forests of West African lagoon country, by way of American zoological gardens and a newspaper syndicate office. Away from the elite places there flourished non-elite views on life, mind, evolution, even the physics of sound. Garner's articles, books, letters, newspaper profiles, notebooks, and manuscripts (almost all of them unexamined before now) preserve the worldview of a sort of person usually left out of standard histories of science—in Garner's case, that of a scientific amateur in the American South during the Gilded Age, coming to terms with the latest theories and technologies.[10] The chapter dwells especially on ideas about evolution, race, and language both absolutely commonplace and crucial to Garner's self-understanding. They included: that evolutionary change was gradual and progressive; that its highest products were humans; that the lowest humans were the savage or barbarous races, such as the Fuegians or the Australian aborigines or the Hottentots; that the highest races were the civilized ones, especially the white European civilized ones; and that human racial highness and lowness showed itself in body as well as mind, including language. For Garner, the playback experiment was so important because it proved just what the theory of evolution, as he understood it, predicted: that just below the most primitive human race, with its primitive language, there was an animal species speaking a still more primitive language.

In part 2, we turn from the world that made the primate playback experiment possible to the world that made it, as a live option for scientific inquiry, virtually impossible. The fourth chapter, "Congo Fever," deals with Garner's expedition to Africa in 1892–93, including the public trial he endured after his return, and from which his research program never really recovered. It was a trial conducted not in the courts but in the press. The prosecutor was Henry Labouchere, an English MP and the owner-editor of a six-penny weekly called

Truth, devoted to high-society gossip and the exposure of frauds in public life. Between 1894 and 1896, on the basis of letters sent to *Truth* from Africa about Garner's activities there, Labouchere worked to show that Garner had lied about what he had done and seen in the African forest. The account offered here of the bizarre and often comical story of the collision of the simian tongue with the new society journalism of late Victorian Britain contrasts the much-contested public version of the expedition with the private version Garner set down in an unpublished manuscript some time thereafter. What I have called Garner's "Fernan Vaz testament," discovered in a sketchbook among Garner's papers, tells an astonishing tale of Catholic conniving ranged against a would-be Darwinian hero. Weighing all the evidence, I attempt to puzzle out what really happened in Africa in 1893, and come to some provisional conclusions. But whatever the truth of *Truth*'s allegations, Garner's testament holds a valuable lesson: that for all the wariness historians now bring, rightly, to the "military metaphor" of conflict or warfare in describing the relationship between science and religion, they must also be sensitive to how this metaphor has influenced people's sense of themselves and their dealings with one another. "Metaphors we live by" is a familiar phrase in the philosophy of language.[11] For better or worse, the military metaphor for science-religion relations has been and remains a metaphor that some people in Western culture live by. Garner was one of them.

After the scandal, the simian tongue and the technique used to discover it virtually disappeared from the sciences. The explanation lies in part with the scandal but also, and more importantly, with the newly emerging professional sciences of humans and animals. The fifth chapter, "The Anthropologists and Animal Language," considers the impact on the animal-language debate of changes in physical anthropology and cultural anthropology. In physical anthropology, the years in which the simian tongue disappeared were the years when those interested in human origins turned their attentions increasingly toward the fossil record. Before the early 1890s, Neanderthal and Cro-Magnon fossils had featured in a limited way in debates on language and evolution. But the available specimens were widely regarded as too near to modern humans in time and form to be of much use in linking apes to humans. Then came the discovery of the Java man. Coinciding almost exactly with Garner's phonographic work of the early 1890s, this sensational find made the fossil record seem a much more promising source of evidence with which to fill the ape-human gap than more intensive work among living apes. At the same time as fossil hominids moved into the ape-human gap, due to changes in physical anthropology, living human races moved out, due to changes in cultural anthropology. In opposition to the evolutionist race-rankers of the nineteenth century, Franz Boas and his students held that evolution had created all peoples as biological equals, with differences arising through the medium of culture. In Boas's view, languages with simple grammars or few abstract words did not reflect a lack of mental capacity on the part of the speakers of those languages. Rather, claims about the highness and lowness of different languages reflected the racial prejudice of those making the claims.

For us, who come after Boas, the racist assumptions of Garner's research, his belief in an evolutionary ladder of language with apes on one rung and the most savage humans on the next rung up, belong to science past. At a recent conference on the evolutionary origins of language, where disagreements were rife over just about everything, there were no papers on the languages of aboriginal Australians or Amerindians as holding clues to the prehuman origins of language.[12] But science past can surprise: a look at Garner's ethnographic writings shows that evolutionist racism and cultural relativism were not mutually exclusive options at the beginning of the twentieth century, however much the Boasians succeeded in making them so later. Garner turns out more generally to furnish fresh vantage points on the professionalizing disciplines that took jurisdiction over the intellectual domain of his experiments. Exploring connections between Garner and a number of better-remembered figures in the anthropology of the period, among them Alfred Haddon in England and Aleš Hrdlička and John Harrington in America, we come away with a more concrete sense both of the diversity of options available and the limited scope these left for experimental, simian-oriented ambitions like Garner's.

The case of psychology, the subject of the sixth chapter, "The Psychologists and Animal Language," is interestingly similar and interestingly different. In psychology too, professionalization gradually put paid to Garner-style inquiries into the natural vocalizations of apes and monkeys. But there was no general turning away among psychologists, in their theorizing or their experimental practice, from research with these animals. The man best known for primatological psychology in the early twentieth century, Robert Yerkes, conducted pioneering studies of simian powers of speech and reasoning. Yerkes was an admiring acquaintance of Garner's, and even sponsored field research during which, in an inconsequential way, some recorded vocalizations were played back to some gibbons. In all these activities, however, Yerkes was in cautiously managed dissent from the legacy of his own teacher, Edward Thorndike. The history of professional animal or "comparative" psychology in the first half of the twentieth century is the history of the takeover of that science by quantitative, laboratory studies of learning in solitary animals set problems in food acquisition, on the model of Thorndike's early experimentalizing of Morgan's canon. The very idea that animals had ideas came to be regarded as belonging to the sentimental, evolution-obsessed nineteenth century. As for the natural behavior of animals, that came to seem no business of the professional student of learning, any more than it did the professional student of hominid bones. By the time comparative psychology found an ideology in behaviorism and a symbol in the maze-running rat, the primate playback experiment had long been an impossibility.

When the idea of the experiment was again entertained seriously, it was within the context of another distinctly twentieth-century science, ethology. The British-born Peter Marler was one of its earliest recruits; his remarkable intellectual biography organizes part 3, on the world that brought the primate playback experiment back into being. The seventh chapter, "Mr. Marler's Spectrograph," takes the stories of ethology and Marler through the 1950s. Forced to compete with comparative psychology, the ethologists presented themselves as correcting deficiencies in the reigning science of animal behavior. To a very large extent, the ethologists defined their science *against* comparative psychology, or, more precisely, against a caricature of that science. In the doctrine that grew up, it was said that psychologists studied animals in order to understand humans, and so wound up misunderstanding both. The rat in the psychologist's maze, the cat in the puzzle box, the monkey in the choice chamber: all found themselves acting under conditions of no relevance to their natural lives. Worse, since these animals could only extract themselves from such artificial experimental setups through trial and error, the psychologist, who never studied animals in nature, concluded that animal behavior, including human behavior, owed nothing to instinct, and everything to trial-and-error learning. The ethologist, it was said, did not make these mistakes. Unlike psychologists, who failed to divest themselves of a human-oriented perspective in their studies of animals, ethologists studied animal behavior on the animals' terms.

This disciplinary contrast with psychology lay at the core of the ethological self-conception. One can find it in the programmatic writings of the European founders, Konrad Lorenz and Niko Tinbergen, and of followers such as W. H. Thorpe, who established ethology at Cambridge. More importantly for understanding how the primate playback experiment became thinkable again, the contrast appears in the early writings of Marler, Thorpe's first ethological student. At Cambridge in the 1950s, Marler worked with Thorpe on an experimental study of song learning in the chaffinch—a study making extensive use of playback of recorded song as well as a new instrument for visualizing and analyzing animal vocalizations, the sound spectrograph. Independently of Thorpe, Marler also developed an interest in the new information theory and its bearing on understanding animal communication systems. Between them, the sound spectrograph and information theory represent a little-studied contribution to European ethology from the American military-industrial complex. This chapter traces the immediate origins of both technologies to work done at Bell Labs during the Second World War, an event that tends, in histories of ethology, to be treated as all hindrance and no help.

The next chapter, "Simian Semantics," follows Marler to America, where he increasingly divided his time, first at Berkeley and then at the Rockefeller University, between lab-based studies of bird vocalizations and field-based studies of primate vocalizations, all the while continuing to theorize about the big issues concerning language, communication, and information. The late 1950s and early 1960s saw a huge surge in primate field studies, thanks in large part to one of Marler's senior colleagues at Berkeley, the physical anthropologist and one-man disciplinary revolution Sherwood Washburn, who believed that observation of living primates—apes and monkeys but also hunter-gatherers—would illuminate the Darwinian agencies at work in the hominid past. Soon Marler was caught up in the primatological enthusiasm. His interest in the vervet alarm calls and their meaning arose at the conjunction of three developments. One was Berkeley graduate student Thomas Struhsaker's discovery of the alarm calls, made in the course of observing and recording the vervet monkeys of Amboseli Reserve in Kenya in 1963–64. The second was the emergence of a neuroscientific consensus, also in the mid-1960s, on the wholly emotional character of vocalizations in nonhuman primates. The third was the stimulus of another discipline, the old rival to ethology, comparative psychology. Starting in the mid-1960s, academic psychologists were engaged in high-profile attempts to teach apes to communicate symbolically—the "ape language projects," as they came to be known. In Reno, Allen and Beatrice Gardner were teaching a chimpanzee named Washoe to use sign language. In Santa Barbara, David and Ann Premack were coaching another chimpanzee, Sarah, in the use of a complex system of plastic icons. Throughout the 1970s, similar projects sprouted elsewhere, involving chimpanzees but also gorillas and orangutans. And in 1975, at the beginning of what would be a period of unparalleled prominence for the projects, David Premack—their most acute and ambitious theoretician—affirmed his view that nonhuman animals in the wild do not communicate symbolically. "Man has both affective and symbolic communication," he declared. "All other species, except when tutored by man . . . , have only the affective form."

Between 1975, when Robert Seyfarth and his wife Dorothy Cheney first contacted Marler as primatological postdocs looking for jobs, and 1980, when the trio's playback research was published, the ape language projects enjoyed their greatest popular cachet. Not since Garner's day had the topics of animal-human communication and the evolutionary origins of language commanded so much scientific and journalistic attention. Although the project psychologists had always had their critics, around 1980 the criticisms started to look fatal, thanks to the recent recantations of one of the project psychologists, Herbert Terrace at Columbia. It was against a backdrop of disillusion with and disarray among the psychological tutors of apes that the Rockefeller team's papers became so celebrated. There was an element of lucky timing here, but it was also true that Marler conceived the vervet playbacks very much as an ethological riposte to the psychological projects. The ninth chapter, "Playbacks in Amboseli," tells how the playbacks came to be done and then to be interpreted as they were, first by the experimentalists themselves, then by the scientific community and popular media. This part of the story is grounded in the second major documentary discovery made during this study: a fairly complete set of the correspondence between Marler and the Seyfarths (as they were known at the time) throughout their years together at the Rockefeller. American-born but educated, like Marler, in Cambridge ethology, the Seyfarths joined his laboratory knowing little about vocal communication and its study. Their expertise was as observers and theorists of social behavior in baboons, and when they left for Amboseli in spring 1977, they were much more excited about the prospects for using the playback technique to explore social relationships than semantic ones. Once in the field, however, they found the alarm-call behavior utterly absorbing. By the time Marler came out to Amboseli in the summer to help with the first playback trials, the couple were hooked. But, as one of the letters they sent to Marler later that year shows, even quite late in their field research, they were not at all persuaded that their results added up to a vindication of the meaningfulness or "semanticity" of the vervet alarm calls. Why they changed their minds is a question whose answer, I suggest, lies as much with the technical details of the experimental work as with the unique scientific ambience in which, post-Amboseli, Seyfarth and Cheney re-examined that work.

It was an ambience, not least, in which the wisdom of Morgan's canon was being actively called into question. That questioning has continued unabated, indeed has intensified and grown widespread.[13] At such times the contest between Morgan's canon and Garner's phonograph—the contest at the analytic and narrative pivot of this book—may be especially worth pondering, for it is at bottom a contest between two permanent and polar positions on the significance of evolution for humankind. At one pole is the belief that evolution has little to teach us about who we are. At the other pole is the belief that evolution teaches us everything about who we are. In returning to the Victorian debate on the simian roots of speech, we return to a debate about problems that, for all that we have learned in the interim, remain very much our problems.

ONE

I

THE LANGUAGE BARRIER

◆　◆　◆

GORILLAS WERE LITTLE MORE than a rumor in Western science when Paul du Chaillu turned up in London in early 1861, stuffed gorillas in tow. He rightly guessed how much interest there would be in these humanlike apes, collected on a recent expedition to western Africa, and showcased them with an impresario's cunning. Public exhibitions, lectures, a book that told thrillingly of gorillas in the wild, even allegations of fraud against him—all ensured that gorillas were seen and discussed as never before. Their public notoriety grew further that spring and summer thanks to a dispute between two of London's most formidable comparative anatomists, the eminent Richard Owen (one of Du Chaillu's patrons) and the up-and-coming Thomas Henry Huxley. According to Owen, the hippocampus minor, a small protrusion on the back wall of one of the ventricles in the human brain, was absent altogether from the gorilla brain, along with a number of other structures. Huxley disagreed, casting doubt on Owen's competence and integrity along the way. Soon the arcana of brain folds and skull angles were spilling onto the pages of the *Athenaeum*, *Punch*, and other popular periodicals. What was ultimately at issue, as *Punch* made explicit in a famous bit of doggerel in May 1861, was the Darwinian explanation of human origins. Was a gorilla but a potential human, given enough time, enough struggling for life, and enough natural selection of the most successful strugglers?[1]

In June, the German-born Sanskrit scholar Friedrich Max Müller declared his verdict, in his final lecture in a series on the "science of language," before a packed hall at the Royal Institution in London:

> Where, then, is the difference between brute and man? What is it that man can do, and of which we find no signs, no rudiments, in the whole brute world? I answer without hesitation: the one great barrier between the brute and man is *Language*. Man speaks, and no brute has ever uttered a word. Language is our Rubicon, and no brute will dare to cross it. This is our matter of fact answer to those who speak of development, who think they discover the rudiments at least of all human faculties in apes, and who would fain keep open the possibility that man is only a more favoured beast, the triumphant conqueror in the primeval struggle for life. Language is something more palpable than a fold of the brain, or an angle of the skull. . . . [For] if this were all, if the art of employing articulate sounds for the purpose of communicating our impressions were the only thing by which we could assert our superiority over the brute creation, we might not unreasonably feel somewhat uneasy at having the gorilla so close on our heels.[2]

Along with the controversies over Du Chaillu's gorillas and the hippocampus minor, Müller's lecture on the origin of language was one of the events that forever linked Darwinism and the apes, well before Charles Darwin himself linked them in *The Descent of Man* (1871).[3] General histories of the debates over Darwinism tend to pass over Müller's lecture,[4] while more specialized studies leave out its connections to the sensational debut of gorillas in scientific London, and to the ensuing debate over the differences between human brains and gorilla brains.[5] For Victorian Londoners, however, Müller's language lecture and the two gorilla controversies surrounding it were of a piece. Thanks to Müller, as one observer later recalled, "[l]adies discussed Grimm's law as cheerfully as they talked of protoplasm, and our Aryan forefathers divided attention with Du Chaillu's gorilla and with a remoter ancestor 'probably arboreal in his habits.'"[6]

Müller (fig. 1.1) went on to become one of the most eminent public men of science in the English-speaking world,[7] and he used this position to keep the language barrier between gorillas and humans conspicuous. His views became well known: that language and reason were mutually entailing; that only humans had them; that this distinction marked a difference of kind between humans and other species; and that no evolutionary bridge spanned the gap between the irrational cries of animals and the rational roots of language, because the roots, the irreducible atoms of language, were fully conceptual, and

Figure 1.1 Friedrich Max Müller (1823–1900), around 1863. From Müller 1902, vol. 1, frontispiece.

must have been so from the beginning. "[N]o process of natural selection will ever distill significant words out of the notes of birds or the cries of beasts," Müller declared.[8] This chapter aims to restore to our picture of the Victorian debates over Darwinism a fuller sense of the impact of Müller's arguments on behalf of an absolute language barrier in nature.

F. Max Müller and the Sudden Origins of the Language Barrier

For anyone familiar with the period, two terms that will come unavoidably to mind in what follows are "catastrophism" and "uniformitarianism." These were invented in the 1830s to describe rival positions in geology. It was said the geological catastrophists explained the surface features of the earth by appeal to causes no longer in evidence, such as huge and sudden floods. By uniformitarian lights, catastrophism was doubly false, involving a false belief about how the world works (that large-scale changes can occur rapidly) and a false belief about how to explain the world objectively (that explanations can appeal to causes no longer in evidence). With some justice, Müller's explanation of the origin of language can be labeled "catastrophist," since it invoked just such a cause: a vanished language-making instinct, arising fully formed in humans in the past, and generating abstract concepts together with the sounds expressing those concepts.

At the opposite extreme from the catastrophists—again, as commonly understood—the geological uniformitarians restricted their explanations to causes still in evidence, such as wind and rain. These have small effects that, over the very long run, can accumulate gradually to produce major changes in the earth's surface. Darwin's gradualist geology came to be regarded as a type specimen of uniformitarian theorizing, and there is more than a passing resemblance to his language-origins account. He explained language as beginning in the past much as it would begin in the present, by the exercise of a power of imitation still common to humans and some nonhuman creatures. Through imitation, the earliest humans would have communicated sense impressions from mind to mind, and gradually there would have arisen languages rich in abstract concepts, together with brains and vocal organs capable of thinking and speaking in conceptual terms.

Questions about causation and explanation are ancient and abstruse. But there were cultural politics at work here as well, by no means irrelevant to understanding the reception of Müller's arguments. In the 1830s, the elite sciences in England were beholden to Anglican Christianity—the state-sanctioned religion, then as now—in several ways. Anglican clergymen filled the scientific societies and the small numbers of scientific professorships at the ancient universities of Oxford and Cambridge. These men drew their salaries from the Church of England, and, in line with traditional Protestant theology, they tended to see the natural world as harmonizing with scriptural authority. A history of the earth that postulated a recent, earth-covering flood—as the Oxford cleric-geologist William Buckland had famously proposed—fitted comfortably with the narrative of Noah's ark. When Darwin's mentor Charles Lyell published his gradualist geology in the early 1830s, he had these cozy, compromising dependencies very much in his sights. By the 1860s, the successes of Lyell and, now, Darwin were celebrated in progressive circles as part of a wider unraveling of old, benighted arrangements.[9]

To the extent that Müller's opponents derided him, as we shall see, as a methodological throwback, it is important to bear in mind this backdrop. We should nevertheless beware taking the dichotomy "catastrophism-uniformitarianism" too much at face value. Even in geology, it oversimplified, making contrasts seem bolder and cleaner than they were.[10] A neat opposition of Müller's vanished-instinct explanation and Darwin's imitation explanation breaks down considerably in the details. Viewed from within, both explanations appear almost as much catastrophist as uniformitarian, as do the rather different explanations to be considered in what follows, from Huxley, the philosopher Chauncey Wright, and other Darwinians in the 1860s. The intermediary position on language adopted in the 1870s by the American Sanskrit scholar W. D. Whitney was, it will become clear, less a synthesis of opposed Müllerian and Darwinian theses than a variation on familiar themes.

Müller and the New Comparative Philology

We begin with F. Max Müller and his Sanskrit-centered science of language. Along with biogeography, animal and plant forms, and electromagnetic phenomena, Sanskrit was one of the central enthusiasms of romantic science in the German lands in the early nineteenth century.[11] As the language of the sacred texts of Hinduism, Sanskrit was the key to an ancient and exotic mythology. At the same time, its widely accepted kinship with Greek and Latin, and its place among the sources of the "Aryan" or Indo-European languages, made Sanskrit part of the heritage of Europe, and so, it was hoped, a repository of clues to the origin and development of the Western spirit.[12] For a philosophically ambitious philologist in the early 1840s, Sanskrit was the obvious specialism, and the young Müller made the language his own, taking his doctorate at the University of Leipzig in 1843. Just in his early twenties, he spent the following year in Berlin, attending the lectures of the comparative philologist Franz Bopp and the idealist philosopher Friedrich Schelling. In search of manuscripts, Müller moved on to Paris, where he began to prepare a major scholarly edition of the Rig-Veda, the oldest of the ancient Hindu texts. Like Bopp before him, Müller's textual pilgrimage eventually took him to London. He arrived in 1846, to work with the incomparable collections of the library of the East India House.[13]

By and large, Britain at that time was not the most congenial place for someone with Müller's talents and interests. Among the rulers of the subcontinent there was little of the Indophilia so widespread elsewhere. Where, for Müller, the Veda was the work in which, as he once wrote, the "bridge of thoughts and sighs that spans the whole history of the Aryan world has its first arch," British commentators were not even sure about the authenticity of Sanskrit, let alone the value of scrupulous editions of Sanskrit literature.[14] Nor was there much regard, in the homeland of empiricism, for what Müller considered the magnificent last arch in that Aryan thought bridge, Immanuel Kant's rationalist masterpiece, the *Critique of Pure Reason*.[15] Yet there were groups in Britain who welcomed the new arrival. He drew important support from the ethnologists, those devout searchers for evidence of the common origin or "monogenesis" of human races. Bopp's work on comparative grammar had proved useful to them in arguing for a single original human stock that had diffused (and, in certain branches, degenerated) to create the different races. The ethnologists remained keen to learn what comparative philology revealed of human unity.[16] Another important constituency was the circle of Liberal Anglican scholars based at Trinity College, Cambridge, including the philologist J. W. Donaldson and the historian and philosopher of the sciences William Whewell (inventor of the terms "catastrophism" and "uniformitarianism"). These men looked favorably on comparative philology as an instance of the historicist and idealist tendencies they generally admired in German thought.[17]

Shortly after coming to England, Müller became a fixture within both groups, thanks to the patronage of Baron Christian von Bunsen, the energetic Prussian ambassador in London. At the meeting of the BAAS in Oxford in 1847, Müller shared a platform with Bunsen and the leader of the ethnologists, James Cowles Prichard. Bunsen also secured financial backing from the East India Company for the preparation and publication of Müller's edition of the Veda.[18] Meanwhile, Müller managed to create an academic role for himself at Oxford. He settled there in 1848 to supervise the printing of the first volume of Vedic texts. Gradually he won over students and dons alike, and in 1854 was elected Taylorian Professor of Modern European Languages. When the Boden Professorship in Sanskrit became available in 1860, Müller was a fellow of All Souls, and one of the top Sanskritists in the world. But the politics of religion at Oxford were exceedingly divisive and complex. There were conservative elements who had never warmed to Müller's enthusiasm for German biblical criticism and Hindu sacred texts. Furthermore, as Müller's rival for the post, Monier Williams, pointed out, the professorship had been founded to promote missionary work in India, in particular the translation of the Bible into Sanskrit—an aim hardly congruous with the leveling tendencies of Müller's scholarship. After a vigorous and public campaign for and against Müller, the crucial vote was taken in December 1860, and Müller lost.[19]

The wound was still raw when, in April 1861, Müller began a series of nine general lectures at the Royal Institution—a second series of twelve lectures followed in 1863—on "the science of language." In his early lectures, Müller eloquently put the case for the philosophical study of language, or, more precisely, the natural-philosophical study of language. He argued that language was no different from plants, animals, electricity, rocks, stars, and the other phenomena treated within the natural or "physical" sciences, in three respects. First, the most important practical benefits of the science of language flowed from the study of language as an end in itself, rather than as a means to practical ends. The student of classical philology aimed at expertise in reading great literature; the student of the science of language aimed at understanding the nature of language as such. For the latter's purposes, the tongue of the Hottentot was equal in value to the tongue of Homer.[20]

Second, language was not "a work of human art," not "a contrivance devised by human skill for the more expeditious communication of our thoughts," but "a production of nature," whose modification occurred in accord with natural laws, operating independently of human agency. "We might think as well of changing the laws which control the circulation of our blood," Müller explained, "or of adding an inch to our height, as of altering the laws of speech, or inventing new words according to our pleasure." According to Grimm's law, for example, sounds changed between Latin and German in a regular and predictable fashion, irrespective of what individual speakers may have wished. Given the law-governed nature of language change, it was thus more appropriate to speak of the natural "growth" rather than the artifactual "history" of language, though, as Müller elaborated, the natural analogue for language was less a growing organism than the regularly and continuously changing crust of the earth.[21]

The third reason the science of language deserved to count among the sciences dealing with "the works of God" rather than "the works of man" was the manner of its progress. According to Müller, observers of the natural sciences such as Whewell and Alexander von Humboldt had noted that there were three stages of progress in a science: first came the empirical or fact-gathering stage; next came the classificatory stage, when the gathered facts crystallized out their systematic relations to one another; and finally came the theoretical stage, when questions about the nature, origin, and meaning of what had been discovered could at last be explored. Müller organized his lessons on language around this threefold scheme, so that the final lecture, about the origin of language, was at the same time a summary of the results of a science just now coming to full maturity.[22] A catastrophic account of the history of knowledge thus climaxed with a catastrophic account of the emergence of language.

Müller's Explanation of the Origin of Language

For Müller, the problem of the origin of language was the problem of explaining how the *roots* of language had come into being. The roots were the elements of language, the smallest units into which words could be decomposed.[23] Each family of languages shared a set of roots, and each set constituted the lexicon, or part of the lexicon, of an extinct ancestral language. Thus, the Aryan languages contained in their roots the words of the aboriginal Aryan tongue, the Semitic languages the words of the aboriginal Semitic tongue, and so on. Like chemical elements, these roots were irreducible. There was no earlier, ancestral language of ur-roots, out of which the root sets themselves had developed, to be discovered. What united the roots was not a common ancestor but a common character. According to Müller, all roots expressed concepts. Furthermore, because reasoning depended on concepts and language, based on roots, was conceptual through and through, language and reason appeared to be aspects of a single phenomenon. In sum, wrote Müller, "[l]anguage and thought are inseparable. . . . The word is the thought incarnate."[24]

It was on the basis of this identity of language and reason that Müller concluded against animals' having language. We know, he argued, that language at bottom consists of roots expressing concepts. We know, thanks to Locke, that nonhuman animals lack concepts ("'the having of general ideas is that which puts a perfect distinction betwixt man and brutes, and is an excellency which the faculties of brutes do by no means attain to'"). Hence it must be that animals do not have language, since their minds, barren of concepts, are ipso facto barren of roots. In Müller's words, "[n]o animal thinks, and no animal speaks, except man."[25]

So how did language-reason begin? Müller rejected the two best-known explanations for the origin of language, the onomatopoeic theory and the interjectional theory—or, as he enduringly renamed them, the "bow-wow" theory and the "pooh-pooh" theory. According to the former theory, language began when people started to imitate sounds in the natural world. In imitation of a particular dog, someone made the sound "bow-wow," found the sound useful for expressing to others the idea of that dog, and so was born the word for *that dog*, which over time generalized into *dog*. According to the latter theory, language began when people started to use the sounds accompanying certain emotional states to refer to those states. Thus the spontaneous cry of pain, "pooh-pooh," became the word for *my pain*, which over time generalized into *pain*. But, Müller argued, the conceptual character of the roots told against such a humble origin for language, for by that character they showed that they had not entered the world as signs associated with particular objects, events, or states of mind. To grasp Müller's logic, it is necessary to share his confidence that the roots were indeed ultimate, that there was nothing more primitive to be discovered via more powerful philological methods. For Müller, if the roots were conceptual, then they must have been conceptual from the start, because otherwise they would not be the roots, but only later developments of the roots. Of course, Müller conceded, languages contained *some* onomatopoeic and interjectional words, but these were small in number and "sterile," each "like a stick in a living hedge." They were "unfit to express anything beyond the one object which they imitate," and "not liable to the changes of Grimm's Law" or any of the other laws that governed true roots.[26]

Disposing of the two false theories of language origin, Müller turned to a third theory, the vanished-instinct theory, which he attributed to the German philosopher Lorenz Heyse, and which came to be known, following Müller's example, as the "ding-dong" theory. According to this theory, just as wood or tin ring with characteristic vibrations when struck, so primeval man, the most highly organized form in nature, rang with roots as new concepts passed through his mind. Müller stressed that the analogy was an imperfect one. But the existence of a now-vanished instinct for making roots "must," he continued, "be accepted as a fact," because "its effects continue to exist":

That faculty was not of [primeval man's] own making. It was an instinct, an instinct of the mind as irresistible as any other instinct. So far as language is the production of that instinct, it belongs to the realm of nature. Man loses his instincts as he ceases to want them. His senses become fainter when, as in the case of scent, they become useless. Thus the creative faculty which gave to each conception, as it thrilled for the first time through the brain, a phonetic expression, became extinct when its object was fulfilled. The number of these *phonetic types* must have been almost infinite in the beginning, and it was only through the . . . process of *natural elimination* . . . that clusters of roots, more or less synonymous, were gradually reduced to one definite type.[27]

In Müller's view, the root sets of the several language families had all come into being independently. This aboriginal diversity of root sets was not, however, of itself evidence for multiple origins of language, much less of human races. On the contrary, as Müller had earlier explained, it was "at least intelligible how, with materials identical or very similar, two individuals, or two families, or two nations, could in the course of time have produced languages so different in form as Hebrew and Sanskrit." He postulated an enormous range of roots among the root set of the first human group. As this group began to diffuse and diverge, and as the new subgroups or clans found themselves living under different circumstances, the original root set fragmented according to local needs and conventions. So thorough and so contingent was the sifting, the "natural elimination" that led to the entrenchment of one root rather than its several synonyms, that only rarely was it possible to find roots common to more than one language family.[28] Such an account fitted well with the oldest account of the origins of language and man—one, according to Müller, about which "we have heard so often from the days of our childhood—'And the whole earth was of one language and one speech,'" but which now "assume[s] a meaning more natural, more intelligible, more convincing, than [it] ever had before." What lent credence to the Tower of Babel story also lent credence to the conclusions of Prichardian ethnology.[29]

There was a great deal in Müller's views on language that harmonized with the ethnological program. We have already encountered the common themes of monogenism and diffusionism. But there was also the theme of degeneration, fallenness, decline from a pristine state in the past.[30] According to Müller, all roots were subject to two general laws: "phonetic decay" and "dialectical regeneration." The latter term referred to the role of regional dialects in partially renewing the ever more depleted resources, phonetic and otherwise, of a language. In the beginning, the root sets were all in the isolating or "radical" stage, and each root was meaningful in its own right. In the Chinese case, the early imposition of a uniform literary language arrested the process of decay that elsewhere leached roots of their power to stand alone. As they decayed, these formerly isolating languages became agglutinating, their words compounded of two or more roots, one of them now utterly meaningless on its own. According to Müller, what he called the "Turanian" languages were good examples of this intermediate stage.[31] In cases of advanced corruption, languages proceeded from the agglutinating to the inflectional stage, wherein all roots ceased to be meaningful outside compounds with other roots. Decline was the fate of all languages. Thus, in Chinese, "we find rudimentary traces of agglutination," and in agglutinative languages, "we meet with rudimentary traces of inflection." Grimm's law testified to the persistence of decay even after the final stage in grammatical development had been reached.[32]

Lest Müller's explanation of the origin of language should seem wholly of its age, it is worth comparing it with a predecessor, that of the German philosopher Johann Gottfried Herder. Herder famously won a prize from the Berlin Academy of Science in 1772 for an essay on the origin of language.

Although Müller in his lectures criticized Herder's essay as a defense of the discredited theory of onomatopoeia, Herder's and Müller's positions were much closer than Müller allowed. It was not the case, for example, that Müller outright denied that the *sounds* of the first words were imitations of the sounds of nature, including man's natural sounds. He insisted rather that, however a root acquired its phonetic shell, what made a root a living germ of language, susceptible of development according to natural laws, was the existence within that outer shell of a concept, implanted by the root-making instinct. Müller explained at length in his second series that he had no quarrel with scholars "who derive all words from roots according to the strictest rules of comparative grammar, but who look upon the roots, in their original character, as either interjectional or onomatopoeic." On this matter, wrote Müller, "I should wish to remain entirely neutral, satisfied with considering roots as phonetic types [pending further inquiry]. . . . Quite distinct from this is that other theory which, without the intervention of determinate roots, derives our words directly from cries and interjections."[33] Herder's position was similarly subtle. "[N]o animal," he declared, "not even the most perfect, has so much as the faintest beginning of a truly human language. Mold and refine and organize those outcries as much as you wish; if no reason is added . . . I do not see how . . . there can ever be human language." For Herder, as for Müller, "language, from without, is the true differential character of our species as reason is from within." Likewise, natural sounds were merely the "raw materials" of language, "the sap that enlivens the roots of language," for "what is to grow must be there as a germ."[34]

Responses to Müller in the 1860s

In the 1860s, Müller's postulation of an extinct root-making instinct did not find favor among most writers on the origins of language.[35] Two criticisms came up repeatedly: Müller's explanation violated canons of sound scientific method, and it presumed the perfect-beginnings scenario of biblical and ethnological thinking against which there was mounting evidence.

In her anonymous review for *Macmillan's Magazine* in 1862, the writer Frances Julia Wedgwood charged Müller with offenses against method. Much of the review was devoted to showing how much more resourceful was the theory of onomatopoeia—also known as the mimesis or imitation theory— than Müller had credited. (Wedgwood was the daughter of Darwin's cousin and brother-in-law, the etymologist Hensleigh Wedgwood, himself an onomatopoeia theorist.)[36] But she also argued that, however many words could in fact be traced to imitative beginnings, the imitation theory should be preferred on methodological grounds. In Wedgwood's view, the history of the sciences showed that progress depended on the expulsion of imaginary barriers from nature. Mechanics and geology had long ago ceased to project "phantasmal limitations," so that even children could now see in falling leaves the same force that moved planets, and in a rain shower a power sufficient to level mountains. "Nature knows no bursts of fitful vehemence followed by intervals of inaction," she wrote. "The laws which preserve are separated by no generic interval from those which produce." For Wedgwood, the example of geology was especially instructive, for the influence of rain showers and other presently acting agencies had "seemed to the geologists of a past generation to occupy as insignificant a place in the mechanism of their science as is taken, in the estimation of Professor Müller, by the imitative principle in the origination of language." But just as geologists had managed "to consign the machinery of 'cataclysms' to the limbo of epicycles in astronomy," so too would philologists, when they learned to hear the origins of language in "the imperfect accents of the child or the savage," and to ignore calls to bestow upon the human race of old "powers different in kind from those it possesses now."[37]

The methodological case against Müller was made still more forcefully in a lengthy essay review four years later in the *Westminster Review*, a periodical long favorable to the progressive cause (in more than one sense) of evolution. Swapping Whewell's and Humboldt's three-stage scheme of scientific progress for the French positivist Auguste Comte's, the anonymous reviewer demoted Müller's explanation from the final stage of science to the intermediate stage. According to the reviewer, in the Comtean scheme, the first, theological stage of science is the time of explanations appealing to divine will. In the second, metaphysical stage, explanations appeal to causes for which there is no evidence. And in the third, positive stage, explanations appeal to causes for which there is positive evidence. Müller's root-causing instinct belonged with the gravity-causing ether, combustion-causing phlogiston, and other bogus entities of the second stage. The vanished instinct was

> a metaphysical entity, which, though a mere empty abstraction, satisfies the mind with a convenient formulized statement of ignorance which it mistakes for cause. . . . This is the turning point of science; this also is the stage reached by philology in our day. . . . [There is] a struggle between the metaphysical theory of language, now almost in sole possession of the field, and its positive rival—a rival for whom the victory is no less certain than for a body of well-disciplined troops engaged against savages. The disproportion of the forces may be immense, the issue of the contest at present most unfavourable; but the result is not even doubtful; for the contest lies between a positive cause and a metaphysical abstraction—between a principle of acknowledged operation to some extent [i.e., imitation], and

a figment of the mere indolent understanding, a product of the "*intellectus sibi permissus*"—the unquestioned, undisciplined prejudice of mankind.[38]

The notion of a "positive cause" had less to do with the positivism of Comte (who wanted science to dispense with causal explanations altogether) than with a distinctively British doctrine, the *vera causa* or "true cause" doctrine. This doctrine emerged in the eighteenth and early nineteenth centuries from commentary on the work of Isaac Newton. In the 1860s, the doctrine's most famous adherents were the astronomer John Herschel, the geologist Charles Lyell, and Darwin, who had taken Herschel and Lyell as his scientific models back in the 1830s. According to the doctrine, there had to be independent evidence for the existence of a cause, quite apart from its ability to explain what needed explaining.[39] Müller's postulated instinct, being, in the reviewer's words, "something of which we have no evidence but the fact which it professes to explain," was thus unacceptable. "The phonetic types [i.e., the instinct-generated roots] explain the origin of language, and they do nothing else," complained the reviewer. "They intervene to fill a gap in the chain of cause and effect, and we know no more of them. Their operation is confined to strictly metaphysical ground." The amount of evidence in favor of mimesis as the origin of language was strictly irrelevant, since mimesis was "the sole *vera causa* yet suggested." Mimesis was "a hypothesis that demands no withering of our primitive instincts, no chasm in the progress of the race, no exceptional agency at work during any part of its existence. . . . it asserts that language originated long ago, just as language would originate today, if any person were isolated among the speakers of a tongue unknown to him."[40]

Quite apart from general concerns about proper method in the sciences, there were also recent empirical findings which, the *Westminster* reviewer continued, lent support to the imitative theory of language origin, or, more precisely, to an onward-and-upward view of human history that predicted such humble beginnings for language. For Müller, man was, as he had written, "in his most primitive and perfect state" when he bodied forth the roots of language in all their magnificent variety. In the beginning was "a period of unrestrained growth,—the spring of speech—to be followed by many an autumn," marred by root cullings and meaning muddles. According to the reviewer, however, the evidence of recently discovered prehistoric settlements in the caves and along the lakesides of Europe told against a golden age of language or indeed of anything else in the human dawn. "What a significance is there in the Lake-dwellings, the *débris* of which yields us so much of our evidence as to prehistoric man! What a tale of insecurity, of mutual distrust and terror, is told in the remnants of those uncomfortable water homes!" The imitative theory alone matched the archaeological evidence for "a condition which we can only describe as degraded as the starting-point of the human race."[41]

Echoing judgments appeared at about the same time in two books championing the imitative theory against Müller's strictures, the Reverend Frederic Farrar's *Chapters on Language* (1865) and Hensleigh Wedgwood's *On the Origin of Language* (1866). Farrar argued that the evidence of comparative philology pointed toward the same conclusion as the evidences of geology, archaeology, and history: that "those primeval lords of the untamed creation, so far from being the splendid and angelic beings of the poet's fancy, appeared to have resembled far more closely the Tasmanian, the Fuegian, the Greenlander, and the lowest inhabitants of Pelagian caverns or Hottentot kraals." Wedgwood argued similarly.[42] It was in the work of the Victorian era's most influential anthropologist, Edward Burnett Tylor, that this shift away from the diffusionist and degenerationist views of the ethnologists was most fully realized during the 1860s. For Tylor, progressive evolution was a feature of nature and culture alike. Cultural progress proceeded in all human groups independently, through the same stages, and upward from the same benighted starting point. Progress was not lockstep, however, and in cultures one found "survivals" of former stages persisting into later ones. Müller's instinct theory was, in Tylor's view, just such a survival—an instance, he wrote in the April 1866 *Quarterly Review*, of "the *a priori* theory of a philosopher, brilliant and subtle indeed, but, to our thinking, ages behind . . . in scientific method."[43]

Tylor put language among the other human arts, thus making it, in his scheme, subject to the same universal pattern of progressive evolution. Furthermore, he argued, since language creation is ongoing, we can see how it happened in the past by observing how it happens in the present. In his own *Researches into the Early History of Mankind* (1865), he had reported seeing the deaf and dumb learn from scratch to express themselves with visual signs such as gestures and pictures. Such displays, he wrote, gave at least "some idea of the nature of this great movement [toward language], which no lower animal is known to have made or shows the least sign of making."[44] As for vocal language, Tylor found support for the imitative origins of words in the fact that, upon seeing a musket fire for the first time, "one savage race after another," from Australia to the Amazon, from the South Seas to South Africa, "named the European musket . . . by the sound *pu*, describing as it seems not the report, but the *puff* of smoke issuing from the muzzle."[45] Imperial encounters thus provided more than metaphors with which to demean Müller (recall the *Westminster* reviewer on "a body of well-disciplined troops engaged against savages"). They also provided evidence with which to refute Müller's instinct theory of language origins.

Charles Darwin and the Gradual Origins of the Language Barrier

Darwin's Early Reflections on Language Origins

Darwin's "transmutation" notebooks, opened in 1837, not long after his return from the prolonged imperial encounter of his years aboard the *Beagle*, show that problems to do with the nature and origins of language were among the earliest problems he addressed as a theorist of evolution.[46] Here Darwin (fig. 1.2) sketched a number of the views he would later defend in the *Descent* (1871) and elsewhere. Acknowledging the existence of a language barrier between man and brute at present, he insisted that the barrier's nature was not such as to preclude man's development out of the brute. "The distinction 'as often said' of language in man is very great from all animals—but do not overrate—animals communicate to each other—[William] Lonsdale's story of Snails, [William Darwin] Fox of crows, & many of insects—they likewise must understand each other expressions, sounds, & signal movements."[47] And again: "Nearly all will exclaim, your arguments are good but look at the immense difference. between man,—forget the use of language, & judge only by what you see. compare, the Fuegian & Ourang & outang, & dare to say difference so great."[48] Other insights bound for later elaboration are present: that certain aspects of language showed origins in music;[49] that, once language existed, it may have enriched a moral sense already arisen from social instincts in primeval humans;[50] that gesture, interjection, imitation may all have been recruited to create meaningful signs with which protohumans communicated ideas from one mind to another.[51] In all of these musings, there is a clear preference, as in all of Darwin's scientific work, for explaining past events in terms of presently acting true causes. "We cannot doubt that language is an altering element, we see words invented—we see their origins in names of People.—Sound of words—argument of original formation.—declension &c. often show traces of origin."[52]

One of the first evolutionist accounts of the origins of language that Darwin read—along with thousands of other men and women—appeared in *Vestiges of the Natural History of Creation*, published anonymously in 1844.[53] "[L]anguage, as the communication of ideas, was no new gift of the Creator to man," wrote the Vestigiarian (whom Darwin rightly surmised was the Ed-

Figure 1.2 Charles Darwin (1809–82), around 1874. Courtesy of James Moore and reproduced by permission.

inburgh journalist and publisher Robert Chambers), "and in speech itself . . . we see only a result of some of those superior endowments of which so many others have fallen to our lot through the medium of an improved or advanced organization." How then did the language-enabling change in bodily organization arise? The *Vestiges* had it that once conditions on the earth had altered to a point of readiness for humankind, then, in accord with the natural laws of development, an apelike creature gave birth to the first human. The larynx, trachea, mouth, and brain of this first human turned out to be so highly organized and coordinated that language followed as a matter of course. "It was unavoidable," the Vestigiarian explained, "that human beings so organized, and in such a relation to external nature, should utter sounds, and also come to attach to these conventional meanings, thus forming the elements of spoken language."[54] For Darwin, with his commitment to the *vera causa* doctrine, such an explanation would have been unacceptable, because it postulated an adapting power in nature for which there was no evidence except the adaptations themselves. As Darwin wrote in his introduction to the *Origin*, an explanation in the style of the Vestigiarian "seems to me to be no explanation, for it leaves the case of the coadaptations of organic beings to each other and to their physical conditions of life, untouched and unexplained."[55]

In the *Origin*, Darwin made a few suggestive parallels between species change and language change (comparing vestigial organs to silent letters, for example),[56] but otherwise avoided the evolutionary origins of language as of all other human powers and faculties, limiting himself to the prediction that, in due course, "[p]sychology will be based on a new foundation, that of the necessary acquirement of each mental power and capacity by gradation."[57] Following the publication of the first series of Müller's lectures, Darwin discussed the book in correspondence with his friends and botanical experts Joseph Hooker and Asa Gray. Like Darwin, Hooker admired all of the book save the catastrophist argument on the origins of language, which he pronounced "fatuous & feeble, as a Scientific argument."[58] Gray, however, considered that he could "hardly think of any publication which in England could be more useful to your cause than this volume is, or should be." "Depend on it," he advised Darwin, "Max Müller will be of real service to you."[59]

Darwin was hard-pressed to see how. "I quite agree that [Müller's book] is extremely interesting," Darwin wrote back to Gray,

> but the latter part about *first* origin of language [is] much the least satisfactory. It is a marvellous problem. I have heard, whether truly or not I do not know, but the book has rather given me the same impression, that he is dreadfully afraid of not being thought strictly orthodox. He even hints at truth of Tower of Babel! I thus accounted for covert sneers at me, which he seems to get the better of towards the close of the book.—I cannot quite see how it will forward "my cause" as you call it; but I can see how anyone with literary talent (I do not feel up to it) could make great use of the subject, in illustration. What pretty metaphors you would make from it! I wish some one would keep a lot of the most noisy monkeys, half free, & study their means of communication![60]

Müller's metaphors were indeed what had impressed Gray, in particular Müller's discussion of what he called the "natural elimination" of superfluous roots. Gray celebrated the "perfect appreciation and happy use of Natural Selection, and the very complete *analogy* between diversification of species and diversification of language" drawn in Müller's *Lectures*. So compelling was the performance that Gray confessed his own skepticism about natural selection weakened thanks to Müller.[61] Until reading the *Lectures*, Gray had believed that evidence of a designing intelligence could be found in the variations upon which natural selection worked. But the analogy with language had helped him to see how variations could arise without design. As Gray elaborated in a subsequent letter to Darwin: "The *use* that I fancied could be made of Max Muller's book . . . is something more than *illustration*, but only a *little* more,—i.e. you may point to analogies of development & diversification of language—of no value at all in evidence in support of your theory, but good & pertinent as rebutting objections, urged against it."[62] It was advice that Darwin would act upon when he came to assemble his own account of the origin of language in the *Descent*. In the meantime, a number of his supporters addressed the problem.

Darwinians on Language Origins in the 1860s

"I rather doubt about man's mind & language" was Darwin's comment, in a letter to Hooker in mid-January 1863, on one of the earliest and also most detailed of these attempts, due to T. H. Huxley.[63] In his last lecture to working men on "our knowledge of the causes of the phenomena of organic nature," Huxley repeated his key argument against Richard Owen: the differences in structure between the human brain and the ape brain were no greater than the differences between the brains of two different species of ape. With no more of a structural hiatus between apes and humans than elsewhere in the scale of nature, there was no reason, Huxley now argued, to suppose that a Darwinian process of structural variation and natural selection, sufficient to bridge the gaps elsewhere in nature, was incapable of bridging the ape-human gap. And yet, Huxley continued, "there is no one who estimates more highly than I do the dignity of human nature, and the width of the gulf in intellectual and moral matters, which lies between man and the whole of the lower creation."[64]

But how could there be an intellectual and moral hiatus without a structural hiatus—especially if, as Huxley claimed, "there is no faculty whatsoever which does not depend upon structure"? Huxley blamed the appearance of contradiction on a widespread misunderstanding. Just because "[f]unction is the expression of molecular forces and arrangements" did not mean that large changes in function require commensurately large changes in structure. The barest of tinkering could turn a working watch into a nonworking watch, and in the same way, the smallest change in anatomical structure—change on the order of the minute variations observed in nature and required by Darwinian theory—could occasion the largest change in physiological function. So it had been, Huxley contended, with the structural changes that transformed the ape brain, and brought into being that which "constitutes and makes man what he is . . . his power of language."[65]

For Huxley, language had set humans apart chiefly because language enabled each generation to transmit its collective experience to future generations. "I say that this functional difference is vast, unfathomable, and truly infinite in its consequences; and I say at the same time, that it may depend upon structural differences which shall be absolutely inappreciable to us with our present means of investigation."[66] He gave a vivid picture of the several muscles, nerves, and organs whose complex coordination alone enabled humans to speak. Were a man to suffer the smallest of modifications at any of several places in his body, Huxley went on, the man could be struck speechless, and "a race of dumb men, deprived of all communication with those who could speak, would be little indeed removed from the brutes . . . though the naturalist should not be able to find a single shadow of even specific structural difference."[67] In his best-selling *Evidence as to Man's Place in Nature*, published in late February 1863, Huxley stated the same argument more briefly, emphasizing now that intellectual power depended, not just on the brain, but also on the sensory and motor organs, "especially those which are concerned in prehension and in the production of articulate speech."[68]

Here was an account of the evolutionary origins of language that blended uniformitarianism (at the level of structure) and catastrophism (at the level of function). Huxley had earlier suggested to Darwin that "you have loaded yourself with an unnecessary difficulty in adopting 'Natura non facit saltum' so unreservedly," adding "I believe she does make *small* jumps."[69] Darwin's skepticism about Huxley's mind-and-language story was of a piece with his skepticism about natural leaps in general. Furthermore, without an account of how speech might have been useful in the struggle for existence, or clarification as to how the several anatomical changes necessary for language were brought into coordination, Huxley's leap into language looked, by Darwin's lights, no less miraculous than the one proposed in the much-criticized *Vestiges*.

At least Huxley presented his views on the origins of language as consistent with the theory of natural selection. Charles Lyell and Alfred Russel Wallace, Darwin's geological mentor and natural-historical collaborator respectively, were much less obliging. "Other animals may be able to utter sounds more articulate and as varied as the click of the Bushman," wrote Lyell in his *Antiquity of Man,* also published in February 1863, "but voice alone can never enable brute intelligence to acquire language." According to Lyell, once the power of language had been implanted in humans, languages themselves began to develop in Darwinian fashion, via a process of variation and selection according to conditions. Nevertheless "we may . . . demur to the assumption that the hypothesis of variation and selection obliges us to assume that there was an absolutely insensible passage from the highest intelligence of the inferior animals to the improvable reason of man." Lyell appealed instead to the occasional but law-governed birth of geniuses. He wrote of "a law capable of adding new and powerful causes, such as the moral and intellectual faculties of the human race, to a system of nature which had gone on for millions of years without the intervention of any analogous cause." To invoke such a law was, of course, to abandon the *vera causa* standard of explanation at the center of Lyell's—and Darwin's—theoretical work.[70]

Some years after Lyell's apostasy came Wallace's. First in a review of one of Lyell's geological works, then in more extended form, Wallace questioned whether the theory of natural selection could explain the origins of a number of human features, among them the remarkable versatility of the human larynx, and the power to formulate and express abstract thoughts. The problem, as Wallace saw it, was that selection could preserve only what was useful in the struggle for existence. Under the savage conditions of life in which the first humans presumably existed, abstract thinking and laryngeal versatility would have been useless, indeed actively harmful, and so selected against. The fact that modern-day savages were nevertheless capable, with training, of producing words as abstract and acoustically subtle as their civilized counterparts showed, in Wallace's view, that something other than selection had implanted distinctively human powers in the human brain. To illustrate this claim, Wallace compared civilized languages, "full of terms to express abstract conceptions," with "savage languages" bereft of such terms.[71] As for "the wonderful power, range, flexibility, and sweetness, of the musical sounds producible by the human larynx, especially in the female sex,"

> The habits of savages give no indication of how this faculty could have been developed by natural selection; because it is never required or used by them. The singing of savages is a more or less monotonous howling, and the females seldom sing at all. Savages certainly never choose their wives for fine voices, but for rude health, and strength, and physical beauty. Sexual selection could not therefore have developed this wonderful power, which only comes into play among civilized people. It seems as if the organ had been prepared in anticipation of the future progress of man, since it contains latent capacities which are useless to him in his earlier condition. The delicate correlations of structure that give it such marvellous powers, could not therefore have been acquired by means of natural selection.[72]

Before Darwin settled down to mount his defense of the power of natural selection against Wallace's critique, someone else beat him to it: the Cambridge, Massachusetts, philosopher Chauncey Wright, writing in the October 1870 issue of the *North American Review*. Wright dwelt in particular on Wallace's arguments about language, the larynx, and abstract thought. What Wallace had failed to understand, Wright argued, was that selection for greater expressiveness in speech might well have generated larynxes *incidentally* capable of producing a wider range of musical sounds. In Wright's view, this solution contained an important and quite general evolutionary principle. As he put it: "There are many consequences of the ultimate laws or uniformities of nature, through which the acquisition of one useful power will bring with it many resulting advantages, as well as limiting disadvantages, actual or possible, which the principle of utility may not have comprehended in its action." In other words, selection for an advantageous new power may have lots of side consequences, sometimes advantageous, sometimes not.[73]

As for abstract language, Wright insisted on the need to appreciate just how momentous a thing it was for the first humans to acquire, in his words, "the one universal characteristic of humanity, the power of language,—that is, the power to invent and use arbitrary signs." For Wright, once humans were able to invent and use arbitrary signs for concrete ideas, the extension of this unprecedented signifying power to abstract ideas constituted a minor innovation. "[A] psychological analysis of the faculty of language," urged Wright, "shows that even the smallest proficiency in it might require more brain power than the greatest [proficiency] in any other direction." As with the larynx, so with the brain: once selection had fixed certain capacities, others came for free. Wright concluded against Wallace's pessimism about the power of natural selection—pointing out, however, that with respect to the evolutionary origins of language, the positive case for selection remained to be seen.[74] Darwin would publish it within a few months.

Language Origins in The Descent of Man *and Later Writings*

In the *Descent*, Darwin set his theoretical sights on the diverse powers of the human mind, including the power that, for Darwin, was "justly . . . considered as one of the chief distinctions between man and the lower animals": man's power of language.[75] In Darwin's view, humans alone were able habitually to produce a variety of well-defined sounds and to connect those sounds to well-defined thoughts; and if such was language, then the mental activities of other creatures were indeed carried out "without the aid of language."[76] But Darwin also insisted on the importance, and at points even the expressive dominion, of the communicative repertoire that humans shared with languageless creatures. Dogs and monkeys expressed what was on their minds through gestures, inarticulate cries, and facial expressions, and so did humans, especially when their minds were full of "the more simple and vivid feelings, which are but little connected with our higher intelligence." Furthermore, "[o]ur cries of pain, fear, surprise, anger, together with their appropriate actions, and the murmur of a mother to her beloved child, are more expressive than any words."[77] Continuity across the language barrier showed equally in the semi-instinctive manner in which humans acquired language, much as birds have an instinct to learn their songs ("an instinctive tendency to acquire an art," as Darwin put it). Just as chicks instinctively chirp away, gradually taking on the song of their species (indeed of their locality) according to the songs sung around them, so human babies take their first steps toward mastering a particular language through an instinctive and universal babbling. In addition, what was learned in each case was not something deliberately invented, but something that had emerged gradually and without conscious intent on the part of its users.[78]

So how did the human "half-art and half-instinct" of language arise?[79] Darwin declared himself on the side of Wedgwood, Farrar, and other scholars who held that "language owes its origin to the imitation and modification, aided by signs and gestures, of various natural sounds, the voices of other animals, and man's own instinctive cries."[80] According to Darwin, the exercise of an instinct for imitation, still present in humans and their near evolutionary kin, led the progenitors of modern humans to speak the first words. Sounds imitative of natural objects and events became words for those objects and events. When these words brought an advantage in the struggle for life, as when the utterance of a sound associated with a predator warned the group of imminent danger, then natural selection acted to preserve and accumulate these new words, along with the creatures able to speak and understand the words. Similarly, sounds imitative of instinctive cries became words for the emotions expressed in those cries. When these words brought an advantage in the struggle for mates, as when the utterance of a sound associated with love or rage made for more successful courtship, then sexual selection acted to preserve and accumulate these new words and word users. And while language grew thus through imitation, natural selection, and sexual selection, the brains and the vocal organs of language users became, through use, ever better at thinking and speaking. Because such effects of use were heritable—Darwin here compared the case of brain-and-vocal-organ modifications to "the case of hand-writing, which depends partly on the structure of the hand and partly on the disposition of the mind; and hand-writing is certainly inherited"—these language-induced refinements of body and mind were transmitted to future generations, where more use brought further refinements. And so, gradually, language came into being.[81]

Darwin broke off from his discussion of the evolutionary origins of language at one point to list a number of parallels between languages and species.[82] Homologies, analogies, correlated growth, effects of use, rudiments, hierarchical classification, dominance and extinction, survival of the fittest (illustrated with a quotation from Müller—"[a] struggle for life is constantly going on amongst the words and grammatical forms in each language"): Darwin catalogued numerous respects in which languages changed according to his theory of how species changed.[83] The point of the exercise was revealed only by what followed. "The perfectly regular and wonderfully complex construction of the languages of many barbarous nations," Darwin continued, "has often been advanced as a proof, either of the divine origin of these languages, or of the high art and former civilisation of their founders." What Darwin was confronting here was the savage-as-degenerate view of the old biblical anthropology, recently defended in Charles Staniland Wake's antitransmutationist *Chapters on Man* (1868). Wake had cited a number of authorities on the existence of languages high on the scale of grammatical perfection among races low on the scale of intellectual perfection. Such mismatches pointed to degeneration from a pristine state, rather than development from a brutish one.[84]

On the basis of the previous parallels, there was, Darwin urged, much to be said for borrowing from the naturalist in order to make sense of phenomena studied by the philologist. When it came to perfection, the naturalist had one way of reckoning it, and the philologist another. Where the philologist rated symmetry above asymmetry in evaluating grammars, the naturalist, according to Darwin, "justly considers the differentiation and specialisation of organs as the test of perfection." True, the grammars of savage tongues exhibited greater symmetry than the grammars of civilized tongues, but this fact, reckoned against the naturalist's scale of perfection, made savage tongues less perfect, not more perfect. Adopt the scale of perfection appropriate to species, and, Darwin concluded, the apparent perfection of the languages of primitive races became predicted imperfection. The burden of the previous parallels was to make this switch to the more favorable scale of perfection seem reasonable. In sum, the aim in drawing parallels was to show that the lower races had lower languages—just as the theory of evolution (as opposed to creationism) predicted.[85]

For Darwin, the evolutionary trajectory of the human species was one in which language, intelligence, and the moral sense all marched onward and upward together, carried within ever-larger brains. The brain was the organ of mind, and as all organs became improved through exercise, so the brain became improved under the demands placed upon it by language. Darwin approvingly cited Chauncey Wright on how the early use of language—"that wonderful engine"—probably propelled the prehuman brain to its present remarkable size (compared to body size) by "excit[ing] trains of thought which would never arise from the mere impression of the senses, and if they did arise could not be followed out."[86] The undoubted fact that dogs dreamed and reasoned their way through problems showed that language was not a necessary precondition for thought as such. Nevertheless, wrote Darwin, "[a] long and complex train of thought can no more be carried on without the aid of words, whether spoken or silent, than a long calculation without the use of figures or algebra."[87] Such long and complex trains of thought pushed humans ever further beyond the mental horizons of the lower animals. "If it be maintained that certain powers, such as self-consciousness, abstraction, &c., are peculiar to man, it may well be that these are the incidental results of other highly-advanced intellectual faculties; and these again are mainly the result of the continued use of a highly developed language."[88] In helping to advance the human intellect, language indirectly advanced the human moral sense, for, as Darwin saw it, moral action depended on the ability to recall and reflect upon instances when strong but fleeting urges were indulged at the expense of weaker but more enduring social instincts. Furthermore, language became the medium whereby human groups agreed to codes of conduct, and then made sure the codes were followed, through expressions of pleasure and displeasure as required.[89]

But for all the distance between language-using humans and the languageless lower animals, language was studded with signs of its origins in the prehuman past. When, just by modulating their voices, orators and poets thereby managed to move their listeners deeply, all involved were experiencing the pull of the deep past, of a time when males and females wooed one another through ever more extravagant feats of vocal acrobatics. It was then, Darwin wrote, that "musical tones became firmly associated with some of the strongest passions an animal is capable of feeling, and are consequently used instinctively, or through association, when strong emotions are expressed in speech." Indeed, the "sensations and ideas" that music and musical speech quicken within us "appear from their vagueness, yet depth, like mental reversions to the emotions and thoughts of a long-past age."[90]

Darwin further explored this and related themes in *The Expression of the Emotions in Man and Animals,* published in 1872.[91] In the new book, Darwin gave a still more vivid sense of the suite of gestural and facial movements that surround and amplify articulate human utterances. In his view, the more primitive "movements of expression . . . reveal the thoughts and intentions of others more truly than do words, which may be falsified." The book aimed to explain in evolutionary terms how such movements came to be expressive of certain emotions rather than others: a smile for happiness, a blush for embarrassment, and so on.[92] More straightforwardly descriptive was an 1877 paper in *Mind*, "A Biographical Sketch of an Infant," in which Darwin summarized observations recorded over thirty years before. The infant in question had obliged its papa by more or less recapitulating the acquisition of language by the species: first squalling; then using facial expressions; then babbling; then laughing; then imitating; then producing articulate but meaningless sounds; then using gestures; and then, at one year old, inventing a word, subsequently elaborated. Reflecting now on the infant's instinctive tendency to speak words with a rapidly rising intonation, Darwin made a connection to a more distant, less human relative: "I did not then see that this [tendency] bears on the view which I have elsewhere maintained that before man used articulate language, he uttered notes in a true musical scale as does the anthropoid ape Hylobates."[93]

W. D. Whitney and the Social Origins of the Language Barrier

Müller's "Lectures on Mr. Darwin's Philosophy of Language"

After Darwin had at last published his views on the evolutionary origins of language in the *Descent* and the *Expression of the Emotions*, Müller again took to the platform of the Royal Institution, delivering three lectures on "Mr. Darwin's Philosophy of Language" in the spring of 1873.[94] Since his last lecture there on Darwinian topics, Müller had largely retreated from the vanished-instinct theory of root origins, insisting by the 1872 (sixth) edition of his *Lectures on the Science of Language* that he had discussed the instinct theory merely to show that there were alternatives to the imitative and interjectional theories then dominant.[95] Müller made plain in his new lectures that catastrophic origins for the roots were entirely consistent with uniformitarian origins for their phonetic shells. Thus "the sounds *cuckoo* and *cock* might [have been] used as the phonetic signs of these two birds . . . [and] if a phonetic sign was required for the singing of more birds, or, it may be, of all possible birds, . . . a filing down of the sharp corners of those imitative sounds, would answer the new purpose." But for all the gradualism in the phonetic aspect of root assembly, the concepts expressed in the roots had arisen *as* concepts, catastrophically, whole.[96] Here Müller was as firm as ever. In stating his case for catastrophism, however, he now turned to a different philosophical idiom. In his 1861 lecture, still smarting from the Boden professorship fiasco, Müller had fitted out his catastrophism with quotes from the eminently English empiricist Locke. The Kantian allegiances that drove the catastrophism had been unstated. More than ten years later, those allegiances were out in the open.

The lectures opened on a note of crisis. According to Müller, the old barriers—between humans and the lower animals, between the organic and the inorganic, between matter and mind—were everywhere said to be crumbling before the assault of materialist and evolutionist science. But science alone could not remove those barriers. Science was tangled up with philosophy. In particular, the belief that the human mind had evolved out of the animal mind presupposed that all knowledge was built up from experience. "If Mr. Darwin is right," said Müller, "if man is either the lineal or lateral descendant of some lower animal, then all the discussions between Locke and Berkeley, between Hume and Kant, have become useless and antiquated," because evolutionism and empiricism stood or fell together. If the empiricists were right—if experience sufficed to supply the contents of the human mind as well as the contents of the animal mind—then the difference between human minds and animal minds was but a difference of degree, and it would have been, as the evolutionists held, a matter of mere "time and circumstances" for nature to transmute "a man-like ape . . . into an ape-like man."[97]

But, Müller urged, Kant had shown that experience was *not* adequate to account for the contents of the human mind. Kant had shown the human mind to be no passive blank tablet. It came equipped with the a priori intuitions and categories through which humans organized their experience of the world. Was it nevertheless possible, as Herbert Spencer had argued, that, in Müller's words, the "genesis of these congenital dispositions or inherited necessities of thought" had begun far lower in the scale of nature, becoming ever more elaborate in the course of evolution? Here Müller adduced the contrary evidence of the abstract roots of language. Only roots, he argued, counted as evidence for the presence of conceptual thought, and since no root had been discovered among the lower animals, there were no grounds for supposing that animal minds were capable of such thought. Furthermore, because the roots were irreducible, and because each one was the sign of a concept, the roots showed that concepts had sprung into being fully abstract. Here comparative philology (the science of language) and Kantian philosophy (the science of thought) converged: words imitative of particular sounds, as signs for ideas of particular objects associated with those sounds, could no more have developed into roots than the signified ideas could have developed into concepts.[98]

Ignorant of the results of the science of language, in the grip of an antiquated understanding of mind and knowledge, Darwin had offered an account of the origins of language built on dogmatic assertion rather than positive evidence. Müller challenged the Darwinians to do better:

> Show me only one single root in the language of animals, such as AK, to be sharp and quick; and from it two such derivatives as asva, the quick one—the horse—and *acutus*, sharp or quick-witted . . . and I should say that, as far as language is concerned, we cannot oppose Mr. Darwin's argument, and that man has, or at least may have been, developed from some lower animal. I do not deny that there is some force in Mr. Darwin's remark, that both man and monkey are born without language; but I consider that the real problem which this remark places before us is to find out why a man always learns to speak, a monkey never. If, instead of this, we say that, under favourable circumstances, an unknown kind of monkey may have learned to speak, and thus, through his descendants, have become what he is now, viz. man, we deal in fairy-stories, but not in scientific research. . . . Certain it is, that neither the power of language, nor the conditions under which alone language can exist, are to be discovered in any of the lower animals.[99]

Müller again defended the most certain conclusion of the science of language: "*There is no thought without words, as little as there are words without thought.*" Language and reason presupposed each other. Those who claimed they were able practitioners of a "new art of ventriloquism," namely, "thinking thoughts without words," were deluded.[100]

When Müller sent Darwin a copy of the lectures (published in *Fraser's Magazine*), Darwin wrote a friendly letter of thanks. "As far as language is concerned," Darwin wrote, "I am not worthy to be your adversary, as I know extremely little about it, and that little learnt from very few books." He confessed he was not at all open-minded about the evidence for and against his views on the origin of language. "He who is fully convinced, as I am, that man is descended from some lower animal, is almost forced to believe *a priori* that articulate language has been developed from inarticulate cries; and he is therefore hardly a fair judge of the arguments opposed to this belief."[101] In the pages of the "Lectures," Darwin scrawled more specific objections. "Monstrous sentence" was his comment on the words-thoughts equation. "One often utterly forgets name of man, animal or substance & yet can think about it definitely.—I am sure that I have when writing forgotten word for complex feeling. (for instance Emulation) & yet have had a definite notion in my head."[102] Later that summer Frances Julia Wedgwood sent Darwin an English version of some criticisms of the "Lectures" from a German periodical.[103] But by far the most damaging review came the following summer from the Ameri-

Figure 1.3 William Dwight Whitney (1827–94), date unknown. From the *Journal of the American Oriental Society* 19 (1897): ii.

can language scholar William Dwight Whitney, based at Yale College, where he was professor of Sanskrit and comparative philology (fig. 1.3). Although, like Müller, Whitney had studied Sanskrit in Berlin, his views on the relations between language and thought were resolutely Lockean. For Whitney, as for Locke, words were merely the instruments of thought. They were signs fixed by convention. A language was not an organism but a social institution. Undoubtedly humans alone communicated ideas from mind to mind using arbitrary signs, and in this sense there was indeed a language barrier in nature; but this barrier was best understood as part of the more general tool-use barrier that separated humans from the other, less intelligent species.[104]

Throughout the 1860s and early 1870s, Whitney poured acid criticism over Müller's work. Müller's vanished-instinct theory of the origin of language, wrote Whitney, "may be very summarily dismissed, as wholly unfounded and worthless. . . . In effect, it explains the origin of language by a miracle, a special and exceptional capacity having been conferred for the purpose upon the first men, and withdrawn again from their descendants."[105] The equation of language with reason and humanness led to "Müller's worst paradoxes, that an infant (*in fans*, not speaking) is not a human being, and that deaf-mutes do not become possessed of reason until they learn to twist their fingers into imitation of spoken words."[106] Müller's translation of the Rig-Veda was in a class of one because "we hardly know a volume of which the make-up is more unfortunate and ill-judged, more calculated to baffle the reasonable hopes of him who resorts to it."[107] Müller's later claim that he had never advanced the vanished-instinct theory himself, but only explored its explanatory power, showed "either disingenuousness or remarkable self-deception; or, perhaps we ought to add, one of the most extraordinary cases on record . . . of failure to make one's self understood."[108] Müller's influence for the good on the science of language was balanced by "nearly equal an amount of harm he has done by inculcating false views and obstructing better light."[109]

It is hard to see just what Müller had done so to irritate Whitney. They seem to have enjoyed cordial if perfunctory professional relations.[110] Perhaps it was enough that the somewhat older Müller had usurped the role of public spokesman for the science of language in the English-speaking world. Whatever the sources of his resentment, Whitney had nursed it for some time before Müller's 1873 attack on Darwinism.

Whitney versus Müller: The View from Down House

Whitney's criticisms were familiar to Darwin. He would go on to quote the jibe about the language-reason paradoxes in the 1874 edition of the *Descent*, adding, with respect to Müller's aphorism about no thought without words and vice versa: "What a strange definition must here be given to the word thought!"[111] Whitney's general approach to language was not wholly congenial, however. One problem was his insistence on the determining role of the will in the growth of language. In Darwin's view, unconscious processes were far more important determinants.[112] In June 1872, Darwin asked Chauncey Wright to adjudicate:

As your mind is so clear and as you consider so carefully the meaning of words, I wish you would take some incidental occasion to consider when a thing may properly be said to be effected by the will of man. I have been led to the wish by reading an article by your Professor Whitney. . . . He argues that, because each step of a change in language is made by the will of man, the whole language so changes; but I do not think that this is so: a man has no intention or wish to change the language. It is a parallel case with what I have called "unconscious se-lection," which depends on men consciously preserving the best individuals, and thus unconsciously altering the breed.[113]

When he received this philosophical assignment from Darwin, Wright was in Britain on holiday; in late August 1872, he sent a long letter in reply. He touched on a number of problems: whether personal agency and respon-sibility extend beyond intended consequences; whether an action freely taken stands outside the causal chains that bind natural phenomena ("I believe that this view is purely fanciful, or at least poetical; but that it is implicitly con-tained in, or lies at the bottom of, such objections as Professor Whitney's"); whether Darwin's theory of unconscious selection for language put Darwin closer to Whitney or to Müller ("the theory as it stands is not, it seems to me, inconsistent even with Müller's views, since it ascribes nothing and denies nothing to variations as a direct cause of changes in species or structures or habits or customs . . . [but] only attributes to them opportunities or the condi-tions for *choice*, and does not deny to them other forms of agency").[114] Later that month, Wright visited Darwin at Down House. Wright continued to work on the problem for some time thereafter.[115]

With the appearance in the July 1874 *North American Review* of "Dar-winism and Language," Whitney's response to Müller's recent Royal Institu-tion lectures, Whitney's relentless public campaign against Müller entered a new phase. According to Whitney, the results of the science of language were irrelevant to the debate about Darwinism. Whenever language began, it be-gan because humans already possessed of distinctly human powers desired to communicate with each other, and set about adapting the materials at hand to satisfy this desire. It was this astonishing capacity to adapt means to ends, rather than specific results of this capacity such as language or machines, that separated man from the lower animals.[116] Were human minds therefore dif-ferent in degree or kind from animal minds? Whitney rejected the question. Indeed "a chasm, not a step, separates us from our nearest inferiors," he wrote, and there were "no steps between the wholly instinctive expression of the animals and the wholly (so far as articulate words are concerned) conven-tional expression of man."[117] But the latter depended on the joint workings of a number of subsidiary capacities. It was an open question, Whitney argued, to what extent these capacities were present but insufficiently developed or coordinated in the lower animals.[118] In his view, between animal cries and human speech "[t]here is neither *saltus* nor gradual transition . . . : no transition, because the two are essentially different; no *saltus*, because human speech is an historical development out of infinitesimal beginnings, which may have been of less extent even than the instinctive speech of many a brute." The Darwinians no less than Müller had erred in viewing speech as part of the human biological endowment. "Speech, like the other elements of our civilization, is the result of our human capacities, not their cause; . . . it trains [man's] mental powers to a higher capacity of labor, but adds no new powers; least of all does it produce modifications of physical structure that look toward the founding of new varieties or species."[119]

On reading Whitney, Darwin immediately wrote to John Knowles, editor of the *Contemporary Review*, and asked if the journal would reprint "Darwinism and Language"—Darwin offered to pay the expenses—to combat the influence among British readers of Müller and a like-minded anonymous reviewer in July's *Quarterly Review*.[120] Knowles politely declined, as reprinting was bad for business, but suggested that Darwin send in something of his own, perhaps with lengthy quotations from "Darwinism and Language."[121] Darwin put the matter to one side for the time being, however, claiming a need for rest far more than for controversy.[122] He reported these editorial negotiations in a letter to Whitney, adding that Whitney's article "seems to me most clearly reasoned, & by far the best argument against Max Müller's views which has ever appeared."[123] But Darwin's own copy of "Darwinism and Language" (sent to him by Whitney) shows that Darwin nevertheless still had problems with Whitney's position. In the *Descent*, for example, Darwin had claimed that, in speaking, early humans had exercised their brains and vocal organs, causing these to become gradually more highly organized. Speech in Darwin's view was thus at once the cause and the consequence of the superior capacities of humans. For Whitney, however, speech was merely a consequence of superior capacities, not a cause. In a marginal query, "wd increase power of brain by use?" Darwin flagged this disagreement over the transformative power of speech in human evolution.[124]

Not long after this flurry of anti-Müllerian maneuvering at Down House, Müller himself came for a visit. He later described his meeting with Darwin:

> Sir John Lubbock took me to see the old philosopher at his place, Down, Beckenham, Kent, and there are few episodes in my life which I value more. I need not describe the simplicity of his house, and the grandeur of the man who had lived and worked in it for so many years. Darwin gave me a hearty welcome, showed

me his garden and his flowers, and then took me into his study, and standing lean-
ing against his desk began to examine me. He said at once that personally he was
quite ignorant of the science of language, and had taken his facts and opinions
chiefly from his friend, Mr. Wedgwood. I had been warned that Darwin could
not carry on a serious discussion for more than about ten minutes or a quarter of
an hour, as it always brought on his life-long complaint of sickness. I therefore
put before him in the shortest way possible the difficulties which prevented me
from accepting the theory of animals forming a language out of interjections and
sounds of nature. I laid stress on the fact that no animal, except the human animal, had ever made a step towards generalisation of percepts, and towards roots,
the real elements of all languages, as signs of such generalised percepts, and I gave
him a few illustrations of how our words for one to ten, for father, mother, sun
and moon had really and historically been evolved. That man thus formed a real
anomaly in the growth of the animal kingdom, as conceived by him, I fully admitted; but it was impossible for me to ignore facts, and language in its true meaning
has always been to my mind a fact that could not be wiped away by argument,
as little as the Himalayas could be wiped away with a silk handkerchief even in
millions of years. He listened most attentively without making any objections, but
before he shook hands and left me, he said in the kindest way, "You are a dangerous man." I ventured to reply, "There can be no danger in our search for truth,"
and he left the room.[125]

Müller met Darwin on the page as well as in person around this time. In
October 1874, the *Westminster Review* published an anonymous commentary on the debate between the two over the evolutionary origins of language.
Just as in the 1860s, the *Westminster* reviewer came down firmly on the side
of Darwin against Müller, adducing along the way a novel but eminently
Darwinian argument. According to the reviewer, only the theory of evolution made it at all intelligible why speech acquisition in children should be a
gradual process. The evolutionist held that each new stage in the acquisition
of speech by the earliest humans or protohumans occurred, in the words of
the reviewer, "not first in the minds and mouths of animals very young or
newly born, but in those of the adult and mature." The reviewer continued:
"Then, like many other variations, [speech acquisition] would have a tendency to be inherited at a corresponding age, and, in course of time, at successively earlier periods of life." This embryological theory of terminal additions was of course Darwin's own, formulated, as the reviewer pointed out,
to explain observations in the embryology of crustaceans. "[I]t must," the
reviewer concluded, "be admitted as at least a remarkable coincidence when
the same theory will equally account for certain phases in the life-history of a
shrimp, and for the fact that infants cannot speak, while both boys and men
are able to do so."[126]

Whitney versus Müller: Public Escalation

Darwin's son George now spoke up in place of his father. Primed for debating mental evolution from his work on the second edition of the *Descent*, George Darwin entered the fray with a brief paper in the November 1874 *Contemporary Review*, "Professor Whitney on the Origin of Language." It was a more or less straightforward report on Whitney's "Darwinism and Language," praising the points made against Müller and fending off the points made against Darwin père. "You have defended me nobly" was Charles's proud verdict.[127] George sent his article to Müller, whose published response, "My Reply to Mr. Darwin," appeared in the January 1875 number of the same journal.

Müller claimed he had been unaware of "Darwinism and Language" until George Darwin's notice. Having failed to obtain a copy thus far, Müller instead had for the first time examined Whitney's own lectures on the science of language, published in 1867 as *Language and the Study of Language*. "I have seldom perused a book with greater interest and pleasure,—I might almost say amusement," wrote Müller. "It was like walking through old familiar places, like listening to music which one knows one has heard before somewhere, and, for that very reason, enjoys all the more." According to Müller, passage after remarkable passage revealed Whitney had consulted the same sources, and had even experienced "the same doubts and difficulties, the same hopes and fears, the same hesitations and misgivings" Müller had experienced in preparing his earlier *Lectures*.[128] Müller padded out his insinuation of plagiarism with unflattering personal remarks quoted from other reviews of Whitney's work: "'That vain man who only wants to be named and praised;' 'that horrible humbug;' 'that scolding flirt;' 'that tricky attorney;' 'wherever I read him, hollow vanity yawns in my face; arrogant vanity grins at me.'"[129]

Müller claimed to be at work on a new book, entitled *Language as the True Barrier between Man and Beast*.[130] Whitney shot back in the April *Contemporary Review* with "Are Languages Institutions?" and later that year with a chapter on the origin of language in a new set of lectures, *The Life and Growth of Language*. The *Contemporary Review* article reiterated Whitney's views on the nature of language and protested the charges against him. Perhaps responding to an intimation of George Darwin's that Whitney was not at heart an evolutionist, Whitney made plain that language placed "no impassable barrier, but only an impracticable distance," between humans and the lower animals, and called for further research into the noninstinctive communicative means of the animals.[131] As he explained in *Life and Growth*,

[w]e need not be surprised to find, in more than one quarter, such methods of communication in use, only limited, and, for lack of the right kind and degree of capacity in their users, incapable of development; and these would be the real analogues of speech, and would bridge the *saltus* of which some are so afraid. If the Darwinian theory is true, and man a development out of some lower animal, it is at any rate conceded that the last and nearest transition-forms have perished, perhaps exterminated by him in the struggle for existence, as his special rivals, during his prehistoric ages of wildness; if they could be restored, we should find the transition-forms toward our speech to be, not at all a minor provision of natural articulate signs, but an inferior system of conventional signs, in tone, gesture, and grimace.[132]

As for the long-ago first words, Whitney, like the evolutionists, favored the "*vera causa*" of mimesis over any explanation "which calls in a special force at the beginning, like a *deus ex machina*, to accomplish what we cannot see to be otherwise feasible, and then to retire and act no more."[133]

While Müller labored away on his new book through the first half of 1875, Whitney twice tried and failed to visit Darwin at Down House, first in May (when illness and the demands of *Insectivorous Plants* caused Darwin to cancel), then again in early September (when the Darwins were away in Southampton).[134] In late September, Müller finished what had become a long essay, now bearing the title "In Self-Defence: Present State of Scientific Studies," and included as the final chapter in the forthcoming fourth volume of his *Chips from a German Workshop*. "I think you will see from what I have stated," wrote Müller in the letter accompanying Darwin's copy of "In Self-Defence," "that Professor Whitney is not an ally whom either you or your son would approve of."[135] The new essay ended with a list of twenty points in dispute between Müller and Whitney, and a challenge to Whitney to assemble a tribunal of three qualified judges to rule on these points.[136] Darwin wrote back only that Müller's criticisms of the evolutionist position "have all been made so gracefully, I declare that I am like the man in the story who boasted that he had been soundly horsewhipped by a Duke."[137]

Near the end of that year George Darwin advised Whitney that "[o]ne does not see the end of this kind of polemic & if it were I myself, I think I should have been rather inclined to allow time to cool the heels of controversy and bring forth the just view of the case."[138] A mutual friend of Whitney and Müller, the American minister Moncure Conway, tried to negotiate a truce, but to no avail.[139] With the publication of Müller's *Chips* volume, the Whitney-Müller debate became more heated, and its public profile more conspicuous. In January 1876, Whitney published a letter in the London *Academy* protesting its favorable review of Müller's volume, adding that Müller himself was welcome to choose the members of the adjudicating tribunal.[140] Two months later, the editors declined to publish a further letter from Whitney on the grounds that, as the *Academy*'s readers learned, "the letter is simply calculated to stir up further bad blood, without contributing a single point towards the final settlement (if such be desired) of the controversy at present outstanding."[141] Whitney succeeded nevertheless in having the letter published (with indignant commentary) both in the London *Examiner* and in the *Nation*. Reports on the developing controversy began to appear in even the smallest-circulation newspapers and periodicals, under titles such as "The Whitney-Müller Controversy," "The Battle of the Sanscrits," and "The Müller-Whitney War." Intrigued by the issues and amused or dismayed by the attitudes of the combatants, the authors of these reports managed to bring the debate about the evolutionary origins of language, and in particular Müller's arguments against the Darwinians, before a much wider and international audience.[142]

Although 1875–76 was the high point of the Whitney-Müller controversy, Whitney kept up a ceaseless tirade, culminating, in the spring of 1892, in an entire book of criticism—and a new spate of journalism about the now aged philological rivals.[143] For his part, Müller ignored Whitney's attacks after the mid-1870s, but the fight against Darwinism was another matter. Early in 1878, Müller published an enthusiastic notice in the *Contemporary Review* of a new theory of root origins, due to the German philosopher Ludwig Noiré, and, in Müller's view, by far the best theory yet formulated. According to the new "sympathic" theory—later known as the "yo-he-yo" or "heave-ho" theory—the roots of language emerged from sounds uttered spontaneously and collectively as the first humans acted cooperatively.[144] In 1887, Müller re-entered the fray with a sprawling summary of his views on the nature, origins, and philosophical implications of language, *The Science of Thought*. The book went on to provoke a lively correspondence in the letters pages of *Nature* over the relations between thought and language, in which, among others, the pro-Darwinian Francis Galton and the anti-Darwinian politician the duke of Argyll debated Müller's arguments.[145]

All in all, Müller managed to keep the debate over the language barrier and its significance on the boil for some thirty years. At the 1891 meeting of the BAAS, he expressed satisfaction at the results of his long campaign. "It required some courage at times to stand up against the authority of Darwin," Müller told his anthropological audience, "but at present all serious thinkers agree . . . that there *is* a specific difference between the human animal and all other animals, and that that difference consists in language."[146]

2

BRAINS AND MINDS ACROSS
THE BARRIER

◆ ◆ ◆

"ONE DOES NOT SEE the end of this kind of polemic," George Darwin had warned. Agreement about the nature and causes of the language barrier separating humans and animals indeed proved elusive in the nineteenth century. But there emerged consensus, and enduring consensus, about two related issues: how language arises in the human brain, and how best to interpret the seemingly rational behavior of animals. This chapter seeks to root this double consensus in the Victorian debate on animal language by showing how it generated doctrines widely taken for granted even now. These doctrines are that language arises through the coordinated activity of several parts of the human brain, and that anthropomorphic interpretations of animal behavior—attributing to animal minds the ideas that would cause humans to behave thus—are unscientific. The doctrine about the human brain is due above all to the physician John Hughlings Jackson; the doctrine about the animal mind above all to the psychologist Conwy Lloyd Morgan.

Jackson's role in producing the brain doctrine is well known. What has been forgotten are the links his contemporaries saw between the problem of locating or "localizing" language in the brain and the problem of showing the brain to be a product of evolution. To restore this evolutionary dimension to the brain-language question, the first part of this chapter pairs Jackson with a less celebrated colleague, the physician Frederic Bateman. Shortly after Darwin's *Descent* was published in 1871, Bateman sent a copy of his recent book *On Aphasia* (1870) to Darwin. Bateman's book was the first in English devoted, in the words of its subtitle, to "the loss of speech, and the localisation of the faculty of articulate language." Along with the book, Bateman sent a letter summarizing his findings and what he took to be their relevance to Darwin's theory of human descent.[1] There was little evidence, wrote Bateman, that loss of speech occurred because of lesions to the region of the brain identified by his friend Professor Broca of Paris as the seat of articulate language. In light of the investigations documented in the book, Bateman continued, "I am tempted to ask whether there be any *cerebral centre* for speech, and whether speech may not be an attribute, the comprehension of which is beyond the limits of our finite minds?" Furthermore,

> this has an important bearing upon the great question you are so laboriously working out, for if the faculty of speech can be traced to no *material centre*, does it not offer an objection to the belief that man has been developed from some lower form? Does it not tend to prove that the possession of this "attribute" is one great barrier between man and animals[?][2]

In 1861, F. Max Müller had publicized the language barrier between brute and man at the same moment that Richard Owen publicized the hippocampus-minor barrier between gorillas and humans. In 1871, it was not the hippocampus minor but the third frontal convolution, Broca's seat of articulate language, on which man's place in nature seemed to pivot. In exploring the origins and development of Bateman's aphasia-based case against evolution, the first part of this chapter builds a new and unfamiliar context for understanding relations between the theory of evolution and the science of brain in the later nineteenth century.

Moving from brains to minds, the second part also seeks to reveal forgotten connections, again through an unlikely pairing. Conwy Lloyd Morgan announced the rule of interpretation he was later to call his "canon" in August 1892 as follows: "in no case is an animal activity to be interpreted as the outcome of the exercise of a higher psychical faculty, if it can be fairly interpreted as the outcome of the exercise of one which stands lower in the psychological scale."[3] From the vantage point of present-day psychology, Morgan's canon stands, in the words of one commentator, as "possibly the most important single sentence in the history of the study of animal behavior."[4] That is because built into the canon is the still-influential view that objectivity and anthropomorphism are mutually exclusive. Referring recently to the American L. H. Morgan's 1868 claim that beavers command engineering know-how, the ethologist Aubrey Manning wrote: "We now know better, and follow another Morgan—Lloyd—whose 'canon' exhorts us never to interpret behaviour using complex psychological processes if simpler ones will serve."[5] Since the latter Morgan and his canon, objective observers have appealed whenever possible to quasi-mechanical processes such as trial-and-error learning and imitation, rather than understanding and purpose-guided planning, to explain how animals come to act in apparently clever ways. What Morgan called "reason" is presumed absent in animal minds until shown otherwise.

So successful did Morgan's canon become that it now takes some effort to see it as anything other than the crystallization of scientific good sense. Historians of comparative psychology have been little help here, tending to present the canon as a more or less inevitable product of a Victorian debate on the nature of instinct, as if, as soon as scientists had started to get serious about understanding animal actions and animal minds, they were bound to subscribe to a skeptical rule of interpretation at some point or other. The desirability of the canon is treated as self-explanatory.[6] In the present historiography, we can thus learn how Morgan's skepticism informed his studies of trial-and-error learning (what Morgan called "intelligence"),[7] or persisted alongside his mind-matter monism,[8] or fed the even deeper skepticism of the behaviorists whom he influenced,[9] but not why Morgan became so skeptical, or recommended others to be so. The second part of the chapter aims to return the canon to its true matrix—the Victorian language-barrier debate—in order to dispel this air of inevitability. We will see that behind Morgan's skepticism about animal reason was F. Max Müller's skepticism about animal language. According to Müller, as we have seen, language and reason went together, and belonged to humans alone. It was Müller, in his 1873 "Lectures on Mr. Darwin's Philosophy of Language," who sparked the first major scientific debate about anthropomorphism in interpretations of animal actions—a debate that eventually resulted in the robustly anthropomorphic writings of Müller's opponent George John Romanes. Morgan's canon in effect turned the central theme of Müller's comparative philology into a reforming rule of method for Romanesian comparative psychology. For Morgan, the canon was needed because animals, lacking language, probably lacked reason.

In the third and final part of the chapter, these stories of the 1870s and 1880s, concerning the brain-and-language barrier (Jackson and Bateman) and the mind-and-language barrier (Morgan and Müller), are tracked through to the early 1890s. Bateman's challenge to the evolutionists and the localizers was, we shall see, less answered than absorbed, in large measure because his central claim—that there is no single seat of speech in the human brain—came to be accepted among students of brain function as just what the theory of evolution predicted; while an examination of Morgan's writings around the time of the canon's 1892 debut reveals how, from Morgan's own point of view, there was nothing at all inevitable about the triumph of his canon.

Brains across the Language Barrier

Speech Localization in the Early Nineteenth Century

Controversy over language and the brain in the first half of the nineteenth century centered on the claims of the Viennese anatomist Franz Joseph Gall as debated in Paris, Gall's adopted home from 1807. If, as was widely believed, the brain was the organ of mind, and if an organ such as the liver had the functions it did in virtue of its internal structure and the needs of the body as a whole, then what, asked Gall, were the functions of the brain? Rejecting the old classification of the mind into abstract subfaculties of the understanding (memory, intelligence, imagination, and so on), Gall looked for functions of the brain that, like the functions of the liver, made sense in view of what creatures needed in order to thrive. In place of the single faculty, memory, Gall postulated different kinds of memory, serving different needs. The first faculty he identified was related to language: the faculty of verbal memory. To discover where in the brain one of these new functional faculties was located, Gall generally looked (in line with eighteenth-century traditions in physiognomy and medicine) for a correlation between an unusually robust expression of a faculty and an unusually prominent bump on the skull, inferring that this faculty had its seat in a cerebral suborgan just behind the bump. He was doubtful about the value of studying damaged brains to learn about brain function, and savagely critical of those who inflicted such damage on purpose—"the mutilators," as he called them, whose "cruel experiments, when they are made on animals of an order comparatively low, are hardly ever conclusive for man." In the end Gall settled on twenty-seven sets of bumps, faculties, and organs, nineteen of which humans shared with other creatures.[10]

Chief among the Parisian mutilators was Marie-Jean-Pierre Flourens. As a mind-matter dualist, Flourens found Gall's claims about mind and brain to be metaphysically suspect. But it was as a physiologist, in particular as a pioneer of the use of surgical removal or "ablation" in investigations of the nervous system, that Flourens criticized those views most effectively. To investigate brain function, Flourens cut out regions of the brains of living animals (principally birds) and observed the consequences on their behavior. Thus he claimed to show that the cerebral hemispheres played no role whatsoever in producing motion; that the strictly intellectual functions of the hemispheres were merely aspects of a single general faculty of understanding; and that these aspects of the understanding were so distributed that each point on the cerebral surface was functionally equivalent to all other points. In a book against the phrenology that grew up around Gall's doctrines, Flourens upheld the unity of the organ of mind as evidence of the unity of the soul and the separation of mind from matter.[11]

Based at the Académie des sciences, Flourens and his research enjoyed great prestige in scientific Paris. Nevertheless, Gall's localization hypothesis remained under active discussion between the 1820s and 1860s, largely though the efforts of Jean Baptiste Bouillaud and his studies of language and its disturbance due to diseases and injuries of the brain. Bouillaud had studied localization with Gall himself, and helped found the Société phrénologique. He nevertheless abandoned the study of bumps for the study of lesions—natural and induced—to show that the hemispheres were indeed involved in motion, and that, "[i]n particular, the movements of the organs of speech are regulated by a special cerebral centre, distinct and independent . . . situated in the anterior lobes of the brain."[12]

Between Gall's precise but poorly supported localization of language, and Bouillaud's imprecise but better supported localization, there took place a much more thorough mapping of the convolutions of the hemispheres.[13] In early 1861, during a debate at the recently founded Société d'anthropologie, the surgeon and professor of medicine Pierre-Paul Broca (founder of the society) presented evidence in support of Bouillaud and the localizers. One of Broca's patients had died after an illness that had destroyed his powers of speech without destroying his powers of thought. Broca inferred from the pattern of damage to the brain tissue and from the clinical history of this "aphemic" patient that the lesion had begun near the posterior end of the third convolution of the frontal lobe. Broca proposed that here was "the seat of the faculty of articulate language." The next few years saw celebrated debates at the Société anatomique and the Académie de médecine on the truth of Broca's claim. As the clinical and pathological data accumulated, Broca began to argue for localization of the language faculty not merely in a single set of convolutions—the two hemispheres being symmetrical in their convolutions—but in the *left* member of that set. Gall and his followers had argued that, while the regions within a hemisphere had different functions, the hemispheres themselves were symmetrical, so that the same functions were represented in the same regions of each hemisphere. Broca was doing away with functional symmetry as well as functional equivalence. He tended nevertheless to describe the faculty of articulate language in terms congenial to Flourens's doctrine that neither hemisphere was involved in movement. Broca's "aphasic" patients (as they came to be known) were diagnosed as having forgotten how to articulate certain words or certain parts of words without loss of other intellectual and motor powers. With Broca's entry into the sixty-year-old debate, the localization hypothesis began to generate attention not merely in France but from men of science and medicine from around the world—including Frederic Bateman.[14]

Bateman, Language, and Brain Function

Born in Norwich in 1824, Bateman (fig. 2.1) first studied medicine in the Norfolk and Norwich Hospital. He continued his education in London, then Paris, and then Aberdeen, where he became a doctor of medicine in 1850.[15] His time

Figure 2.1 Frederic Bateman (1824–1904), date unknown. From the *Lancet*, 20 August 1904, 567.

in the hospitals of Flourens's Paris in the mid-1840s was the beginning of what would be a strong and lifelong connection with French medical science. (His first book, published in 1849, was a translation of a French essay on the treatment of cholera.)[16] His education completed, Bateman moved back to Norwich, where he became house surgeon to the hospital and set up in general practice. In the early 1860s, he returned briefly to Paris, and was soon caught up in the excitement over Broca's claims. A junior colleague later recalled that he "frequently saw [Bateman] in the wards of the Hôtel-Dieu, La Charité, and other hospitals, and noted at the time the zeal and energy with which he followed the practice and teachings of the men who had contributed to raise the Parisian school to the then zenith of its fame." Bateman took a particular interest in the work of Armand Trousseau, professor at the Hôtel-Dieu, and one of Broca's most formidable critics. Back in Norwich, Bateman gave up his general practice to become, in 1864, physician to the city hospital, where he would preside for the next thirty-one years. In 1865, he saw his first aphasic patient at Norwich, a fifty-one-year-old waterman named William Sainty, who had suddenly and without any other impairment lost his ability to say more than "Oh dear! Oh dear!" Bateman wrote up the Sainty case for the *Lancet* and, beginning with Trousseau's lectures, entered into a thorough review of the literature on aphasia, the results of which he published as six articles in the *Journal of Mental Science* from January 1868. (These articles in turn became the six chapters of *On Aphasia*.)[17]

In August of that year, Norwich hosted the annual meeting of the BAAS. At Bateman's invitation, Broca addressed the biological section on aphasia and the localization of articulate language.[18] Sharing the bill was the London physician John Hughlings Jackson, already well known for his studies of the nervous system and its disorders, including aphasia. The *Lancet* looked forward to "the opportunity of immediately comparing the best English and the best French views on the pathology of this remarkable disease."[19] The session did not disappoint. Broca mounted a spirited defense of his localization of language in the third frontal convolution of the left hemisphere.[20] But the *Lancet* nevertheless detected a general turning away from "that coarse and mechanical view of the function of speech which regarded it as a small and separate portion of the brain."[21] This turning away from Broca was at the same time a turning toward Jackson. At Norwich, Jackson insisted on the complex nature of language, and the weakness of inferences made about language and brain function from lesion-induced aphasia. He reported cases where the ability to form propositions and the ability to swear were affected separately. He also pointed out how often aphasia did not occur in isolation from other intellectual and motor impairments.[22]

In the general discussion that followed, Bateman sided with Jackson. Of the twenty-seven cases of speech disruption Bateman had examined, a mere five, he reported, confirmed Broca's localization of the language faculty. Indeed, five cases showed no brain lesion whatsoever.[23] But others spoke up in defense of Broca. Among the most formidable was Carl Vogt. Distinguished for his anatomical work, notorious for his materialism and socialism, the Geneva-based Vogt was a tireless advocate of *Darwinismus*. He joined Joseph Hooker, Alfred Russel Wallace, T. H. Huxley, and the physicist John Tyndall in arguing the transmutationist case in religiously conservative Norwich that summer.[24] In support of Broca, Vogt offered his observation that in apes a distinct third frontal convolution in the left hemisphere was, as the *Lancet* emphasized, "*entirely absent.*"[25] Around this time or shortly thereafter, Bateman received a letter from Vogt describing his findings and their significance in more detail. Bateman quoted at length from the letter in the final chapter of *On Aphasia*, commenting that "these views of Professor Vogt are not very generally known in this country, and I need hardly allude to the extremely important bearing they have upon the issues in question":

> In Man, the third frontal convolution is extraordinarily developed . . . [while] in the Ape . . . the third frontal convolution is but slightly developed. . . . To show the bearing all this has upon the seat of speech, I would refer to the Microcephali who do not speak—they learn to repeat certain words like parrots, but they have no articulate language. Now, the Microcephali have the same conformation of the third frontal convolution and of the central folds as Apes—they are Apes as far as the anterior portion of their brain is concerned, and especially as far as regards the environs of the fissure of Sylvius [i.e., the region of the third frontal convolution]. Thus, Man speaks; Apes and Microcephali do not speak; certain observations have been recorded which seem to place language in the part which is developed in man and contracted in the Microcephali and the Ape; comparative anatomy, therefore comes in aid of M. Broca's doctrine.[26]

There is some evidence that Charles Darwin, at least, was independently familiar with Vogt's conclusions. In 1867, Darwin had received from Vogt a copy of his massive *Mémoire sur les microcéphales ou hommes-singes*, summarizing its contents in a letter. Of the brains of microcephalous idiots, Vogt wrote: "I arrive at the conclusion that [their] abnormal conformation is an atavism which leads back towards the point of separation of the two stocks, men and apes—but that this point of separation is no longer represented in the present scheme of things."[27] Darwin later used Vogt's *Mémoire* to buttress the argument for the imitative origins of language in the *Descent*. Citing Vogt, Darwin wrote of the evidence of "the strong tendency in our nearest allies, the monkeys, in microcephalous idiots, and in the barbarous races of mankind, to imitate whatever they hear."[28]

Just after the cited passage in Vogt's *Mémoire* was an entire section devoted to "articulate language."[29] Discussing the case of a highly intelligent
woman whose power of speech was diminished after an apoplectic attack,
Vogt wrote:

> If I am not mistaken, it is precisely [the] faculty of combination [of sounds, letters,
> and signs into words] which, following the admirable work of M. Broca, resides
> in the posterior part of the left superciliary formation. The observations in which
> certain words and certain categories have been conserved shows even, in my opin-
> ion, that this faculty possesses so to speak its store which can be destroyed entirely
> or in part. Now, if I compare with these facts the language of animals, it seems
> to me that the apes [and] the microcephalous do not speak because the faculty of
> combination and the store of the third convolution is missing in them.[30]

Did Darwin share Vogt's and Broca's belief that the highly developed third
frontal convolution was the anatomical and evolutionary seat of speech? On
the specific claim for the third frontal convolution, Darwin was silent. But
on the more general claim that *some* part of the brain was devoted to speech,
there is evidence of agreement. The evidence comes from Darwin's marginal
notes on two works. The first was the German zoologist Ernst Haeckel's
Natürliche Schöpfungsgeschichte, published in 1868, the year after Vogt's
memoir. In Haeckel's book, Darwin read of the arguments of the comparative
philologist August Schleicher, Haeckel's friend and colleague at the University
of Jena. In addition to promoting an analogy between language change and
species change, Schleicher had defended the view that language had evolved
gradually along with the brain and speech organs.[31] Introducing Schleicher's
arguments, Haeckel wrote: "The origin of human language must, more than
anything else, have had an enabling and transforming influence upon the mental life of Man, and consequently upon his brain."[32] In a note on the sentence,
Darwin scribbled: "Remember a special part of the brain for speech."[33] Darwin made a similar comment in the margins of an 1871 paper by W. D. Whitney. "*Organ for Language* in the Brain; but this may be a result"—a result,
that is, of the increased demands placed on the evolving brain by evolving
speech, as opposed to the view (supported by Whitney) that the human brain
was fully formed before humans spoke the first words.[34]

Bateman versus Darwin

So Bateman probably bore old news when he wrote to Darwin in 1871, of
the relevant passage in *On Aphasia*, "you will see that Carl Vogt's dissections
on the ape tend to support the theory of my friend Professor Broca."[35] In the

book, Bateman presented a broad sampling of the literature on aphasia and related topics, along with a historical overview, detailed reports of the cases he had observed at first hand, and tips on diagnosis and treatment.[36] He concluded that none of the specific proposals for a language center in the brain (fig. 2.2) had passed empirical muster, and doubted whether such a center ever would be found.[37] There was, after all, more than one form of language; more than one form of language impairment; more than one area of the brain where damage induced such impairment; and more causes of such impairment than just brain damage.[38] Perhaps, Bateman suggested, investigators of aphasia should attend less to gross structural damage than to subtle changes in the

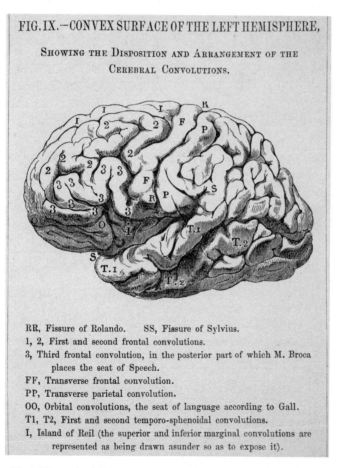

FIG. IX.—CONVEX SURFACE OF THE LEFT HEMISPHERE,

SHOWING THE DISPOSITION AND ARRANGEMENT OF THE
CEREBRAL CONVOLUTIONS.

RR, Fissure of Rolando. SS, Fissure of Sylvius.
1, 2, First and second frontal convolutions.
3, Third frontal convolution, in the posterior part of which M. Broca
    places the seat of Speech.
FF, Transverse frontal convolution.
PP, Transverse parietal convolution.
OO, Orbital convolutions, the seat of language according to Gall.
T1, T2, First and second temporo-sphenoidal convolutions.
I, Island of Reil (the superior and inferior marginal convolutions are
    represented as being drawn asunder so as to expose it).

Figure 2.2 The left hemisphere of the human brain, engraving from a cast sent to Frederic Bateman by Paul Broca around 1870. From Bateman 1877, 127.

brain's thermal or chemical or electrical state.[39] In Bateman's view, however, whatever the state of affairs in the aphasic brain, or in the healthy brain, there would nevertheless remain the problem of specifying just what the faculty of speech or articulate language *is*, and the possibility that this problem, like the problem of the soul, defeated the limited reach of human understanding.[40]

In the *Descent*, Darwin had appealed to the evidence of aphasia to show that language was rooted in the brain, and therefore could have evolved in tandem with the brain. "The intimate connection," he wrote, "between the brain, as it is now developed in us, and the faculty of speech, is well shown by those curious cases of brain-disease, in which speech is specially affected, as when the power to remember substantives is lost, whilst other words can be correctly used."[41] Darwin's source was the Edinburgh physician John Abercrombie's book on the intellectual powers—a favorite during the 1830s.[42] Aside from the suggestion that Darwin was wrong about evolution, Bateman's letter was highly complimentary of Darwin's treatment of language. How right of Darwin to insist that there was more to language than speech! Bateman singled out for praise Darwin's conjecture that in the distant past "we might have used our fingers as efficient instruments."[43] Bateman himself had once observed just such a "language of *signs*," he reported. He referred Darwin to the pages of *On Aphasia* describing an extreme form of "pantomimic language," made up entirely of mimicked or echoed sounds and gestures, to which an aphasic woman Bateman had seen in La Salpêtrière had been reduced.[44]

To judge by Darwin's annotations and by his later use of *On Aphasia*—no doubt to Bateman's chagrin—in the *Expression of the Emotions* and in the second edition of the *Descent*, Darwin was most interested in Bateman's clinical reports.[45] In the *Expression*, Darwin cited the pantomimic case as further evidence for the existence of a human instinct for imitation—an instinct that, Darwin now argued, had helped to build inarticulate as well as articulate language:

> It is perhaps worth consideration whether movements at first used only by one or a few individuals to express a certain state of mind may not sometimes have spread to others, and ultimately have become universal, through the powers of conscious and unconscious imitation. That there exists in man a strong tendency to imitation, independently of the conscious will, is certain. This is exhibited in the most extraordinary manner in certain brain diseases, especially at the commencement of inflammatory softening of the brain, and has been called the "echo sign." Patients thus affected imitate, without understanding, every absurd gesture which is made, and every word which is uttered near them, even in a foreign language.[46]

In the second edition of the *Descent*, Darwin again referred to the Parisian pantomime, to much the same end. He also embellished the old sentence on aphasia and the brain to take account of Bateman's greater variety of cases in which "substantives" were lost with brain disease or trauma.[47]

By the time this second edition appeared in 1874, Bateman had turned his tentative and private queries about the truth of evolution into bold and public statements. In March 1872, Bateman read a paper before the Victoria Institute in London, "Darwinism Tested by Recent Research in Language."[48] The institute had been founded several years before for the purpose of defending Scripture from the predations of "pseudo-science,"[49] and Bateman insisted that his case against evolution was motivated not by religious convictions but by respect for the evidence.[50] Neither human dignity nor belief in a Creator was threatened by the theory of evolution.[51] Nevertheless, in his view, the evidence of language showed the theory to be false. This evidence was of two sorts. The first sort was the evidence for language as a difference of kind between humans and other species. Here Bateman praised F. Max Müller and criticized Darwin, in particular Darwin's account of imitation. As Bateman saw it, the point about the Parisian pantomime was that her power of imitation was entirely separate from her power of speech. Furthermore, though parrots and monkeys had been imitating away for millennia, they had yet to evolve anything like language.[52]

The second sort of evidence was the evidence against a speech center in the human brain. "I have shown, and on the most indisputable authority," Bateman told his audience, "that persons could talk when the *presumed* seat of speech was invaded by an enormous tumour, completely disorganized by disease, or destroyed by a pistol-shot!"[53] The relevance of these conclusions, outlined in Bateman's letter to Darwin a year before, was now spelled out in full. Bateman argued as follows. The case for an evolutionary origin of language depended on a structural analogy between the human brain and the ape brain. Suppose the third frontal convolution were indeed the seat of speech. Then the more rudimentary version of the structure in the ape brain might reasonably be considered the seat of a more rudimentary version of speech. Out of these rudiments of brain and therefore speech in the ape could well have evolved their more elaborate human counterparts. However, concluded Bateman, as the third frontal convolution "has *not* been proved to be the seat of speech in man, the Darwinian argument from analogy falls to the ground, and speech remains a barrier the brute is not destined to pass . . . whilst the common belief in the Mosaic account of the origin of man is strengthened."[54] So localization and evolution fell together, and with them materialism, for, as

Bateman had hinted in his aphasia treatise, the immaterial faculty of language was not the same thing as its material instruments in the brain.[55]

Back from the metropolis, Bateman published a letter in the *Eastern Daily Press* of East Anglia summarizing his paper to the Victoria Institute. Thus began a much-discussed public controversy lasting into the summer.[56] One of Bateman's letters-page opponents argued that articulate language was not even a universal human feature. Bateman conceded that if there were indeed humans without language, the theory of evolution would be strengthened. But there were no confirmed reports of such humans.[57] Darwin's own encounters with supposedly languageless Fuegians (capable only of henlike clucking) turned out, on inspection of the relevant passages in the *Beagle Narrative*, to show that Fuegians did possess the faculty of speech, albeit little developed. Thus they were able to learn English, as had the Fuegians aboard the *Beagle*. "Bring a Fuegian to England, and give him time, and he will talk," wrote Bateman. "Put a monkey under training for any number of years, and he will never evince the slightest capacity for the acquisition of language."[58]

Other participants in the debates over human evolution, meanwhile, were beginning to make use of the new studies on aphasia. In his 1873 Royal Institution lectures, Müller appealed to Jackson's work to support a distinction between emotional language (which humans shared with other creatures) and rational language (which humans alone enjoyed). For Müller, the existence of this divide counted against an evolutionary origin for rational language, hence for humankind.[59] For Darwin, however, the evidence of aphasia pointed in precisely the opposite direction. It showed that rationality and language were separate phenomena, and thus need not have emerged together in a single catastrophic leap. "Persons with aphasia are certainly intelligent," Darwin scrawled on his copy of Müller's "Lectures," "& yet they cannot utter any words, the vocal organs being still perfect, & if rational thought depended absolutely on language, how can they be intelligent?"[60] In the October 1874 *Westminster Review* article on Müller and Darwin, the anonymous reviewer countered Müller's claims about aphasia and evolution with a discussion of the recent work of the physiologist David Ferrier at the West Riding Lunatic Asylum:

> Now Professor Ferrier, following up the researches of Hitzig and Fritsch, has recently been experimenting on the brains of various animals, and on that of the monkey among others, and "the part," he tells us, "that appeared to be connected with the opening of the mouth and the movement of the tongue was homologous with the part affected in man in cases of *aphasia*." This must surely be admitted . . . as at least a very curious coincidence, easily explicable, indeed, on the evolution theory, but not easily on any other. A certain intimate connexion between the

organs of speech and a particular portion of the brain appears to be essential to man's use of rational language. In the course of evolution, therefore, such a connexion would need to be established before rational language could be acquired; and this very intermediate stage, where the connexion exists, although the further acquisition of speech has not yet been attained, is now presented to us in the monkey, a creature of all animals, beyond contradiction, in general appearance and physical conditions, most closely similar to ourselves.[61]

In 1877, Bateman intervened in this debate once more, arguing his aphasia-based case at book length in *Darwinism Tested by Language*. As before, the "Darwinism" tested was not so much Darwin's theory about how species evolve but the more general claim *that* species evolve. Against this general claim, Bateman adduced several of the stock objections: the fossil record showed little or no gradual change, present-day creatures still fitted Aristotle's descriptions, and so on.[62] "The evolutionists," wrote Bateman, "deal largely in the subjunctive mood,—the *may* and the *might*—and on purely hypothetical premises, they attempt to found conclusive arguments."[63] But at the center of his case remained the earlier argument about language, apes, brains, matter, and mind.[64] Although *Darwinism Tested by Language* carried an admiring preface from the dean of Norwich, Bateman again protested against the accusation of "using Scripture to refute Darwinism," insisting that his was a strictly scientific case against evolution and its corollaries, localization and materialism.[65]

A few years later, Bateman attempted to shore up that case with the results he obtained in the course of his work at the Eastern Counties' Asylum for Idiots in Colchester.[66] The Colchester asylum was one of five such rural institutions established in the middle of the century for the "training" of the mentally deficient.[67] Bateman judged the asylum received less than its share of charitable funds due to widespread misunderstanding of the nature of idiocy and the prospects for its treatment. To correct this misunderstanding, and solicit funds for the institution, he lectured to philanthropic audiences on his work at the asylum. An address he gave in Norwich in January 1882 was subsequently published as a small pamphlet, *The Idiot: His Place in Creation and His Claims on Society*.[68] Here Bateman argued that the key to improving the unfortunate estate of the idiots was to be found among the conclusions of his earlier inquiry into language, evolution, and the brain (to which he referred his readers): one had to take care "not to confound the instrument with the person who possesses those organs [i.e., that instrument]."[69] Trapped within the damaged bodies gathered at the asylum were souls no different from those of the physically more fortunate. In Bateman's view, there was no other explanation for the successes of the training regime at the asylum. The "sociable, affectionate, and happy" Colchester idiots were thus incontrovertible evidence of the true relation of matter to mind.[70]

Minds across the Language Barrier

Amtsberg's Pike and the Debate over Anthropomorphism

Speaking, in his 1873 lectures, of the new, "positive" science of the mind-brain, Müller flagged two major problems. First, its practitioners, dazzled by their success in distinguishing afferent from efferent nervous tubes, in locating the exact destinations of messages sent from the exterior nerves to the brain, had mistakenly presumed that "nothing were wanting but a more powerful lens to enable us to see with our own eyes how, in the workshop of the brain, as in a photographic apparatus, the pictures of the senses and the ideas of the intellect were being turned out in endless variety."[71] Second, inappropriate though it was in the case of the human mind, the positivists' commitment to observable matters of fact was being dispensed with altogether when it came to animal actions, whose interpretation most called for positivist rigor. "[I]f there is a *terra incognita* which excludes all positive knowledge," Müller declared, "it is the mind of animals. We may imagine anything we please about the inner life, the motives, the foresight, the feelings and aspirations of animals—we can *know* absolutely nothing." Is there any way at all to glimpse what a mollusk experiences (to use Müller's example)? Perhaps mollusks no more experience their world than rocks do. Or is the inner life of a mollusk rich beyond the wildest dreams of mere humans? "It may be so, or it may not be so, for there is no limit to an anthropomorphic interpretation of the life of animals."[72] To show just how much of animal activity could be explained without trusting to "anthropomorphic analogies," and without supposing, against the evidence of the science of language and "the strict rules of positive philosophy" alike, that animals, like humans, generalize, conceptualize, and reason about the world, Müller offered the example of Amtsberg's pike.[73]

According to Müller, who took the story from a published lecture by the Kiel zoologist Karl Möbius, a Mr. Amtsberg of the German town of Stralsund had conducted the following experiment.[74] He inserted a pane of glass between a pike and a number of smaller fishes sharing the tank with the pike. From time to time the pike tried to attack the smaller fishes, ramming itself into the glass pane. Over the next three months, these unsuccessful attempts at predation grew fewer and fewer, and finally ceased altogether. Six months later the glass pane was removed, and the pike was able to swim freely among the smaller fish. The pike refrained entirely from attacking these familiar fish, though it swallowed up new fish with such regularity that soon the pike outgrew the tank, and the experiment thus came to an end. What did the pike's apparently merciful treatment of the familiar fish reveal about the inner life of the pike? In Möbius's view, as Müller translated, "the pike acted without reflection." The pike's instinct was to attack its prey. But sufficient experience of the invisible barrier had led the pike to recollect pain whenever it caught sight of its prey, and to experience "the sad impression that it was impossible to reach the prey." The result was the suppression of the instinct to attack. Möbius concluded that the pike was thus "like a machine, but like a machine with a soul, which has this advantage over mechanical machines, that it can adapt its work to unforeseen circumstances, while a mechanical machine can not. The pane of glass was to the organism of the pike one of these unforeseen circumstances."[75]

Darwin registered his disagreement on his copy of Müller's "Lectures." "When it is objected," Darwin wrote at the back, "that we can know nothing of what is in mind of an animal, so it may be said of any savage, whose language we do not understand.—we can judge only by action & have our doubts."[76] Amtsberg's pike in particular prompted a rather gnomic comment: "*Reason*, like association—in contrast with Pike & pane of glass."[77] What could Darwin have meant? It was one thing to contrast reason on the one side with the mere association of sensations or ideas—and association machines such as Amtsberg's pike—on the other side. This was precisely Müller's contrast. But what could Darwin have meant by contrasting reason and association *together* on the one side and the pike on the other side? Darwin went some way toward explaining in his discussion of reason in the second edition of the *Descent*. According to Darwin, reason was different from instinct, and both were different again from the association of ideas or sensations. But, Darwin continued, reason and association were "intimately connected." He then summarized the pike experiment, commenting thus:

> If a savage, who had never seen a large plate-glass window, were to dash himself even once against it, he would for a long time afterwards associate a shock with a window-frame; but very differently from the pike, he would probably reflect on the nature of the impediment, and be cautious under analogous circumstances. Now with monkeys . . . a painful or merely disagreeable impression, from an action once performed, is sometimes sufficient to prevent the animal from repeating it. If we attribute this difference between the monkey and the pike solely to the association of ideas being so much stronger and more persistent in the one than the other, though the pike often received much the more severe injury, can we

maintain in the case of man that similar difference implies the possession of a fundamentally different mind?[78]

In other words, what the pike shows is not how much animals can achieve without reason or reflection, but how much reason or reflection *depends upon* powers of association far greater than those of the pike. For Darwin, reason was not a mental power existing in splendid isolation from all other powers of the mind; rather it arose at the intersection of several such powers—of association, of observation, of inference—increasing as these increased, and decreasing as these decreased, up and down the scale of organization. At the top of the scale, Darwin explained, his infant child, "before he could speak a single word," had manifested powers of association far in advance of even the cleverest dog Darwin had known.[79] In his "Biographical Sketch," Darwin commented further: "What a contrast does the mind of an infant present to that of the pike . . . !"[80]

As for Müller's claim that objectivity about animal minds began with the presumption that animal minds were nothing like human minds as far as general ideas, concepts, and reason were concerned, Darwin argued for precisely the opposite point of view, at one point framing what we might anachronistically call *Darwin's canon*:

> If one may judge from various articles which have been published lately, the greatest stress seems to be laid on the supposed entire absence in animals of the power of abstraction, or of forming general concepts. But when a dog sees another dog at a distance, it is often clear that he perceives that it is a dog in the abstract; for when he gets nearer his whole manner suddenly changes, if the other dog be a friend. A recent writer remarks, that in all such cases it is a pure assumption to assert the mental act is not essentially of the same nature in the animal as in man. *If either refers what he perceives with his senses to a mental concept, then so do both.*[81]

The writer in question was a Mr. Hookham, who, in May 1873, published an open letter to Müller in the *Birmingham News*, in response to Müller's Royal Institution lectures.[82] A better-known opponent of Müller on animal minds, also cited by Darwin in the second edition of the *Descent*, was the distinguished man of letters Leslie Stephen. When Stephen's 1872 essay on "Darwinism and Divinity" was republished a year later in his book *Essays on Freethinking and Plainspeaking*, Stephen inserted a brief response to Müller's "Lectures." For Stephen, the "capacity to understand is as good a proof of the presence of vocal intelligence, though in inferior degree, as the capacity to speak. A dog frames a general concept of cats or sheep, and knows the corresponding words as well as a philologer."[83] In Stephen's view, "Kant and

Hume must fight out their quarrel" over other grounds. As for the language barrier, that "grammatical dike" Müller had interposed between humans and the animals, it was not "destined to hold back the deluge any better than its predecessors."[84]

Whitney too, in his "Darwinism and Language" of 1874, took up the case against Müller on what animal minds were like and what could be known about them. "We believe," wrote Whitney, "that the horse sees green, and tastes water, and feels pain, as confidently, and on nearly the same grounds, as we believe that our neighbor does the same." Of course, animals cannot report on their inner states as humans can, Whitney continued. But human speech is neither exclusive nor infallible as evidence as to other human minds. Human speech, in Whitney's view, needs to be "both supplemented and controlled by that same observation of conduct under conditions which is all we have to rely upon in the lower animals." What of animals having general ideas? Whitney emphasized what a humble thing a general idea could be. Its formation need depend on no more than "the power of being so impressed by a thing in the assemblage of its qualities that on seeing another like it we recognize it as being like, and expect the same acts or effects on it." If a dog consistently acts appropriately on encountering a man, then there are no grounds for doubting that the dog "possesses such germs of the faculty of generalizing as are distinct only in degree from those which I possess," or that the dog, if it had language, "would as certainly say *man*, or something equivalent, and would apply it as correctly, as any of us do."[85] A German critic of Müller likewise argued against Müller's extreme skepticism. If animals acted as people did when people were fearful, hopeful, suspicious, expectant, and so on, what reasons could there possibly be to deny that animals too experienced fear, hope, suspicion, anticipation? None. Here is still another canon, much like Darwin's: "when signs have been ascertained to denote things in one place, they shall be supposed to denote them in another till proof is shown to the contrary."[86]

One of the most remarkable responses to Müller's views on animal minds appeared in the 1874 *Westminster Review* article on Müller versus Darwin. The anonymous author professed surprise that Müller held the case of Amtsberg's pike to demonstrate how different were the animal and the human minds. True, the pike did not act out of mercy when it spared its smaller neighbors, but neither did humans act out of mercy when they acted mercifully:

> How does [Amtsberg's] result differ from that which we aim at in the treatment of the dangerous classes, the pikes of human society? . . . The whole analysis of the pike's education might be fitted to the analysis of human education in the

sphere of morals and personal prudence, so that of the loftiest actions done in obedience to the dictates of conscience, we should have to say that they were "not based on judgement," but due only to "the establishment of a certain direction of will in consequence of uniformly recurrent sensuous impressions." It is an old doctrine in regard to human character that by training and habit our wills do become so set in one direction that morally we lose the power of choice and *cannot help* acting in the manner which has become habitual to us. That a pike should be like a man in this respect will be no great matter of surprise to an evolutionist.[87]

The author went on to speculate about how such a gap as there was between animal minds and human minds came into being. On one side of the gap, language had enabled humans to accumulate knowledge, and thereby to improve their state immensely. On the other side, the first language users would have turned their newfound power against the animals, driving the most humanlike ones to extinction, and jealously guarding the source of their power from the rest. The same went on at present:

> Practically language has been a weapon of war, an instrument for gaining the mastery. How little anxious men will have been in past days to impart it to other animals may be understood from what has happened in the parallel and analogous case of reading. Though a man incapable of reading is only half a man compared with his fellows who have that capacity, still it is only with serious misgivings and the greatest reluctance that the governing classes of our own country have been induced to concede education in this respect in any secure and substantial form to the classes beneath them. To substitute an instructed mind for a machine-like pair of hands was thought dangerous to the exigencies of farm-labour and domestic service, and many religious and kindly persons, fully if unconsciously believing in the divine right of the upper ten thousand, would gladly find some expedient, if they could, by which the children of the poor might be taught how to read the Bible and Catechism, without learning how to read any other literature which can only make them "uppish," "above their places," discontented with the state of life to which they have been called, and thorns in the sides of their "betters," instead of obedient machines, to work and work on without comment and without reproach.[88]

Embattled though he was, Müller had allies in this debate. The anti-Darwinian anatomist St. George Jackson Mivart took a position on language and on anthropomorphism strongly reminiscent of Müller's. In an anonymously published 1871 *Quarterly Review* article on the *Descent*, Mivart had charged Darwin with failing to explain what most needed explaining about the putative evolutionary origins of language. "It is not," wrote Mivart, ". . . emotional expressions or manifestations of sensible impressions, in whatever way exhibited, which have to be accounted for, but the enunciation of distinct deliberate judgements as to 'the what,' 'the how,' and 'the why,' by definite articulate sounds; and for these Mr. Darwin not only does not account, but he does not adduce anything even tending to account for them."[89] In Mivart's slim volume on the comparative anatomy of primates, *Man and Apes* (1873), he argued for a distinction between two kinds of soul: the human one, "capable of articulately expressing general conceptions," and the nonhuman one, with no such capacity. As, in Mivart's view, body and soul were "so essentially and intimately related," and "as no natural process would account for the entirely different [human] kind of soul" emerging out of the animal soul, it followed that "no merely natural process could account for the origin of the body informed by it."[90]

Two years after Müller's "Lectures," in an April 1875 article "Instinct and Reason" in the *Contemporary Review*, Mivart turned to theological notions once more, contrasting the opposite errors of *"theological anthropomorphism"* and *"[b]iological anthropomorphism."* Where earlier theologians had gone too far in speaking of God as if God's attributes were just like human attributes, modern theologians were now going too far in speaking of those attributes as if they were nothing at all like human attributes. What was needed was to see that God's attributes could be analogous to human attributes—"the old, safe *via media* of the schoolmen." For Mivart, the silence of brutes presented an intellectual challenge much like the silence of God. "Since we are unable to converse with brutes," wrote Mivart, "we can but divine and infer from their gestures, motions, and the sounds they emit, what may be the nature of their highest psychical powers." But where the rise of science had propelled theologians from extreme anthropomorphism to extreme antianthropomorphism, scientific observers of animals had made precisely the opposite journey, from the Cartesian error of supposing that "animals are nothing but wonderfully complex machines" to the Darwinian error of supposing that "there is a substantial identity between the brute soul and the soul of man."[91] Drawing on his 1871 review, Mivart gave an example of Darwin's own anthropomorphic excesses, his "hasty attribution of human qualities to brutes, on account of certain superficial resemblances":

> What praises of the patient fidelity of the bird sitting on her unhatched progeny do we not meet with, and yet this constancy is promoted by something very different from maternal tenderness! In truth, a multitude of branching arteries and veins furnish such an abundance of blood to the bird's breast as to cause it to seek in the contact of the eggs a refreshing sensation. Cabanis and Dugès tell us that if a capon be plucked in that region which is naturally bare in a sitting hen, and if an irritating substance be applied to the part so stripped, then not only will the local inflammation cause the capon to seek the contact of eggs and to sit, but even to act maternally to the young when they come to be hatched.[92]

George John Romanes and Comparative Psychology

Whatever Darwin's anthropomorphism, it was modest next to that of his successor in the defense of mental evolution, the physiologist George John Romanes. The 1880s were Romanes's decade, when his articles and books defined the state of the art in comparative psychology—a science that, as late as 1878, could "hardly be said to exist," as Huxley observed.[93] Like Mivart, Romanes wrestled to reconcile science and faith, and reached for an analogy with theology to defend his view of the legitimate role of anthropomorphism in understanding animal minds. In his introduction to *Animal Intelligence* (1882), Romanes wrote: "Just as theologians tell us—and logically enough—that if there is a Divine Mind, the best, and indeed only, conception we can form of it is that which is formed on the analogy, however imperfect, supplied by the human mind; so with 'inverted anthropomorphism' we must apply a similar consideration with a similar conclusion to the animal mind." Of course the human conception of, say, the insect mind would be imperfect. But it would not be *too* imperfect, Romanes urged, so far as the theory of evolution was true (and it was). For the evolutionist, according to Romanes, "there must be a psychological, no less than a physiological, continuity extending throughout the length and breadth of the animal kingdom."[94] The book itself was a compendium of evidence bearing on the mental lives of creatures from protozoa to monkeys. Here was the empirical basis upon which Romanes would erect his theory of mental evolution in *Mental Evolution in Animals* (1883) and *Mental Evolution in Man* (1888). Here too were accounts of canine Euclids and simian Archimedes, reasoning their way to elegant solutions to tough practical problems, just as humans would. As Romanes wrote in *Mental Evolution in Animals*, "there is no difference *in kind* between the act of reason performed by the crab and any act of reason performed by a man."[95]

In his theoretical as well as his practical work, Romanes took aim at the language barrier, and Müller's claims on its behalf. Much of *Mental Evolution in Man* was devoted to showing how human language and human conceptual thought could have evolved out of their less advanced animal counterparts. Against Müller, Romanes argued that the Sanskrit roots, while referring to general ideas, referred to ideas at the lowest level of generality. "Scarcely any of them," Romanes wrote of the roots, "present us with evidence of reflective thought, as distinguished from the naming of objects of sense-perception, or of the simplest forms of activity which are immediately cognizable as such."[96] Between percepts and concepts, Romanes identified an intermediate species: what he called "pre-concepts," or "recepts."[97] So far as the Sanskrit roots expressed recepts rather than concepts, the roots constituted, not a leap to full conceptual thought, but a step in the direction of full conceptual thought.[98] Furthermore, "the witness of philology" itself testified, on Romanes's exhaustively evidenced account, to the first humans commanding a far more primitive form of language than Müller had suggested. The roots must have come from a relatively late period in the transition from brute to man, argued Romanes, as many roots had pastoral referents ("to dig," "to milk," and so on). In the beginning, he concluded, were not roots, but an undifferentiated speech-protoplasm, spoken by a creature who was less Darwin's manlike ape than Haeckel's apelike man, the *Homo alalus*, the "speechless man," largely human but lacking language.[99]

Romanes's critique of Müllerian philology formed one part of his attempt at, in his words, "bridging the psychological distance which separates the gorilla from the gentleman."[100] Observations and experiments with a chimpanzee in the Regent's Park zoological gardens formed another part. According to Romanes, the high intelligence of Sally, resident in the menagerie nearly six years, explained her remarkable powers of language comprehension. When told to put a piece of straw in one of several straw meshes distributed around her cage ("the one nearest your foot," "the one next to the key-hole"), Sally would dutifully carry out her instructions. By rewarding her with fruit for correct responses, Romanes taught Sally to fetch just the number of straws he asked for, up to five straws. It was, in Romanes's view, "difficult to overrate the significance of these facts":

> The more that my opponents maintain the fundamental nature of the connection between speech and thought, the greater becomes the importance of the consideration that the higher animals are able in so surprising a degree to participate with ourselves in the understanding of words. From the analogy of the growing child we well know that the understanding of words precedes the utterance of them, and therefore that the condition to the attainment of conceptual ideation is given in this higher product of receptual ideation. Surely, then, the fact that not a few among the lower animals (especially elephants, dogs, and monkeys) demonstrably share with the human infant this higher excellence of receptual capacity, is a fact of the largest significance. For it proves at least that these animals share with an infant those qualities of mind, which in the latter are immediately destined to serve as the vehicle for elevating ideation from the receptual to the conceptual sphere: the faculty of understanding words in so considerable a degree brings us to the very borders of the faculty of using words with an intelligent appreciation of their meaning.[101]

So much for understanding words. What about uttering them? In a fuller report on Sally, published in 1889 in *Nature*, Romanes presented a vivid account of Sally's vocal repertoire:

the only attempts that she makes by way of vocal response are three peculiar grunting noises—one indicative of assent or affirmation, another (very closely resembling the first) of dissent or negation, and the third (quite different from the other two) of thanks or recognition of favours. . . . By vocalizing in a peculiar monotone (imitative of the beginning of her "song"), [her keepers] are usually able to excite her into the performance of a remarkable series of actions. First, she shoots out her lips into the well-known tubular form (depicted in Darwin's "Expression of the Emotions" . . .) while at the same time she sings a strange howling note interrupted at regular intervals; these however, rapidly become shorter and shorter, while the vocalization becomes louder and louder, winding up to a climax of shrieks and yells, often accompanied with a drumming of the hind feet and a vigorous shaking of the network which constitutes her cage. The whole performance ends with a few grunts.[102]

In January 1891, lecturing before the Philosophical Society of Glasgow on the relations between thought and language, Müller responded to Romanes's chippings away at the language barrier. According to Müller, for all Romanes's eminence as a biologist, he was out of his depth when it came to language and philosophy. Thus, Müller contended, "the very fact that no animal has ever formed a language" was for Romanes but "an unfortunate accident." No facts seemed to count against what Romanes himself had called "the immense presumption that there has been no interruption in the developmental process in the course of psychological history." Indeed, for the most part Romanes dealt not with true language at all, observed Müller, but with baby talk and ape howls. As a student of true language, appreciative of the nature of the phenomenon to be explained, and of the hopelessness of onomatopoeia to do the explaining, Müller proclaimed his refusal "to argue with [Romanes] or any other philosopher either in the nursery, or in the menagerie."[103] Published subsequently in the *Monist*, Müller's lecture drew a vigorous response from Romanes. Of the Sanskrit roots, Romanes wrote: "archaic though they be in a philological sense, in a phylological sense they are things of yesterday. . . . This has to be inferred from observations in the 'menagerie,' as distinguished from research in the library." That "Prof. Max Müller expressly refuses to give me the pleasure of his company where the best materials for studying the really 'primitive' condition of the sign-making faculty are to be met with" had, in Romanes's view, made a rapprochement between the sciences of language and of life virtually impossible.[104] Commenting on this impasse in a further article, the *Monist*'s editor, Paul Carus, argued that the way forward lay with seeing how differences in kind could arise by degrees. Allow for this, and, wrote Carus, the two great scholars of language and life respectively could be seen more to "complement than refute each other."[105]

Peacemaking was not the aim of the American philologist and psychologist E. P. Evans in his 1891 *Atlantic Monthly* article "Speech as a Barrier between Man and Beast." According to Evans, it was not comparative philology but comparative psychology that would discover how and why the human mind differed from the animal mind. "The philologist," wrote Evans, "who recognizes in the roots of language the Ultima Thule beyond which he dare not push his investigations, confesses thereby his incompetency to solve the problem of the origin of language, and must resign this field of inquiry to the zoöpsychologist."[106] Evans was equally impatient with the skepticism about animal reason that Müller and others had promoted alongside the presumed lack of animal language. One writer had argued that, as dogs were unable to talk about their inner states, "[t]here is no reason to suppose, because the burnt dog shuns the fire, that it perceives any relation between it and the pain of being burnt." It may be, Evans conceded, that the sight of the fire merely calls up the associated pain state in the dog's mind. For Evans, however, the evidence for the perception of cause and effect in the mind of a burnt dog was no better or worse than for the perception of cause and effect in the mind of a burnt child. "The misfortune of dogs in not being endowed with articulate speech," concluded Evans, "is greatly aggravated if it renders them liable to have such elaborate philosophy as this mouthed over them."[107]

Morgan's Reform of Romanesian Comparative Psychology

In the end, Müller's miserliness in attributing higher powers to animal minds triumphed over Romanes's generosity, thanks in large measure to the success of the rule known as Morgan's canon. Conwy Lloyd Morgan's scientific career began amidst the widespread and wide-ranging discussion of anthropomorphism that followed Müller's 1873 Royal Institution lectures. Morgan (fig. 2.3) started at the Royal School of Mines in London, where he trained in the late 1860s and early 1870s as a mining engineer. Seated at dinner one night next to the school's most famous professor, T. H. Huxley, Morgan impressed Huxley enough to win an invitation to stay on as a research student after graduation. A year in Huxley's laboratory led to a succession of short-term teaching and engineering jobs. In 1884, after a longish stint in South Africa, Morgan took up a professorship in geology and zoology at University College, Bristol, where he would remain for the rest of his long career, becoming principal of the college (later university) in 1887. As a geologist and zoologist, he was, he later recalled, "no more than a tolerably conscientious hireling."[108] But comparative psychology engaged his interests and abilities to the full. Throughout the 1880s, he worked to make the new science his own.

Figure 2.3 Conwy Lloyd Morgan (1852–1936), around 1900. Courtesy of Special Collections, University of Bristol, and reproduced by permission.

Morgan first developed his views on the animal mind in a series of papers written between 1882 and 1886, largely in response to Romanes's work.[109] Romanes had claimed that animals have abstract thoughts—that the concept of "good for eating," for example, passed through the mind of a dog as the dog sniffed at a biscuit. But, Morgan pointed out, no less an authority than Locke had claimed that "the having of general ideas is that which puts a perfect distinction betwixt man and brutes, and is an excellency which the faculties of brutes do by no means attain to." As we have seen, this quotation had formed the centerpiece of the argument on the origin of language in Müller's widely read *Lectures on the Science of Language*, and Morgan cited

"Prof. Max Müller and other living thinkers" as advocates of the Lockean position.[110] After contrasting Romanes's views with Müller's, Morgan distinguished different kinds of abstraction. His most important distinction was between abstraction-as-elimination and abstraction-as-isolation. When a dog sees a biscuit, Morgan argued, the image thus impressed on the dog's mind immediately triggers associated expectations of smell and taste. The result is a perceptual construct in which the visual, olfactory, and gustatory features of the biscuit so dominate that all other features are effectively eliminated. The dog's mind has executed abstraction-as-elimination, and the dog comes to believe that the biscuit is indeed "good for eating." But can a dog reflect on the quality of "good for eating" in isolation, independently of an apparently good-for-eating object? Morgan's answer was no. Abstraction-as-isolation was the privilege of human minds, because, he continued, only humans had isolation-enabling language.[111] "By means of language and language alone has human thought become possible," wrote Morgan. "This it is which has placed so enormous a gap between the mind of man and the mind of the dog. . . . Through language has the higher abstract thought become possible."[112]

The human monopoly over language and truly abstract concepts was, Morgan added, a boon in all domains of knowledge save one. A science of the animal mind from the inside would never be possible, he argued, because, lacking language, animals would never be able to report on their own introspections. No animal could verify the "ejective" inferences (that is, inferences based on human mental life) of the would-be animal psychologist, so no knowably reliable description of the animal mind could develop from those inferences. Worse, Morgan wrote, "such is the extraordinary complexity of the human mind—a complexity largely due to the use of language—that we may well suppose that any conception we can form of animal consciousness is exceedingly far from being a true conception."[113] Such considerations, he noted, prompted extreme skepticism about animal minds. They had led Müller, notoriously, to declare (and here Morgan quoted Müller) that "according to the strict rules of positive philosophy we have no right to assert or deny anything with respect to the minds of animals." But, argued Morgan, such extremity was unwarranted so long as one accepted on other grounds that humans and other species were evolutionary kin. In Morgan's view, the evolutionist justifiably believed that animals were conscious, because "[a]nimals have inherited brain-structures in many respects similar to those possessed by man; and there is no reason for supposing that in them no psychoses [mental states] run parallel, or are identical, with their neuroses [brain states]."[114] So far as animals displayed associative learning, the evolutionist could even claim to know a little about the content of animal consciousness:

we may say that consciousness . . . at a very early stage of evolution became, so to speak, polarised into pleasurable and painful; that those actions which were associated with pleasurable feelings were more frequently performed than those associated with painful feelings; that those organisms in which there was an association between right action and pleasurable feelings would stand a better chance of survival than those in which the association was between wrong actions and pleasurable feelings; and that finally those organisms in which conscious adjustments of all orders were more perfectly developed would be the winners in life's race.[115]

But, Morgan insisted, to infer more than this about the minds of creatures in whom "the ratio of the senses" was often very far from the human ratio was a hazardous business, like using "mirrors of varying and unknown curvature" to study the heavens. A suitably cautious inquirer might glimpse in the human mind the mental vestiges of descent from the lower animals, but a true science of comparative psychology needed to be otherwise founded. Like comparative anatomy, comparative psychology needed to deal as much as possible in publicly verifiable statements about observable structures. It needed to forgo Romanes's rich interpretations of animal behavior in favor of systematic investigation and fact gathering. "Let us, therefore," Morgan urged, ". . . stick to the objective study of habits and activities, [of] reflex, instinctive and intelligent, making use of ejective inferences as sparingly as possible."[116]

It was no mean feat for Morgan to bring Müller and evolutionism together thus. Here the example of Morgan's former teacher Huxley was no doubt influential. In later life, Morgan recalled how he had once asked Huxley what distinguished his belief about a leap in mental evolution from the similar belief of the detested anti-Darwinian Mivart. According to Morgan's fragmentary notes, Huxley reached to "speech and language" to clarify his position, insisting there was "no evidence of *jump* either in laryngeal, mouth, or brain structure," that "neuroses" and "psychoses" were at all times tightly correlated, and that the "child passes from animal stage to man stage *continuously*." As with children, so with species: a process of continuous change at the level of structure brought discontinuous change—from mutism to speech—at the level of function.[117]

A correspondence began to flow in the late 1880s between Morgan and the greatest Victorian skeptic about animal language and reason, Müller. In September 1887, Müller wrote to Morgan to thank him for his letter and (probably) a copy of Morgan's *Springs of Conduct: An Essay in Evolution* (1885), the first chapter of which analyzed the close links between language, concepts, and the human mind. "It is such a pleasure—as when travelling in a foreign country—to meet someone who speaks and understands one's own language," wrote Müller. "As soon as I had read your letter, I recognised you as a countryman in philosophy—as one who had tramped the same tracts of thought, and was pushing to the same goal." Müller went on to praise Morgan's discussion of abstraction and language.[118] The echo of at least one more letter (or perhaps a conversation) occurs in the pages of Morgan's first major scientific work, *Animal Life and Intelligence* (1890–91). Describing an experiment of Sir John Lubbock's, in which the removal of the leader from a group of ants had caused the rest of the group to turn back to the nest, Morgan added in a footnote: "Professor Max Müller suggests to me that perhaps the ants were frightened."[119]

In *Animal Life and Intelligence*, Morgan returned to the dog-and-biscuit example, affirming that, while the animal mind certainly had the power to construct a *"predominant,"* that is, "a perceptual construct with eatability predominant," the absence of language in animals put "a conceptual isolate or abstract idea of eatability" beyond their mental grasp. Morgan added that "this capacity of analysis, isolation, and abstraction constitutes in the possessor [i.e., in humans] a new mental departure, which we may describe as constituting, not merely a specific, but a generic difference from lower mental activities." He explained that he preferred the phrase "generic difference" to the more familiar "difference in kind" to signal his belief that novel features such as language and abstraction were nonetheless products of standard evolutionary processes (surveyed at length earlier in the book). So long as this evolutionary origin was granted, wrote Morgan, "I am prepared to follow Professor Max Müller in his contention that language and thought, from the close of that [brute] stage onward, are practically inseparable, and have advanced hand-in-hand."[120]

His discussion of reason brought Morgan even closer to Müller's position. In Müller's 1887 book, *The Science of Thought*, Müller had defined reason as "neither more nor less than the faculty, or if we dislike that word, the act of handling abstract concepts." (Müller's epigraph: "No Reason without Language, No Language without Reason.")[121] Now Morgan argued similarly that the term "reason" ought to be reserved for that analytic, conceptual, isolating thought that language alone bestowed on its users:

> I repeat, then, that the introduction of the process of analysis appears to me to constitute a new departure in psychological evolution; that the process differs generically from the process of perceptual construction on which it is grafted. And I hold that, this being so, we should mark the departure in every way that we can. I mark it by a restriction of the word "intelligence" to the inferences formed in the

field of perception; and the use of the word "reason" when conceptual analysis supervenes. Whether I am justified in so doing, whether my usage is legitimate or not, I must leave others to decide. But, adopting this usage, I see no grounds for believing that the conduct of animals, wonderfully intelligent as it is, is, in any instances known to me, rational.[122]

Jackson, Morgan, and the Tools of Consensus

Language, the Human Brain, and Jacksonian Neurology

During the 1880s, the case for mental continuity between humans and animals had its doughtiest champion in George Romanes. In *Mental Evolution in Man*, he defended cerebral continuity as well, disputing the now-faded arguments of Frederic Bateman for a difference in kind between ape brains and human brains:

> Since Dr. Bateman wrote, a new era has arisen in the localization of cerebral functions; so that, if there were any soundness in his argument, one would now be in a position immensely to strengthen "the Darwinian analogy" [between ape brains and human brains]; seeing that physiologists now habitually utilize the brains of monkeys for the purpose of analogically localizing the "motor centres" in the brain of man. In other words, "the Darwinian analogy" has been found to extend in physiological, as well as anatomical detail, throughout the entire area of the cortex.[123]

As Romanes reported, a new era had indeed arisen in the localization of cerebral functions since Bateman read his paper at the Victoria Institute in the early 1870s. The *Westminster* reviewer heralded the new era's first experimental achievements, the discovery of sensorimotor centers in the cortex via electrical stimulation of animal brains by the physiologists Gustav Fritsch, Eduard Hitzig, and David Ferrier. But the Darwinian analogy between human brains and ape brains had not so much been vindicated in the new era as transformed. Darwin's evolutionism, as we have seen, was tied to a picture of the brain derived from Franz Gall: a mosaic of well-demarcated cerebral seats for the different mental faculties. When Bateman attacked the theory of evolution, he argued that it predicted the existence of such a seat for the faculty of articulate language—"a special part of the brain for speech," as Darwin had written in his note on Haeckel-Schleicher, and as Darwin's ally Carl Vogt had argued publicly.[124] If no such seat existed, Bateman reasoned, then no apelike ancestor to humans could have carried in its brain the precursor to the seat. On Bateman's showing, there was no seat of speech in humans, hence no grounds for believing that the human brain had an evolutionary origin. Pointing to the evidence from clinical pathology, Bateman charged the theory of evolution with making a false prediction. Pointing to the evidence of comparative anatomy, Vogt had answered that the prediction had been borne out after all.

It was John Hughlings Jackson's distinction to reject the terms of this debate. At Norwich in August 1868, Jackson urged a principle of cerebral localization owing less to Darwin than to that other great Victorian theorist of evolution, Herbert Spencer. In the Spencerian-Jacksonian view, what were localized were not faculties but impressions and movements. Like the spinal cord out of which it had evolved, the brain, cortex and all, was but a machine for associating and coordinating sensorimotor impulses. It followed that the "language faculty" in an evolved brain just was the cells and fibers representing the movements of speech.[125] It did not follow, however, that these cells and fibers were packaged up in a single speech center. On the contrary. Because the muscle groups needed for speech (for motor control of the lips, tongue, and so forth) were needed for other, and evolutionarily more basic, activities, speech centers were effectively distributed all over the brain. Thus, as Jackson argued at Norwich, "speech resides in each part of the brain," yet "there are points—probably in Broca's convolution—where the most immediate processes are *specially* represented."[126] For Jackson, this motor view of speech explained why damage to Broca's convolution often brought not only speech impairment but paralysis on the right side of the body. And the brain-wide distribution of these speech centers explained why such damage did not disrupt speech completely—why, as Jackson reported at Norwich, Broca's aphasics often swore even though they could not put sentences together. Jackson later argued that the stubborn persistence of oaths and other emotionally charged expressions even revealed something of the evolutionary origins of language: the more highly evolved and more fragile sensorimotor centers of the left hemisphere (the intellectual, voluntary, "proposition" regions of the speech network) when damaged lost control over the less highly evolved and more robust sensorimotor centers of the right hemisphere (the emotional, involuntary, "oath" regions of the network).[127]

In effect, Jackson traded Darwin's faculty-and-seat evolutionism for Spencer's sensorimotor evolutionism, and thereby saved the phenomena of aphasia on evolutionary terms. The analogy between ape brains and human brains was now a Spencerian analogy, and the theory of evolution predicted just what Bateman had found: the absence of a single seat of speech in the human brain. Throughout the 1870s, the new sensorimotor approach moved from strength to strength, so that, by the late 1880s, as Romanes observed, it had swept the field.[128] There was still plenty to disagree about within the new framework. Was language more of a sensory or a motor process? What exactly was happening in Broca's region—and, from 1874, in Wernicke's region, linked to sensory as opposed to motor aphasia? Did apes and microcephali have well developed or poorly developed versions of these regions? Such questions excited much debate and research.[129] But that debate and research was now conducted within a larger consensus about how the human brain had come into being and how it ought to be studied. Evolutionism had entered into the core of the science of brain function, whose investigators now busied themselves mapping the sensorimotor constitution of the human brain. Monkeys and electricity became tools of the trade—tools it made sense to use only so far as the human brain had indeed evolved.

As the new orthodoxy took hold, Bateman's views on the brain continued to attract attention and admiration, if little allegiance.[130] A revised and vastly expanded edition of his aphasia treatise (1890) was awarded a major prize from the Académie de médecine. In 1892 he was knighted. Meanwhile, the appropriation of his work for the evolutionary cause, begun by Darwin himself, continued apace, as in E. P. Evans's 1891 *Atlantic Monthly* article. For Evans, the phenomena of aphasia had shown what a permeable thing the language barrier was. Surveying a number of striking cases, Evans at one point considered a case from Bateman's treatise, recorded by Trousseau. A professor was reading one evening when suddenly he found himself unable to comprehend his text. Alarmed, he rang for his servant. When the professor tried to explain his predicament, he found himself unable to talk, though still able to control his tongue and vocal organs. When the servant brought paper and pen, the professor found himself unable to write, though still able to use his hands. When at last a physician arrived, the professor rolled up his sleeve and pointed to his arm. All took this to be a sign of the professor's wish to be bled, which he promptly was. Right away the professor was able to manage a few words, and twelve hours later was fully recovered from the episode. But, for Evans, the significance of this case could be seen only by the light of another. He continued:

> An orang-outang that had once been bled on account of illness, not feeling well some time afterwards, went from one person to another, and, pointing to the vein in his arm, signified plainly enough that he wished the operation to be repeated. In this instance, the orang, not being endowed with articulate speech owing to the rudimentary condition of a convolution of the brain, expressed his ideas just as the Frenchman did, who had been temporarily deprived of the faculty of articulate speech owing to the suspension of function of the same convolution of the brain. The process of reasoning was identical in both cases. The idea of recovery from

sickness was associated with the act of venesection as the result of experience. In short, the man reverted for the time being to the condition of the monkey. How then should it be deemed a thing impossible for him to have risen out of such a condition?[131]

Language, the Animal Mind, and Morgan's Canon

The Harvard philosopher and psychologist William James had no doubts about evolution or the role of Broca's convolution in motor aphasia. But he did have doubts about animal language and animal reason, and in his epochal *Principles of Psychology*, published in 1890, he expressed them eloquently.[132] Like Morgan, James recommended suspicion of animal reason on account of the absence of animal language. James even cited Morgan "[o]n the possible fallacies in interpreting animals' minds."[133] But James owed his point of view less to Morgan or F. Max Müller than to Chauncey Wright, a fellow Cambridge philosopher and, as we have seen, Darwin's defender against Wallace's apostasy over the human brain. For Wright, again, language was sign making, and the power to make signs at will was what chiefly separated the human mind from the animal mind. On the basis of this insight, Wright had gone on to argue that uniquely human self-consciousness was a consequence of this uniquely human power of intentional sign making. In his *Principles*, James agreed, arguing that the sign-making power itself flowed from more basic powers of dissociation and association. In James's view, humans alone were able fully to dissociate the idea of a character from the idea of the thing instantiating that character, and then to associate similar ideas with one another.[134] Brute minds were confined to the routine and the here and now. "[I]f the most prosaic of human beings could be transported into his dog's mind," wrote James, "he would be appalled at the utter absence of fancy which reigns there." Accordingly, stories of "animal sagacity" could "as a rule be perfectly accounted for by mere contiguous association, based on experience."[135]

Morgan reviewed James's book for *Nature*, and was soon using Jamesian concepts and terminology in his own work.[136] In August 1892, first in London, at an international congress on experimental psychology (where he introduced his new rule of method), and then again in Edinburgh, at the annual meeting of the BAAS, Morgan presented a distinctly James-tinged account of the gap between animal minds and human minds.[137] Altering slightly James's famous "stream" metaphor, Morgan now spoke of a "wave of consciousness," with its crest of full or "focal" consciousness, and its trough of "marginal" consciousness.[138] He argued that what James had called a "'fringe of relation'" surrounds each object that we humans apprehend in the course of "the simple psychical life of external perception." As we turn our attention from object to object, the relations between objects—spatial relations, relations of similarity and dissimilarity, and so on—fleetingly register at the margins of consciousness. For Morgan, this "feeling or sensing of relations" was crucial to practical skills in humans and animals alike. But he insisted on "a great difference in practical experience between a relation dimly felt and a relation perceived or cognized." To focus consciousness on relations themselves seemed to require the use of introspection and reflection, and there was no evidence that these were available to animals.[139] The new distinction between awareness of relations and perception of relations mapped neatly onto the old distinction between perceptual construction and conceptual isolation. As before, in Morgan's words at Edinburgh, "it is well to restrict the words 'reason' and 'rational'" to the higher and exclusively human process. "Animals," he affirmed, "are certainly intelligent; they may be rational."[140] The retention as well of a key role for language was signaled in a letter Morgan published in *Nature* a few weeks after the BAAS meeting, at the beginning of September 1892:

> The power of cognizing relations, reflection and introspection, appear to me to mark a new departure in evolution. But whether, as I am at present disposed to hold, the departure took place through the aid of language coincident with, or subsequent to, the human phase of evolution; or whether, as other observers and thinkers believe, it took place, or is now taking place, in the lower mammalia or in other animals, is a matter for calm temperate, and impartial discussion founded on accurate, and, as far as possible, crucial experiment and observation.[141]

There is a note of hesitation here that is new in Morgan's claims about animal language and animal reason. Within a year, he was expressing more than hesitation. Here is the conclusion to Morgan's article "The Limits of Animal Intelligence," published in August 1893:

> I have expressed my opinion that, in the activities of the higher animals, marvellously intelligent as they often are, there is no evidence of that true perception of relationships which is essential to reason. But this is merely an opinion, and not a settled conviction. *I shall not be the least ashamed of myself if I change this view before the close of the present year. And the distinction between intelligence and reason will remain precisely the same if animals are proved to be rational beings the day after tomorrow.*[142]

What proof of animal language and animal reason pressed from so near a future in 1892–93? The next chapter ventures the following answer. When Morgan wrote those surprising sentences, an American student of primate language and primate reason, Richard Garner, was away in the French Congo.

Garner's plan, as Morgan and much of the rest of the world knew, was to capture the "speech" of wild gorillas and chimpanzees—the missing links of language—on the cylinders of the Edison phonograph. Garner and Morgan both attended the BAAS meeting in Edinburgh in 1892, and Morgan subsequently wrote a review in *Nature* of Garner's book about his research. So Garner was almost certainly among those "other observers and thinkers" Morgan had in mind as disagreeing with him about language and reason in nonhuman animals. Furthermore, in Garner's view as in Morgan's, a phonograph cylinder full of remarkably humanlike ape language would have "proved" apes "to be rational beings." In the summer of 1893, Garner was due back from his phonographic expedition among the apes "before the close of the present year," perhaps "the day after tomorrow."

3

PROFESSOR GARNER'S
PHONOGRAPH

✦ ✦ ✦

WHEN RICHARD GARNER announced the discovery of a "simian tongue" in June 1891, in the English *New Review*, *Punch* responded with a poem, "Simian Talk":

> PROFESSOR GARNERS, in the *New Review*
> Tells us that "Apes can talk." *That's*
> nothing new;
> Reading much "Simian" literary rot,
> One only wishes that our "Apes" could *not!*[1]

Much the same witticism appears in the notebooks of the anti-Darwinian man of letters Samuel Butler the next year:

> In his latest article (Feb. 1892) Prof. Garner says that the chatter of monkeys is not meaningless, but that they are conveying ideas to one another. This seems to me hazardous. The monkeys might with equal justice conclude that in our magazine articles, or literary and artistic criticisms, we are not chattering idly but are conveying ideas to one another.[2]

Soon the American satirical magazine *Puck* was wondering whether "the rage fur doilect stories" would extend to tales "related in the doilect av quadhrupids or two-legged apes."[3] In the autumn of 1892, following news of Garner's departure for the French Congo, equipped with a cage to protect him from the gorillas and chimpanzees whose speech he was to learn, *Punch* responded once more, comparing those far-off apes with anti-Gladstonian Tories in their Pall Mall clubs:

> IN THE MONKEY-HOUSE;
> *Or, Cage versus Club.*
> PROFESSOR GARNER goes to the Gaboon
> To garner Monkey talk; a dubious boon! . . .
> Why hang, Sir, up a tree, in a big cage,
> To study Simian speech, which in our age
> May be o'erheard on Platform or in Pub,
> And studied 'mid the comforts of a Club?
> And yet perchance your forest apes would shrink
> From Smoke-room chat of apes who *never*
> But cackle imitatively all round,
> Till their speech hath an automatic sound. . . .
> "Voluntary and deliberate," their speech,
> "Articulate too"—those [forest] Apes! Then could they teach
> Their—say *descendants*—much. Does Club or cage
> Hear most of rabid and unreasoned rage?
> "Apes' manner of delivery shows" (they say)
> "They're conscious of the meaning they'd convey!"
> Then pardon, GARNER! Apes, though found in clans,
> Are *not*, of course, political partisans.
> Tired of the Club-room's incoherent rage,
> One pines for the Gaboon, and GARNER'S cage.
> For what arboreal ape *could* rage and rail
> Like him, with fierce Gladstonophobia pale,
> That Smoke-room Simian, though without a tail![4]

Richard Garner and the experiments that made him famous in the early 1890s are the subjects of this chapter. It tells first how the language barrier became a problem for Garner, how the phonograph became a solution, and how his contemporaries got word of that solution, so effectively that Victorian wits were able to use Garner to comic effect in their broadsides at human culture. From there the chapter turns to consider Garner's theoretical and experimental program in some detail, as well as his plans for transporting that program from the American zoo to the African forest. In its final part the chapter surveys a range of contemporary responses to Garner's claims. Among the patrons of Garner's research was the phonograph's inventor, Thomas Edison, collaborating with Garner on new phonograph designs for his upcoming expedition. Among men of science, W. D. Whitney declared Garner a genius, while Conwy Lloyd Morgan judged him a tender-minded amateur.

What one made of Garner's phonographic results depended not on whether one was an evolutionist, but whether one accepted Garner's view that his results were crucial for vindicating the theory of evolution. From the mid-1870s, the truth of evolution ceased to be a burning public issue, even in the American South, Garner's home. Most of Garner's audience were evolutionists like him.[5] There were exceptions: an 1894 book about Garner's work, for example, sought to combat his influence and defend the scriptural account of the origin of language. But most readers and listeners seem to have been impressed with the phonograph and Garner's manipulations of it. Indeed, no one seems ever to have queried Garner's own understanding of how the phonograph worked, or the extent to which that understanding influenced his interpretation of his results; though, as we shall see, Garner was much concerned to show that his phonograph manipulations merely amplified extant features of monkey utterances, and did not distort those features or introduce new ones. In short, there was a great deal of consensus within the debate over Garner's claims. The main sticking point was whether evolutionists could accept the existence of a language barrier. Garner said no, but many other evolutionists had said yes, Darwin and Romanes included.

Garner's extreme position—even Darwin was not evolutionist enough by Garner's lights—is doubly revealing. First, unlike the other evolutionists we have encountered thus far, Garner was not concerned so much to *explain* the origins of language as to show that language had no origins, that language existed in higher and lower forms at all points on the scale of nature. His lack of engagement with debates about imitation, the third frontal convolution, the possible usefulness of language in the struggle for existence, and so on, in part stems from a framework of assumptions within which those topics had no significance. Second, and again unlike the other evolutionists, Garner was a true outsider. He acquired his assumptions and his aspirations at a great distance from the intellectual and social milieu in which we have so far traveled, and from which much of the historiography on the evolutionary debates never departs. In seeing Garner's experiments as he saw them, we adopt the perspective of someone who bought the books, listened to the lectures, read the periodical pieces generated in that other milieu, and wanted to make a contribution of his own.[6]

Richard Garner versus the Language Barrier

The Making of a Self-Made Scientist

Richard Lynch Garner was born in 1848 in the town of Abingdon, in the Blue Ridge Highlands of southwestern Virginia (fig. 3.1).[7] Although the region was poor, rural, and isolated, Abingdon was its commercial and cultural center, and as such grew prosperous and even genteel. Its historic downtown still boasts a number of large and elegant buildings from the mid-nineteenth century, when Abingdon bustled with shopkeepers, bankers, doctors, lawyers, and merchants trading in salt, wool, flax, livestock, and other products from the surrounding farms, mines, and mills.[8] There were churches, schools, and

Figure 3.1 Richard Lynch Garner (1848–1920), around 1891. Courtesy of the National Anthropological Archives, Smithsonian Institution (Garner Papers, box 7), and reproduced by permission.

halls for lectures and theatrical performances.[9] Garner's father, Samuel, supported his large family as, at one time or another, a horse and cattle trader,
a mill owner, proprietor of a foundry, and a plow and tin supplier. He sent
Richard to private school from the age of eight to prepare for the ministry. In
later life the failed minister recalled an early but silent skepticism about religion. "I was surrounded on all sides with the bulwarks of orthodox religion
and any one, old or young, who would have confessed to the sin of unbelief
would have been anathematised from every pulpit and fireside in the land."[10]
As Garner entered adolescence, hostilities between North and South intensified. When war broke out, Garner's father (who had owned several slaves)
and older brothers served in the Confederate army, as did Garner himself,
though only in his teens. Fighting at times with Forrest's Third Tennessee
Mounted Infantry, later celebrated in the South for its raids on Union territory, Garner served in the battles of Bull's Gap, Cheek's Cross Roads, and
Morristown, emerging unharmed, though spending a good deal of time in
Union prisons.[11]

After the war, Garner studied from 1865 to 1867 at the Jefferson Academy for Men, in Blountville, Tennessee, taking courses in divinity and medicine. He took up schoolteaching, a profession he pursued in the region for
the next fourteen years. Soon with a small family of his own, Garner supplemented his income as he could. One of his more colorful ventures outside the
schoolhouse involved crossing the Missouri and Kansas plains on foot, riding
broncos through southern Colorado, fighting Apaches in their northern lands,
and capturing and breaking wild ponies to sell in the Texas markets of Sher-
man and Paris.[12] It was in the course of his work as a teacher, however, that
Garner first encountered the scientific controversy over the nature and origins
of the language barrier.

The occasion was a meeting about the latest developments in phonetics.
Far from the dry-as-dust subject it has become (for nonspecialists), phonetics in the nineteenth century had tremendous intellectual and social cachet.
As the goals of mass literacy and efficient communications had spread, so
had awareness of the limitations of the conventional alphabet. The phonetic
dream was to replace that hodgepodge with a system of simple, accurate, and
self-explanatory visual analogues for the sounds of human speech. Just by
looking, people would know exactly how to pronounce a word. Such a system
would be a boon to education, especially for the deaf. It would help adults
learn a foreign language. It would promote more efficient communication by
stamping out regional differences in pronunciation. It would help the imperial nations govern their colonies better, by ensuring swifter learning of the
languages of colonizers and colonized alike. As Alexander Melville Bell, the

Scottish elocutionist and inventor of a widely acclaimed "physiological" phonetic system (depicting not sounds but the associated positions of the tongue,
throat, and other parts of the vocal apparatus) wrote in his 1869 book *Visible
Speech*, with "such a medium of self-interpreting letters, the . . . foundation is
laid, and the Linguistic Temple of Human Unity may at some time, however
distant the day, be raised upon the earth."[13]

Phonetics was Garner's entrée to the debate on the evolutionary origins
of language. He later recalled his pivotal encounter with the language barrier
as follows:

> As a young school master, I once attended a Teachers' Institute where I took
> part in a discussion on phonetics in the course of which I innocently mentioned
> the sounds of animals as rudiments of speech—This was instantly and violently
> assailed as rank heresy and I barely escaped being immolated as an infidel. In de-
> fence of my assertion, I promised to adduce the evidence at some future meeting
> and stimulated by this tirade of opposition I resolved to study those sounds in a
> methodic manner and try to learn the speech of animals. This may be regarded as
> the first note in my researches.
>
> At that time I could find no literature on the subject except the negative ipsi
> dixits of those microsophic authorities who have filled whole libraries with soph-
> istry. There were no precedents to guide my researches, and I was left to my own
> resources to devise methods and invent means of experimentation.
>
> For some years my studies were only casual and incoherent; but the goal
> at which I aimed was not obscured by the universal negation and derision that I
> had to combat. Although my progress was slow and my data nebulous, I was not
> discouraged and firmly believed in my own ability to solve the riddle of animal
> speech.[14]

What Garner knew of evolution he probably learned from books, newspapers, magazines, and lectures during the "vogue of Spencer" that swept
Gilded Age America.[15] He read Spencer, Darwin, Tyndall, the anthropologist
Rudolf Virchow, the physicist and physiologist Hermann Helmholtz. Press
reports on the row between Müller and Whitney in 1875–76 carried Müller's
already famous views to a much wider popular audience, and Garner later recalled Müller among several authorities he had read who insisted, in Müller's
famous phrase, that language was "the one great barrier between the brute
and man."[16] In the 1880s, Garner turned from teaching to business, achieving
fitful success in real estate, stock sales, and other enterprises. In 1884, his long
doubts about the language barrier were confirmed at the zoological garden in
Cincinnati. He claimed that after listening for a while to the chatter of a group
of monkeys he was able to predict the behavior of a mandrill who shared their
cage. Further translation proved elusive, however.[17] Though he continued this

research as he could,[18] he soon turned his attention instead to the origins of
writing, in particular to the enigma of the Maya glyphs, which he examined
at the Smithsonian whenever business took him to Washington.[19] Attempting
in 1888 to advance his studies, Garner asked the curators to furnish him with
photographs from several of the collections. His frequent entreaties at last
exasperated the Smithsonian officials: "This man is a weariness to my flesh,"
complained one in an internal memo. "He seems to be pegging away at something, but he never publishes anything."[20] Garner indeed had a novel scheme
in the works. As he later described it:

> Dr. Alexander Melville Bell has shown, in his work on "Visible Speech," that the
> organs brought into use in the production and modification of sounds must work
> in harmony with each other; hence it is that by a study of the external forms of
> the mouth the movements of all the organs used in making any sound can be de-
> termined with such certainty that deaf-mutes can be, and have been, successfully
> taught to distinguish these sounds by the eye alone. And it was by such a method
> that I set out to read the temple inscriptions from the ruins of Palenque, some
> years ago. . . . The main feature of those glyphs, by which I was guided, was the
> outline of the mouth, which the artist had sought to preserve and emphasise at the
> cost of every other feature, and by this process I found to my satisfaction some
> ten or twelve sounds or phonetic elements of the speech used by these people; but
> not knowing the meaning of the sounds in that lost tongue, I did not attempt to
> verify them, but when I find the time to devote to them I believe I can accomplish
> that.[21]

It was probably on one of his trips to Washington that Garner first saw in the
cylinder phonograph a solution to his problems with the monkey language.[22]
As he later recalled, "At last came the Edison phonograph. I was being ship-
wrecked, when this wonderful machine saved me."[23]

The Edison Cylinder Phonograph

Thomas Edison had come upon the idea for his wonderful machine unawares,
in the course of work on telegraph and telephone improvements in his Menlo
Park, New Jersey, laboratory in 1877.[24] The first phonographs basically had
two parts: a thin metal disk or "diaphragm" with a stylus attached to one
side, and a grooved, cylindrical drum, wrapped with a sheet of tin foil, and
mounted on a crankshaft. Unlike the phonographs and gramophones that followed, the pioneer phonographs could record as well as reproduce sounds. To
record, the user aimed the sound to be recorded (sometimes through a horn
or "trumpet") at the disk while working the hand crank. As the disk vibrated,
the sound etched its pattern via the stylus onto the rotating tin foil. The result

was a record of the sound. To reproduce the recorded sound, the user fitted a smoother stylus and once again cranked the handle, letting the stylus ride over the tin-foil record as it rotated. Now the record set the stylus in motion, which set the disk in motion, which set the surrounding air in motion so as to recreate the original sounds. At first the mere existence of such a machine was cause for astonishment. But the sound quality was terrible, and the records themselves wore out after just a few plays. Soon the novelty wore thin, even for Edison, who soon moved on.[25]

For the next ten years, Edison turned his prodigious energies to the more lucrative projects of manufacturing incandescent lamps and distributing electrical power.[26] In 1887, he returned to the phonograph with a literal vengeance. His competitors had begun (behind his back, as he saw it) to patent improved versions of the device. Out of Edison's belated burst of enthusiasm came the so-called perfected phonograph. It was notable for a number of innovations, but two in particular. First, in place of the humble hand crank of the early phonographs, there was a heavy electrical motor, often powered by a lead-acid battery and regulated by a rather cumbersome governor. Second, in place of the tin foil, wrapped around a horizontally shifting drum, there was a wax-covered cardboard cylinder, slipped onto a nonshifting spindle or "mandrel." Wax recordings made at more uniform speeds sounded better, recorded longer, and endured more robustly—hundreds of plays were now possible.[27] "Here is a pretty little machine," wrote one admirer, "a little bundle of iron nerves and sinews imbued through electricity with the great principle of life, 'the soul of the world,' chained by Edison's genius and forced to obey our behests."[28] When *Punch* brought Michael Faraday back from the grave in 1891 to pronounce his pleasure at Science's progress, the perfected phonograph had center stage, surrounded by the paraphernalia of cylinder containers, an enormous battery, and other examples of Edisoniana, including a chart of telegraph cables and a telephone (fig. 3.2).

The phonograph was a technological triumph, but, despite massive publicity, it remained a commercial underachiever. In Europe and Britain, public expositions and demonstrations were really the only places where most people encountered the phonograph.[29] In the States, phonographs were available for rental or purchase in several regions, but they were just too expensive for most people to have in the home.[30] Edison was keen to see phonographs put to practical use in offices, for taking dictation or sending messages.[31] Although the phonograph met with only limited success in this regard,[32] Washington proved a welcome exception. "During the session of Congress between fifty and sixty machines were used in the Capitol alone, by Senators, representatives, officials, etc.," reported the April 1891 issue of the *Phonogram*, house

A SCIENTIFIC CENTENARY.

Faraday (returned). "Well, Miss Science, I heartily congratulate you; you have made Marvellous Progress since my Time!"

Figure 3.2 The Edison cylinder phonograph celebrated in *Punch*, 27 June 1891, 309.

organ of the new industry. "Every department of the Government is now supplied. In the homes of men of wealth and culture the phonograph is already a fixture."[33] Outside those homes, the phonograph was slowly becoming a fixture as well, in railroad stations, drug stores, saloons, and even new "phonograph parlors." These sometimes unsavory places began to feature the forerunners of the jukebox, the nickel-in-the-slot phonograph, where a nickel bought a few minutes of the latest John Philip Sousa march or Stephen Foster ditty.[34]

The anthropologist Jesse Walter Fewkes announced a new scientific use for the phonograph in May 1890. Fewkes was one of a new, professional breed of anthropologist, working full-time for the United States government under the auspices of the Bureau of American Ethnology at the Smithsonian.[35] Writing in *Science* "on the use of the phonograph in the study of the languages of the American Indians," Fewkes described "experiments" he had made with the phonograph in preserving the speech, songs, folklore, and rituals of the Passamaquoddy Indians of Maine. "What specimens are to the naturalist in describing genera and species, or what sections are to the histologist in the study of cellular structure, the cylinders made on the phonograph are to the student of language," he wrote. No longer was the ethnologist hostage to an imperfect ear, a faulty memory, hopeless systems of notation, limited time, or, indeed, the extinction of the speakers of a language. "There are stories, rituals, songs, even the remnants of languages which once extended over great States, which are now known only to a few persons," wrote Fewkes. "The phonograph renders it practicable for us to indelibly fix their languages, and preserve them for future time after they become extinct or their idiom is greatly modified or wholly changed."[36]

Was Garner aware of Fewkes's innovations? No doubt they were much discussed among the Smithsonian anthropologists Garner was pestering for access to the Maya glyphs. Be that as it may, in November 1890—a couple of months after Garner's debut experiments at the National Zoological Park, latest addition to the Smithsonian imperium—the anatomist Frank Baker, the park's acting director, received a cylinder in the post. Baker and his staff gave the cylinder a spin on the office graphophone (a commercial rival of the phonograph, produced by telephone inventor Alexander Graham Bell, son of Melville Bell). They tried the cylinder one way round, then the other, but unintelligible noise was the result each time. As Baker later wrote, "I came to the conclusion that it represented an entire carnivore house in full blast about feeding time." No letter of explanation or instruction came. At last someone suggested that the mystery cylinder might have been recorded at a speed outside the range of the office machine. Baker went to the headquarters of the Columbia Graphophone Company to find out. Several attempts with various models gave equally poor results. At last one of the older machines, faster than its descendants, was tried, and the roars and bellows became articulate speech.[37]

The cylinder carried a dictated letter from Garner. Very probably Baker had been among the scientific witnesses to Garner's first attempt at phonographic inquiry among the monkeys, and the letter aimed to bring Baker up to date on Garner's progress since, with reports on more recent experiments with the phonograph at the zoological gardens of Cincinnati and Chicago. Although much more work was needed, Garner was even more convinced of his eventual success. And would Baker kindly let Garner know when that celebrated explorer and habitué of gorilla country, Mr. Henry Stanley, was due to arrive in town? In December, Baker sent a letter of reply, telling Garner that, due to bafflement over his cylinder, the Stanley visit had come and gone without him. But Baker allowed that he found the account of Garner's research thus far interesting, and advised him to take more care in the future in his efforts to communicate by cylinder.[38]

Garner's phonographic investigations, and his campaign to secure publicity and support for them, appear to have continued apace. In April 1891, the *Phonogram* carried a small notice about "novel experiments with the phonograph":

> A learned professor at the Smithsonian Institute has recorded on the phonographic cylinder the chatter of monkeys; and, after careful practice of the sounds, he finds that, by repeating them, he can make himself understood by the animals. Sounds expressive of fear, cold, hunger, and other sensations common to the human race and its four-footed brethren, have alone been recognized. It is asserted that there are forty different sounds in the Simian vocabulary.[39]

Spreading the Word

Garner gave a full account of his research in the *New Review* article of June 1891, entitled "The Simian Tongue." "In coming before the world with a new theory," Garner announced,

> I am aware that it may have to undergo many repairs, and be modified by many new ideas. On entering the world of science, it begins its "struggle for life," and under the laws of "the survival of the fittest" its fate must be decided. I am aware that it is heresy to doubt the dogmas of science as well as of some religious sects; but sustained by proofs too strong to be ignored, I am willing to incur the ridicule of the wise and the sneer of bigots, and assert that "articulate speech" prevails among the lower primates, and that their speech contains the rudiments from which the tongues of mankind could easily develop; and to me it seems quite possible to find proofs to show that such is the origin of human speech.[40]

Punch's doggerelist took notice, as we have seen. So did William James. In his copy of his *Principles of Psychology*, next to the sentence "Man is known again as 'the talking animal'; and language is assuredly a capital distinction between man and brute," James inked in a note: "Cf. R. L. Garner, the Simian Tongue, (New Review, June 1891)."[41] On 6 June, the English *Spectator* celebrated Garner's article under the heading "Apes and Men." For the *Spectator*, Garner's phonograph was an emblem, not of the ties that bound apes and men, but of the free will that scientists exercised in examining the laws of nature, laws that governed apes and all else nonhuman. Where Garner as the human investigator had freely chosen how to direct his attention in conducting his experiments, his simian subjects attended only to that which excited them and their desires. Hence the "whole series of experiments, and the wonderful instrument which rendered them possible, are a triumph of this one great fundamental root of all science, the act of voluntary attention."[42]

Why proclaim the discovery of the simian tongue in the *New Review*? In several respects, it was not an obvious choice. While old lions such as Huxley, Spencer, and Wallace still held forth on the nature of life and evolution in the *Nineteenth Century* and other nonspecialist journals, empirical work in the natural sciences now appeared largely in journals devoted to the natural sciences, indeed to particular natural sciences.[43] Even among the nonspecialist journals, the *New Review* was neither especially successful nor well known for its coverage of science. It was founded in 1889 by Archibald Grove, a young Liberal MP who hoped the review's combination of low price and serious content would fill a gap in the periodical market. While articles on science and technology appeared from time to time, politics and the arts were the mainstays of the *New Review*. Garner's articles appeared alongside work by Henry James, Émile Zola, and a host of lesser but still considerable Victorian lights, such as Frederic Farrar; the novelists Grant Allen, Olive Schreiner, and Eliza Lynn Linton; and the mythographer Andrew Lang. But sales were stubbornly low. Rather than filling a gap, the *New Review* seems instead to have fallen into one, being neither as high-caliber as the more expensive intellectual journals, nor as inexpensive as the low-caliber illustrated magazines.[44]

The placing of "The Simian Tongue" in the *New Review* was probably the work of Samuel Sidney McClure, Garner's literary agent. As proprietor of a major literary syndicate, the New York–based McClure in the early 1890s was always on the lookout for fresh writing talent. He had excellent contacts in the English publishing world, and traveled constantly between England and the States, negotiating terms all the while.[45] His antennae for what would sell were remarkably good, and he seems to have become Garner's agent from early days.[46] It was McClure who turned the obscure Virginian into the talked-about "Professor Garner." The list of outlets where McClure flogged Garner's writings reads like an atlas to a vast newsprint continent: the *Atlanta Constitution*, the *St. Louis Post Dispatch*, the *Detroit Tribune*, the *Minneapolis Tribune*, the *Chicago Tribune*, the *Boston Herald*, the *Philadelphia Inquirer*, the *Montreal Star*, the *Montgomery Advertiser*, the *Cleveland Leader*, the *Louisville Courier Journal*, the *Washington Post*, the *Albany Telegraph*, the *Buffalo Courier*, the *Toledo Blade*, the *Kansas City Times*, the *New Orleans Picayune*, the *Chattanooga Times*, the *Richmond Dispatch*, the *Syracuse Herald*, the *Springfield Republican*, and the *Cleveland Plain Dealer*.[47] The market was huge, and the well-connected McClure knew how to exploit it. For McClure, Garner represented an investment worth cultivating, a colorful scientist and (under McClure's guidance) soon-to-be explorer who could write vivid prose about his studies in the controversial field of human origins.

In large part thanks to McClure's canny promotional efforts, Garner became news. Shortly after June 1891, according to a reporter for the *New York World*, "the entire scientific world knew of [Garner's] doings," and before long, the "Simian tongue, Mr. Garner's name for his discovery, ha[d] gone all over the world and . . . been written about in every known language save in the simian tongue itself."[48] Certainly Garner was being listened to and taken seriously. Barnet Phillips, a reporter sent to Central Park to profile Garner for *Harper's Weekly*, described Garner's phonograph as opening a new route into the prehuman past. "Farther back than man, Mr. Garner is looking for the first understood syllable," wrote Phillips. "If we accept the Darwinian theory, and should Mr. Garner's life work be accomplished, the advocates of evolution will have found a new and strong argument in their favor."[49] Introducing Garner at a meeting of the Nineteenth Century Club in the assembly rooms of Madison Square Garden the following February, the club's president remarked, in the words of a reporter for *Scientific American*, "that Darwin's 'Descent of Man' was the most important scientific work since the 'Principia' of Newton, and that Mr. Garner's brilliant researches were well calculated to sustain the views introduced by Darwin."[50]

Garner followed up "The Simian Tongue" with two more installments in the *New Review*, in November 1891 and February 1892. He published a number of other articles about his research, in monthlies such as the *Forum* and the *Cosmopolitan*—American counterparts to the *New Review*.[51] At McClure's suggestion, Garner wove these articles together, along with much new material, into a book, *The Speech of Monkeys*, published in the summer of 1892 in New York and London by Charles L. Webster and William Heinemann, respectively. The advances came to about $400. This was all the money Garner earned from the book, in large part because Webster went out of business shortly after the book came out.[52] In the first half, Garner described his phonographic and psychological investigations among a cast of simians including Jokes, Jennie, Banquo, Dago, McGinty, Pedro, Puck, Darwin ("a quiet, sedate, and thoughtful little monk, whose grey hair and beard gave him quite a venerable aspect"), Mickie, Nemo, Dodo, Nigger, Nellie, Dolly Varden, and Uncle Rhemus. Many of these, he wrote, were his "little friends." Indeed, he "regarded the task of learning the speech of a monkey as very much the same as learning that of some strange race of mankind, more difficult in the degree of its inferiority, but less in volume."[53] The second half of the book was more speculative and theoretical, relating Garner's views on the nature and evolution of speech, thought, and life, with remarks on the phonograph and other instruments used in his investigations.

A Closer Look at Garner's Experimental and Theoretical Program

The Phonograph Experiments

Some of those who observed Garner in action have left descriptions of his experimental technique. Here is Garner at work at the zoological garden in New York's Central Park in December 1891 (fig. 3.3):

> [Garner] attaches a new wax cylinder to the machine and then speaks into the trumpet, saying "Cylinder number one, New York, December 9, 1891." This is, of course, recorded on the wax, and will always be repeated when put into the phonograph.
>
> His plan is to collect monkey words, noting their apparent meaning. By this time one Park policeman and two or three keepers are standing spellbound in the monkey house, and when the experimenter calls for something eatable for the simians all hands hurry to help him. Some pieces of carrot are brought, carefully concealed in paper. The monkeys must not see them until the machine is set going and the trumpet is pointed in their direction.
>
> Mr. Garner bites off a little bit of carrot from a slice and holds it up. "Wh-u-u-h!" he says. A little chap darts forward and makes the same sound, getting the carrot as a reward.
>
> Did you ever try to get a monkey to talk into a phonograph? Don't ever attempt it unless you have lots of time. The little scamps utter their cries freely enough, but the shrewdest of lawyers could not guard himself more carefully against going on record. They won't talk into the trumpet. When they talk into it it is when the wheels aren't moving, so that their remarks are lost. The experimenter's patience is stupendous. He moves the phonograph and the cumbersome pedestals of boxes and ladders on which it rests half a dozen times over to some promising monkey whose garrulity he thinks he can catch on the fly. . . .
>
> The experiments were brought to a close at about half-past ten o'clock [a.m.]. The work had been going on for three hours, but so absorbed and interested were we all that it did not seem half as long. Six wax cylinders were loaded altogether, and as this was all Mr. Garner had, proceedings were discontinued necessarily. No further work could have been done anyway, as a big crowd began to swarm into the monkey house, the report having got out that some exciting things were going on there.[54]

Garner used the phonograph to address three basic questions: first, about the sounds and meanings of simian utterances; second, about variation within

PROFESSOR GARNER.
FROM A PHOTOGRAPH BY BRADY, WASHINGTON.

PROFESSOR GARNER'S EXPERIMENTS AT CENTRAL PARK.—DRAWN BY F. S. CHURCH.—[SEE PAGE 1030.]

Figure 3.3 Richard Garner with his phonograph in Central Park, New York City, December 1891, and (as imagined by the artist) in West Africa. Illustration by F. S. Church, from *Harper's Weekly*, 26 December 1891, 1036.

the tongues of different species (for, in his view, there were as many "simian tongues" as there were ape and monkey species); and third, about the arrangement of these tongues into a hierarchical series. For purposes of translation, Garner recorded the often rapid and, he claimed, subtly modulated utterances of the monkeys, and then repeated these, on the phonograph or, with lots of practice, on his own lips, and observed the effect of these sounds on other monkeys. In his first "Simian Tongue" article, he described how he had thus translated the capuchin words for food, drink, sickness, storm, and alarm. He also gave detailed instructions for how to pronounce these words (which turned out to be composed mostly of vowels).[55] An opportunity to go beyond translation and examine variation arose at Central Park. Garner was there doing the experiments chronicled above when a shipment of rhesus monkeys arrived from abroad. At Garner's request, the foreign monkeys were kept separate to ensure that no communication passed between them and the local rhesuses. When Garner recorded the local word for "salutations" and repeated it to the new monkeys, their excited response showed, he believed, that the word was their word too.[56] Most important for Garner's evolutionary claims was the task of hierarchy building. He argued that the whole of life could now be seen as a great chain of expression, with "one unbroken outline, tangent to every circle of life from man to protozoa, in language, mind, and matter."[57] For Garner, all creatures had the means for expression, commensurate with their physical and mental development generally—the expressive powers of mammals, for example, correlated closely with such features as the "craniofacial angle" and the "gnathic index"—and with the demands their mode of life placed on them.[58] He had discovered the language of spider monkeys to be "almost as inferior to that of the brown Capuchin as the brown Capuchin's appears to be below the Chimpanzee's, and as the Chimpanzee's appears to be below the lowest order of human speech."[59]

But were these simian languages ancestral to human languages? In what sense were the simian "languages" truly languages at all? Garner justified his claims with a brief argument: "To reason, [simians] *must think*, and if it be true that *man cannot think without words*, it must be true of monkeys: hence they must formulate those thoughts into words, and words are the natural exponents of thought."[60] It helps to distinguish three implicit and separate premises here. First, humans and monkeys are, in some way, related. Second, evolution is uniformly progressive—that is, features present in a later stage of the process are only more fully developed versions of features present at an earlier stage. Third, speech and reason always go together. This last Müllerian premise appears several times over in the argument above, and elsewhere in Garner's writing from this time as "speech is materialised thought" and "words are the body of which thoughts are the soul." (Compare Müller's own "the word is the thought incarnate.")[61] In Garner's view, if humans and monkeys are related, if evolution is uniformly progressive, and if speech and reason go together, then monkeys ought to speak and reason to a lesser degree than humans do. To Garner's satisfaction, at least, the phonograph showed he was right. He had found in the "*monophones*" of monkey speech, in which "each idea seems to be couched in a single word of one syllable and nearly, indeed, of one letter," the homogeneous rudiments that later differentiated, first into the "few score of words" and "small range of sounds" of the tongues of savages, and then into the fully heterogeneous tongues of the civilized races.[62] Where the most advanced humans used complex sounds to transmit complex ideas between complex minds, monkeys used simple sounds to transmit simple ideas between simple minds, but for each group, words served the same basic function. Simian words were no different in kind from human words.[63]

Understanding what Garner used the phonograph for is helpful for understanding what he actually did with the machine, and why.[64] One thing he did was to inspect the phenomena he was studying more closely. Using the phonograph, Garner was able to fix the aural blur of monkey utterances, slow them down, and analyze them. In his words, "to magnify the sounds as I have shown it can be done, allows you to inspect them, as it were, under the microscope."[65] He reported that playing cylinders at speeds much lower than the speed of recording made it possible to "detect the slightest shades of modulation . . . [and t]he slightest variation of tension in the vocal chords."[66] Garner magnified the features of monkey utterances still further by using two phonographs in conjunction. While phonograph A repeated a monkey utterance at reduced speed, phonograph B recorded that slowed-down utterance. Then the B record was itself played at a reduced speed.[67] In this way the phonograph indeed functioned something like a microscope: it compensated for impoverished senses by putting the world into sharper focus. "The rapidity with which these creatures utter their speech," Garner wrote, "is so great that only such ears as theirs can detect [the] very slight inflections" upon which meaning depended.[68] Of the rhesus tongue, he claimed: "The 'u' sound is about the same as in the Capuchin word, but on close examination with the phonograph it appears to be uttered in five syllables very slightly separated, while the ear detects only two."[69] If there was indeed more to monkey utterances than met the ear, such as nuances of meaning or subtle phonetic features, and if these previously undetectable features did indeed show some kind of continuity with the roots of human speech, then the phonograph was bringing new facts to the debate about the evolutionary origins of language, as Garner claimed.

In some cases Garner did more than just inspect phenomena with the phonograph. He also used it to purify phenomena. The phonograph enabled Garner to repeat monkey utterances back to monkeys with hitherto unobtainable precision. Furthermore, he was able to screen out or suppress the visual and even (he believed) psychic cues that accompanied the utterances of live monkeys, and so to determine with certainty whether monkeys really did respond to the utterances as such.[70] Typical of his reports is the one on Dago, in Chicago. According to Garner, one of the utterances recorded while Dago was sitting by a window seemed to be a complaint about bad weather. Garner continued: "this opinion was confirmed by the fact that on a later occasion, when I repeated the record to him, the weather was fair; but when the machine repeated those sounds which he had uttered at the window on the day of the storm, it would cause him to turn away and look out of the window."[71] Garner also wrote of the value of the phonograph in removing unwanted features of a sound. By playing a cylinder in reverse, he reported, he was better able to isolate a sound in its fundamental state. "By this means we eliminate all familiar intonation, and disassociate it from any meaning which will sway the mind, and in this way it can be studied to advantage."[72] Looking again to the emblematic devices of the scientific revolution, the phonograph in these instances functioned less like a microscope than an air pump, enabling intervention in, rather than mere inspection of, the course of nature. But note too an important difference between the phonograph and the air pump. The air pump was located in, indeed helped to define, a specific site, the laboratory. Garner had no phonographic laboratory. Garner and phonograph went where the monkeys were. And monkeys, it seems, were just about everywhere: on the street corners where immigrants played hand organs, on the stages of traveling shows, in the homes of the well-to-do, and in the ever-burgeoning zoological parks and gardens.

One of the most intriguing uses to which Garner put the phonograph involves neither inspection nor purification so much as the creation of phenomena. Garner regarded as some of his most important findings phenomena that, in a rather straightforward way, he created, by manipulating the sounds he had recorded. Here is a sample of Garner the creator of phenomena:

> By the aid of the phonograph I have been able to analyse the vowel sounds of human speech, which I find to be compound, and some of them contain as many as three distinct syllables of unlike sounds. From the vowel basis I have succeeded in developing certain consonant elements . . . from which I have deduced the belief that the most complex sounds of consonants are developed from the simple vowel basis, somewhat like chemical compounds result from the union of simple elements. Without describing in detail the results, I shall mention some simple experiments which have given me some very strange phenomena.[73]

What is described here is something like this. Garner claimed that monkey utterances were composed almost entirely of vowels. Playing recordings of human vowels at different speeds and so on, he had found, first of all, that these vowels could be broken down to still simpler vowels (of the sort one found in monkey utterances); and second, that the recorded vowels could be manipulated to produce the more complex consonants found only in human utterances. Here we have the use of apparatus to create phenomena that had never before existed. No one had ever noticed the kinds of regularities Garner was producing with his cylinders and phonograph. Moreover, for Garner, these and similarly created phonographic phenomena were *evolutionary* phenomena: they revealed the evolutionary continuity between human speech and monkey speech.[74]

But what guarantee was there that Garner's phonograph manipulations legitimately magnified, purified, and extended the deliverances of nature? How could one be sure that the subtleties "revealed" in a slowed-down recording of an utterance, say, were in nature waiting to be discovered, and not meaningless artifacts of the process of slowing down? Garner needed a theory of the phonograph to answer such skeptical questions. In *The Speech of Monkeys*, a fair number of pages expound his theory in detail. "At a given rate of speed," Garner wrote in a discussion of his slowing-down technique,

> I have taken the record of certain sounds made by a monkey, and by reducing the rate of speed from two hundred revolutions per minute to forty, it can be seen that I increased the intervals between what is called the sound waves and magnified the wave itself fivefold, at the same time reducing the pitch in like degree. . . . in this process all parts of the sound are magnified alike in all directions, so that instead of obtaining five times the length, as it were, of the sound unit or interval, we obtain the cube of five times the normal length of every unit of the sound. . . . From the constant relation of parts and their uniform augmentations under this treatment, it has suggested to my mind the idea that all sounds have definite geometrical outlines, and as we change the magnitude without changing the form of the sound, I shall describe this constancy of form by the term contour.[75]

Notice the emphasis on the *preservation* of features present in the original recording: "all parts . . . are magnified alike in all directions," "constant relation of parts," "uniform augmentations under this treatment," "we change the magnitude without changing the form." Garner presented a theory of the phonograph that is at the same time a defense of his uses of the phonograph to find out about monkey speech and its relations to human speech. According to Garner, phonograph manipulations of recordings introduced no distortions. He even reported finding this conclusion confirmed in the visual displays of the phoneidoscope—a simple device, about as old as the phonograph itself, in which different sounds caused different patterns to form on a soapy film.[76]

The Wider Program

Important though it was as a window onto the simian mind, the simian tongue was not the sole window. On the basis of a number of psychological experiments, Garner claimed to show that the minds of monkeys had at least the beginnings of ideas of color, number, and quantity. To find out about color ideas, Garner presented monkeys with a number of differently colored items—candies, marbles, bits of ribbon—and checked whether the monkeys consistently favored some colors over others. In general, he concluded, monkeys preferred green things, perhaps, he surmised, because of an association with some green food of which they were fond.[77] To assess mathematical skill, Garner devised two tests. One test involved presenting monkeys with two small platters, one with a single nut on it, the other with several nuts. When the monkeys consistently chose the platter with the larger number of nuts on it, no matter whether Garner held the platter in his left hand or in his right hand, Garner concluded in favor of the presence of an idea of quantity and perhaps even of number. The other test assessed how high monkeys could count. Garner gave a monkey three marbles to play with for a while, then put the marbles in a small wooden box with a hole in the side, just big enough that the monkey could reach in and pull out one marble at a time. Next Garner hid one of the marbles and put only two in the box. After recovering two marbles, the monkey continued for some time to search in and around the box, but gave up after a while and started playing with the two marbles. Again Garner repeated the process, so that now there was only one marble for the monkey to recover from the box, and again the monkey searched about. Increasing the number of marbles to four, Garner was less sure that one was missed; but that one was missed from three he was in no doubt.[78] Of his subjects he concluded:

> I do not think that they have any names for numbers, colors, or quantities, nor do I think that they possess an abstract idea of these things, except in the feeblest degree; but as the concrete must have preceded the abstract idea in the development of human reason, it impresses me that these creatures are now in a condition such as man has once passed through in the course of his evolution; and it is not difficult to understand how such feeble faculties may develop into the very highest degree of strength and usefulness by constant use and culture. . . . In brief, they appear to have at least the raw material out of which is made the most exalted attributes of man.[79]

Conwy Lloyd Morgan—not someone Garner would have read— had distinguished a third category, "intelligence." This was the domain of trial-and-error learning, between instinct and reason. Garner distinguished only instinct and reason, and then argued the distinction was not a sharp one. Reason, for Garner, was the power "to think methodically and to judge from attending facts."[80] To test for reason in animals, one needed to put the animals on their own resources under novel circumstances.[81] "When a monkey examines the situation and acts in accordance with the facts, doing a certain thing with the evident purpose of accomplishing a certain end, in what respect is this not reason?" he asked. In his view, a monkey that figured out how to escape from its cage was exercising reason no different in kind from human reason. Wherever reason began in the scale of nature, Garner concluded, "it is somewhere far below the plane occupied by monkeys." Garner was as chary of dividing emotion from reason as of dividing instinct from it. Indeed, reason and language alike were rooted in the emotions and their expression. "The uniform expression of the emotions of man and Simian," he wrote, "is such as to suggest that, if thought was developed from emotion, and speech was developed from thought, then the expressions of emotion were the rudiments from which speech is developed."[82]

Garner professed surprise at the sins of error and omission in Darwin's discussion of language in the *Descent* and elsewhere.[83] Certainly Garner rejected any notion of a language barrier between humans and simians. But Darwin's assessment of that barrier had been much qualified, and Garner himself had made modest claims for the simian tongues he surveyed, and for his own ability to learn them. His vocabulary of capuchin monkeys, for example, still included only words for food, drink, weather, love or friendship, alarm, and summons.[84] Monkey conversations were usually limited to utterance and reply (often the same sound).[85] Where Darwin and Garner differed most was in their accounts of how language had evolved. Darwin speculated a great deal about how language would have aided early humans in the business of survival and reproduction, and therefore how the power of language would have increased under natural and sexual selection. Although Garner wrote that language "is essential to all social order, and no community could long survive as such without it," for the most part he was silent on the *usefulness* of language.[86] Rather, for Garner, the capacity and desire for expression were fundamental properties of vital matter, indeed matter as such, so that evolution amounted to the gradual, progressive unfolding of expressive potential.[87] The more complex the conditions of life under which a creature lived, the greater the demands on expressive powers, and the more those powers, through exercise, developed.[88] As speech developed, concrete ideas became more abstract, sound became more important than gesture in communication, and the vowel basis of utterances became elaborated with ever more prominent and diverse consonants.[89]

Garner emphasized how small were the increments in expressive power as one moved up the natural scale of expression. He presented his "law of cranial projection" (as he later called it) as predicting, more or less accurately, the vocal powers of creatures from hissing reptiles to eloquent Europeans.[90] He was more cautious about the order in which the various positions in the scale had been filled. Perhaps humans had evolved from apes; perhaps apes had degenerated from humans; perhaps both had emerged and gone their separate ways from a single protostock, neither ape nor human.[91] But, he asked, supposing "man derived his other faculties from such an ancestry, may not his speech have been acquired from such a source? If the prototype of man has survived through all the vicissitudes of time, may not his speech likewise have survived? If the races of mankind are the progeny of the Simian stock, may not their languages be the progeny of the Simian tongue?"[92] The phonograph had opened vistas on the past, but also, for Garner at least, on the future:

> Standing on this frail bridge of speech, I see into that broad field of life and thought which lies beyond the confines of our care, and into which, through the gates that I have now unlocked, may soon be borne the sunshine of human intellect. What prophet now can foretell the relations which may yet obtain between the human race and those inferior forms which fill some place in the design, and execute some function in the economy of nature? A knowledge of their language cannot injure man, and may conduce to the good of others, because it would lessen man's selfishness, widen his mercy, and restrain his cruelty. It would not place man more remote from his divinity, nor change the state of facts which now exist. Their speech is the only gateway to their minds, and through it we must pass if we would learn their secret thoughts and measure the distance from mind to mind.[93]

From the Zoo to the Forest

It was not only Garner's experiments with monkeys, marbles, and phonographs that excited interest. He had big plans, and he was talking about them almost from the beginning. "I am now trying to arrange for a trip to interior Africa to visit the *troglodytes* in their native wilds," he wrote in the second installment of "The Simian Tongue," "and if my plans (which are all practicable) can be arranged, I agree to give to the world a revelation which will rattle the dry bones of philology in a wholly new light." He added that Edison himself was helping to modify his invention—"the only thing which makes these researches possible"—for the upcoming expedition.[94] More details appeared in the New York press in December. Readers learned that a seven-foot-square cage made of metal bars would be Garner's forest home, in which he would sit with phonograph and camera to obtain a full record of the speech and habits of gorillas and chimpanzees in the wild:

> It is certain that in all the history of the world there has never been such a combining of the wild and adventurous with the civilized and scientific as that which Mr. Garner has planned. Think of a man seated in a cage of steel wire work. . . . His cage is in a dense African forest, and the gorilla nearby is making the air ring with comments on the stranger's appearance, delivered in his own Simian dialect. At the same time the phonograph, silent and infallible, is taking down the gorilla's remarks. Up in the fork of a tree a camera, which Mr. Garner operates by touching an electric button, is taking instantaneous pictures of the gorilla as he talks, and Mr. Garner, by telephone and by electric bell, is in communication with friendly forces located at a convenient distance.[95]

In the words of one New York journalist, "If Jules Verne had put such an idea into a novel it would have excited great interest. As an actual part of scientific experiment in real life it has aroused more interest than anything that has come to the surface in a very long time."[96] Barnet Phillips from *Harper's* concurred:

> It all seems like a Jules Verne excursion into the animal kingdom; but with a man as a directing spirit who will go to Africa, taking with him all those scientific implements which have positive and practical effectiveness, much may be expected. No one can know what Mr. Garner may not accomplish. He may advance only by one footprint into the realm of the long past, where all has been heretofore hazy, confused, indistinct. Certainly he is a brave man who has the courage to try and solve nature's greatest mystery.[97]

Garner at first expected he would leave for Africa as early as December 1891 or January 1892. In the event, he departed New York for Liverpool in early July 1892, and Liverpool for Libreville, West Africa, in mid-September 1892.[98] Meanwhile he published two articles about what he expected to do in Africa and how he expected to do it, in the *North American Review* (June 1892) and the *New Review* (September 1892). The cage, camera, and phonograph were just the beginning of the list of matériel. There would be chemicals and tools, needed to preserve plants and animals. There would be electrical light "globes." There would be deadly weapons: a rifle and revolver, with two thousand rounds of ammunition; a kind of modified crossbow, firing arrows whose tips had been modified to discharge prussic acid upon impact (guns were too noisy and disruptive, too apt to scare away all the wild animals and people); a spear similarly modified; and what Garner called his "masked battery," a device for spraying concentrated ammonia at a moment's notice. The cage itself was capable of being electrified. Of course Garner would work alone, but stationed nearby he would have the servants, carriers, and guides he needed to transport and maintain all this equipment. To ingratiate himself to the natives, he would be carrying a letter of introduction to the chief of Lukalela in the Middle Congo, in the form of a phonogram from the explorers Henry Stanley, Paul du Chaillu, and E. J. Glave, one of Stanley's lieutenants.[99]

The phonograph was at the heart of the enterprise. Using the instrument, Garner expected to learn the tongues of the anthropoid apes, in much the way he had learned the tongues of their simpler kin. But he promised to do much more. He would also record and analyze the tongues of the tribesmen, and compare these with each other and the ape tongues. "Granted that I have got to the bottom of monkey talk, my task would be but half accomplished," Garner told the *Harper's* reporter, Phillips. "I will have but forged a single link in the chain. I want another. I propose taking down the speech of the lowest specimens of the human race—the pygmies, the Bushmen . . . the Hottentot cluck and click." Garner continued: "If there be family resemblance, structural relationship, between the Rhesus monkey, the chimpanzee, and the lower grades of humanity, there may be correlation of speech, philological kinship, and then—and then—the origin of man's talk might be found."[100] Garner also planned to study how speech developed in individual apes and natives. Through daily recordings, he would chart this development with precision, allowing him, for example, "to ascertain whether or not the development of consonant sounds follows the same law with them [apes and natives] as with children of the Caucasian race." As well as translating ape words, he would attempt to teach the apes new words, for if "it can be shown that these animals can acquire new sounds and ascribe to them new meanings, it will indicate a capacity which has always been denied them."[101]

His inquiries into speech were central, but not exclusive. With his camera, he would assemble a photographic record of the forest at different times of day. (He intended to send photos of the mouths of speaking apes to Melville Bell for analysis.)[102] He would contribute much of ethnological interest by securing songs, rites, and ceremonies on wax cylinders and photographic plates. He would study the domestic life, habits, and forms of worship of the natives, the better to compare them with their kinsmen in the States "who have grown up under the influences of civilization."[103]

So much effort, and at great risk, but for what? Though Garner planned to use his skills as an ape philologist to negotiate "a very limited commercial treaty" with the apes, he was more interested in what successful negotiations would reveal about ape minds than in whatever commercial gain might thereby accrue. Ape business acumen would "indicate in them a feeble appreciation of values and the law of exchange," and would "go far to prove the intelligence and reasoning power of these creatures." Garner proclaimed his goal to be knowledge—"the philosopher's stone" with which "the alchemy of speech will be complete."[104] His final *New Review* article concluded on a note of brave defiance:

> Many who are called men of science spend their lives in acquiring knowledge of what others have done, and these men of letters are usually regarded as the high priests of science; but they are really the laymen. . . . I would rather blaze one mile through the forests of science than travel forty leagues along the paved and lighted thoroughfares of learning. I would rather find the bones of one new and unknown genus than possess a whole museum of familiar forms; and so I feel about the "Simian Tongue." I would rather find the primal truth concerning it than master all the modern tongues of men; and to accomplish such a feat I am willing to forego the comforts of civilization, the endearments of home, and the blessings of plenty, and take upon myself the hardships, undergo the privations, and endure the toil of such a journey that I may give to the world the "open sesame" by which to pass the gates of speech.[105]

Responses to Garner's Work

Garner and Edison

At McClure's suggestion, Garner drafted an appeal to prospective patrons in November 1891.[106] Garner reckoned the undertaking would cost fifteen thousand dollars. In return for contributions toward his research, Garner promised vaguely "to try to reward you by the results of my expedition." The following April, a rather different document was circulating among patrons. Headed "In the Interest of Science," it told of Garner's taking out a ten-thousand-dollar life insurance policy against his dying "in wresting this secret from the deadly marshes of that dark coast." The one hundred people who individually lent Garner the one hundred dollars he requested for expedition expenses would become beneficiaries of this policy in the event of his death. Should he make it back, the loans would be repaid over the course of three years of writing and lecturing. McClure signed on, as did Melville Bell, former president and now presidential candidate Grover Cleveland, James Bailey of Barnum and Bailey Circus, the Washington politico John Hay, and several others. "The plan you propose," Cleveland wrote to Garner, in a letter then circulated with Garner's petition, ". . . seems to me to be a manly and straightforward one, and I certainly hope that you may succeed in that direction."[107] Among those who declined were the *Harper's* artist F. S. Church, the Peabody Museum curator Frederic W. Putnam, and a Mrs. Wilcox, who wrote: "Mrs. Wilcox is much interested in the projects of Mr. Garner: but she has two nieces to educate—in the *human* language and fair to clothe in *modern* garb—. . . so she will be obliged to let the monkeys chatter in an unknown tongue during her mortal career she fears."[108]

Among the most important backers of Garner's phonographic expedition was Edison himself.[109] As early as the summer of 1891, letters from Garner to Edison had broached the desirability of certain minor modifications to the phonograph, the possibility of a trip to Africa for further phonographic research among the gorillas and chimpanzees, and concerns about costs and proprietary rights ("on my return I should want to use the phonograph in lectures, to deliver to an audience the original sounds secured in the wilds of Africa. . . . I should not want to be barred the use of it, whether by any territorial limits or by reason of prices demanded for it").[110] Edison already knew of Garner's work, and in a sense had anticipated it. In 1890, the inventor had set down his notes for *Progress,* a Vernesque novel set in the future. Among the innovations in manufacture, transportation, and warfare Edison envisioned was an "experimental station of the international Darwinian Society at Para on the Amazona," where there were "a great number of educated beings derived from the interbreeding and assiduous cultivation and care of the higher anthropoid apes. Two species being of the eleventh generation were capable of conversing in English."[111] Edison's personal secretary, Alfred Ord Tate, wrote to Garner at the end of August 1891 that Edison had been following Garner's work with interest and that, provided the Edison United Phonograph Company (which had rights to African distribution) gave its blessing, "Mr. Edison thinks that he will be able to fit you out with some good apparatus by the time you are ready to go abroad. He can provide you with a phonograph that will be twenty times more sensitive and accurate than any yet made . . . [with which] it will be possible to take a fifteen minute continuous record."[112]

Garner obtained the permission of Edison United, and wrote back to Edison in December with design suggestions for the modified phonograph. Garner's most pressing problem was with simultaneous recording and reproducing. "Very often when I am repeating the cylinder to [the monkeys], one or more of them will respond at different parts of the cylinder," he explained in a letter. "I have been unable to make records of these responses. Using two phonographs I have to labor under the difficulty of either recording the sounds of the first phonograph or lose entirely the sounds of the monkey." He proposed a solution: a phonograph with two cylinder spindles on it instead of one, with both spindles

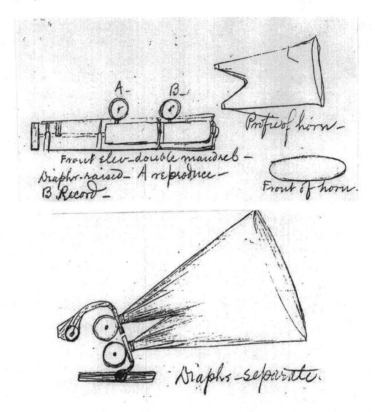

Figure 3.4 Richard Garner's designs for a double-spindle phonograph. Sketches in Garner's letter to Thomas A. Edison, 21 December 1891, Edison Papers, part 3, reel 131, frames 868–70. Courtesy of the Thomas A. Edison Papers, Rutgers University.

sharing a single horn, but with one spindle dedicated to recording, and the other to reproducing (fig. 3.4). According to Garner, this double-spindle phonograph would not suffer the same problems as two phonographs working together, because the sound waves would pass each other coming from opposite directions, and so would not interfere. Garner included diagrams of the proposed device. He also asked that the new phonograph run on a kind of clockwork windup mechanism, which would be much lighter and more efficient than batteries. Garner's letter concluded with a rather impolitic postscript: "Can you not construct a cylinder similar to those used on the graphophone [of Edison's rival Bell]? Something to stand rougher usage and be less in weight?" Above the letter Edison scrawled: "Say to him that he can come to the lab & I will give him an expert for a day to experiment & learn just what he wants."[113]

A notice of Garner's upcoming expedition appeared in the January 1892 issue of the *Phonogram*, as one item among many in "Mr. Edison's forecast" for the future of his instrument: "An expedition is also forming, whose mission will be to carry the phonograph to South Africa for experimenting among the monkeys. If, as scientists have asserted, we are descended from the Simian race, and the conductors of this mission succeed in establishing communication with it and discover a genuine vocabulary, we shall then owe to the phonograph the obligation of teaching us our 'Mother Tongue.'"[114] Elsewhere in the issue Garner and his expedition were treated at greater length ("The Phonograph to Aid in Establishing the Darwinian Theory").[115]

Edison and the phonograph both were now based in the northern New Jersey town of West Orange. Here, almost next door to each other, were the Edison laboratory complex and the Edison Phonograph Works.[116] At the time there was much interest in combining phonographs and telephones. Edison had invented the carbon button microphone that made Bell's telephone workable, and held numerous patents relating to the telephone.[117] Telephone-phonograph combinations were another item in the January forecast: "phonograph operators are now in the habit of connecting this instrument [the phonograph] with the telephone, and when a telephone operator is absent at a time when a message is sent to him, the phonograph cylinder will take a record of this message and keep it until the absent clerk returns. Its importance as a factor in the work of sending telephonic messages long distances [should also be noted]."[118] Garner had been planning to take telephones with him all along, but around this time he too became excited about the prospects of a telephonic phonograph. In mid-February, according to one account, he told an audience at Madison Square Garden in New York City of a "fine phonograph with telephone attachment . . . being constructed specially by Mr. Edison for Mr. Garner."[119] There was no mention of the clockwork double-spindle phonograph. Some time later Garner described the new telephonic phonograph in detail:

> I shall have attached to the diaphragm of my phonograph [a] telephone wire, which can easily be carried to any point in the forest, within a few miles of my cage, at which point will be attached another telephone concealed in a tin cone, all of which will be painted green, in order that it may be easily concealed in the leaves or moss. Having arranged this at a suitable point, and placed in front of it some bait, effigy, mirror, or other inducement, the place will be watched with diligence, and when an ape may be found within a proper distance if he should utter any sounds they will be transmitted over the telephone wire to the phonograph, where they will be recorded.[120]

It is not clear whether the telephonic phonograph was considered an alternative or an auxiliary to the clockwork double-spindle phonograph. Although no mention of the latter appears after December, there are a couple of reasons for thinking that, for a time, both devices were being discussed. First, using a telephone cable to increase the distance between the recording phonograph and the recorded utterance would not obviously have solved the problems of simultaneous playback and recording. Second, when, in April 1892, the secretary of Edison United wrote to Edison for some idea of how much they should charge Garner for the new machine, Edison wrote at the bottom: "Mr. Ott [John T. Ott, probably the expert assigned to Garner] thinks that the apparatus proposed would be uncertain & there is no assurance that the thing will work satisfactorily if it is done. He advises two phonographs." Here the "apparatus proposed" seems to be the double-cylinder phonograph.[121]

However long the telephonic and clockwork double-spindle phonographs survived as designs, there is no evidence that either device was built before Garner left for Africa in July 1892. Garner put the blame on greed. In *The Speech of Monkeys*, sent to the publishers in early June, he praised the phonograph and its scientific potential, but then complained: "In many ways the use of this machine is so hampered by the avarice of men as to lessen its value as an aid to scientific research, and the Letters Patent under which it is protected preclude all competition and prevent improvements."[122] On the eve of his departure for England, Garner told the press he intended to bring two phonographs, while the July *Phonogram* ran an article on the expedition that included a description of the telephonic phonograph.[123] It may be that Garner intended to pursue his plans with the phonograph people in England. A letter that July from the Edison United secretary to the head of the English branch, Stephen Moriarty, informed Moriarty that Garner was on his way.[124]

Garner and American Theorists of the Language Barrier

Garner reported once sending a letter to F. Max Müller challenging his views, but failing to get a reply.[125] That Müller's work in particular was under threat from Garner's phonograph was clear to Garner's contemporaries. Watching Garner at work in Central Park, an onlooker ("who wanted to flaunt his erudition," commented Phillips, the *Harper's* reporter) was said to have remarked: "And what's to become of the Max Müller business about Sanskrit? As a lover of dogs, I never quite forgave him sitting down on the imitative source of language, and killing the bow-wow theory."[126] As for Müller's American rival, Whitney, he testified in the summer of 1892, in a letter to W. H. Page, editor of the *Forum* (in which Garner had published): "I have met Professor Garner personally, and can say of him that he is a genius. His observations are certainly of a very high degree of interest, scientifically as well as to the general public. . . . I am confident that very important results will follow [from] his labors."[127]

In the United States, Whitney was one of three highly accomplished linguists who participated in the Victorian debate on the language barrier. The other two were the anthropologists Daniel Brinton and Horatio Hale. Brinton was one of the commentators when Garner spoke at Madison Square Garden in February 1892 (the Columbia philologist E. D. Perry was also commenting).[128] Garner later recalled a less-than-enthusiastic response on the part of the experts. "I distinctly remember the attitude of . . . [the two professors, who] virtually voiced the sentiments of the philological world in maintaining that animal sounds were not speech and that they were inarticulate."[129] Based in Philadelphia, Brinton worked for most of his life as a physician. He was prominent in several learned societies, and published over an astonishing range: on North, South, and Central American languages and literatures; on race and racial affinities; on mythology and folklore; on archaeology and ethnography; on anthropology as a science; on modern poetry; on the prospects of spelling reform. The Maya glyphs were a particular specialty. His appointment as professor of American archaeology and linguistics at the University of Pennsylvania in 1886 made him the first university professor of anthropology in the United States. But it was a professorship without money and, for lack of students, without much opportunity to teach.[130]

In his views on race, language, and the brain, Brinton was close to Darwin. Both held that, while humans alone had cerebral convolutions sufficiently developed for articulate language, humans had nevertheless evolved from apelike beings; and the different human races and their languages could be ranged on a scale from the lowest to the highest. In 1888, at a meeting of the American Philosophical Society, Brinton read a paper called "The Language of Palaeolithic Man," in which he attempted to infer the features of the earliest human speech from the features of the most primitive languages existing at present (mainly aboriginal American languages). He concluded that the Paleolithic tongue had been grammarless, tenseless, prepositionless, and conjunctionless, with "words [that] often signified logical contradictories," and a phonetics "so fluctuating . . . so much depend[ent] on gesture, tone, and stress, that its words could not have been reduced to writing."[131]

Hale and Brinton admired each other's work.[132] But Hale was of an earlier generation. As a young Harvard graduate in the 1840s, he had spent four years as a member of the Wilkes Exploring Expedition, learning the languages and studying the customs of aboriginal peoples throughout the Oregon Territory and the South Pacific, and publishing the results in a series of widely acclaimed volumes. Then, for thirty years or so, he led the quiet life of a prosperous lawyer and businessman in Canada. By the time Hale returned to fieldwork (among the Iroquois) in the late 1860s, the intellectual world had gone evolutionary. So, at some point, did Hale.[133] But Hale's evolutionism was a remarkable hybrid of the old and new. Not for him the rankings of races and languages as higher and lower. In Hale's view, the theory of evolution predicted no such hierarchy, and indeed no such hierarchy existed. In an 1888 lecture in Toronto on "the development of language," Hale argued that Darwin too had come to this conclusion:

> The doctrine of evolution, whose importance I would in no way depreciate, has, in reference to the intellectual powers of the human race, been strangely misapplied, to such an extent as to lead to serious errors. The misapplication, it must be said, began with Darwin himself; but he, with that noble candor which distinguished him, admitted and corrected the mistake, in which some of his followers still persist. We know how frankly and fully, near the close of his life, he withdrew, on better information, the opinion which he had originally expressed of the low intellectual and moral character of the Fuegians. By just implication, this reversal of his opinion will apply to all savages—for the Fuegians have always been ranked among the lowest of the low. On further consideration, it becomes apparent that this final judgment of the great investigator of nature was in strict accordance with the law of evolution. It is certain that there has been, from one geological age to another, a steady though somewhat irregular increase in the size and complexity of the brains of vertebrate animals. But this increase appears to occur in the transition from one species to another. When a species is once established, there is no evidence (as I am assured by high zoological authority) to show that any material change in the quantity or quality of its brain occurs from first to last. When "speaking man" appeared as a new species on the world's stage, the size and power of his brain was fixed, once for all. There are variations in different races, as there are differences in this respect among children of the same parents; but the variations do not pass certain defined limits, and are constantly tending, as Mr. Galton has shown of the human stature, towards the general average.[134]

Hale's appeal to the work of the anthropometrist, mathematician, and evolutionary theorist Francis Galton is revealing. Galton was making catastrophism respectable for evolutionists. He had proclaimed important limits to what gradual selection of variation could achieve. Generation after generation, the same range of variation appeared in natural populations, with selection occasionally trimming the extreme forms (the tallest or the shortest, say), but leaving the mean forms untouched. In Galton's view, evolution was a matter of shifting means, of creating new points of stability, around which the usual bell curve of variations would form anew. These shifts to new means did not happen gradually, and could not be engineered. They happened in a single leap, and they happened by chance, spontaneously, unpredictably. Such a view was, of course, anathema to Darwin. In the 1850s, it savored of *Vestiges*. But in the 1880s, it was cutting-edge.[135] Hale's making Darwin an honorary Galtonian is intriguing. Were tales of a death-bed conversion to saltationism making the rounds among Darwin's followers, as tales of a death-bed conversion to Christianity were making the rounds among his religious opponents? Hale's image of speaking man's sudden irruption on the world stage was not at all a Darwinian image. But in putting Darwin's name to it, Hale was putting in motion the process of dissociating belief in evolution from belief in a language hierarchy among human peoples.

In the 1888 paper as elsewhere, Hale offered no explanation for how or why the language barrier between humans and other creatures had evolved. What he proposed was that the first humans to appear on its far side were no different in mind or language than humans in the most advanced civilizations. All humans, as humans, were born with one and the same language-making instinct, and would thrive or not according to circumstances. Where Brinton attributed the barest rudiments of speech and thought to the earliest humans, Hale gave them, in his words, "not a mere mumble of disjointed sounds, framed of interjections and of imitations of the cries of beasts and birds . . . [but], like every language now spoken anywhere on earth by any tribe, however rude or savage, a full, expressive, well-organized speech, complete in all its parts." Hale went so far as to declare in favor of Müller's instinct theory of language origins.[136]

For Hale, in sum, the language barrier between humans and animals was conspicuous.[137] But Hale was no dogmatist. Writing in the *American Anthropologist* in 1892, he praised both the substance and style of *The Speech of Monkeys*. He looked forward to the light Garner's work would throw on Broca's claims about the brain no less than on Darwin's claims about human origins:

> One of the most notable of Broca's discoveries in physiology was that which found the seat of the faculty of language in a portion of the third or lowest of the three frontal convolutions of the human brain. This convolution is also found, though imperfectly developed, in the anthropoid apes, but not in the other quadrumana or in any other of the lower animals. In the only experiment which Dr. Garner was able to make with two chimpanzees in the Cincinnati collection, whose utterances he took down in the phonograph, he "was able to discern in a careful study of one cylinder as many as seven different phones, all of which come within the scope of the human vocal organs." There seems, therefore, good reason to expect that in

the African trip which he contemplates making, with the view of continuing his investigations, and especially of studying the gorilla and chimpanzee side by side in their native wilds and recording the sounds they utter, he will be able to obtain many novel facts of great scientific value. Those who desire either to confirm or to confute the opinion expressed by Darwin, that a cry of alarm uttered by "some unusually wise ape-like animal" to warn his fellow-monkeys of expected danger was "the first step in the formation of a language," will await the result of Dr. Garner's further inquiries with much interest. The caution and good judgment which he has thus far displayed in his researches will give special weight to any facts which he may report, and the phonographic records will afford indubitable evidence of their truth.[138]

The Reception of The Speech of Monkeys

Reviews of *The Speech of Monkeys* appeared in a range of other publications, including *Nature, Science, Popular Science Monthly,* the *American Naturalist,* the *Spectator,* the *Dial,* the *Critic,* the *Nation,* the *Athenaeum,* the London *Times,* the *New York Times,* the *New-York Daily Tribune,* the *Brooklyn Daily Eagle,* and the Dutch journal *De Gids.*[139] There were translations into German and Russian.[140] Some readers were unimpressed, the London *Times* and *Popular Science Monthly* reviewers among them, with the former declaring the book "a delightful addition to the literature of the 'clever pet' order," but of little scientific value.[141] After the *New York Times* printed a sympathetic review, a reader wrote in to suggest that, as Garner's work was taken up, "we may thus create a class of persons capable of speaking monkey talk, and, within the limits of their vocal powers, teaching such animals human speech." Might this put the simians at risk of debasement, as they began not just to speak with but as humans, and so to lie and to speak hypocritically?[142] Other responses were more positive, especially about the use of the phonograph. The *Spectator* reviewer emphasized the importance of the manipulative possibilities the phonograph afforded: "By recording the monkey notes on the drum, and then spinning the machine at a slow rate, the sounds are analysed, and modulations detected, and vowel sounds resolved, in a way hitherto impossible."[143] And Garner's German translator, the zoologist William Marshall, remarked in his afterword that until Garner "all analyses [of the matter of speech] had not been carried out methodically. Ultimately such studies produced no more than unproved and arbitrary assumptions, while the document at hand produces facts based on experiment. To have employed the phonograph in this area was a stroke of brilliance."[144]

The reviewer in the *Critic* noted a convergence between Garner's results and those of George Romanes:

A comparison of Mr. Garner's experiments with those of Prof. Romanes seems to offer a curious confirmation of the opinion of the latter with regard to the superior mental advancement of the anthropoid apes. Mr. Garner found by an ingenious experiment that one of his most intelligent monkeys could distinguish numbers certainly as high as three and probably as high as four, but apparently no higher. Prof. Romanes gives equally clear proof that the chimpanzee in the London Zoological Garden could count as high as five. It may therefore be reasonably supposed that the language of these great man-like apes will be at least as far advanced as their arithmetical capacity beyond that of their humbler simian brethren. In the African trip which Mr. Garner has now in view he may reasonably hope to determine this and many other points of curious interest, in connection with the study which he pursues with so much persevering industry and describes with such felicity.[145]

Romanes's critic Conwy Lloyd Morgan was less favorable. Reviewing the book for *Nature*, not long after he and Garner had both attended the BAAS meeting in Edinburgh (to be discussed in chapter 4), Morgan deplored the "anecdotal style." Too many of the claims, he wrote, "savour of the prattle of the parlour tea-table rather than the sober discussion of the study." When Garner stated, for example, that "all mammals reason by the same means and to the same ends, but not to the same degree," and that it was only the "siren-voice of self-conceit" that kept men from admitting this truth, he showed, in Morgan's estimation, utter lack of sound training in psychology. Only through such training would Garner gain "the right of expressing a scientific opinion on this difficult question." Morgan argued further that Garner's results showed monkeys to have more modest mental powers than Garner ascribed to them. The elements of the simian tongue were no more impressive than the chick sounds Morgan himself had recently been investigating. Morgan reported that within their first week baby chicks commanded a five-sound repertoire: a "cheep" of contentment, a "churr" of danger, and so on. He allowed that these sounds were intentionally emitted by the chicks, and conveyed to fellow chicks some kind of "intimation," sometimes of an inner emotional state, sometimes of an external object. In Morgan's view, Garner's detailed record of his difficulties in translating monkey words showed that these too lacked more than the most general "suggestive value." The capuchin word for food, for example, seemed "mainly expressive of a craving for something." The rest of the simian tongue appeared likewise "emotional in [its] nature." However ineptly, Garner was nevertheless, in Morgan's judgment, "working on the right lines, namely, those of experiment and observation in close contact with phenomena," and he wished the traveler to the apes "all success in the prosecution of his inquiry."[146]

A number of writers praised the use of the phonograph but lamented Garner's interpretations of his phonographic results. Writing in the *Dial*, the American psychologist Joseph Jastrow commented: "The author had the happy idea of studying the chatterings of monkeys by recording them in a phonograph . . . but it was one thing to have a good idea, and another to be possessed of the proper ingenuity, patience, and scientific habits, to carry it out."[147] Although he complained of Garner's illogical conclusions and misjudged emphases, the anti-Müllerian E. P. Evans enthused about the prospects for phonographic inquiry. "Mr. Garner's superiority to his predecessors in this department of linguistic research consists in the greater excellence of his material rather than of his mental equipment. The possession of the phonograph alone gives him an immense advantage in this respect, by enabling him to record and to repeat the utterances of monkeys with perfect accuracy."[148] Evans was particularly eager to learn what would come of Garner's work in Africa. From two of Evans's publications in the 1890s:

> Armed with this scientific weapon of phonetic precision and all the instruments and appliances which modern invention has placed at his disposal, he may perhaps completely conquer a province of investigation hitherto but partially explored, and, by making important contributions to zoöglottology and working out a system of alphabetical signs for the language of the anthropoid race, become the Cadmus of the simian world.[149]
>
> . . .
>
> Mr. Garner is now in the wilds of Africa, amply equipped with the instruments requisite for his personal safety and for the pursuit of his investigation in the habitat of the gorilla and the chimpanzee; and the fruits of the studies which he is carrying on under so favourable circumstances will be awaited with interest. *Nomen et omen*: may he return with a well-filled garner! Brehm remarked nearly twenty years ago: "The language of apes may be called quite rich; at least every ape has at its command a great variety of tones for the expression of different emotions. Man also learns the significance of these sounds, which are difficult to describe and still more difficult to imitate." Indeed without the aid of the phonograph it would have been impossible to determine their exact nature and to reduce them to a phonological system. This is an interesting illustration of the far-reaching and beneficent influence of great inventions. The phonograph may yet render as valuable service to philology by extending the field of linguistic research as the microscope has rendered to medicine, and especially to bacteriology.[150]

Perhaps most remarkable among the responses to *The Speech of Monkeys* was a book-length rebuttal, *The Speech of Man and Holy Writ*, published anonymously in 1894 in London.[151] Its author aimed to defend the scriptural account of the Adamic origin of language against Garner's claims. "If there be one faculty of humanity, especially difficult to reconcile with the theory of evolution of mankind from the brutal ape, it is the special power to utter articulated sounds, illimitably and consecutively, an attribute not possessed by any other creature, but which constitutes the speech of man." On Garner's own showing, the author argued, the utterances of monkeys were completely unlike human words. The phonograph had captured noise, the simian equivalent of feline mewing. Furthermore, Garner had failed utterly to trace any of the roots of human language back to simian antecedents. What Garner had produced was not a book of science, but "a collection of mere tales of monkey tricks, [presented] as scientific evidence to establish the unsavoury and incredible theory, that man should see and acknowledge his progenitor in the gibbering ape."[152] No, men had come into the world as Scripture taught, perfect in speech as in all other things, according to the will of the Divine Creator.

The Speech of Man and Holy Writ fits perfectly into views of the "warfare between science and religion" so hoary that few historians take them seriously today. Yet the book exists. Was it read? The copy I examined had its pages uncut. But someone bothered to write the book, and a commercial firm believed in it enough to publish it. We sometimes forget that, for all the caveats, the warfare between science and religion was just that for some people, rightly or wrongly. Richard Garner came to believe his work on the simian tongue was one of the most serious casualties of science-religion warfare. It is time now to examine his case.

TWO

4

CONGO FEVER

✦   ✦   ✦

RIGHT FROM THE BEGINNING, Garner's search in the forests of West Africa for the facts about apes, language, and evolution became the stuff of fiction. Early in 1893, "with apologies to Mr Rudyard Kipling and Professor Garner," *Blackwood's Edinburgh Magazine* published a doggerel fantasy in which the simians of the forest reclaimed Garner as one of their own from his "prison of twisted wire." As one of the simian liberators narrated:

> We surrounded the wire-twisted fortress,
>   And, stretching forth welcoming paws,
> We seized on the bars of his prison
>   And wrenched them asunder like straws.
>
> With eager and loving caresses
>   We drew forth our trembling brother;
> And as he crept out, midst a deafening shout,
>   He was clasped in the arms of his mother!
>
> How she cuddled and fondled her darling,
>   How she wept o'er his features so pale;
> How she bathed him in koko to strengthen his hair,
>   How she sewed on a *beautiful tail*,

It boots not to tell, but, as seasons rolled on,
     And he drank in his pure native air,
His speech it returned and his hair it grew long,
     And our food he delighted to share.

And now through the forest he springs with his kin,
     Having wedded our chieftain's daughter;
For one of our poets wrote ages ago
     That "Bluid is aye thicker than water."[1]

A more enduring fictional tribute came in 1901, when Jules Verne published *The Aerial Village* (*Le village aérien*).[2] As in so many of Verne's previous works, the narrative followed a band of plucky male adventurers on their journey through a strange and exotic land—here the West African interior.[3] After facing elephants on the rampage and other terrors, the protagonists at one point come upon a large cage by a river. The cage is covered with lianas and empty but for a few rusting supplies and a notebook (fig. 4.1). The name on the notebook, Dr. Johausen, is "a revelation."[4] When last heard from, the doctor had been heading into the interior to observe apes in the wild from his cage. His purpose had been to solve the great riddle of the origins of language. As Verne reminded his readers, an American, Professor Richard Garner, had become well-known for a similar plan some years before.[5] But the missing links of language had slipped Garner's grasp, and Garner himself had been swept away in a tide of accusations. Verne summarized them:

> Professor Garner had had made a portable iron cage, which he had taken into the forest. If he were to be believed, he had lived there three months, most of that time alone, and had been able to study the apes in their natural condition. The truth is that the prudent American had simply erected his metallic house within twenty minutes' walk of [a] mission, and quite near to its spring, in a place to which he gave the name "Fort Gorilla." He had even slept in it three consecutive nights. Devoured by the swarms of mosquitoes, he could not hold out any longer, so he had taken down his cage and asked from the Fathers of the Holy Spirit a hospitality which they readily gave him. At last, on 18th June, leaving the Mission for good, he had returned by way of England to America, taking as the only souvenir of his journey two tiny chimpanzees who obstinately refused to talk.[6]

Through Verne, fiction thus gave rise to fiction, as Garner's supposedly embellished account of his expedition—an expedition journalists described beforehand as something out of a Verne novel—became the spur to the Garneresque labors of the fictional Dr. Johausen. We turn now to examine the mix of fact and fiction in the controversies surrounding Garner's expedition of 1892–93. The first part of the chapter will set the scene, recounting Gar-

Figure 4.1 Professor Johausen's empty cage. Illustration by G. Roux, from Verne 1901, 93. Courtesy of the Bibliothèque nationale de France and reproduced by permission.

ner's experiences in Britain in the summer of 1892 and the state of affairs in West Africa at the time of his arrival there later that year. Next we consider Garner's public accounts of his expedition, from his sole field dispatch to his 1896 book on the expedition, and the responses of the transatlantic press. Especially important in shaping those responses was the English journalist and politician Henry Labouchere, who pressed the charges of fraud later immortalized in *The Aerial Village*. The third and final part of the chapter introduces Garner's private account of the expedition and the controversies surrounding it. This survives in a testimonial Garner wrote in his notebooks while in Africa around 1909. Never published, Garner's Fernan Vaz testament (as I call it) tells a remarkable story of intrigue, adventure, betrayal, even attempted murder. Garner blamed his public disgrace in large part on the fanaticism of one of the Holy Spirit missionaries, who, according to Garner, sought to disrupt and discredit Garner's expedition in an attempt to keep him from establishing the truth of the theory of evolution.

The 1890s were the age of the military metaphor for science's relationship with religion. Large numbers of people read John W. Draper's *History of the Conflict between Religion and Science* (1876) and Andrew Dickson White's *History of the Warfare of Science with Theology in Christendom* (1896) as reliable guides to the scientific past. Since then, generations of scholarship have revealed how much distortion there is in the military metaphor popularized in these tomes and the anticlerical essays of T. H. Huxley. We now read these works not as histories but as emblems of the divided culture that gave rise to them.[7] We recognize a sociological dimension to their rhetoric and distortions, at a time when men of science were displacing men of the cloth in positions of authority throughout public life and public institutions.[8] Quite apart from its functional role in advancing the professional interests of scientists, however, the military metaphor affected the way people experienced themselves and their world. Whatever really happened during Garner's 1892–93 expedition—and what follows is in part an attempt to tease the facts from the fictions—his Fernan Vaz testament matters as a record of a life lived and contemplated through the military metaphor. But there is also a more disconcerting possibility. If, on balance, we conclude that Garner told it like it happened in his testament, then his tale is an almost pure case of science-religion warfare, all talk of metaphor notwithstanding.

From Liverpool to Libreville

Preparations and Frustrations

Shortly after his arrival in Britain in the summer of 1892, Garner was invited to speak before the BAAS, meeting that year in Edinburgh. There was no more auspicious platform from which to address men of science in Britain or, indeed, the world. On the morning of 4 August, Garner presented himself in the rooms of the anthropological section. The previous year, F. Max Müller himself, as president of the section, had affirmed that an impassable language barrier separated man and beast. Now Garner had come with, he claimed, experimental proof that no such barrier existed. But his moment slipped away.

He learned upon arrival that his address had been scheduled not for that day (Thursday) but the following Monday. Garner explained that he was not able to wait, due to prior engagements. He offered to leave his paper for discussion in his absence, but the section officials refused, and Garner returned to London paper in hand. His name remained on the program, however, and a large crowd (perhaps including Conwy Lloyd Morgan, who lectured on Friday about "The Method of Comparative Psychology") gathered on Monday to hear him speak.[9]

"We can easily understand that the members of the British Association at Edinburgh were disappointed when Mr. Garner did not appear to read the paper on the speech of monkeys which he had promised," wrote the *Athenaeum* reviewer of *The Speech of Monkeys*, some months after the event. The reviewer thought the book so poorly argued and badly written that a "slight cross-examination in public might have enabled [Garner] to make clear what really are the tangible results to which he has attained."[10] Others also registered their disappointment over Garner's lecture. A vivid account of what transpired in the anthropological section appeared in the "Girls' Gossip" column of *Truth*, the London weekly edited by Henry Labouchere. Taking the form of a letter to "Dearest Amy" from her "loving cousin MADGE," the gossip of 18 August featured a letter from a reader, Ursula, who wrote about the Edinburgh meeting:

> What appeared to be the most inviting lecture for the general public was that announced to be given by Professor Garner on the "Language of Monkeys." He is the scientist who devised the happy plan of studying monkeys, both great and small, amid their natural surroundings in the forests, entrenched the while himself within a strong cage of steel bars, but after scrambling and pushing for a place in the crowded hall, we were all much disappointed when the usual telegram . . . was handed in to inform us of the Professor's inability to be present.[11]

Readers of the journal *Natural Science* were alerted to how Garner's "not altogether credible" excuse had aroused suspicions in Edinburgh.[12] Garner defended his reputation in an angry letter to the *Times* a few weeks after the meeting. Rumors to the contrary notwithstanding, it was, Garner insisted in his letter, neither disrespect for the BAAS nor unwillingness to submit his results to public scrutiny, but the most trivial misunderstandings and schedule conflicts that had forced his absence. "I only consented to read the paper with the desire to aid in the laudable work of that great school of science known all over the earth," he protested, "and I fully appreciated the honour of appearing before it, and deny any want of respect for that august body."[13] The London correspondent of the *New York Times*, Harold Frederic, came to

Garner's defense, blaming Garner's unfair treatment at the hands of the officials on his "being a plain, unpretentious, practical worker instead of a dry, polished plant," adding that, after his letter appeared, Garner had received "a large number of expressions of regret and annoyance from British men of science."[14]

Money was a problem that summer, and would remain a problem. Garner's insurance scheme had attracted far fewer than the one hundred backers he had hoped for. He nevertheless managed to purchase what he believed he needed to survive nine months alone in the forest.[15] What he did not manage to purchase was, of all things, a phonograph. Edison's representatives in Britain, led by the colorful Colonel George Edward Gouraud, had "boomed" tirelessly on behalf of the phonograph for years. But their responsibility for managing all the foreign patents had left them with virtually no resources to put the machine in the shops. Worse, Gouraud's replacement, Stephen Moriarty, seemed less concerned to sell phonographs than to fleece Edison. In 1892, Garner's lack of funds met with Moriarty's refusal to let Garner have a phonograph at a reasonable price.[16] The *New York Times*'s Frederic gave a sympathetic account:

> Prof. Garner expects now to sail on his monkey-language quest from Liverpool to Gaboon on Sept. 14 by the steamship Maidi. Delay was caused here by the inexplicably wooden-headed conduct of the people who own the phonograph rights in Europe; they refuse to allow him to take one of their instruments on his mission. Many other complaints have been brought me heretofore concerning the methods that distinguish this concern, which seems to divide its energies between declining itself to sell phonographs and striving to prevent other folk making them. But there has been nothing quite so stupid and narrow as its treatment of Garner. However, it is unlikely that this dog-in-the-manger policy will avail to spoil the professor's plans.[17]

On the eve of his departure, Garner himself wrote in the same newspaper of his frustrations with the English phonograph men. He had intended to record the speech of a Regent's Park orangutan at regular intervals, and so to find out whether the vocal powers of the orang progress according to the same laws governing human vocal powers. "It is the first opportunity," Garner declared, "that has ever been afforded science to do such a thing, and it may be the last for many years to come." But, he continued,

> the avarice of a few men makes science hide her head in shame, while they strangle her babes and cut off her posterity. The use of the phonograph in England is inhibited for all purposes, and I shall therefore abandon this one very important step merely to avoid delays in my journey which might result from it, but not through

any compunction or scruples that I would have in using a machine whose own-ership belongs to the public. But science will, nevertheless, suffer the loss of this experiment.[18]

In mid-September Garner boarded a steamer at Liverpool, bound for West Africa, and without a phonograph.[19] Arriving in Libreville in early November, Garner wrote of his predicament to Edison. For Garner, the British phono-graph men were damnable less for their greed than for the short-sightedness of their greed. As Garner explained, a phonograph expedition in Africa would be good for the phonograph business. "I cannot understand why they should strive so hard to thwart my efforts as a new and lucrative market for phonograph[s] would be the certain result of my success with it here. Every cylinder recorded here would be studied all over the civilized world by phi-lologists, linguists and students of acoustics." Garner asked Edison if it was possible to secure a phonograph on affordable terms.[20] In January 1893, Gar-ner's literary agent McClure sent Edison a letter on Garner's behalf, offering to come out to the lab to discuss Garner's phonographless state. At the bot-tom of the letter Edison wrote: "Garner is the most impracticable man extant. He never arranged to get a phonograph. He should have spent 2 or 3 days here & got one & learned its peculiarities instead of that he simply talked & did nothing."[21] McClure's efforts notwithstanding, no phonograph arrived while Garner was in West Africa in 1892–93.[22]

Before leaving England, Garner had written to Walter Page, editor of the *Forum*, to ask Page's help in finding a better agent. For reasons he did not disclose, Garner had lost confidence in McClure's ability to secure the fees Garner felt his writings merited. "I am aware," Garner wrote, "that my figures are not the same as those of the average spring rhymer, but I am also aware that my stuff is not the same in quality." He had been paying McClure ten percent of the proceeds for placing the articles, and would offer the same to the new agent.[23] However, a few months later, Garner's wife, Margaret, wrote to Page asking him *not* to do as Garner had asked. She explained that Garner had decided to drop McClure out of anger over something McClure had said to her. In her view, it was all an unfortunate misunderstanding, the more so because McClure, she wrote, "understands more about Mr. G's busi-ness than anyone else does," and really should remain in charge of Garner's affairs in the United States.[24]

Meanwhile, Garner was exploring ways of generating income besides writing. In December 1892, while still in Libreville, he wrote to F. W. True at the Smithsonian to find out how much the museum would pay for specimens collected in the Gaboon. Gorillas, buffalo, bush deer, bush cats, bush pigs, monkeys, butterflies, beetles, ants, birds, rodents, even manatees: all could be procured, preserved, and packaged up for delivery. "I do not mean by these inquiries," Garner wrote, "to rush into the Nat. Museum some day with a cargo of junk and demand the prices you may mention, but my purpose is to get a bird's-eye-view of values to what I may be able to collect on this trip."[25] True responded that a gorilla skeleton or skin would fetch between one and two hundred dollars, an African manatee half that amount.[26] In addition to animal collecting, Garner continued to pursue outright solicitation from eminent men of means. At some point during his time in Africa, he posted a bond for twenty pounds to T. H. Huxley, asking for a loan of that amount in return. Apparently the bond never arrived, for when Garner, back from Africa, asked to have it back, the aged Huxley wrote: "Your letter is unintelligible to me. I have never received a letter from you, and know nothing of the bond of which you speak."[27]

The French Congo in the Early 1890s

In preparation for his trip, Garner had sought out a number of old Africa hands, including Henry Stanley, Stanley's former lieutenant E. J. Glave, and Paul du Chaillu. Du Chaillu's advice would have been most relevant, as he had traveled so extensively in the equatorial forests where Garner was headed, in what is today Gabon, on the western coast (fig. 4.2). Then as now, it was a land covered with dense tropical rain forest,[28] with the Ogowé River, running roughly east to west, serving to connect the interior to the coast. Along the length of the Ogowé were a number of important settlements, including Lambaréné (where Albert Schweitzer would set up his famous hospital in the 1910s) and Njole. The Ogowé emptied near the coastal towns of Port-Gentil and Cape Lopez, and, farther to the south, in the Fernan Vaz lagoon, also known as Lake Nkomi. The most important town, Libreville, was located on the Gaboon estuary, to the north of the river. Sparsely distributed throughout the region were some forty different tribes, including pygmies and cannibals (the Fan). In certain respects, the place was much the same in the 1890s as it had been in Du Chaillu's day, in the 1850s and 1860s. There were now, however, vastly more Europeans and Americans in the region. These fell into three basic groups: colonial officials from France, which, since the Congress of Berlin in 1885, had governed the region as a district of the French Congo, with its capital at Libreville; traders, principally from Britain, Germany, Portugal, and the United States; and missionaries, from France (Catholics) and the States (Protestants). Each group had a foothold in the Fernan Vaz region, where Garner was to spend much of his time.[29]

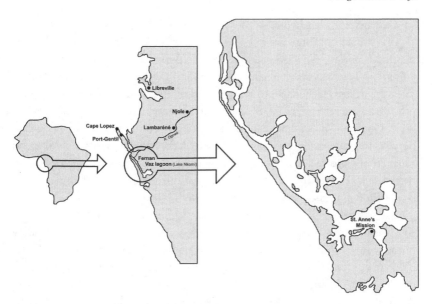

Figure 4.2 The Fernan Vaz region of West Africa, at three levels of magnification. Designed by Alex Santos, University of Leeds, after the maps in Kingsley 1897, xxi; and Gaulme 1974, 396.

As Garner dealt with members of each group, it is well to consider each in its own right. We begin with the colonial officials. During the 1880s, a drive to consolidate French colonial power led to the establishment of increasing numbers of government posts, penetrating ever deeper into an increasingly well explored interior. Libreville was the administrative as well as commercial center of the colony, and as such the first stop for travelers from abroad. Garner spent over a month there after his arrival in late autumn 1892, gathering information about the region and its inhabitants. "I was kindly received by the governor and others, and assured of any aid that they could render me," Garner reported in his first and (so far as I know) only published dispatch from the Congo, which appeared in the September 1893 issue of his agent's new enterprise, *McClure's Magazine*. "They manifested great interest in my work and anxiety for its success."[30]

Garner wrote little more about his initial encounter with officialdom in the French Congo. Mary Kingsley's testimony is much fuller. Niece of the evolutionist minister and man of letters Charles Kingsley, Mary Kingsley traveled up and around the Ogowé shortly after Garner's departure. In her classic narrative *Travels in West Africa* (1897), she detailed the administrative formalities with which a foreign visitor of the time needed to comply.[31] Kingsley wrote that, having dealt with the Customs House at the port town of Glass (she left her revolver behind rather than pay for the license), she walked the short distance to Libreville up a road lined with a telephone wire and newly planted coconut palms, to the office of the directeur de l'Administration de l'intérieur. Left waiting on "a fine stone-built verandah," she and her companion were at last "ushered into a cool, whitewashed office, [to] find a French official, clean, tidy, dark-haired, and melancholy, seated before his writing table. Courteously bidding us be seated, he asks our names, ages, and avocations, enters them in a book for future reference, and then writes out a permit for each of us to reside in the colony, as long as we behave ourselves." From there she walked to the Palais de Justice, for further ordering of papers.[32]

Although France was the colonial power in the region, the French did not dominate its trade. Two of the biggest firms, John Holt and Hatton & Cookson, were British. These and the many other smaller firms operated trading houses called "factories," where natives brought palm oil (used in soap), rubber, ivory, ebony, timber, and other raw materials for export, in exchange for goods such as cloth, rum, tobacco, and gunpowder.[33] From about the 1870s, the traders ran steamboats over the navigable parts of the Ogowé and its associated rivers and lakes, though the principal means of transport remained the canoelike pirogue. Like the colonial officials, the traders began pushing out from main centers such as Libreville. Soon lone traders such as the later famous Hatton & Cookson agent Trader Horn (born Aloysius Smith in Lancashire) were crisscrossing ever more remote parts of the interior, leaving new subfactories in their wake.[34]

Garner wrote little of trade in his 1893 *McClure's* article, remarking only that a native had once offered him a fine gorilla skull for some rum or tobacco, but that, "not having either of these current articles of trade, I could not make the purchase." Garner does not seem ever to have turned his energies to trade proper, in contrast to Kingsley, who was a trade enthusiast, so shrewd in her rubber dealings that other traders accused her of swindling the natives.[35] Kingsley noted that the use of "trade English" was widespread throughout the region. "On the south-west Coast," she reported, "you find individuals in villages far from the sea, or a trading station, who know it [trade English], and this is because they have picked it up and employ it in their dealings with the coast tribes and travelling traders."[36]

The use of English was fostered by traders but also by missionaries. American Protestant missionaries had vied against French Catholic missionaries in the region since the 1840s. The Americans had missions in Libreville, Lambaréné, and Njole, and Garner, a Protestant American himself, probably spent a good deal of time among his countrymen in the French Congo, chatting on steamer decks and mission verandas. Indeed, 1893 was an excellent year for gossip within the Protestant community, divided at the time over a proposal to forbid single male missionaries (and, clearly, one of their number in particular, Robert Nassau) from hiring native women to do domestic work.[37]

Nevertheless it was the French missionaries of Fernan Vaz who received a brief mention in Garner's 1893 article. Early that year Garner had left Libreville for Fernan Vaz. "After arranging here for a sojourn of a few months," he wrote, "I placed . . . most of my heavy effects in the custody of the Ste. Anne Mission, and began a journey up the Ogowé."[38] He also mentioned that one of the missionaries at St. Anne's, Père Buléon, claimed to have twice observed gorilla families where the father sat still while the others scurried about feeding him plantains.[39] Joachim Buléon and his fellow missionaries belonged to the Congregation of the Holy Ghost. Known as Holy Ghost Fathers or Spiritans, they had founded St. Anne's only a few years before, in 1887. When Garner arrived, the mission compound housed women missionaries as well, the Sisters of the Immaculate Conception, who ran a school for native girls as the fathers ran a school for native boys. The missionaries' relations with English traders and French officials were cordial. Their relations with the local Nkomi tribespeople were better than cordial: at the end of 1892, the Nkomi chiefs gathered in the mission parlor and granted Buléon the title of *Ozoungé*, or savior.[40]

In short, the many officials, traders, and missionaries, not to mention explorer-naturalists such as Garner and Kingsley, gave the French Congo in the early 1890s a sizeable population of Europeans. When Garner arrived in 1892–93, with enough canned food to last him through nine months alone in the bush, he found a place where it was easier than expected to get out to the bush, and harder than expected to be left alone once there. We shall return to the problem of solitude in the discussion below of Garner's Fernan Vaz testament. As for the ease of transport, both Garner and Kingsley left Libreville on a small trading steamer, the *Mové*. Garner traveled in January or February, at the height of the wet season ("We have terrific lightning here now," Garner wrote to Edison), and Kingsley in late May, at the beginning of the dry season.[41] Perhaps Garner left England when he did, with no phonographs, because he believed Fernan Vaz to be (in the words of his 1893 article) "the true habitat of the gorilla," and because he knew the vessel that bore his cage and supplies could reach Fernan Vaz only in the wet season.[42] Be that as it may, he probably traveled in comfort. Here is Kingsley's description of life aboard the *Mové*:

> She is a fine little vessel; far finer than I expected. The accommodation I am getting is excellent. A long, narrow cabin, with one bunk in it and pretty nearly

everything one can wish for, and a copying press thrown in. Food is excellent, society charming, captain and engineer quite acquisitions. The saloon is square and roomy for the size of the vessel, and most things, from rowlocks to teapots, are kept under the seats in good nautical style.[43]

What Happened on Garner's 1892–93 Expedition?

Dispatches, Reports, and Lectures, 1893–94

As described in his 1893 *McClure's* dispatch from the field, Garner's "mission to Africa" had two phases: first, several months of traveling in and around the Gaboon and the lower Ogowé; then several months of observing gorillas and chimpanzees from his cage in Fernan Vaz.[44] According to Garner, after more than a month based in Libreville, he steamed into Fernan Vaz, stowed his equipment at the mission of St. Anne, and headed up the Ogowé. Over the next few months he traveled as far upriver as Njole, over two hundred miles from the coast, before heading back to Fernan Vaz. He wrote of the splendor of the forest flora and fauna—palms, ferns, vines, flowers, monkeys, hippos, bright-feathered birds—and the squalor of native villages such as the Orungo village of M'biro. "I do not know what the name signifies," he wrote, "unless it is *mud*."[45] Garner was generally curious about native words for things, and in his *McClure's* dispatch frequently gave the native name for a place or an animal alongside the English name.

Back in Fernan Vaz, as Garner narrated for *McClure's* readers, he set about preparing to make scientific history. "On reaching Ste. Anne I selected a site for my cage and erected it at once," Garner reported. "It is located in the heart of the jungle, a trifle more than a mile from any human habitation, and I named it Fort Gorilla. It is in a spot where nothing but the denizens of the bush has any cause to come." He had chosen a spot near a plantain grove, where "the gorilla feasts with the gusto of a charter member of the Gourmand Club," and had so far seen a small number of gorillas and chimpanzees strolling about. He heard them still more often, noting a resemblance between certain chimpanzee sounds and certain native words, and the ease with which the novice confused gorilla howls with the cries of a large local bird. He related much about the habits of the apes: their gait, their speech, their centerless wanderings through the forest, their male-dominated social hierarchies, even their rituals. (Like the natives, the chimpanzees were said to dance to the beating of drums; although Garner had not seen the dance of the chimp, he reported being shown "a peculiar spot of sonorous earth," a clay-peat mound of irregular shape, which, he surmised, could well have served as a ritual drum.)

He also related much of ethnographic interest. He explained that African kings were much less powerful than their European counterparts; he told the myth of the Galoi people concerning the origins of apes and humans; he commented on the increasingly dominant role of that "great tribe of cannibals," the Fan (Garner called them the Pangwe), in the trade life of the region.[46]

The problem of the origin of speech was not at all central in the 1893 McClure's article. Nor did Garner mention the phonograph experiments he had intended to carry out in the bush. Instead he wrote of a new line of inquiry. He had recently acquired a young chimpanzee, Moses ("because he was found in a papyrus swamp of the Ogowé"), and had housed him in a little cage of his own next to Garner's. "I am trying to teach Moses to speak English," Garner wrote, "but up to this time he has not succeeded. He tries to move his lips, as I do, but makes no sound. However, he has only been in school a very short term, and I think he will learn by and by."[47] In mid-November 1893, some six weeks after the appearance of his McClure's dispatch, Garner returned to Liverpool. With Garner were two chimpanzees, Aaron and Elishiba (the two Verne later wrote about as "obstinately refus[ing] to talk"), but they died soon after arrival. (Moses had died in Fernan Vaz.)[48]

Notwithstanding Garner's own representation of his activities in McClure's, newspaper accounts of his return featured the phonograph and the simian tongue as prominently as ever. It is unclear whether Garner misled journalists about having had a phonograph on his expedition, or whether the error of a careless journalist on the scene then became embedded in the story. The New York Times reported that Garner had used the phonograph to make records of his many ape visitors over the 101 days spent locked in his cage.[49] One of the newspaper's writers commented: "Their remarks will doubtless be ground out again for the benefit of learned societies, in the wheezy and asthmatic tones into which the phonograph converts all sounds." For all Garner's talk of success, the writer went on, his claims for the wild noises on his cylinders were so incredible that none but the least learned of societies would take him seriously. "Prof. GARNER'S discovery . . . will excite the deepest interest in the Society for Psychical Research, and in such circles as are left of the American Spiritualists, who will receive the phonographic reproductions of the chatter of apes with the same reverent belief that they accord to messages from the other world." Tapping a comic vein opened long before, the writer concluded: "In a different spirit, but with equal zeal, will the revelation be received and exploited by the realistic novelists of France, to whom simian literature and conversation may be expected to open a new and most attractive and congenial field."[50]

On 19 February 1894, at Prince's Hall in London, Garner gave his first public lecture about his time in the cage and his successes with Moses. Garner brought his cage along, fitted out just as it had been in the Congo.[51] Afterward the persistent phonograph vied with the cage for newsprint. In James Payn's *Illustrated London News* notebook item, the phonograph dominated his speculations about why Garner had so little to report of the simian tongue:

> If . . . [the apes] discovered that the Professor was using a phonograph, their behaviour is explicable at once. It is not a very delicate proceeding to take down the conversation of one's companions, as it were, in indelible ink. How should we like it, if a visitor thus set down for reproduction our own observations? Some of them would not well stand transplanting. Conceive a phonograph that has been left in the dining-room, being taken upstairs to amuse the ladies, with a "gentleman's story" in it! It is fair to the gorillas to say that no narrative of this nature seems to have been communicated to the Professor. He had other experiences that showed a divergence from conversational humanity; for example, he found a lady gorilla "too astonished to speak." This seems incredible, but it must be remembered that she came upon him in the middle of a forest shut up in an iron cage—a spectacle which, to say the least of it, must have been unexpected.[52]

Cage and phonograph both featured in *Punch*'s doggerel response to the lecture:

> TO A GORILLA GIRL.
> *(By a Disciple of Professor Garner.)*
>
> MAID of Afric, kindly stay,
> From my cage I wish to say
> Words of thine—not said with ease—
> Looking like a cough or sneeze,
> Or a cipher telegram,
> *Hxerrg ztti hnnwpflb srth kkqam!**
>
> Goodness knows how one should sound
> Words where vowels don't abound;
> I should hurt my throat or lungs
> If I tried these monkey tongues,
> Feeble linguist that I am!
> *Hxerrg ztti hnnwpflb srth kkqam!*
>
> By that lip, which thus can speak
> With a most appalling shriek;
> By that rather hairy face,
> Full of troglodytic grace—
> Thy complexion is not sham;
> *Hxerrg ztti hnnwpflb srth kkqam!*

This my phonograph will say
In a sentimental way,
Thy pronunciation seems
Far beyond the wildest dreams
Of a MEZZOFANTI, ma'am;
Hxerrg ztti hnnwpflb srth kkqam!

Maid of Afric, when I'm gone,
Think of me, sweet, all alone
In the London Prince's Hall,
With my talk, a trifle tall,
And my priceless phonogram,
Hxerrg ztti hnnwpflb srth kkqam!

*These words, in the Gorilla language, are translated by some authorities, "Oh my eye! Ain't she a stunner, and no mistake?" and by others, "Waiter, bring me a cocoa-nut and mashed bananas."[53]

But it was the cage that mattered to Henry Labouchere, editor of *Truth*, and it was Labouchere's response that would prove decisive.[54] On 1 March 1894, under the title "A Few Questions for Professor Garner," Labouchere wrote of his disappointment at having to miss Garner's lecture at Prince's Hall. "I consoled myself," he wrote, "with the reflection that I should find Tuesday morning's papers full of the Professor's astounding contributions to Darwinism, and the whole world talking about this adventurous pilgrim of science." But the reports he had read were on the whole dismissive. They were also oddly bereft of details, especially concerning Garner's exact movements in the Congo. Like all thinking men, Labouchere wrote, he wanted to know as much as possible about "what gorillas have to say for themselves." But his interest had other motives as well. According to Labouchere, a reader of *Truth* in the Congo had taken it upon himself to send certain letters to the editor that, no doubt, amounted to "a cruel attempt to damage the reputation of a great scientific pioneer." Nevertheless, Labouchere added, it would be best if Garner put all doubts to rest with an account of his exact itinerary in the jungle.[55]

Two weeks later, Labouchere reported, there was still no word from Garner. "[T]he Professor is, I suppose, so deeply immersed in the mysteries of the simian language that he has no time to look at newspapers, and does not even notice a contents-bill when he walks about the streets." More was the pity, because others *had* noticed, and were themselves supplying the missing accounts—but, according to Labouchere, they "all omit . . . to mention when and where the Professor was in the cage in the bush." "There ought to be no room for doubt," he added, "about the basis on which our knowledge of the gorilla tongue is to rest."[56]

Labouchere's Challenge

Today Henry Labouchere, Radical MP (fig. 4.3), is best remembered as a Victorian agitator for the abolition of the House of Lords (contemporary car-

Figure 4.3 Henry Du Pré "Labby" Labouchere (1831–1912), 1887. From Pearson 1936, frontispiece.

toons pictured him as Guy Fawkes). In his own day, however, "Labby" was equally famed for his editing of *Truth*.[57] Founded in 1877, *Truth* was among the most profitable enterprises in what Matthew Arnold derisively called the "new journalism." Unlike what Arnold esteemed as the elevating, serious, trustworthy, difficult journalism of old, papers such as Labouchere's aimed merely, as Labouchere cheerfully admitted, to fill "the interstices of conversation."[58] The form of *Truth* fitted its undemanding contents: large pages precut and stitched together for ease of reading on train journeys. The package was a success right from the start. One contemporary observer wrote that *Truth* became "perhaps the only paper ever published which more than paid its expenses from the first number."[59]

By Arnold's lights, *Truth* was the lowest of the low, a variety of the new journalism known as "society journalism" or "personality journalism." It was a gossip sheet, chronicling who was doing what at court, in parliament, in high society, in the church, the military, the theater, music, finance, and so on. There was politics, but it was a politics of personalities, the emphasis being less on issues than on characters and character. In addition to its gossip, presented in short, easily digested paragraphs, *Truth* was read for its tireless pursuit of shady charities, corrupt officials, and other abusers of the public trust. "People bought *Truth* with a desire to see who was 'going to get it next,' who or what institutions would be marked down for exposure," recalled one reader. In spite of, or perhaps because of, constant libel suits brought by those accused in *Truth* of defrauding the public, *Truth* by 1887 was selling some thirty thousand copies a week.[60]

Garner was perfect *Truth* fodder, in two respects. First, as Ursula described so vividly in her "Girls' Gossip" letter about the 1892 BAAS meeting where Garner nearly spoke, public scientific lectures such as Garner's were social occasions, bringing people together and replenishing their stocks of conversation filler. Garner's fascinating but accessible research in Darkest Africa, his larger-than-life personality and ambitions, could not have been better designed to meet the social needs of *Truth*'s readers. Second, Garner had made a number of public promises about what he would do in Africa, and *Truth*'s readers were vigilant in keeping such figures to their promises. Indeed, where most periodicals quarantined readers' letters in a special "Letters" section, *Truth* integrated them into the regular columns, thus creating a sense of *Truth* as a collaborative enterprise, jointly created by a community of readers and writers.[61] By the 1890s, that community stretched far beyond London. "The world has shrunk down in the last few years to such a degree," marveled Labouchere in March 1894, "that a philosopher can hardly go and sit in a cage in a Central African jungle without being dogged by a reporter and worked up into copy for the London press."[62]

What Labouchere wanted out of Garner were details and documentation of his activities in Africa:

> When the Professor's lecture was announced I thought that my time had come. "Now," said I to myself, "we shall get the whole truth from the Professor himself. We shall hear precisely when he reached Gorilla-land, where the cage was fixed up, the date when he first entered it, and the date when he finally left. We shall get all the details of the transportation of the cage to the distant depths of the jungle, we shall know who assisted in the work, and so on; and with this, we shall have photographs of the denizens of the jungle, phonograms of their utterances, shorthand notes of their views on the cage and its contents, and abundance of circumstantial detail."[63]

When Garner arrived back in New York in late March 1894, reporters from the *New York Times* and other papers found the Professor keen to clear up misunderstandings of what he had and had not done in Africa. There had been no phonograph, Garner insisted. The cage had stood about one mile from the lake in Fernan Vaz. Natives from a nearby village had helped him transport it and erect it, and he had occupied it between 25 April and 6 August (whether continuously or just from time to time, Garner did not say). For a short while, he had retained a native boy as a servant, but the boy deserted him after hearing the terrifying shrieks of a gorilla. All in all, Garner said, the trip had cost him two thousand dollars—probably about as much as he had raised.[64] In "Gorillas and Chimpanzees," an article whose publication in *Harper's Weekly* coincided with Garner's return to the United States, there were photographs of the gorilla and chimpanzee skulls Garner had collected, together with some "drawings from photographs," including one of Garner standing before Fort Gorilla (with rifle, pith helmet, and spear-clutching native companion).[65]

In the *New York Times*, Garner's homecoming was front-page news, reported under the heading "Thinks Well of Gorillas." So favorable was American coverage generally that a few weeks later, Labouchere was stirred to protest. "America hails her adventurous son and his home-made American cage," Labouchere sneered. For all their gullibility, however, American reporters at least had pressed Garner for more details than their unimpressed English counterparts, for which Labouchere was grateful. He again asked that Garner provide independent corroboration of his hundred-plus days in the cage. Labouchere also called attention to an apparent discrepancy between what Garner had said in England and in the States about shooting gorillas. At Prince's Hall, according to press reports, Garner had stated that, far from attacking his cage, gorillas were hardly to be seen in its vicinity. Yet, wrote Labouchere, the *New York Recorder* reported that though Garner "did not find his cage a complete protection against the attacks of gorillas . . . it was strong enough to keep them at bay until he could shoot them." Again, Labouchere wanted details: how many apes did Garner shoot, what had happened to the bodies, and so on. Labouchere concluded: "Even if the Professor cannot see that an answer is due to himself, the Press, I think, will see that an answer is due to the public, to say nothing of the cause of scientific research."[66]

In May 1894, there were still further questions for Garner, but from a new source. According to a report in the *Figaro*, entitled "Le cas du Docteur Garner" and later summarized in the *Nation*, Garner had been the subject of a lecture given before the Société de géographie. The speaker, the explorer Jean Dybowski, had just returned from a government-sponsored expedition to the French Congo. Dybowski had passed through Fernan Vaz after Garner's departure, and had also stayed at the mission of St. Anne. There he learned from one of the missionaries, as the *Nation* put it, that Garner had "evidently preferr[ed] the society of the monks to that of the monkeys." In Paris, Dybowski told the following story. According to the missionary, whose account had been confirmed by Garner's servant boy Rozoungué, Garner had set up his cage within earshot of the mission bells, had spent three nights in the cage, and had seen no apes during that time, though he subsequently bought some short-lived ones. At one point during his three months at Fernan Vaz, Garner did leave the mission, joining Father Buléon on a journey to the Eshira tribe of the interior. But two days after setting out, Garner complained of severe pain in his legs. He was duly hammocked off to a nearby factory, run for the Tomlinson firm by a man named Sheridan, who, over the next two months, nursed Garner back to health, after which Garner left the region. Such was the sum of Garner's time among the apes. In the opinion of the *Figaro* writer, though Garner had succeeded for a short time in being taken seriously as an amateur comparative philologist, it was now abundantly clear he was a mere humbug ("un simple fumiste").[67]

Provoked by Labouchere's insinuations and Dybowski's allegations, Garner in the summer of 1894 set out, as Labouchere noted in *Truth*, on "the war path." Garner was furious, telling one American paper that Dybowski was a "second-hand clothing dealer," who, by attacking a more famous man, was trying to drag his own name out of obscurity. As for Labouchere, his "attack is too mean and contemptuous to be answered in any other way" but with a slap in the face or a good knocking-down.[68] But Garner was out of his league when it came to sparring with Labouchere, who wrote:

> It has been a question with me whether I should have an iron cage erected at this office, fitted up with cameras and phonographs in order that upon the arrival of the bellicose philosopher I may retire into my citadel armed with a Winchester rifle

and Garnerise Garner. Failing this, it is my intention to procure an active young gorilla, and chain him up at the front door, for from all I hear the Professor will never trust himself within reach of an animal of that species if he can help it.[69]

So far the charges against Garner had all been made at secondhand, by unnamed English traders via Labouchere, and by the missionaries at St. Anne's via Dybowski. In December 1894, the pages of *Truth* carried what purported to be a letter from one of the missionaries, with commentary by Labouchere, under the title "Garner's Unpaid Bill." According to the writer of the letter, "H. T.," the events of spring and summer of the previous year were much as reported. At the mission, wrote H. T., Garner had enjoyed "a good bed and a good table," the latter always supplied with a bottle of claret. "Shall we add," asked H. T., "that the number of bottles was not usually limited to one, and that the study of the simian language developed [in Garner] a most astonishing amount of thirst?" Reprinted with the letter was Garner's IOU for 500 francs to the mission for its hospitality—never paid.[70]

H. T. was especially concerned to defend the reputation of Father Buléon. In his July column, Labouchere had quoted Garner in an American newspaper as follows: "This fellow, Dybowski, who claims to be an explorer, but of whom I never heard, has evidently conspired with the Jesuit priest, Buléon, to do me lasting harm because I denounced this fellow Buléon for deserting me in the jungles."[71] But, wrote H. T., the missionaries at St. Anne's were not Jesuits, a label thrown at them to cast doubt on their testimony. Nor would most people regard the hospitality shown Garner as tantamount to desertion in the jungle.[72] Case closed, concluded Labouchere:

> Were it necessary, I could easily enlarge upon the self-evident inconsistency of the various accounts that have been put forward by him [Garner], or on his authority, as to his experiences with the gorillas in the cage; upon the ridiculous farrago of twaddle which he has offered the American public as the result of his "wonderful experiments with gorillas and chimpanzees while learning their language in the heart of the African forest" (*vide New York World*); and, above all, his truly ludicrous failure to adduce a single discovery that he has made regarding the language of apes or gorillas in a state of nature. But to spend time and labour over these points, after the above exposure of Garner's doings during his sojourn in Africa, seems like killing the slain.[73]

Garner's Gorillas and Chimpanzees

Garner's book about his time in the French Congo, *Gorillas and Chimpanzees*, was not published until near the end of 1896, and then only in Britain.[74]

In the book, Garner did not mention the controversies swirling around the events of the 1892–93 expedition, nor did he attempt to defend his integrity or the truth of his version of events. Indeed, the book is surprisingly impersonal in many respects, not least when compared with Mary Kingsley's *Travels*, published at more or less the same time, about travels through the same part of the world. Garner began with an old-fashioned lesson in the comparative anatomy of monkeys, apes, and humans, and ended with an essay on the proper treatment of apes in captivity. In between, the focus was on gorillas and chimpanzees, Garner first summarizing what was known about each species of ape, then chronicling his interactions with and observations of certain of its members, along with the relevant testimony of natives, traders, and hunters.

Not much is said of evolution, though what is said is revealing: "The common opinion that man has descended from or is related by consanguinity to a monkey is silly and absurd. Science has never taught such folly, nor advanced any theory from which such a conclusion could be justly deduced."[75] But of course science had taught as much.[76] He explained himself a little further along, in a discussion of the peculiar gait of the chimpanzee. For Garner, the chimpanzee was a creature caught in a transitional state, neither a quadruped nor a biped, and the human habit of swinging arms in alternating fashion when walking was but a vestige of this transitional mode of locomotion. "Such a fact," he wrote, "does not show that he [man] was ever an ape, but it does point to the belief that he has once occupied a like horizon in nature to that now occupied by the ape, and that having emerged from it, he still retains traces of the habit."[77]

On his own account, at least, Garner's investigation of the speech of apes was by no means as barren as Labouchere had suggested. Among the discoveries reported in *Gorillas and Chimpanzees* were nine words of the chimpanzee tongue, plus the first human word spoken by a chimp.[78] In his translation efforts, Garner proceeded much as he had when working with the phonograph. "The chief purpose of my living among the animals being to study the sounds they uttered," he wrote, "I gave strict attention to those made by Moses. . . . By constantly watching his actions and associating them with his sounds I learned to interpret certain ones to mean certain things."[79] The limitations of this primitive method meant, for example, that Garner could not even repeat all of Moses's sounds, let alone translate them.[80] The phonograph would have obviated this dependence on Garner's limited powers of mimicry and memory.[81] The absence of the phonograph also led Garner to introduce a new notational system, much like Alexander Melville Bell's "visible speech," using English punctuation marks variously configured to represent volume, initial sounds, and degrees of glottal closure. Thus a loud "feu," the chimpanzee's breakthrough human word, became ']('').[82]

To Garner's chagrin, the Congo gorillas rarely emitted any utterances for him to transcribe.[83] "One special trait of the gorilla which I wish to emphasise is that he is one of the most taciturn, if not quite the most, of any member of the simian family. This fact does not appear to confirm my theory as to their high type of speech, but it is a fact so far as I observed, although the natives say that they are as loquacious as the chimpanzee."[84] Headaches beyond those presented by phonographless translation made work with even the most gregarious of the chimps, the "kulu-kambas,"[85] very difficult for the would-be lexicographer:

> these apes do not answer the call when they can see the one who makes it, and they do not always comply with it. In this respect they behave very much the same as young children, and it may be remarked that one difficulty in all apes is to secure fixed attention. This is exactly the same with young children. Even when they clearly understand, sometimes they betray no sign of having heard it. At other times they show that they both hear and understand, but do not comply.[86]

Besides Moses, Garner's ape companions included fellow kulus Aaron and Elishiba (whom Garner brought to Liverpool in late autumn 1893, to the famous animal dealer Dr. Cross, in whose care they soon died) and a gorilla, Othello (who, like Moses, died in the Congo), although Garner instructed only Moses in human speech.[87] "It was never any part of my purpose to teach a monkey to talk," Garner wrote, "but after I became familiar with the qualities and range of the voice of Moses, I determined to speak a few simple words of human speech."[88] Of the words tried, including "mamma," "wie," and "nk-gwe" (the native Nkami word for "mother"), only "feu" took. Moses pronounced it about as well as most foreigners, Garner judged. Garner even taught Moses to sign his name ("X").[89] As for the presence of a Catholic mission, its bells within earshot, Garner reported leaving young Moses at a nearby but unnamed mission for safekeeping while Garner traveled on a three-week journey to the Eshira country. (There is no mention of the leg pains and long recuperation reported by others.) According to Garner, the missionary in charge of Moses passed him along to a native boy at the mission, who kept Moses tied up outdoors through heavy dews, winds, and chilly nights, so that, on Garner's return, the first chimp to cross the threshold to human language was dying.[90]

Garner described his cage, its contents, and location in detail. As the accompanying illustrations showed (see fig. 4.4), Fort Gorilla was a cube, six-and-a-half-feet on each side, made up of twenty-four square panels of steel wire lattice joined together with hinges and rods, the whole, in Garner's

Figure 4.4 Richard Garner and his servant boy at Fort Gorilla, 1893, "Starting for a Stroll."
From Garner 1896, 23.

words, painted "dingy green, . . . [so that] in the forest it was almost invisible among the foliage," mounted on posts two feet above the ground. The top was covered with bamboo leaves and, when it rained, canvas curtains; the bottom was covered with thin boards. Furnishings included a fold-up bed, a fold-up camp chair, a board that doubled as a table, a swinging shelf, and cases full of clothing, bedding, medicine, guns, ammunition, tools, a camera, and supplies. A small kerosene stove served for making coffee and soup, and heating canned meats (eaten with crackers, on plates cleaned by the ants overnight).[91] Here "one could see without obstruction on all sides, and yet feel a certain sense of safety from being devoured by leopards and panthers." Garner included a vivid portrait of a composite day in the cage:

> [The gorilla] is now within a few yards of the cage, but is not aware of my presence. He plucks the tendril from a vine, smells it, and puts it in his mouth. He plucks another and another. I shall note that vine, and ascertain what it is. Now he is in a small open space, where the bush is cut away, so as to afford a better view. . . . He comes nearer. Now he has detected me.[92]

"In this novel hermitage," Garner wrote, "I remained for the greater part of the time for one hundred and twelve days and nights [i.e., close to four months] in succession, watching [these animals] in perfect freedom following the pursuits of their daily life."[93] Garner refers several times to, for example, "my abode in this desolate spot," "my long and solitary vigil," "that solitary gloom," "the place where I had so long lived in my cage."[94] But more frequent are details that suggest not only constant human contact but, by their sheer volume, the absence on Garner's part of any sense that such contact compromised his claims and needed to be concealed. Garner wrote of newspapers, letters, and contracts being delivered to the cage; of a trading boat entering the lake; of a village "a mile or two away," whose inhabitants agreed at Garner's behest not to shoot within a half-mile of the cage, to avoid scaring away his subjects; of "the monotony . . . often [being] relieved by going out for a day or two at a time, or hunting on the plains, a few miles away."[95]

To Labouchere's immense consternation, *Gorillas and Chimpanzees* received a small number of respectful notices in the British press and elsewhere. Of Garner, the "Munchausen of Monkey-land," Labouchere wrote: "In the description he gives [in *Gorillas and Chimpanzees*] of life in the jungle at Fernan Vaz . . . there is not a word which might not have been as easily and graphically penned by anyone who had resided, as Garner undoubtedly did, in comfortable quarters at a mission station, and passed a few hours in excursions into the forest."[96] Notwithstanding the friendly reviews and other signs of confidence in Garner and his enterprise,[97] Labouchere's campaign against Garner had indeed taken its toll. Even E. P. Evans, otherwise sympathetic to Garner, read his blustery replies to the allegations against him as "a tacit admission of their correctness."[98] In November 1896, the *New York Times*'s Harold Frederic reported on "Myths about Monkey Talk Exploded" in England. Thanks to Labouchere's attentions, wrote Frederic,

> Mr. R. L. Garner, whose professorship in simian tongues used to be talked about a good deal, is having a rough time here with his new book. . . . When Garner

passed through London on his way to Africa I saw something of him and tried to help him, but while he was absent officials just home from the Gold Coast told me that he was a fraud. When Garner returned his yarns quite confirmed this view.[99]

Two Later Versions

Garner's Version: The Fernan Vaz Testament

Among Garner's papers there is a notebook or sketchbook bearing the title *The Record of Idle Moments*, probably written in 1909, about sixteen years after the events in question. The contents page lists two main entries: "Monkey Talk: What the Monkeys have to say about Henry Labouscheister and His Criminal Chums," and "The True Version of the Same with the Correct Names of the Star Liars: Ananias Labouchere, Hypocrite Buléon, and their two chief Muffets, Allstink Wolfgang and Humble-Bug Machlichspittle." The rest of the *Record* consists of sketches with titles such as "Tinker Ben," "Teedie Bunson's Lesson," and "The Gypsy Fortune-Teller." These are mild minifictions of the most conventional sort. They could not be more different from the ferocious diatribes they follow.[100]

"Monkey Talk" takes place in the monkey house at the Zoological Gardens in London on a rainy Sunday afternoon. The narrator, a long-absent student of the monkeys, enters the monkey house and, finding it deserted, approaches the cages. Much to his satisfaction, the monkeys recognize him and immediately greet him as a friend. No sooner have greetings been exchanged than Spud, an intelligent and gossipy old capuchin, turns to his companion and says: "This is the gentleman about whom I was telling you this morning." Speaking to the narrator, Spud continues: "I was talking to my friend about you this morning and wondering where you were. Have you seen the scandalous statement or insinuations about you that old Peggy Labonshack has made about you?" There ensues a long mock of an editor of indeterminate sex who slanders men of achievement just to sell more copies of his vile newspaper. Next the group discuss the two traders who were the sources of the slander in question (itself never identified). "Well," comments a mangabey who had known the traders back in West Africa, "of all the human beings with white faces that I have ever known on my coast, those two are about the shabbiest lot that I have ever seen and that is saying a good deal, for I must add that we get about the last scrapings of men that ever wore trousers." Taking his leave of Spud, the narrator heads out to the garden, sits down, and composes this epitaph: "Under this pile of reeking dung, lie the bones of

Labouchere: Through life he lied with pen and tongue, And now his rotting form lies here—A worthy tomb for the snarling cur: For such was the stink-pot Labouchere. As dust to dust in fitting urns, From muck you came: to muck return."[101]

Next is Garner's Fernan Vaz testament, purporting to chronicle the true events fictionalized in "Monkey Talk." This prefictionalized version carries the title "Notes and Comments on the Affairs involving one Buléon, a Mission Priest of Fernan Vaz, Henry Ananias Labouchere, a Slum Editor of a Gutter Sheet in London, and some of their Allies, Representatives and Emissaries." In a preface to both versions of events, Garner wrote:

> Some years ago a certain editorial scab in London known as Henry Annanias Labouchere, who works off a filthy little sheet of libellous slander that floods the slums and gutters of White Chapel and Shore Ditch every week—at a penny a throw—attempted to malign my character by publishing a lot of innuendoes, which are always the weapons of cowards and blackmailers. . . .
>
> I do not charge the said . . . Labouchere with being the original author of the slanders in question, for I am aware that he only lent himself as a sewer to the pitiable outcasts who conspired with him to defame a man as punishment for an imaginary grievance for which not one of the guilty cowards had the manhood to complain to the supposed offender.[102]

Garner identified two of these pitiable, cowardly, grievance-nursing outcasts. The first was John McLaughlin, an assistant trader for John Holt.[103] After his trip up and down the Ogowé, but before his steamer journey back to Fernan Vaz, Garner stayed with McLaughlin in Cape Lopez, in late March or early April 1893. (Of the Ogowé trip that interrupted his time in Fernan Vaz, Garner wrote that he had gone reluctantly and at the insistence of a trader he had met.) According to Garner, McLaughlin was a lout, given to mistreatment of the natives and a tendency to stroll naked on the public beach when drunk, shouting for the attention of the women. Still more loathsome to Garner, it seems, was McLaughlin's lying about his religious views and background. He claimed to be English, but turned out to be Irish—the better, Garner surmised, to get along in an English firm. As part of his masquerade, McLaughlin professed his commitment to, in Garner's words, "the principles of an agnostic and evolutionist," denounced all religions, and told scurrilous stories about the missionaries.[104] But McLaughlin was a practicing Catholic, as Garner claimed to have discovered one afternoon while the trader was out.

> In his absence I remained at the station and spent the time in reading some old papers. The house boy at length asked me if I feared Mr. McLaughlin. I was a bit surprised at such a question, and asked the boy why I should be afraid of him.

In reply he said that Mr. McLaughlin was a witch, and could bewitch anyone—I demanded to know why he thought so, and in response he pointed to an old clock on a shelf and said that behind it Mr. McLaughlin kept an "Umburi" (MBURI) that was very powerful. Curious to know what it was that inspired such fear in the boy, I told him to show me what it was, and after assuring him that I would protect him from its effects he started to show me what it was. But being still afraid to touch it, he removed the clock to one side, and behind it was a Romish prayer book, which he told me he had several times seen Mr. M. making some kind of fetish with.[105]

With hindsight, however, Garner decided it was not McLaughlin's religious beliefs that had led him to write slanderous letters about the agnostic evolutionist Garner. No, what sealed Garner's fate, in his retrospective view, was his taking exception to McLaughlin's ostentatiously severe views about the Irish:

One Sunday afternoon this man denounced the Irish in such unmeasured terms that I volunteered some remarks in their defense. His condemnation of the land laws and his explanation of them were so strongly prejudiced that they were ludicrous and showed utter ignorance of the land laws of Ireland. Without the slightest feeling or show of passion on my part I ventured to say that he was certainly not informed upon that subject. This gave him mortal offence which however he did not betray at the time but bottled up and nursed to fruition. Just as such low minds always do, he resorted to secret revenge, and knowing of this muck-monger, Labouchere, and his filthy little penny sheet, he immediately wrote a scurrilous letter to him which was made the basis for all that black-mailer's vile calumnies about me.[106]

McLaughlin's alleged fellow informant and coward was also an employee of John Holt. Garner first met Alfred Wolfgang at Lambaréné, where Wolfgang ran the Holt station. Garner and Wolfgang subsequently saw each other several times, in the Ogowé region and back in Liverpool, Wolfgang's hometown. According to Garner, Wolfgang always showered him and his work with praise. Wolfgang had no reason to betray Garner. But Wolfgang was of weak character, easily bent to McLaughlin's conspiratorial will.[107]

How did Garner know that McLaughlin and Wolfgang were the culprits behind Labouchere's allegations? "It was little more than a year [later]," Garner wrote, when "I learned from [Wolfgang's] own employer, who had shamed him for the act, that he had been the chief ally of the mendacious McLaughlin, and both had been the pliant feeble tools of the monumental hypocrite, Buléon." Indeed, Buléon is the blackest villain in Garner's tale (fig. 4.5).[108] Garner first met Buléon on the steamer from Libreville to Fernan Vaz early in 1893. As Garner tells it, from the first, Buléon took a strong interest

Figure 4.5 Joachim Buléon, date unknown. Courtesy of the Archives Générales, Congrégation du Saint-Esprit, Paris (D 6024), and reproduced by permission.

in Garner's research and welfare. In a friendly manner, the priest offered a great deal of information about the native apes, warmly insisting the naturalist stay at the mission while in the region. Garner recalled explaining that he had not expected to find any white men at all, and had come amply supplied with everything he would need for the next nine months. But Buléon insisted, refusing on principle Garner's offer to pay for whatever accommodation he might from time to time enjoy at the mission.

Following this first meeting, Garner deposited his cage and supplies with Buléon at the mission. At the invitation of a trader, Garner spent the next few months traveling up and around the Ogowé (during which time he met McLaughlin and Wolfgang). Returning to the mission in April 1893 to retrieve his things, Garner found Buléon once again implacable in his wish to play the host. With the help of another priest, Garner managed to set up his cage, albeit, as Garner wrote, "on a remote corner of the mission concession about two kilometers from the mission." He continued: "It was a favorable site, and with all subsequent experience I am convinced that it was one of the best that I could have found within that region anywhere." But it was also close to Buléon, who often came to the cage himself or else sent messengers to invite Garner to spend the night or Sunday at the mission.[109]

In Garner's version, Buléon's interventions up to this point were at worst distracting. In mid-June 1893, however, they took a darker turn. Together, Buléon and Garner set out (at Buléon's urging, and with a great many native carriers) on 18 June into the region of the Eshira people, to the east. They traveled upriver in the mission canoe for several days, then headed into the forest. Awaking with fever on 22 June, Garner took some quinine and started off again, but found himself too weak to keep up. Soon he was far behind the main caravan, led by Buléon.[110] "Near noon," Garner writes, "I became quite exhausted, and I began to realize that my colleague was playing me falsely, although I could not then foresee his intentions nor understand his methods."

> But he made almost continuous use of the native language in all his conversations with guides, interpreters and carriers, quite contrary to his usual custom, for ordinarily he was very rigid in requiring everyone under his authority to speak French. This fact, coupled with certain others which had an air of mystery about them, first caused me to suspect that he was trying to conceal something from me. But I had not the remotest idea what it could be.
>
> I struggled along alone, barely able to trail the caravan, for there was scarcely a visible sign of a road except at the crossings of swamps or jungle patches, and finally I yielded to the inevitable and sat down on the side of the trail. I was conscious that I was alone and practically helpless. I knew that the last carrier had long since passed me and was far ahead, but there was no possible means of my communicating with anyone, and I finally surrendered myself to my fate.
>
> I lay down upon the ground and within a moment became unconscious—but before doing so I had the presence of mind to tie my handkerchief to the muzzle of my rifle, with a few words on a slip of paper, and attach it and set the gun where the signal might be seen.
>
> The time I remained in this place was oblivion to me, for I was unconscious of anything or even my own existence, and any wild beast of the forest could have claimed me with impunity. In this condition I lay until I was aroused by someone pulling at my sleeve and holding me in an upright position.[111]

Garner's savior was a carrier who had also fallen behind. The carrier was soon joined from the other direction by two interpreters who had left the caravan to look for Garner. They had not been sent by Buléon. According to Garner, they reported that Buléon had wanted to press on without Garner. It was the interpreters who insisted on rescue, and because one of them was a young chief of an especially powerful family, with whom Buléon needed to stay on good terms, the priest had capitulated. The interpreters brought the ailing professor to the main camp. But soon the caravan was off again, with Buléon promising to send men and a hammock along to fetch Garner from the village of Molemba, two days' march from the camp.[112]

Four days later no one had arrived, and Garner determined to press on. Walking slowly from weakness, surviving only on some packets of desiccated soup he found in his coat pockets, Garner and a small servant boy made their way to Molemba. After five days' walking, they arrived at another, smaller, village, still a day's walk from Molemba, but a suitable resting place, where the travelers bought chicken and some manioc (a potatolike vegetable—still a staple food in Gabon).[113]

When Garner arrived at last in Molemba, he discovered that Buléon had made none of the promised arrangements. The next day Garner pushed on to yet another village, where he stayed with a white trader (neither the trader nor the factory is named), until, about a week later, and two weeks since his desertion, Garner intercepted Buléon's caravan on its way back to Fernan Vaz. Walking with Buléon once again, Garner "took occasion to shame him for his dastardly conduct. Soon after I returned, he asked that all our differences should be forgotten, and desired to renew our former terms of friendship. I acceded to his proposal, but with the reservation of mind that I would never again trust him in anything."[114]

There had been nothing unambiguously hostile in Buléon's actions, even on this hellish journey. The misfortunes that befell Garner could all be explained as innocent accidents, or due to simple misunderstandings, even to selfless friendship and kindness. According to Garner, however, Buléon at several points revealed the true motives that drove him:

> To put it in the most charitable language that truth will admit of, this priest acted the part of a trickster and a traitor from beginning to end of all my relations with him and in the end had the unblushing audacity to admit to me that *it had been his purpose from the beginning to defeat my purpose in studying the gorillas and chimpanzees, because it was a denial of the catholic* [sic] *religion and that it was his duty as a priest to do anything he could to thwart anything that tended to prove the doctrines of evolution.* Not once but at least half a dozen times he made this statement to me in person—but in each instance he qualified his remarks by

the voluntary assurance that he would not resort to any indirect means of discrediting my work. . . . The sole purpose of this treacherous and two-faced priest was to try to discredit my work in the hope of counteracting it. As a priest and a catholic he was afraid of it—just as thousands of zealots are afraid of any truth that appears to conflict with their narrow creeds.[115]

The Historian's Version

What to make of Garner's tale of Catholic cunning and press opportunism? An assessment of the Fernan Vaz testament is difficult for two reasons in particular. First, there is evidence of anti-Catholic feeling on Garner's part before he went to the Congo in 1892–93. A collection of poems Garner published privately in 1891, *The Psychoscope*, shows a strong commitment to naturalism and a strong hostility to religious (specifically Catholic) dogmatism. "Meditations," about the "voiceless tomb" that is "the only goal / which all men win," is representative:

> Each claybound cell, thatched with the grassy sod,
>   Is a cathedral, where no pampered priest,
> With up-raised chalice to his frowning god,
>   Offers his flattery and sues for peace.
> No pealing anthems echo through its arches,
>   No burning tapers from its chancel shine,
> No choral train in solemn order marches
>   With alms or incense to its tearful shrine.[116]

The strong anti-Catholicism of this and other passages in *The Psychoscope* no doubt has several sources. Garner's South was overwhelmingly Protestant, and though Garner professed little affinity for the religious beliefs of his neighbors and family (the Garners descended from French Huguenots), suspicion of things Catholic was part of the cultural background. Moreover, the debates over Darwinism in the United States in the later nineteenth century occurred at a time when Catholic immigration and anxieties over its effects were rising steadily. In the much-mooted warfare between science and religion (especially as presented in John W. Draper's book), the warring religion, foe of Columbus, Galileo, and now Darwin, was emphatically Catholic.[117]

The second reason for caution is that the Fernan Vaz testament was written sixteen years after the events recounted, and probably under extreme circumstances, physical and psychological.[118] Other unpublished manuscripts arguably written at around the same time as the testament paint a bleak picture.[119] "The inquisition was not more trying to the patience and possibly not more to the endurance of its victims than the situation I have occupied most of the time for the last fourteen months," Garner wrote in one manuscript, entitled "What Next?" "Today the last particle of my manioc has finished. I have not one cent on earth to buy anything with. I have no meat or fish, nor anything to buy either with. . . . at this moment [I am] on the verge of starvation."[120] In another manuscript, "A State of Mind and a State of Facts," he wrote of being too poor even to conduct his scientific work: "I have been so pressed for the means of subsistence that I have not been able to buy or to keep on hand a specimen of either race [gorilla or chimpanzee] to study or to experiment with. Even if I had one I should not be able to find food for it, although there are millions of bananas and plantains rotting every day within easy reach of me, and could be had by the ton at 5 cents a bunch."[121] Under conditions so extreme, Garner may have found it consoling to put down the story of the first expedition with himself cast as hero.

Given the anti-Catholic biases Garner brought to his Fernan Vaz testament, and the difficult circumstances under which he probably composed it, we need to treat the testament with utmost wariness. That said, there is at least some corroborative evidence for certain aspects of Garner's tale. Independently of Garner and Labouchere, for example, we know that Buléon did attempt to discredit Garner's work, although the exact nature of the attempt is unclear. In March 1895, the *Figaro* carried a report about a meeting at the Museum d'histoire naturelle in Paris, where Buléon had spoken about Garner's true activities in the Congo "before an elite audience composed of professors, scientists, explorers, [and] missionaries." This audience, it was said, heard how Garner had arrived at the mission laden with research equipment, but then settled in to write his book about the apes without having observed any. When the missionaries protested the dishonesty of what Garner was doing, he dismissed their concerns, explaining that he was only out to make money, and that, so long as nobody could gainsay the accuracy of his account, his book would sell whether or not he actually observed what he wrote about. "It is fair to say," concluded the reporter, "that if this blunderer Garner, instead of pressing on as far as the Congo, had contented himself with writing his book at the British Museum or the Bibliothèque National, he could have spared himself all this awkward misadventure and still made lots of money."[122]

In his Fernan Vaz testament, Garner questioned whether this reported meeting had ever taken place. Most likely, Garner surmised, Buléon had invented the whole story, then passed on his invention to an unscrupulous newspaper reporter. Who could believe such a meeting had ever taken place, Garner asked? The published report—he does not identify it as being in the *Figaro*—"began with a lie which was intended to give color of importance to the occasion. It began by saying that 'at a mass meeting of scientists, missionaries and

explorers,' as though any such incongruous meeting could have the slightest
signification for anyone, for the oil and water here described never mix in mass
meetings—and as evidence that the scheme was concocted for the purpose, the
whole proceedings reported were directly devoted to an attack upon me by this
man Buléon."[123] Another of the missionaries also attempted to debunk Garner
in the French press. Monsignor Alexandre Le Roy, the bishop with responsibility for the Fernan Vaz region at the time of Garner's expedition, had visited
St. Anne's mission in June 1893, and had journeyed with Garner and Buléon
into the Eshira country. In May 1895, two months after the *Figaro* article
appeared, Le Roy published two articles about Garner in the French science
journal *Cosmos*, "The Language of the Monkeys" ("Le langue des singes").
According to Le Roy (from whose article, perhaps, Verne took the story), Garner had managed to spend but three days in his cage, and to observe but one
gorilla, before mosquitoes sent him scurrying back to the mission.[124]

So in part the missionaries corroborated the Fernan Vaz testament, at
least in their eagerness to discredit Garner through the press.[125] More corroborative still is the book Garner wrote after the first expedition, *Gorillas
and Chimpanzees*. In many ways this book is just the opposite of what an
attention-seeking, money-grabbing fraud should have produced. Far from repeating old dramatic stories from the small available literature on apes in the
wild, *Gorillas and Chimpanzees* was one of the first books of any consequence
to call into question the popular image of African forests teeming with ferocious apes.[126] Garner lambasted the dramatizers:

> Almost every yarn told by the novice is quite the same in substance and much the
> same in detail as those related by others. It seems that most of them meet the same
> old gorilla, still beating his breast and screaming just as he did thirty years ago.
> The number of gun-barrels that he is accused of having chewed up would make
> an arsenal that would arm the volunteers. What becomes of all those that are at-
> tacked by this fierce monarch of the jungle? Not one of them ever gets killed, and
> not one of them ever kills a gorilla. Does he merely do this as a bluff and then
> recede from the attack? Or does he follow it up and seize his victim, tear him open
> and drink his blood as he is supposed to do? How does the victim escape? What
> becomes of the assailant? Who lives to tell the tale?[127]

Accordingly, Garner's own observation reports are modest, both in number
and excitement. Most often (and it was not very often), an ape hove into view,
silently looked around, silently caught sight of Garner in his cage, and walked
away.[128]

The Fernan Vaz testament shows that Garner long felt angry over the allegations made against him. These allegations were personally wounding, of

course. But they were wounding to Garner's claims and career as well. Former admirers such as Harold Frederic and E. P. Evans had written Garner off as a fraud, as we have seen.[129] Chapters 5 and 6 examine the remarkably sudden disappearance within professional science of the simian tongue and the phonographic techniques used to discover it. Garner regarded this collapse in his personal and professional fortunes as a tremendous injustice.

Obviously we can never know with certainty what happened on the 1892–93 expedition. But neither should we forfeit the opportunity to ask what, in light of the record assembled here, we should make of the allegations against Garner. Let us consider the five main ones in turn.

1. *Garner set up his cage not deep in the forest but within earshot of the St. Anne's mission bell.* This is true. But Garner never pretended otherwise. Before the expedition, he had told reporters that "by telephone and electric bell" he would be "in contact with friendly forces located at a convenient distance." In his *McClure's* dispatch from the field, Garner wrote that Fort Gorilla was "a trifle more than a mile from any human habitation," that a dog from the mission was a visitor, that "[w]ithin a few feet of my cage [there] is a small, rough path cut through the bush to mark the boundary of the mission lands." After the expedition, in *Gorillas and Chimpanzees*, he studded his account with references to picking up his post from time to time, dropping Moses off at a nearby mission, asking the residents of a village not to shoot their guns and scare off the apes, and so on.[130] Nor is there any reason to doubt Garner's claim that two kilometers from the mission was as good a place as any for observing apes, not least for purposes of contact with (putatively) friendly forces in case of danger. I suspect what happened is the following. Labouchere received letters from Fernan Vaz impugning Garner's credibility. Press reports of Garner's comments on his return to Britain in November 1893 and during his February 1894 Prince's Hall lecture indicated that life in his forest cage had been tough going. Labouchere took the least charitable reading of these reported remarks, accused Garner of lying about how isolated he had been, and challenged Garner to prove otherwise.

2. *Garner spent only three days in his cage.* This allegation is surely false. The craven fraud about whom Buléon spoke in Paris in 1895 should have produced a book crowded with sightings of apes doing extravagant things. It is because Garner had so *little* to say about what he saw from his cage that his really spending time in the cage seems likely. Furthermore, we have independent reports from journalists about Garner's persistence and patience in pursuit of monkey speech. Add to this all the time Garner subsequently spent in Fernan Vaz—some ten years over the next quarter century—and it is hard

to believe that, after three days, he was so afraid and uncomfortable that he fled his cage for good. He never claimed to have spent 112 days in the cage without interruption. Again, Labouchere, for his own reasons, seems to have read into Garner's reported remarks the most extreme claim possible, then challenged Garner to prove the extreme claim was true.

3. *Garner lied about having shot gorillas.* With this allegation we see most clearly how Labouchere turned ambiguous press accounts to his advantage. After quoting from an American report that Garner had found his cage "strong enough to keep gorillas at bay until he could shoot them," Labouchere accused Garner of contradicting himself, as Garner had told the British there had been no gorilla attacks whatsoever. If Garner could not keep this part of his story straight, Labouchere insinuated, then the rest of his story was equally suspect. But the statement on which Labouchere had built his case was not at all clear. Garner was not reported as having claimed he had shot gorillas. He was reported as having found his cage better protection against gorillas than he had expected. Garner could well have made this discovery without experiencing a gorilla attack, much less ending one with a gunshot (which he was loath to do anyway, for fear of scaring off the apes he wanted to observe). Or Garner may just have said his cage was stronger than expected, and the American reporter then added the detail about shooting gorillas to make for a more colorful story.[131]

4. *Garner journeyed two days into the Eshira country, then spent two months convalescing at the Tomlinson factory.* In the Fernan Vaz testament, Garner does claim to have become ill not long after starting out for the Eshira country. But there his account of the trip, and the account given by the explorer Dybowski on behalf of the Fernan Vaz missionaries, go their separate ways. I find the sheer volume of detail in Garner's version, and the fact that he put down that version not for others' eyes (there is no indication it was ever intended for publication) but for himself, persuasive. Nor does Garner's version read to me like ex post facto revision-as-wish-fulfillment. It was not heroism but happenstance that saved Garner in Garner's version.

5. *Garner left an unpaid IOU at St. Anne's mission.* In the Fernan Vaz testament, Garner explained how his IOU came to be written:

> In a moment of generosity I offered to contribute something more than I had done to the mission funds and it was my purpose to offer him [Buléon] a contribution of one hundred francs. When I mentioned it to him he asked me if I meant $100—or 100 francs. From the nature of the question and his manner of asking it I felt convinced that he regarded the offer as a very niggardly contribution and without higgling or hesitating I agreed to make it $100. But not having that sum

in my possession at the time I gave him a due bill for the amount, and in the face of the same I plainly stated that the sum mentioned was to be paid when I received my remittance from America.[132]

This, to me, rings true. I doubt Garner could have made up the detail about the hundred francs instantly inflating to a hundred dollars when there was a risk of him seeming a cheapskate—even in the eyes of the detested Buléon. I do not know what "remittance" Garner was referring to, but presumably it never came. Garner may not have paid his bill, but, it appears to me, he told the truth even about that. In short, doubts about Garner's veracity emerge weakened, not strengthened, from an examination of the unpaid-bill allegation.

Putting to one side questions about what really happened in 1892–93, tantalizing though these are, there remains the fact that Garner really interpreted his expedition as a battle in the ongoing war of science against religious prejudice and bigotry. The Fernan Vaz testament is the work of someone who inhabited the military metaphor, who took his bearings from within it. No doubt Buléon told Garner the Bible was true and therefore the theory of evolution false. (The author of *The Speech of Man and Holy Writ* believed as much; why not a French Catholic missionary in West Africa?) When things started going wrong for Garner back in Britain and the United States in 1894–96, Buléon's skepticism about evolution might well have taken on a new significance. Rightly or wrongly, Garner may have begun to suspect dark intent behind what had seemed at the time mere accidents or irritations. It all made sense to him once he saw that Buléon had acted the inquisitor to Garner's Galileo. "To be snarled at by mangy curs is not pleasant," Garner wrote in his testament, ". . . but it does not deter me from my course in searching for truth."[133]

5

THE ANTHROPOLOGISTS AND
ANIMAL LANGUAGE

◆ ◆ ◆

GARNER'S TANGLE WITH the missionaries during and after his 1892–93 expedition was not to be his last. When, ten years later, he published an account of the French Congo's native peoples, championing their codes of conduct as better suited to their modes of life than the codes that white men, in their "ignorance" and "egotism," often sought to impose, a missionary reader took exception.[1] According to the critic, writing in a Liverpool newspaper in June 1902, Garner was absolutely right to urge greater understanding of native ways, but wrong to suppose that greater acceptance would follow automatically. On the contrary, seeing native ways as natives saw them was a route to more effective missionizing:

> It will no doubt please Mr. R. L. Garner to know that lectures on primitive customs and belief among aboriginals are being organised at Cambridge. If we [missionaries] were not convinced that we are right, and if we did not regard ourselves with favour, we should never become missionaries. Mission work as a matter of course implies the subversion of what we regard as bad institutions. The civil and moral code of one tribe of mankind may differ widely from that of another and yet both be right; but mission work carries with it a conviction that it is for the

spreading of light in darkness, and for the righting of wrong. Slavery, fetishism, human sacrifices, and so forth seem to us customs to be contended against in every possible way. Yet it must be confessed . . . that in any case the more that can be known as to the inwardness of native custom and belief the better it will be for the civilised nations seeking to wield influence over them. . . . It is desirable to understand, because that is the way to know how to successfully extirpate. To be content with the ability to interpret certain customs from the standpoint of those who practise them would be to accept stagnation as a permanent condition.[2]

This chapter and the next concern the wider scientific changes that, in addition to the scandal described in chapter 4, contributed to the disappearance in the early years of the twentieth century of Garner's great innovation, the primate playback experiment. The complaint about Garner quoted above furnishes a useful point of entry, in particular the mention of lectures being organized in Cambridge. An announcement about them had recently appeared in the *Journal of the African Society*, on the page opposite Garner's offending article on native institutions. Alfred Cort Haddon, leader of a recent Cambridge anthropological expedition to the Antipodes and university lecturer in ethnology, was, it said, due to give a four-lecture evening series on primitive customs and beliefs "with a view to interesting missionary students in the subject." It was hoped the course would lead on to a lengthier and more systematic one in the near future.[3] And so it did: the 1902 Michaelmas term at Cambridge saw the debut of Haddon's course "Social and Religious Institutions of Primitive Peoples" for students reading art and archaeology.[4] Thus did Haddon establish a place in the curriculum in one of Britain's ancient universities for the scientific study of savage cultures—a major stepping-stone toward the subject's professionalization. It remains for historians of anthropology to do justice to the role of missionaries in making that happen.[5] Here the salient point is something even more surprising: the connection that a member of Haddon's audience made between Haddon's efforts and Garner's. Understanding that connection, and how it came to seem surprising, is the burden of this chapter, on the anthropologists and animal language.

We begin with Haddon, since the Cambridge expedition that he led marked a deliberately professionalizing transition out of nineteenth-century anthropological traditions. Literally and figuratively, Haddon represented the mix of sciences out of which the professional anthropologies of the early twentieth century emerged. Next the chapter considers the survival into that era of a main feature of Haddon's approach, an evolutionary analysis of human cultures, as exemplified in the ethnographic writings of Garner himself and of his remarkable protégé, the phonograph-toting John Harrington, expert in the languages of the American Indians. The cultural evolutionism of

Garner and Harrington was, however, increasingly the exception. These were the years when physical anthropology and cultural anthropology came to inhabit separate professional spheres, with evolutionism becoming the near-exclusive possession of the former. The remainder of the chapter tells how these two anthropologies came to coalesce around their divergent ambitions, conceptions, and ideologies. In physical anthropology, the dominant themes became measurement, race, and evolution, contemplated from within a museum full of skulls and bones. In cultural anthropology, meanwhile, the dominant themes became the self-consciously contrasting ones of "participant observation," culture, and history, as contemplated from within a tent pitched at the edge of a native village. Neither of these disciplinary shapes proved congenial to the primate playback experiment, though for very different reasons.[6]

At almost exactly the moment Garner first traveled to the apes, the first truly apish hominid fossils, belonging to the "Java man," were discovered. The result was the gradual but rapid displacement of the tropical jungle by the fossil record as the preeminent place to track human ancestry. It would take a revolution—the rise of the "new physical anthropology" of Sherwood Washburn in the mid-twentieth century, discussed in chapter 8—before physical anthropologists would again see the observation in the wild of the behavior of living apes and monkeys, including their communicative behavior, as part of the job description. Even Aleš Hrdlička, from 1910 to 1930 the leading physical anthropologist in America and a fan of Garner and his work, never became more than an incidental observer of simian behavior, in the field or elsewhere. Hrdlička's career will serve to anchor the analysis below of the emergence of playbackless physical anthropology. Where Hrdlička and his descendants anatomized the animal origin of language, his cultural-anthropological contemporaries disowned the problem altogether. For them, evolutionist anthropology of the Garnerian sort was founded on a huge racist mistake, since no human language was any closer to the apes than any other. Typically, and correctly, anthropologists of this view credited the abandonment of an evolutionary, hierarchical scale of human races—the scale in which Garner conceived his experiment, in which it found its rationale (as vindicating a Darwinian prediction)—to the German-born American anthropologist Franz Boas. The final part of the chapter examines the intellectual means by which Boas and his students wrenched human language and culture out of the accustomed evolutionary framework, putting what they saw as enormous distance between these and anything even the most advanced animals produced. It was the Boasian success that turned a project like Garner's into a cultural-anthropological untouchable.[7]

Traditions and Transitions in
Turn-of-the-Century Anthropology

"Even now," wrote Haddon in his 1898 book *The Study of Man*, "the scope and significance of Anthropology have scarcely been recognized." Too often, he complained, one part of the science was taken to stand for the whole. Setting out his own inclusive vision, he made use of a disciplinary map recently proposed by Daniel Brinton, professor of linguistics and folklore in Philadelphia (and, as we have seen, one of Garner's critics). Brinton had identified four main component sciences: "somatology," "ethnology," "ethnography," and "archaeology." Somatology, Haddon explained, was what the British had tended to call "physical anthropology" and Continental researchers simply "anthropology." Its core was anatomical and anthropometric, but it also embraced embryology, heredity, criminal anthropology, the biologies (including diseases) of the different human races and anthropoid species, and the experimental psychology of sensation and perception. After somatology—the name never much caught on—came ethnology, a kind of cultural counterpart, taking in studies of social organization, technology, religion, language, and folklore. Where somatology and ethnology dealt with humankind as a whole, ethnography dealt with particular human groupings. Its concerns were the taxonomy and historical geography of specific peoples—in Haddon's words, "the classification of peoples, their origin, and their migrations"—as discerned in both the biological and the cultural data. Fourth, and furthest from Haddon's own interests, there was archaeology, covering the geology of the human epoch and "palethnology" as well as the preserved material cultures of past communities and civilizations.[8]

Haddon put this vision into practice that same year, as leader of a celebrated Cambridge expedition to the islands of the Torres Straits, between Australia and Papua New Guinea. The straits had been the site of his own initiation into anthropology a decade earlier. A Cambridge-educated embryologist with a Dublin chair in zoology, he had made the journey to study the region's coral reefs and associated marine life, but found himself captivated instead, as he afterward explained in an evening lecture at the Royal Institution, by "the manners and customs of a people small in number but rich in interest," who "thirty years ago were naked, unknown savages, who to-day are British subjects, and who in a very few years will have lost the last remnants of their individuality, and possibly ere long will practically cease to exist."[9] Launching himself on the business of professional retooling, he turned first to the more anatomical side of anthropology, helping to found Ireland's first anthropometric laboratory and, a few years later, relocating to Cambridge to

lecture in physical anthropology (though he kept the Dublin chair and continued to teach there).[10] From the start he took manners and customs too as part
of his remit, perhaps in emulation of the holistic, biocultural approach of an
admired friend, the Scottish polymath Patrick Geddes.[11] An early expression
of Haddon's synthetic ambitions was his proposal in 1892 that the specialist
anthropological, antiquarian, and folklore societies in Britain coordinate efforts on the problem of Britain's racial composition and history. The result
was the Ethnographic Survey of the United Kingdom, run under the auspices
of the BAAS. Survey business took Haddon and whatever Cambridge students
he could find—including the young physicist Ernest Rutherford—into the
surrounding villages, where they inventoried everything from skull dimensions to children's games.[12]

Through all this activity, Haddon's fascination with the peoples of the
Torres Straits and New Guinea was undimmed. As the survey, incomplete,
wound down in the late 1890s, he began drumming up finances for an an-
thropological expedition to the straits, making sure the new venture would
have all the comprehensiveness of the survey but none of its fatal amateurism.
With the University of Cambridge as the major financial backer, the team
Haddon assembled set off in the spring of 1898. Besides Haddon, there was
William H. R. Rivers, a physician-turned-psychologist now running experimental psychological laboratories in Cambridge and London; William Mc-
Dougall and Charles Myers, Cambridge students of Rivers's and, like him,
medically trained; Charles Seligmann, a medical pathologist; Sidney Ray, a
schoolteacher who was also a leading expert on the languages of the region;
and Anthony Wilkin, a Cambridge student with experience in archaeology
and photography.[13] In a prospectus in *Nature*, Haddon described the division
of labor. His own responsibility was to be the recording of physical characters. Rivers, McDougall, and Myers, wrote Haddon, "will initiate a new departure in practical anthropology by studying comparative experimental psychology in the field." Toward this end, the expedition would bring along not
just anthropometric instruments but psychological ones, including, Haddon
noted, two phonographs, "to record the native songs, music and languages."
Seligmann was to be the team naturalist, identifying the plants and animals
used in native life (though once there he became an industrious anthropometrist as well). Ray would take care of the languages, and Wilkin, in addition
to photographing everything, and even filming dances and the like, would
investigate native sociology.[14]

Arriving in the straits in April, Haddon's crew headed for the comparatively remote eastern islands, choosing the fertile and well-populated Mer
(also known as Murray Island) as their first base of operations (fig. 5.1). They

Figure 5.1 Charles Myers on Mer with his cylinder phonograph, recording a ceremonial song performed by local men, 1898. Courtesy of the Museum of Archaeology and Anthropology, University of Cambridge (Haddon Collection CUMAA P.950.ACH1), and reproduced by permission.

took over a disused mission building and turned it into a combined anthropological, psychological, and photographic laboratory. There they measured the islanders' heads and heights, tested their vision and hearing, quizzed them about their genealogies, asked them to draw scenes from their lives. Ray fixed vocabularies on paper, Seligmann songs on wax, Wilkin legendary sites and stones on film. Ceremonies, children's games, cures for diseases, constellations, crafts of all kinds: the notebooks and collections swelled as the expedition team pursued Haddon's all-embracing agenda, on Mer and then coastal New Guinea, the western islands of the Torres Straits, and, after the expedition left the straits in November, Borneo.[15] Not long after, with the fieldwork completed, a protracted scheme of publication began. Appearing between 1901 and 1935, the reports from the expedition eventually filled six large tomes, dedicated to general ethnography; physiology and psychology; language; arts and crafts; the western islanders' sociology, magic, and religion; and the eastern islanders' versions of the same.[16] But the synoptic thrust of the project could already be glimpsed in 1900, in a two-part paper that Haddon published in the *Geographical Journal*, "Studies in the Anthropogeography of

British New Guinea." Here Haddon attempted to show how the distribution of head forms along part of the southern coast of New Guinea tracked the distribution of customs, arts, and crafts closely enough to allow tentative proposals as to the kinds of people there and the history of their interaction.[17]

Concerned to document traditions and transitions among faraway peoples, the Cambridge expedition was itself both traditional and transitional. It took the established repertoire of nineteenth-century anthropological methods and added to it the methods of nineteenth-century experimental psychology. It gave new life to the Tylorian notion that, as Haddon put it in *The Study of Man*, the "theory of evolution throws a bright and far-reaching light on the problems of Anthropology," while at the same time breaking decisively and influentially with Tylor's "armchair" mode of inquiry.[18] Even the evolutionism comprised traditional and transitional elements. Taught at Cambridge to understand developing embryos as miniaturizing the evolutionary past, Haddon perceived the childishness of savages and the savagery of children as expressions of the same state. Rivers shared this recapitulationism.[19] But for neither man did it betoken uncomplicated racism. Haddon, for instance, insisted that there was no unqualified sense in which one human race was nearer to the apes than another, since all races were specialized in different ways.[20] For Rivers's part, he argued that crudeness of vocabulary in the savage's language—in the case that concerned him, the paucity of terms for shades of blue among the Torres Straits islanders—can reflect not inborn deficits (for tests revealed the islanders were perfectly capable of telling apart different kinds of blue) but cultural preferences (savages, he surmised, take no aesthetic interest in nature, and blue is a boringly ubiquitous natural color in that part of the world anyway).[21] Both arguments, as we shall see, were put again, more famously, by the antievolutionist Boas. But what made Haddon's and Rivers's evolutionism characteristic of its moment above all was the debt to Herbert Spencer. In his 1895 book *Evolution in Art: As Illustrated by the Life-Histories of Designs*, Haddon proposed, in Spencerian fashion, that the shifting balance between conserving and innovating forces determined the evolutionary trajectory of a design.[22] Rivers, who had trained under Spencer's neurological apostle, John Hughlings Jackson (discussed in chapter 2), explained in the Torres Straits reports that the islanders' superior performance on certain perceptual tests was part and parcel with their savagery, in that, in his Spencerian words, "[i]f too much energy is expended on the sensory foundations, it is natural that the intellectual superstructure should suffer."[23]

For all that the Cambridge expedition managed to redefine professional anthropology as a field rather than armchair endeavor, Haddon's design for the expedition—the global disciplinary vision behind it—did not much survive its debut in the Torres Straits. Although Haddon himself retained interests in the physical anthropology and evolutionary biogeography of race, his intellectual descendants in Britain's universities tended to give biology a wide berth, concentrating near-exclusively on topics he had grouped under "ethnology."[24] (A similar fragmentation with a similar outcome occurred in America among students of Boas.) And though the comparative psychology of civilized versus savage thought-worlds remained a major talking point, the psychological instruments used on Mer did not, with one exception, become part of the standard fieldworker toolkit.[25] The exception was the cylinder phonograph.[26] From the 1890s to the 1920s and beyond, cylinder phonographs traveled wherever anthropologists went. In the United States, for instance, some fourteen thousand cylinders of American Indian songs, tales, and speech accumulated in the archive of the Smithsonian Institution, thanks to fieldwork sponsored by the Bureau of American Ethnology. For the most part, the instrument became part of the taken-for-granted recording repertoire, little commented upon in published reports. Indeed, as anthropologists came increasingly to idealize "the field" as a place of absorption in and imaginative communion with exotic cultures, the notion of sitting subjects down in front of a phonograph horn and commanding them to perform seemed embarrassingly artificial. But before professional ideology overtook it, the phonograph made stars of several American fieldworkers, such as Frances Densmore and Alice Cunningham Fletcher.[27] Among their number was John Harrington, protégé of Richard Garner.

Before turning to Garner and Harrington, we should note in passing that the abundance of ethnological phonographs has an important corollary for the primate playback story: throughout the years of its virtual scientific extinction, the experiment remained a technical possibility. From around 1900, the most popular commercial machines were, it is true, reproduction-only devices. But wax-cylinder phonographs and, later, disk-cutting machines could always be purchased. Cumbersome though they were, these found their way into the field. American ethnologists and European comparative musicologists used them extensively.[28] Among zoologists, the best-known proponent was Ludwig Koch, who, from his childhood in Germany in the 1890s to his long career at the BBC, specialized in the recording of birdsong in the wild.[29] But there were others. In the 1910s, the French scientist Louis Boutan used the phonograph in Garnerian research into gibbon language. In the 1920s, Raymond Ditmars, reptile specialist at the Bronx Zoo and a longtime friend and supporter of Garner's, was engaged in experimental work with a macaque. A month after Garner's death in 1920, a New York newspaper reported that an "old style cylinder phonograph was employed by Dr. Ditmars in an experiment in recording monkey speech and then observing the animal's reactions to the reproduction of its own voice."[30] In the 1930s, Koch collaborated with Julian Huxley, a distinguished observer and theorist of bird behavior (among many other things), on *Animal Language*, a lavish "sound book" combining text, photographs, and records of animals in zoos, mainly in Regent's Park and Whipsnade. As Huxley explained, after recording, they assayed "the animals' reactions to their own sounds as recorded here. Mr. Koch has played these from the records to most of the animals used. The results were rather surprising. Some animals became excited, while others showed no reactions at all." Among the animals thus studied were a mandrill, a mangabey, and a baboon.[31]

Language and the Evolutionary Study of Human Culture

Although Garner's research program disappeared from the sciences, the man himself did not. After *Truth*'s attack in 1894–96, there were, it is true, no more lecture invitations from the BAAS, no more reviews in the leading scientific journals. Never again would Garner command scientific attention of the scale or seriousness as that he enjoyed in the early 1890s. Still he remained a presence. Over the last twenty-five years of his life—he died in January 1920, aged seventy-one—he made five more trips to West Africa for research purposes.[32] All told he spent a full decade's residence among the gorillas and chimpanzees (and missionaries and traders and colonial officials and native tribespeople) of Fernan Vaz. His public profile remained high. He published frequently in popular newspapers and magazines, on anthropological topics and much else. He collected gorillas, chimpanzees, and other creatures for the New York Zoological Park in the Bronx and the National Museum in Washington.[33] When not in Africa, he mounted lecture tours, speaking to packed halls about his work, his adventures, and the shame of French colonial rule in West Africa.[34] Garner died a famous man, with friends and admirers among the scientific elite. But his innovative uses of the phonograph were little taken up and soon forgotten. The reasons are complex, and best appreciated from the perspectives of those new professionals who respected Garner without, however, seeking to do as he had done.

Garner as Ethnographer

Apes and monkeys were never Garner's exclusive research concern. His evolutionism drove him toward a broad comparative perspective, taking in much of the animal kingdom, humans included. The native peoples of the Fernan Vaz region were of special interest, being, as he saw it, so close anatomically and mentally to the higher apes. Garner attended closely to native languages, customs, and legends. He did not produce a great deal of straight ethnography, but what he did produce deserves attention, for it links him firmly to that transitional, Torres Straits moment in anthropology, when the fieldworker merged with the theorist, the scientific instrument came to the fore, and egalitarian respect for the worldview of the Other suffused the evolutionist picture of a human racial scale.

Let us consider first the article mentioned at the outset, "Native Institutions of the Ogowe Tribes of West Central Africa," published in the new *Journal of the African Society* in 1902.[35] Here Garner described a number of native practices, offering, in the words of his subtitle, "an interpretation of their meaning as viewed from the standpoint of the native philosopher" (fig. 5.2). Too often, Garner wrote, arrogant white observers never tried to understand native practices from the perspective of the natives themselves. "The few white men who go among these people do not study the meanings of native customs, but on the contrary, they deride and condemn them as heathen rites and denounce those who practice them as a lot of idiots." In their fervor

Figure 5.2 A gathering in Garner's West Africa, date unknown, found among Garner's papers. Courtesy of the National Anthropological Archives, Smithsonian Institution (Garner Papers, box 7), and reproduced by permission.

to extinguish "heathen" practices as fast as possible, these whites, according to Garner, never learned to appreciate how well suited those practices were to the minds and conditions of life from which they sprang. The native practice of witchcraft was a case in point. Ignorant of the natives' own understanding of witchcraft, confident of the childish irrationality of native ways, white observers, Garner explained, had interpreted witchcraft in terms of Western religion, with its notions of a supreme being, of lesser gods and devils, of worship and devotion. But such notions, Garner claimed, were wholly inappropriate. Witchcraft among the Ogowé tribes, he wrote, "rests upon a tangible, physical basis, which makes it far more rational than similar follies believed and practised by our own proud race."[36]

Evolutionist race-ranking and cultural relativism came together seamlessly in Garner's account of Ogowé witchcraft. For the natives, Garner reported, witchcraft was a matter of putting nature to work, nothing more. But the knowledge of how to do this, of which "medicines" had which effects, was secret knowledge. Witches were powerful beings not because they were supernatural beings, but because they were knowledgeable beings, knowledgeable about natural causes and processes as others generally were not. Thus the witches alone knew which plants contained insomnia-inducing substances, and which plants contained sleep-inducing substances. "Their belief [in witchcraft] is absolutely materialistic," Garner concluded. Indeed, such had to be the case, for "the native mind cannot conceive of the power apart from the medicine. His powers of abstraction are too feeble to grasp such a thought."[37] A similar blend of concreteness and pragmatism lay behind native uses of charms and amulets. Garner presented the native reasoning behind the wearing of a leopard claw to ward off danger in the jungle:

> The logic of this belief is found in the fact that the person wearing it must have conquered the leopard . . . , and if he was strong enough to overcome and disarm a leopard, then all other leopards should beware of him and all other things that fear leopards should avoid him. Another inference is that he is in alliance with the leopards, and therefore if a leopard see him wearing this charm he thinks "that man is a friend of my people and I must not harm him," and with this reflection he goes on his way and leaves the man, and if any other animal see this sign he fears the man because he is the friend and ally of the leopards. In other words, the chief source of its virtue is in deceiving the observer.[38]

Like Garner himself, it seems, the Ogowé peoples regarded the creatures of the jungle as potentially good reasoners, though, crucially, not excellent ones. And as we have seen, Garner even urged tolerance of, indeed respect for, native views on right and wrong, as of all other native cultural institutions.

"The civil and moral code of one tribe of mankind may differ very widely from that of another and yet both be right," he wrote. Yes, some of the native codes "would not be consistent with our code of ethics, but we must interpret them from the standpoint of those who practise them."[39]

Upsetting as Garner could be to missionary sensibilities, he could be just as upsetting to the sensibilities of the would-be missionized, as one of his unpublished papers, "The Phonograph among the Savages," illustrates.[40] As ethnography, this later piece—written around 1907—contrasts interestingly with the 1902 "Native Institutions of the Ogowé Tribes."[41] The narrative voice is first person rather than third person; the natives are particular individuals in specific places; and what is observed are responses to Garner and the scientific instrument with which he was most closely identified, the phonograph. "The Phonograph among the Savages" tells what happened when Garner took his phonograph (and his interpreter, and his carriers) on a trip through several villages in the interior. In one village near the coast, he wrote, the chief, a man named Obundu, died of a sudden illness shortly after speaking into the phonograph horn. "I was not openly accused of being responsible for his illness," Garner wrote, "[but] it was not difficult to see that his people entertained a lurking suspicion that the phonograph might, in some mysterious way, be the source of his malady."[42]

These suspicions followed the phonograph up the river. According to Garner, panicked chaos ensued when he played a record of Obundu's voice at another village. "They insisted that no man's voice could live after the man himself was dead and that unless Obundu was present his voice could not be. They demanded that I should take the 'box,' as they called it, and leave the town at once." No amount of explanation reassured the locals. Later Garner learned of claims that Obundu's face had been seen looking out of the phonograph horn. In another village, seeking to entertain the villagers rather than "tempt their credulity by reproducing the voice of the dead man," Garner played records of women from another tribe, singing and laughing boisterously. When the record finished, the chief, irritated, asked Garner to play a record of the same women crying. It was disrespectful, Garner learned, for a woman to behave so when a man, their master, was present. Crying was the sound of the conquered, and much more appropriate in women. In a third village, Garner's request to record a witch doctor's song was turned down for fear "the 'box' would find out his 'secret' and betray it to the [other] witches, which would destroy his power over them. In a confidential manner he assured me that he knew the 'secret' of the 'box' and what made it talk, at the same time promising me that he would not expose it to the people unless he heard of its bewitching someone."[43]

The Ethnographer as Garnerian

For one of the most dedicated anthropological fieldworkers of the first half of the twentieth century, Richard Garner set the model. Born in 1884, raised in southern California, John Peabody Harrington (fig. 5.3) was in his teens when he read about Garner's adventures among the apes and monkeys.[44] As Harrington later recalled:

Figure 5.3 John Peabody Harrington (1884–1961), with three Cuna Indian informants, 1924. Courtesy of the National Anthropological Archives, Smithsonian Institution (Neg 4305-A), and reproduced by permission.

I had loved the study of languages when a boy, but my first resolve to spend my life in the field of ethnology and linguistics came from reading when in high school R. L. Garner's two books on the speech of monkeys and apes, and telling of his experiences afar in the jungles of Africa. Shortly after reading these books and making them a portion of my life, a certain Mr. Edwards brought to Santa Barbara, California, where I resided at the time, a show consisting mainly of a chimpanzee and a Congo negro who talked his native language. I remember yet how I stood for hours fascinated by the chimpanzee. It was the first time I had seen anything of the kind. And what few little sounds the largely silent animal made were learned and imitated. Later I got to visit a number of times and to study the chimpanzee owned by Dr. Knowles of Pasadena, California.[45]

Harrington's path to a career as a full-time, salaried anthropologist began as precariously as would Aleš Hrdlička's, and ended in the same place, the Smithsonian. Taking his undergraduate degree in modern and classical languages at Stanford in 1902, Harrington entered into postgraduate work in anthropology, linguistics, and phonetics at universities in Leipzig and Berlin. He had already developed his passion for American Indian languages (he had studied these as an undergraduate with the Berkeley anthropologist and Boas student Alfred Kroeber), and before finishing his degree returned to California to begin linguistic fieldwork. For the next ten years, he scrounged for teaching jobs and carried on with his research. He wrote some guidelines on learning a new language for the fourth edition (1912) of the BAAS's anthropological handbook, *Notes and Queries*, recommending phonographs as well as other devices for fieldworkers unable to find reliable native informants.[46] Harrington's considerable skills brought him to the attention of the chiefs at the Bureau of American Ethnology, who hired him as a field ethnologist in February 1915. Around the time of this appointment—Harrington remained with the Smithsonian for the next forty years—he played host to his boyhood hero Garner in Los Angeles:

It was . . . in 1915 that I had the privilege of meeting Dr. Garner in person. Already an employee of the Smithsonian Institution at the time and stationed at the Southwest Museum, Highland Park, between Los Angeles and Pasadena, California, I learned through officials of the Museum that Dr. Garner was coming to lecture at Los Angeles. It was my great pleasure to show him the Southwest Museum, and to then attend his lecture on apes, which was held down town in Los Angeles, and especially after the lecture it was a privilege to go with him to his hotel room where I remained until two o'clock in the morning[,] examining his pictures and hearing his first-hand anecdotes and general information about a subject which proved indeed to be a life interest with me.[47]

The lecture that evening was probably a slightly updated version of "Studies of the Great Apes at Home," one of three that Garner had advertised in promotional materials for a lecture tour mounted a few years before. (The other two lectures were "The Empire of Darkness," about Africa and its peoples, and "Child Life of the Jungle Folk," a minilecture for children.) Harrington was dazzled. "When I met Garner in Los Angeles," he recalled, "I was thrilled with his personality and the wealth of knowledge and experience which he carried and imparted. He was a true naturalist if ever there was one. . . . He had all the traits of a forceful and successful teacher, a good lecturer, and true scientist."[48]

Though Harrington was never much given to theorizing about the origins of even individual languages, much less language as such, his conception of language was profoundly evolutionist. Notes that survive from his linguistics lecturing in the early 1910s show how concerned he was to relate human language to the expressions and psychologies of animals.[49] He quoted the great psychologist Wilhelm Wundt, founder of modern experimental psychology, with whom Harrington had studied in Germany: "If psychologists of today, ignoring all that an animal can express through gestures and sounds, limit the possession of language to human beings, such a conclusion is scarcely less absurd than that of many philosophers of antiquity who regarded the languages of barbarous nations as animal cries."[50] Nor, as Harrington's future wife Carobeth Tucker discovered, was his evolutionism strictly an intellectual matter. Not long after Garner's visit, in the summer of 1915, the nineteen-year-old Tucker was one of the students in Harrington's summer school course on linguistics at the San Diego Normal School. "For almost two years," she later wrote, "I had been bringing home from the public library musty, outdated, inconceivably dull, and mostly incomprehensible tomes on paleontology, anthropology, and related subjects. . . . I had discovered *Evolution*, I had discovered *Science*, I had come to believe that there were those who spent their lives in pursuit of absolute truth, and I wanted above everything to belong to that elite band."[51] One day Harrington told his students to read and transcribe phonetically something of their own choosing. Tucker chose a poem, an epigraph to one of those musty books on evolution. One line of the poem, as she later remembered, ran: "And Thou, O Sea, great mother of my soul." Harrington gave her top marks, and asked her to see him after class. "He confided that the subject I had chosen pleased him immeasurably," she wrote. "He, too, was enthralled by the mystique of evolution (although I don't think he learned or cared to learn anything of genetics)."[52] Language and evolution remained leitmotifs of their peculiar courtship, Harrington at one point raising "the possibility of bringing up a child without ever talking to it or allowing it to hear human speech in order to see what sort of language it would evolve." For a time Tucker played along, until she realized Harrington was not kidding. He really wanted to do the experiment. Luckily (for their children, if not for science), he regarded the experiment as too impractical to attempt.[53]

Like Garner, Harrington the evolutionist combined racial prejudice with an intense interest in how other races understood and talked about the world. The prejudice was much in evidence when he took his bride-to-be to see a group of Igorote natives at a Philippine exhibit in San Diego. "I remember the dim light, the shoddy, carnival atmosphere, and the tiny dark people, pathetic as caged animals," she recalled. "Harrington wanted me to listen carefully to their speech and observe their 'primitiveness.'"[54] But Harrington's vivid sense of the antiquity of primitive languages, of the need to document these languages before they went extinct, drove him to work among their speakers with unequalled zeal and thoroughness. For Harrington, each day away from the field was a potential disaster for science, because tomorrow the data might not be there to collect.[55] He filled warehouses with his field notes, publishing as little as he could, roaming endlessly, obsessively, friendlessly, from one Indian group to the next.

In the 1930s, now in middle age, Harrington's thoughts turned back to Richard Garner. As Harrington explained to Garner's son, Harry, in a letter in 1937, a near-fatal illness had brought on a reassessment of priorities:

> You know when a person lies on his back for six weeks with his body reduced to a skeleton, he sees the visions and sees the recondite things, like Ghandi [sic] does in his fastings. One considers past, present and future. It was in my case a mid-life mark, a taking stock of life's store, and I never distinguished so clearly what is the rubbish and what are the worth-whiles. It became more and more apparent to me as I lay there alone and aloof that to get out a book on your father's life and discoveries was one of my principal worth-whiles, and I resolved to do it giving it the precedence over everything else with exception of holdovers the odds and ends of which must be finished up. . . . [As a boy,] I resolved that my life was going to be dedicated to the speech of primates, and to an experiment of raising children without hearing human speech, and of studying them daily in every new development of self-invented language. I was possessed by this idea—and still am. I was resolved that I would forsake marriage, and everything else that ties men down, for the experimentation necessary over a period of years. . . . All through my college work the speech of apes and of savages was my chief interest, and in Germany I took Wundt's course in psychology, which went into that very sort of thing. I remember how I went to call on Professor Wundt (he was the greatest

living psychologist) at his residence, at his sprechstude (hour of consultation), and asked him if apes have language and if raising children without their hearing language would make possible study of the genesis of language. He told me of the work of Garner, and [that] the experiment of my dreams would be sure of extraordinarily important results, but that it would take a martyr to carry it through. I had already heard of Garner. When he came to Los Angeles to lecture about 1915 I made an appointment with him and spent the afternoon and half the night with him. I also spent time with Mr. Cherry, African explorer and ivory hunter, who was retired and living at Santa Ana, California. I plunged into Indian language work, but did not find in it the primitive in speech that I craved to explore, the Indian languages are old Asiatic tongues more comparable to Chinese and to Sanskrit then they are to anything primitive. Even Australia does not have the primitive, when it comes to language.

All these things raged through my mind when I lay on my cot in Emergency Hospital, in the winter of 1936–7, and a new resolve arose. Time is shooting ahead by leaps and bounds. Whatever I do while I am still in the form of a human being on this old planet, the surface of which is moulding with what we call life, will have to be done in a hurry. There is no time to wait. Your father was the discoverer of the fact that apes and monkeys have speech, and that is the most important discovery that will ever be made in the field in which I am most interested. And he was from start to finish a Smithsonian man. Therefore it behoves some official of the Smithsonian Institution, such as myself, to write his life and work, and I am especially adapted to do this because I believed in your father's discovery from the first, and am most sympathetic to set his life forth in the light that it deserves to be seen in. I also loved your father's ways, personally, and was one with his entire outlook, his naturalist, investigating point of view. He has never been given the lasting niche that he deserves in the hall of fame, and I am going to give it to him in the form of writing his life and discoveries as from one within the Institution which he loved. All future work on the speech of apes and primitives will go back to him, he was the Columbus in this field. . . . I shall be proud to be the man in the Smithsonian best fitted to set him forth in the depictation in which I see him. . . . This book will make his work complete, and will put it on record in the properly authorized way.[56]

The same urge to preserve and document the vanishing past led Harrington to secure what remained of Garner's papers. It was Harrington who deposited them in the archives of the Smithsonian, along with a characteristically incomplete draft of a brief biography of Garner. "His Bible was the book of nature," Harrington wrote of his long-gone hero, "and from it he read through first-hand experience without bias the story of ever new experiences and found that they built themselves into a unity in his mind. There was no disagreement within the truth of experience. . . . Going through deprivations and hardships and dangers of disease, Garner devoted his life to the accumulation of new facts along a line unexplored and promising."[57]

Language and the Evolutionary Study of Hominid Anatomy

We turn now to consider the first of the two major changes that deflected professional anthropologists from seeking the origins of language in the vocalizations of nonhuman primates. Before the discovery of the Java man in 1891–92, many evolutionists doubted the fossil record would be much help in filling the apparent gap between apes and humans: hence the very strong interest in showing the ape-human gap to be narrower than commonly supposed. As we have seen, in the 1860s, T. H. Huxley produced arguments about how small changes in apish anatomical structures could bring about large, language-enabling changes in the function of those structures. In the 1890s, Garner produced experimental proofs that apes used words structurally and functionally continuous with human words. An important exception to this general trend was Horatio Hale (fig. 5.4). When in 1892 he presented Garner's work to fellow anthropologists in a favorable light, Hale had already been thinking about the evolutionary origins of language for some time, and the prospects for filling the ape-human gap, which he took to be as wide as it seemed, with the not-very-apish Neanderthal and Cro-Magnon fossils then known. A summative lecture of Hale's along these lines will serve as a point of departure in what follows, as we attempt to track the changing relations between the fossil record and the simian tongue, from before the discovery of Java man to the emergence, in the first half of the twentieth century, of a new class of experts, trained in the finding and interpretation of hominid fossils. One of the leading figures in this new subdiscipline of physical anthropology—and in the narrative to come—was Aleš Hrdlička. As we shall see, his correspondence with Garner reveals much about the significance of ape studies within the new discipline, and the significance of the new fossil discoveries within Garner's own understanding of his achievement.[58]

Neanderthals, Cro-Magnons, and the Language Barrier

In 1886, in his address as vice president of the anthropological section at the Buffalo meeting of the American Association for the Advancement of Science, Hale surveyed the fossil evidence accumulated thus far as it bore on the problem of the origin of the power of speech. As he emphasized, from the 1850s onward, discoveries of fossil human bones and manmade artifacts in deposits

HORATIO HALE

Figure 5.4 Horatio Hale (1817–96), date unknown. From *American Anthropologist* 10 (January 1897): plate 1 (between 27 and 28).

containing extinct animals had forced dramatic revisions in estimates of human antiquity. There was now wide consensus that two different kinds of humans had inhabited prehistoric Europe. One kind, the Neanderthals, were held to be the earlier inhabitants; and fossil evidence of their presence in Europe turned up first, in the German lands, in the late 1850s. (The fossils were discovered in the Neander valley, or *tal*—hence the name.) More than ten years later, a number of skeletons excavated at a site in France led to the identification of the second and distinct kind of human, named the Cro-Magnons (again in honor of the site). As Hale noted, most investigators agreed the Cro-Magnons were more similar to modern humans than the Neanderthals were, but there was much disagreement about just how dissimilar the Neanderthals were. Huxley, for example, in his *Evidence as to Man's Place in Nature* of 1863, had argued that the range of variation in the skull measurements of modern humans was such that the Neanderthals fell well within that range.[59] In Hale's view, there was an immense divide between the apelike Neanderthals and their remarkably unapish Cro-Magnon successors. For Hale, the power of speech could have arrived only with the appearance of the latter.

Hale gave three arguments for a Cro-Magnon origin of language. First, citing the work of Gabriel de Mortillet, professor at the École d'anthropologie in Paris and taxonomist of the tool-making cultures of prehistoric Europe, Hale argued that Neanderthals had lacked the tongue control needed for articulate language. The key piece of evidence here was a fragment of Neanderthal jawbone recovered from a cave at La Naulette, in Belgium, in 1866 (fig. 5.5). Several features of the bone—its thickness, its tooth-socket patterns, its

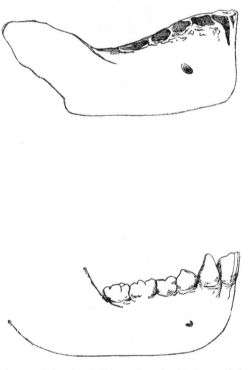

Figure 5.5 The Naulette jaw (*above*) and the lower jaw of a chimpanzee (*below*). From Hartmann 1885, 120.

backward-projecting chin—suggested a being intermediate between humans and apes.[60] But what mattered most was the lack of two pairs of small bumps, known as the "genial" or "mental tubercles," in the center of the inner jawbone. In modern humans, the muscle connecting the tongue to the skull inserts into these tubercles. According to De Mortillet, this unique anatomical arrangement enables an unsurpassed degree of tongue control, which in turn enables the subtle maneuverings of articulate language. Apes lack the genial tubercles, and do not speak. The Neanderthal owner of the Naulette jawbone lacked genial tubercles. Hence Neanderthals did not speak either. De Mortillet published this finding in 1883.[61] Around the same time, Hale reported, an expert on dentistry, Dr. Robert Baume, examined the Naulette jawbone alongside another Neanderthal jawbone, discovered in 1880 at the Schipka cave in Moravia, and came to the same conclusion about the speechlessness of the Neanderthals.[62]

In addition to the evidence of Neanderthal jawbones, there was also the evidence of Neanderthal skulls. Like their jawbones, Neanderthal skulls showed an unmistakable brutishness in several respects, in particular the protruding brows and receding foreheads. Even Huxley had admitted as much. But the matter of cranial capacity was more controversial. Huxley had pointed out that the skull found in the Neander valley in 1856 had a capacity comparable to the skulls of modern-day Polynesians and Hottentots. True enough, Hale conceded.[63] But, for Hale, the crucial issue was less the *quantity* of cranial capacity than its *distribution*. It was the distribution of capacity in the Neanderthal skull that showed the brain housed therein to have lacked the speech center present in the brains of modern humans in the third convolution of the left hemisphere of the frontal lobe.

[T]he greater portion of the capacity is in the posterior part of the skull. The narrowness and depression of the forehead are remarkable, and exceed anything known in the skulls of existing races. The height of the forehead depends, of course, upon the development of the frontal lobe of the brain. The frontal lobe is made up, as regards its height, of three folds, or convolutions, termed by anatomists the first, second, and third frontal convolutions. These convolutions lie one above the other, the third being the lowest. This third convolution is somewhat thicker than the other two, and adds therefore, in general, somewhat more to the height of the forehead than either of the others. Its absence, or almost entire absence, from the brain, would produce just such a depression, or extraordinary flatness, as we find in the foreheads of these ancient skulls. . . . This third frontal convolution is sometimes called "Broca's convolution," from the fact that the distinguished French physiologist, Dr. Paul Broca, was the first to localize the faculty of language in it. . . . If this convolution were absent from the human brain, or were only present in a rudimentary form, as in the anthropoid apes, man would be incapable of speech, and the height of his forehead would be greatly diminished. We should have, in fact, the precise difference which exists between the frontal portion of the Neanderthal . . . skull, and that of the average skull of the present race of men.[64]

Where Neanderthal jawbones and skulls evidenced the absence of speech, their Cro-Magnon counterparts evidenced its presence, Hale argued. Cro-Magnon jawbones projected forward. They contained genial tubercles. The lofty foreheads of Cro-Magnon skulls would have allowed the full development of all the convolutions of the frontal lobe, including the third convolution.[65]

A third argument for speech beginning with the Cro-Magnons looked to the archaeological record. The Neanderthals had left behind stone hand axes remarkable above all for their crudeness and uniformity. As Hale pointed out, some observers had wondered at the difference between these primitive tools and the far more advanced tools in use among modern-day savages. The difference was indeed impressive; and its explanation, for Hale, was obvious: where the later tools had been designed within speech-saturated minds, "the earlier implements were the production of beings whose minds were in the undeveloped state that must necessarily characterize men who had not yet attained the power of speech."[66] Again, Cro-Magnon artifacts could hardly have been more different. Here, in stone, but also bone and ivory, were beautifully rendered works of art. Endowed with speech, and with all the intellectual power that speech brings in its wake, the Cro-Magnons had carved and engraved astonishingly lifelike representations of mammoth, elk, reindeer, and other creatures of the Pleistocene.[67]

So how did the speaking Cro-Magnons emerge? In Hale's view, the Cro-Magnons were the direct, modified descendants of the Neanderthals. As we have seen, however, Hale supposed the evolutionary process generally, and the emergence of "speaking man" in particular, to have been less a gradual, Darwinian process than a saltatory, Galtonian process. For Hale, the modifications that brought forth Cro-Magnon speech from Neanderthal speechlessness probably took place not over a great number of generations, but within a single generation, indeed within a single family. Citing Huxley's precedent, Hale emphasized that so sudden a leap into language would not have required a great leap in bone, brain, or muscle structure. The vocal organs, for example, were organized to such an extent even among the apes that these creatures "can utter cries of some sort, and some of them can make a variety of sounds." Surely the Neanderthals had been capable of at least as much. As for the rest of their anatomies, the changes which took place when the speaking children were born to the speech-less pair were in the greater development of the cerebral convolution in which the faculty of language resides, in the new direction given to the under part of the lower jaw, which now projected forward instead of receding, and in the increased volume and strength of the genio-glossal muscles, which by their action developed the genial tubercle, and gave at once greater size and more freedom of movement to the tongue. These changes, though so important in their results, were really slight compared with the changes in a case of polydactylism [i.e., the birth of a child with six-fingered hands to normal parents]. The chief alteration was, of course, that which took place in the brain. It was simply the enlargement of a fold of that organ; but its effect was prodigious, and has transformed the globe. This enlarged fold was the seat, not merely of the faculty of language, but of many other faculties, all of which showed at once the effect of their newly acquired power.[68]

A New Kind of Fossil and a New Kind of Anthropologist

In 1888, the Dutch anatomist and physician Eugène Dubois assessed the same Neanderthal skulls and jawbones adduced by Hale and drew the opposite conclusion. Far from proving that Neanderthals had occupied an interme-diate position between apes and men, these fossils, in Dubois's judgment, proved "nothing more than . . . the existence in Europe, during the diluvial [i.e., Pleistocene] period, of low-level yet still never really pithecoid races, not even markedly inferior to the lower races of today."[69] At the time, Dubois was hunting for evidence of the true ape-men. To find that evidence, he had made his way to the Dutch East Indies, employed in the Dutch army as a medical officer.[70] His 1888 article, published in a local journal of natural sciences, outlined his reasons for believing the fossils would be found in the Indo-Malay archipelago.[71] Soon two engineers and fifty prisoners were car-rying out excavations under Dubois's command. In September 1891, digging on the banks of the Solo River in Java, near Trinil, Dubois's men turned up a molar bearing strong apelike features. An apelike skull cap and humanlike thigh bone followed, in October 1891 and August 1892, respectively.[72] These fossils, according to Dubois, were remnants of the true missing link. Modify-ing Haeckel's preemptive label of *Pithecanthropus*, Dubois named this van-ished ape-man *Pithecanthropus erectus*.[73] What Garner, on the other side of the world and at the very same moment, attempted to accomplish with the phonograph, Dubois's men accomplished with shovels and sieves: the partial filling of the gap between apes and humans.

By the turn of the century, there was wide consensus that Dubois's "Java Man" was indeed a new kind of object, a fossil hominid of undoubted antiq-uity, showing a degree of apishness without parallel among living humans.[74]

Less often noted, but no less important, was the emergence as well of a new kind of investigator. Before Dubois, there had been geologists and paleontologists, digging for fossil vertebrates and invertebrates; there had been prehistoric archaeologists, digging for hand axes and figurines crafted by early humans; there had been anthropologists, measuring skulls from the various human races and ordering their angles and volumes into progressive hierarchies; but no one had set out deliberately to dig up fossil bones linking apes and humans. The Neanderthal and Cro-Magnon fossils had been incidental discoveries.[75] Nor, before Dubois, was there much interest among evolutionists in vindicating the postulated ape-human link with fossil proofs.[76] Dubois's discovery redirected evidential ambitions toward the fossil record. In the process, fossil hominids took over from living apes as the chief evidence for and emblems of human evolutionary origins.[77]

Apes still mattered, of course.[78] But there grew a keener sense among physical anthropologists and others that apes could not be regarded as unproblematic placeholders for the earliest human ancestors. In a soon-classic survey of the accumulating fossil evidence, Hrdlička in 1914 insisted that humankind's fellow primates had for a long time followed evolutionary trajectories all their own. "The various actual species of primates lower than man," he wrote, "may in a sense be viewed as by-products of his own evolution, partly perhaps as his distant cousins, descendants from some of the old primate stocks, or as the retarded and aberrant relatives, unable or not called upon by their environment to keep up with his progress, and slowly modifying more or less sui generis."[79] On the matter of the evolutionary origins of language, it was skulls and jawbones, not simian utterances, that came to dominate. New fossil finds at Heidelberg in 1908 and La Quina in 1911, for example, reopened the question of speech before the Cro-Magnon era.[80] Hrdlička in his report speculated that articulate language might well have begun before the close of the Tertiary period.[81]

As one of the preeminent physical anthropologists in the post–Java man generation, Hrdlička was at once a product and begetter of the changes that turned his science in the early twentieth century into a professional, ever more fossil-oriented enterprise (fig. 5.6).[82] Born in Bohemia in 1869, he left at the age of thirteen and traveled with his father to New York City, where the two took up work in a tobacco factory, Aleš educating himself as he could at night school. Eventually deciding on a career in medicine, he earned a medical degree and ran a private practice until, in the mid-1890s, he landed a plum research position in a hospital for the insane in Middletown, New York. These were the days when Cesare Lombroso, the Italian criminal anthropologist, famous for his theory that criminals were evolutionary throwbacks in body and

Figure 5.6 Aleš Hrdlička (1869–1943), early 1930s. Courtesy of the National Anthropological Archives, Smithsonian Institution (Neg MNH 31,513), and reproduced by permission.

mind, enjoyed his greatest influence; and Hrdlička later recalled that it was his reading in the scientific literature from France and Italy on, in his words, "criminals and other abnormals," that introduced him to anthropometry.[83] Soon subjecting hospital inmates to an ambitious measurement program, the young physician swiftly came to the attention of the staff at the new Pathological Institute in New York City. When they offered him a job as house anthropometrist, he accepted. But before starting in his new role, he decided to seek training in anthropometry from its masters in Europe. Thus it happened that in 1896, at the height of debate over Dubois's Java finds, Hrdlička was in Paris, studying at the École d'anthropologie with one of the major contributors to that debate, Léon Manouvrier.[84]

At first, the new interest in human antiquity and its fossil evidence was but one of several souvenirs Hrdlička brought home from Paris, along with a firmer grounding in anthropometric technique and a fresh resolve to turn himself from a medical anthropometrist into an academic anthropologist, on the model of Manouvrier (widely regarded as Broca's successor).[85] Back in New York, Hrdlička began repositioning himself. The first task was to become attached to a good anthropological museum. He started with one of the best, the American Museum of Natural History, managing between 1898 and 1902 to accompany a number of museum-sponsored expeditions to Mexico and the Southwest. There he carried out anthropometric research among the Indian populations, seeking the data on human normality without which, in his view, his data on human pathology were uninterpretable. He also began to publish descriptions of prehistoric skeletons and parts of skeletons, from Indian burial mounds and other sites. Throughout, Hrdlička was unsalaried, relying largely on his wife's inherited income to finance this shift of ambition. In 1903, the gamble paid off. At the Smithsonian, where anthropology had tended to mean ethnology or archaeology, a new head of anthropology succeeded in establishing the Division of Physical Anthropology at the National Museum of Natural History, with Hrdlička, as assistant curator, in charge.[86] Under his stewardship, lasting into the early 1940s (he was appointed curator in 1910), what started as a small and disordered collection of human crania and skeletons grew into one of the most comprehensive in the world. Its excellence was a matter of vital public as well as scientific concern, or so Hrdlička argued in a 1908 address, "Physical Anthropology and Its Aims," published in *Science*. Even to get the facts right about variation in just one bone of one human group required, he explained, the careful study of the remains of hundreds of individuals. Massive as the undertaking was, it was just a beginning; for without similarly large-scale inquiries for all other parts of the body, over the whole range of human races, living and extinct, exposed to the widest range of environments, no confident generalizations about human variation could be made. At stake, he continued, was knowledge of human evolution: where it had been, where it was going, and how it might usefully be directed from here.[87] (Eugenics, he later claimed, was but "applied anthropology.")[88]

Inseparable from Hrdlička's building of the Smithsonian's collection were his efforts as patron and practitioner of physical anthropology. The well-stocked museum of his dreams needed, as he pointed out, lots of qualified anthropologists if its potential were to be realized. As he envisioned them, the new recruits would be drawn from medical programs, educated in their new specialism with up-to-date textbooks, and able thereafter to consult and contribute to dedicated research journals.[89] With characteristic energy, Hrdlička over the next decade or so proceeded to fill out this brief. He turned the Smithsonian into an important training ground, his influence radiating outward as those he taught themselves became teachers.[90] He published a textbook that rapidly established itself as the standard source on anthropometry; the third, posthumous edition was entitled *Hrdlička's Practical Anthropometry*.[91] At the end of the First World War, he founded the discipline-defining *American*

Journal of Physical Anthropology, serving as its first editor as well as a major contributor.[92] Meanwhile, his own researches on "the most ancient skeletal remains of man"—the title of his 1914 report—were having a notable impact. The quality of his surveys of the accumulating paleoanthropological evidence, much of it examined in person, became authoritative for American readers in much the way that Arthur Keith's did in Britain and Marcellin Boule's in France.[93] As an original interpreter of that evidence, moreover, Hrdlička ventured two well-remembered synthetic theories. On the origin of anatomically modern humans, he argued that they were in direct descent from the Neanderthals. On the peopling of the Americas, he argued that it had occurred relatively recently, over a now-vanished land bridge between Siberia and Alaska, so that no truly ancient human or hominid fossils would be found in the Americas. (The latter argument has fared rather better than the former.)[94]

In a vision of physical anthropology so osteological and, increasingly, paleontological in focus, there was scant room for the study of the living nonhuman primates, and virtually none for the study of their powers of reasoning and speech. Yet no one with Hrdlička's evolutionary interests could have been completely indifferent to apes and monkeys, still less someone who worked for decades at the Smithsonian, within easy walking distance of the zoo whose simians launched Garner's career. Indeed, in Hrdlička's 1908 address, he described primatological study as "the vestibule to the space occupied by man's natural history and . . . indispensable to the understanding of man's past and continued evolution, collectively and in every particular."[95] Unsurprisingly, from time to time, Hrdlička engaged in a spot of primatological observing himself—once, for instance, spending a couple of hours in a London zoo to see whether orangutans walk on their knuckles or with open palms (the former, he found).[96] But by and large he stuck with bones, leaving behavior to others. One of them was Garner.

Farewell to the Apes

"I wonder if you, too, have forgotten us here in Africa," Garner wrote to Hrdlička in July 1918.[97] Accompanied by a young taxidermist from the Smithsonian, Charles Aschemeier, Garner, now seventy, had arrived in Fernan Vaz over a year before to lay the groundwork for a major collecting expedition. The head of the expedition, and its financial backer, was the Philadelphia businessman and big-game hunter Alfred M. Collins. As Collins explained to a reporter in 1916, the main object of the Collins-Garner expedition was to collect gorillas for the National Museum, and as no one knew the habits and language of the gorillas better, Garner would be a tremendous ally in this undertaking. The plan was for Collins and a fourth man, the Boston naturalist Charles Furlong (in charge of taking motion-picture footage of the expedition), to join the advance guard of Garner and Aschemeier in the spring of 1917.[98] But with the entry of the United States into the European war then raging, Collins and Furlong devoted themselves to the war effort, leaving Garner and Aschemeier to make do in Fernan Vaz. Fortunately for Garner, Aschemeier turned out to be an expert marksman, and had shot some two thousand animals by the time Garner sent his letter to Hrdlička.[99]

Most of the June 1918 letter was devoted to observations about diseases afflicting the local primates, the nocturnal habits of some chimpanzees trooping about near Garner's house, and general remarks on the mental affinities between humans and apes. Where Garner in the 1890s had dismissed the idea that humans and apes were literal kin, he now embraced it, describing his own endeavors as the psychological counterpart to Hrdlička's anthropological work ("Your studies of physical anthropology of course must convince you of the undeniable consanguinity of men and apes, as animals derived from a common source; and my studies of the psychic unity of the two races convince me of their mental affinity as different editions of the same text").[100] Hrdlička wrote back that he considered Garner's letter valuable enough to merit publication in the new *American Journal of Physical Anthropology*. But Hrdlička wanted more from Garner, as he explained:

> I must earnestly entreat you . . . to write more on your exact observations on the chimpanzee and other apes. Put down every minute observation of anatomical, physiological, and psychological nature, as well as all possible details about diseases. You do not know how valuable these observations are from one of your honesty and experience as well as ability. Write in preference to collecting, which you can readily leave to the younger man [Aschemeier]. Do not imagine that you are going to remember things, or live forever to complete what may seem to you only a partial story. Don't shoot the apes, or at least not those in your neighborhood, for the living are far more valuable than the dead. Befriend them, learn to know them still more than you do, and then tell me all about them, just as if we were sitting with each other as we used to. I will publish in the best form every valuable particular that you send and you may be sure that credit will be given where it is due. . . . [A]bove all take scrupulous care of your health. The war is going grand and the world will soon be good enough again to live in, so keep your health and vigor.[101]

Garner and Aschemeier did not return to the States until June 1919. Upon arrival in New York City, Garner spoke to a reporter about how his work among the living apes fitted with the work of others among fossil humans.

"My work has always been misunderstood and misinterpreted," Garner told the reporter.

> I suppose it's because no one but myself can take a monkey seriously. I am earnestly seeking to bridge the intellectual hiatus between the ape and man. Through thousands of experiments, I am learning step by step the points of difference and of similarity which exist between the mental qualities of human beings and their animal prototypes. I study the reasoning power, the instincts, the family lives and habits of apes; so that if no physical link is discovered between them and man, I may at least prove, by properly describing and classifying the intellectual phenomena of the simians, how very near they are to being flesh of our flesh, bone of our bone, mind of our mind, soul of our soul.[102]

So far as this statement reflects Garner's considered opinion of the value of his work at this time, and another comment to the same reporter shows that Garner regarded the fossil record as the place where the missing physical link might one day turn up ("seven skulls . . . found buried hundreds of feet beneath the [surface of the] earth, each marking a gradually ascending stage of evolution, . . . [provide] almost conclusive proof that Darwin's theory may be regarded as a scientific principle"),[103] then the newspaper headlines that greeted Garner on his return, such as "Missing Link Reported Found—Monkey Said to Talk Like Man,"[104] showed that his work indeed continued to be thoroughly misunderstood and misinterpreted.

According to these reports, Garner had discovered an ape more humanlike than any hitherto known to science, with a language remarkably similar to that of the human natives, and had shot the creature dead. A minor and rather bizarre controversy ensued over who actually shot this living missing link. Aschemeier claimed he had done it, and furthermore that Garner had learned to converse neither with this particular ape nor with any other ape.[105] For at least some observers, Garner's noninvolvement in the shooting of so humanlike a being was to his credit.[106] On 11 June, Hrdlička wrote to welcome Garner back and encourage him to start writing more articles on the great apes for Hrdlička's journal. "The recent distasteful newspaper occurrence must not ruffle you in the least," Hrdlička added. "The opinion here is either that the young man did not say what he was reported to have said, or that he suffered with a bad attack of brain fever."[107]

January 1920 found Garner in Bristol, Tennessee, convalescing at the house of a relative. He dictated a letter to Hrdlička, telling him, "you shall have that manuscript at the earliest possible date, or in lieu thereof an invitation to my funeral." Garner also sent along two of his books. "They are the last of the mohicans; and I don't know whether another copy can be purchased for love or money, or stolen by any system of strategy." There was still an audience for new books by Garner, or so the publishers reckoned, Garner wrote. Harper and Brothers wanted a new book on the anthropoid apes, while the Century Company wanted a book on travel, elaborating some of the material to be published soon in their monthly magazine, the *Century*.[108] Hrdlička responded right away, recommending some good doctors. "I have read your larger book from cover to cover, and was envious of the part taken up by the index and blanks," he wrote. "I shall be sorry indeed to return the book when you call for it. It is the most honestly and genuinely written book I have ever had in my hands. That is all I will say." Hrdlička wished his friend in closing a happy New Year—"happy in the knowledge that you have many true friends who thoroughly appreciate both your work and the sterling spirit in which you have always imparted its results."[109]

Garner wrote back that Hrdlička's letter had cheered an ailing old man. As soon as he was well enough, Garner wrote, he was to set off for Cuba, where he was to visit the ape colony of Madame Abreu (described five years later in Robert Yerkes's book *Almost Human*). At the moment Garner was working up the article on apes for Hrdlička's journal. "I have made out a complete census, including every Chimpanzee, male and female, living or dead, that I have ever known, used, or experimented with, and I mean to make the information in this article as exact as your scrutiny would attempt; and while I am delaying it far beyond the time limit allowed me, I am doing it for the good of both of us and the welfare of the Anthropoid races, as well as the rest of mankind, but try and have faith in my promise . . . [and] believe that you will eventually get it."[110] Garner's ape census, never published, was to be the last of his contributions to science. En route by train to Florida, not far from his birthplace at Abingdon, Virginia, Garner became ill, was taken from the train, and died soon after. Obituary notices for the "monkey man" appeared in newspapers throughout the country and abroad.[111] A two-sentence notice appeared in Hrdlička's *American Journal of Physical Anthropology*, saluting Garner as "an indefatigable worker" who had "contributed greatly to our knowledge of the life and habits of the apes."[112]

Hrdlička wrote a fuller obituary for his friend ten years later, as an introduction to a small, limited-edition collection of Garner's letters about his Abingdon boyhood, *Autobiography of a Boy*. (Edison owned a copy.) While celebrating Garner's life and character, Hrdlička at the same time intimated something of the personal and professional isolation, the sense of broken-backed failure, in which Garner lived after (the reference is oblique) that first, disastrous expedition to West Africa:

Somewhere in his early dreams and visions he [Garner] had partaken of the nectar of a genius, and henceforth he was unlike other humans. He saw clearer and felt more. . . . When visiting civilization, Garner was an "innocent abroad." Unassuming, shy, unselfish—but what a boon companion to those where he felt genuine friendship and understanding. On such occasions he would open [up] and for hours hold spell-bound his solitary listener. Could only those conversations, spiced constantly from that well of his humor, [have] been recorded. Had he become suspicious, however, of any stenographer nearby, or note taking, he would have lost his originality and spontaneity; and he could not be persuaded to apply himself to writing down what he told. When he did write, however, especially in his letters to intimate friends, he showed a natural high ability which, under more favorable circumstances, would surely have led him to fame as a writer. Regrettably he had no great urge in that direction. He was a keen and patient observer. But either as a professional author or a professional scientist, he could not have been a greater joy to his friends than just a "Garner." His scepticism of religion, as of anything else, was good natured, never malicious, for he knew no malice. His humor-tinged "virulence" of it must have been connected with some chapters of his life about which he kept silence.[113]

The Boasian Extraction of Language from Nature

No famous facts about language stem from the Hrdličkian tradition in American physical anthropology. At most it offered the student of language new fossil finds with which to fill the gaps in an evolutionary story long familiar in outline—a story of enlarging brain cases and genial tubercles from chimpanzees to Neanderthals to modern humans.[114] The Boasian tradition in the cultural anthropology of the same place and period makes for a sharp contrast. It changed the way educated people generally thought about language, and in directions that made it much harder to see how the power of language had evolved. Banished was the notion, largely taken for granted up to then, that the world contained an evolved racial hierarchy, extending from the savage, near-simian bottom to the civilized top, with language advancing in tandem with culture and anatomy.[115] Gone too was the virtue that, in Garner's eyes at least, accrued to someone who took the trouble to understand simians and savages on their terms, the better to demonstrate how their languages and mentalities were fitted to their positions in a natural scale. In Boasian anthropology, the task of on-their-terms understanding and the task of shoring up the evolutionary picture were held to be mutually exclusive. Transmuting old creationist heresy into new scientific orthodoxy, Boas and his students and admirers argued that no human language was more primitive than any other; indeed, that the most technologically primitive community might well sustain a language in certain respects more refined than the languages of the civilized—a claim made memorable in a famous (if long-since garbled) teaching about Eskimo words for snow. What held for languages, moreover, held equally for cultures and anatomies: all changed independently of each other and independently of up-from-the-ape evolution. The goal became the demonstration, not of highness and lowness, but of equality, as part of a comprehensive campaign to expunge nature from accounts of human achievement—a campaign as much moral and political as it was intellectual and professional.[116]

The Case against Primitive Languages

Boas was not, of course, the first in the post-Darwinian sciences to decouple language from evolution. There are oft-noted affinities between his position and that of W. D. Whitney, for instance.[117] An even closer antecedent is Horatio Hale.[118] Beginning in the mid-1880s, as we have seen, Hale had denied that evolution had produced a linguistic hierarchy among living human races. Yes, the power of language was an evolved faculty, but, for Hale, it did not follow that different languages should fall along a complexity continuum. He believed there was a unique moment in the human past when the power of language sprang into being, more or less whole—around ten thousand years ago, he reckoned—and that all modern humans were equally its beneficiaries, such that even now, when children of whatever race, however savage, become isolated, they spontaneously recreate language in the full. Boas knew and admired Hale's work on these themes. The two men spent a couple of hours in conversation at the 1886 meeting in Buffalo where Hale gave his paleontological overview of language's emergence.[119] Two years later, now in correspondence in connection with Boas's fieldwork among the Indians of the Pacific Northwest (the BAAS had agreed to fund the work on condition that Hale supervise), Boas published a notice in *Science* about Hale's theory of the origin and development of language—"one of the best ever suggested in regard to this difficult problem," Boas later described it.[120] He went on the next year, in his first major attack on the evolutionists, to criticize claims about phonetic highness and lowness upon grounds that Hale had earlier staked out. Where the likes of Daniel Brinton held that the sounds of the languages of primitive peoples were indeterminate, unstably "alternating" between what were stably distinct sounds in evolutionarily advanced languages, Boas, echoing Hale, argued that the appearance of indeterminacy was just that, an illusion arising from the European listener's tendency to assimilate unfamiliar sounds to the sounds of European languages.[121]

The convergence of views is striking—there are even signs that Boas came, like Hale, to prefer Galton's brand of evolutionism to Darwin's—but they point less toward the elderly Hale's influence on Boas than toward the imprint of the German world that formed him.[122] Before setting foot in America in 1884, at the age of twenty-six, Boas (fig. 5.7) was already well disposed toward anthropological work in an antiracist, culturally relativist, evolutionarily indifferent mode. As a Jew growing up in Bismarckian Germany, he had felt the sting of racial prejudice at first hand. As a child of a left-liberal family, he had assimilated the egalitarian values of the thwarted revolution of 1848. As a devotee of fashionable neo-Kantian philosophy, and even more so as a trained laboratory physicist, he had developed a keen sense of the gap between the world as it really is and the world as an enculturated human mind constructs it. And as a disciple of the antiracist physiologist and physical anthropologist Rudolf Virchow, he had learned how to demolish the evidence and arguments of the race rankers.[123] If none of these attitudes exactly endeared Boas to the anthropological establishment in his adopted country, it made room for him nevertheless, providing a succession of jobs and commissions that culminated in 1896 in a dual appointment at the American Museum of Natural History and Columbia University.[124] By this time his sense of the scope of his science was as inclusive as that of Haddon, in many ways his British counterpart. In 1898, the year of the Torres Straits venture, Boas gave the BAAS a summary report on the Pacific Northwest research, which, with its expert discussions of everything from verb forms to head types to the cultural meaning of totem animals, suggested nothing so much as a one-man expeditionary team.[125]

Again like Haddon but even more so, Boas was a determined professionalizer. His arrival in America coincided with the spread of the German research ideal through America's universities. With Columbia as his base—he quit the American Museum of Natural History in 1905 over their refusal to desist from evolutionizing the displays—he set out to remake American anthropology after the image of his own, PhD student by PhD student.[126] Foundational to his reforms, as much for its conclusion as for the style of reasoning it deployed, was his case against primitive languages. The case was put most fully in his 1911 classic, *The Mind of Primitive Man*, in which he gathered together and expanded on his work of the previous quarter century. The book was above all an assault on the "racial prejudices" (the title of his first chapter) that, as Boas saw it, erroneously led civilized people to regard the noncivilized as inferior in body and mind, stuck at the lower stages of some uniform

Figure 5.7 Franz Boas (1858–1942), around 1893. Courtesy of the American Philosophical Society (APS Neg 715) and reproduced by permission.

evolutionary sequence. Again and again, Boas showed that judgments of highness and lowness in human anatomical, cultural, and linguistic traits turned on biased selection and interpretation of evidence. Take anatomy. Choose brain size as the criterion of highness, and big-brained Europeans rose to the top of the scale, farthest from the animals. But, Boas urged, there was little evidence that brain size and mental power correlated in any straightforward way. Brain size was a criterion biased in favor of Europeans. Choose another, equally valid criterion, such as hairlessness, and Europeans slid to the bottom of the scale.[127] When it came to cultural attainment or linguistic complexity, race rankings were, Boas continued, if anything more suspect, as the relevant data often came from prejudiced observers. To observe without bias, he wrote, "the student must endeavor to divest himself entirely of opinions and emotions based upon the peculiar social environment into which he is born." In Boas's view, even when they knew the native language, missionaries and traders were generally unreliable observers, as their agendas made them hostile or indifferent to much that mattered to the observed. Boas used the case of primitive language to drive home the need fully "to enter into the inner life of a people" before rushing to the race scale.[128]

For Garner, as we have seen, the apparently brutish features of primitive languages had constituted a warrant for searching out humanlike features among animal languages. On Boas's estimation, such searching had proved fruitless, the warrant bogus. Nowadays, Boas reported, "little or no diversity of opinion exists" as to the importance of "organized articulate language" as a distinction between animals and humans. "Although means of communication by sound exist in animals, and although even lower animals seem to have means of bringing about co-operation between different individuals," wrote Boas, "we do not know of any case of true articulate language [among animals] from which the student can extract abstract principles of classification of ideas."[129] Moving from animals to humans, Boas restated his case against evolutionist phonetics, and added to it arguments against evolutionist semantics.[130]

Among the allegedly low features of primitive languages was what John Wesley Powell, the Spencerian chief of the Bureau of American Ethnology in the nineteenth century, had called (following others) "holophrasis"—the use of a single word to express a multitude of different ideas. Like organic form, linguistic meaning was supposed to have evolved from homogeneous muddle to heterogeneous clarity. When primitive languages modified a single term to express what civilized languages expressed via several independently derived terms, the former were exhibiting their lowness, the latter their highness.[131] "Here," warned Boas, ". . . we are easily misled by our habit of using the classifications of our own language, and considering these, therefore, as the most natural ones, and by overlooking the principles of classification used in the languages of primitive people." In Boas's view, people classified the world according to local interests and needs. To illustrate, he introduced what would become a famous case: the existence among the Eskimo of a number of independently derived words for different forms of what English speakers referred to, crudely, as "snow."[132] "Thus it happens," concluded Boas, "that each language, from the point of view of another language, may be arbitrary in its classifications." Indeed, "every language may be holophrastic from the point of view of another language. Holophrasis can hardly be taken as a fundamental characteristic of primitive languages."[133]

A parallel point applied to the supposed absence of abstract words from primitive languages, which evolutionists had long celebrated as bearing out the truth of their theory. What primitive peoples lacked, Boas argued, were not abstract words but occasions on which such words were required or appropriate. It was not high intelligence but a peculiar culture that spurred some human beings to create abstract words:

> Primitive man, when conversing with his fellow-man, is not in the habit of discussing abstract ideas. His interests center around the occupations of his daily life; and where philosophic problems are touched upon, they appear either in relation to definite individuals or in the more or less anthropomorphic forms for religious beliefs. Discourses on qualities without connection with the object to which the qualities belong, or of activities or states disconnected from the idea of the actor or the subject being in a certain state, will hardly occur in primitive speech. Thus the [American] Indian will not speak of goodness as such, although he may very well speak of the goodness of a person. He will not speak of a state of bliss apart from the person who is in such a state. He will not refer to the power of seeing without designating an individual who has such power. Thus it happens that in languages in which the idea of possession is expressed by elements subordinated to nouns, all abstract terms appear always with possessive elements. It is, however, perfectly conceivable that an Indian trained in philosophic thought would proceed to free the underlying nominal forms from the possessive elements, and thus reach abstract forms strictly corresponding to the abstract forms of our modern languages.[134]

For Boas, different levels of cultural or linguistic achievement bespoke different histories, not different mental abilities. The concrete cast of the primitive's mind derived from the conditions of primitive culture, not from innate limitations in the architecture of the primitive's brain. To prove the point, Boas reported a linguistic "experiment" he had once conducted with an Indian companion on Vancouver Island. The language of this man's people was indeed bereft of free-floating abstract terms. To show the man could understand such terms, Boas modified the native word meaning "my pity for you" to express merely "pity"; the word meaning "his love for him" to express merely "love"; and so on. "After some discussion," Boas recalled, "I found it perfectly easy to develop the idea of the abstract term in the mind of the Indian, who stated that the word without a possessive pronoun gives good sense, although it was not used idiomatically."[135] Boas went so far as to suggest that it was civilized languages, not primitive languages, that suffered from impoverished semantic resources. "[T]he grammatical categories of Latin, and still more so those of modern English, seem crude when compared to the complexity of psychological or logical forms which primitive languages recognize, but which in our speech are disregarded entirely."[136]

The Animal-Human Divide in Boasian Anthropology

It is a commonplace of American intellectual history in the twentieth century that between roughly the First World War and the Vietnam War, nurture beat nature as the intellectual elite's explanation of choice for why humans behave as they do. These were the decades of "the triumph of culture in American social science," as Carl Degler has put it, when the anthropologists, psychologists, and sociologists stood firm against the false science of the eugenicists and (often though not always the same thing) the racists. The culturalist triumph can be exaggerated, and the reasons for it are complex, but what is undoubted is that Boas in *The Mind of Primitive Man* and related works defined a program for the study of language and culture with broad appeal for a new generation of aspirant professionals and their audiences.[137] Some of the new professionals studied with Boas himself; others felt his influence at a distance. Among the latter was Leonard Bloomfield, whose hugely successful efforts to professionalize the science of language melded Boasianism with the other great social-scientific culturalism of the day, behaviorism (about which more in the next chapter). With Boas as his model, Bloomfield in the 1920s took to the field in order to undertake on-their-terms descriptive surveys of the languages of American Indians. He thus inaugurated what became, for decades, a rite of passage for trainee linguists, none of whom would have dreamed of reporting that a tribe had been found with a remarkably lowly grammar or phonology.[138] In the wider anthropological sphere, meanwhile, Boas's students were taking over the societies, the journals, and the university departments, forging a robustly Boasian consensus about the aims and methods of the science.[139] (Those anthropologists who did not fit, left; hence, for instance, Hrdlička's founding of his physical anthropology journal.)[140] Unsurprisingly, like Boas himself, when his students turned to the question of the relationship between human language and animal communication, they stressed the huge difference between them.

A few passages from the writings of two especially distinguished Boasians, Alfred Kroeber and Edward Sapir, will serve to illustrate. Kroeber was Boas's first doctoral student at Columbia and—lest Boasianism seem too tidy—the man who, in the early years of his long tenure at Berkeley, introduced John Harrington to the study of the languages of the Californian tribes. In 1917,

Kroeber published a soon-famous paper in the *American Anthropologist*, on "the superorganic." Although the phrase came from Herbert Spencer, Kroeber argued here that biological evolution and its laws had nothing whatever to do with human cultural evolution. Human attainment had long ago ceased to be a matter of biological inheritance. Consider, Kroeber suggested, the contrast between human languages with "the so-called language of brutes." Let a cat raise a pup, wrote Kroeber, and the pup will soon be barking just like any other pup, however raised. But let a Chinese parent raise a French baby, and the baby will soon be speaking Chinese as fluently as any other baby raised by Chinese parents. Kroeber acknowledged that birds had to learn their songs at least in part, but denied that birdsong was a matter of accumulating and changing cultural traditions in anything like the human sense. As for human shrieks, moans, and other inarticulate, emotional, animal-like cries, these stood outside language proper. "To deny that something purely animal underlies human speech, is fatuous," Kroeber argued, "but it would be equally narrow to believe that because our speech springs from an animal foundation, and originated in this foundation, it therefore is nothing but animal mentality and utterances greatly magnified."[141]

Kroeber made the point still more emphatically in his enormous survey text *Anthropology*. In the 1948 edition, he offered the following on animal communication:

> Communications of a sort they [the social insects] undoubtedly have; but these, like the noises of all animals other than man, are not language, except metaphorically—something like the language of flowers, or the language of machine guns, which can also be "understood." As here used, the term "language" properly denotes a system of audible symbols able to communicate objective facts. A bird's chirp, a lion's roar, a horse's scream, a man's moan express subjective conditions; they do not convey objective information. By objective information we mean what is communicated in such statements as: "There are trees over the hill," "There is a single tree," "There are only bushes," "There were trees but are no longer," "If there are trees he may be hiding in them," "Trees can be burned," and millions of others. All postinfantile, nondefective human beings can make and constantly do make such statements, though they may make them with quite different sounds according to what speech custom they happen to follow. But no subhuman animal makes *any* such statements. All the indications are that no subhuman animal even has any impulse to utter or convey such information. It is doubtful whether it possesses any concept as generalizing or abstract as "tree," "bush," "burn." This seems to hold as essentially for dogs and apes—or for that matter for parrots—as for insects.[142]

The much more linguistically oriented Sapir propounded a similar view in his influential 1921 textbook on language. Sapir included its uniquely human status in his definition of language as "a purely human and non-instinctive method of communicating ideas, emotions, and desires by means of a system of voluntarily produced symbols."[143] In his 1933 article on language in the *Encyclopaedia of the Social Sciences*, Sapir gave a fuller account of how, as he saw it, the capacity for language had emerged as part of a more general symbolic capacity among the ancestors of modern humans, most likely as an aid in solving problems. He held out no hope of scientists today glimpsing the beginnings of speech in savages or simians, since the former spoke fully complex languages and the latter used vocalizations in a purely nonsymbolic way. But, he continued, there just might be clues in psychological studies of symbolic behavior in children and, especially, apes:

> It is probable that the origin of language is not a problem that can be solved out of the resources of linguistics alone but that it is essentially a particular case of a much wider problem of the genesis of symbolic behavior and of the specialization of such behavior in the laryngeal region, which may be presumed to have had only expressive functions to begin with. Perhaps a close study of the behavior of very young children under controlled conditions may provide some valuable hints, but it seems dangerous to reason from such experiments to the behavior of precultural man. It is more likely that the kinds of studies which are now in progress of the behavior of the higher apes will help supply some idea of the genesis of speech. . . . About all that can be said at present is that while speech as a finished organization is a distinctly human achievement, its roots probably lie in the power of the higher apes to solve specific problems by abstracting general forms or schemata from the details of given situations; that the habit of interpreting certain selected elements in a situation as signs of a desired total one gradually led in early man to a dim feeling for symbolism; and that in the long run and for reasons which can hardly be guessed at the elements of experience which were most often interpreted in a symbolic sense came to be the largely useless or supplementary vocal behavior that must have often attended significant action.[144]

In other words, what became language started not when something ancestral to modern humans instinctively used its voice to communicate with others of its kind, but when it intelligently used its mind to solve a problem whose solution depended on a grasping of the problem situation as a whole. Thus did Sapir, loyal Boasian, extract language from nature and, at the same time, the problem of language's origin from anthropology. He passed the problem instead to another freshly professionalized scientific discipline, comparative psychology. It was no accident that he described the rudimentary symbolic capacity he attributed to apes in the way he did. Just as well-trained cultural anthropologists of the mid-twentieth century took for granted a conception of their science which Boas and other, like-minded figures had to fight for, centered on a notion of culture as a nature-free realm, so the well-trained comparative psychologists of the era took for granted a certain conception of their science, centered on setting problems that animals under experimental conditions might learn to solve. Indeed, just as modern cultural anthropology in America spread outward from Columbia in the late 1890s, so too did modern comparative psychology. The latter traveled even farther, shaping the work done as far away as Manchester and Tenerife—as it happens, work with apes, to which Sapir must have been referring in his discussion. And just as the dominant conception of cultural anthropology put paid to the primate playback experiment in that discipline, so too did the dominant conception in comparative psychology, to which we turn next.

6

THE PSYCHOLOGISTS AND
ANIMAL LANGUAGE

◆ ◆ ◆

How does a monkey think? There are two ways that scientists are to-day using in order to find out—by observing monkeys and by experimenting on them. One is passive; the other active. The former and older method is loose and inconclusive; the latter new method is exact and accurate. It has the advantage, too, that it can be repeated by anyone. Such is the method of the new school of experimental comparative psychologists. It is not animal training. The animal trainer can do all sorts of wonderful things by patience and skill, but they prove little more than how patient and skillful the trainer was. What the scientists want to see is the mind which Nature herself has given the monkey.[1]

"TO-DAY" WAS 1909, the year of the Darwin centenary. In and around the sciences, the urge to reckon progress since the *Origin* became irresistible. From astronomy to anthropology—we will later encounter Boas's effort on the latter's behalf—there were inventories, judgments, calls to arms. For the general reader interested in psychology, the pickings included the journalist Arthur Reeve's "Men and Monkeys: Primates," published in *Hampton's Broadway Magazine*.[2] Beginning with a brisk survey of the findings of anatomists and paleontologists on physical continuity between men and monkeys (the connecting chain of fossils, wrote Reeve, "now lacks only minor 'missing links' to be complete as a chain of evidence"), the article dealt in the main with the question of mental continuity and the men whose experiments—as the extract above suggests—heralded an answer.[3]

Three of those profiled have since entered the psychological pantheon. From the vantage point of the present, Edward Thorndike's was the ur-achievement. Based at Columbia (where Boas taught him statistics), Thorndike had shown how to study animal learning experimentally and quantitatively, by timing animals as they figured out, over successive trials, how to open the doors of "puzzle-boxes" to get food rewards (Thorndike, Reeve commented, "approached the mind of the monkey through his stomach"). As Thorndike interpreted them, his experiments showed that nonhuman animals lack even the low level of understanding needed to learn by imitation. They got by with a minimal form of trial-and-error association learning: the "stamping in" of feelings that led to successful impulsive movements, the stamping out of feelings that led to unsuccessful ones.[4] While Thorndike is remembered much as Reeve portrays him, two other profilees, John Watson and Robert Yerkes, had their defining work still ahead of them in 1909. Not yet the founder of "behaviorism," Watson, then at Chicago, was noteworthy for monkey experiments largely confirming Thorndike's deflationary findings. The Harvard-based Yerkes too, though he went on to establish the first major primatological laboratory in the United States, was lauded for showing experimentally that mice, like monkeys, learn only by doing, not by watching, much less by reasoning.[5]

Other scientists discussed in Reeve's survey of the new science of the simian mind are now more familiar to historians than to psychologists. There was Leonard Hobhouse, an English journalist turned sociologist; Melvin Haggerty, a student of Yerkes's; and Andrew Kinnaman, a member of a lively group of experimental comparative psychologists at Clark University in Massachusetts. All three had done experiments that turned up evidence seemingly contrary to Thorndike's, suggesting that monkeys and apes did have at least rudimentary capacities for imitation and reasoning.[6] Even historians of psychology, however, would be surprised to find that the largest part of the article dealt with the life and work of one Richard Garner. A nonentity in today's historiography, Garner in 1909, though past his prime, was still more famous than Thorndike would ever be. A portrait of Garner in full jungle regalia, seated in his cage, was one of only two portraits illustrating the article (the other was of Darwin). For Reeve, Garner's work as translator of simian speech, educator of simians in human speech, and observer of simian emotional life had done most to demonstrate how similar were the human and monkey minds.[7] In Reeve's view, Garner and Thorndike marked out polar tendencies within the new experimentalist spectrum, for "Garner had a tendency, perhaps, to read too much into the actions of animals; Thorndike to reduce all animals to mere instincts and associations."[8]

A history of experimental comparative psychology that remembers Garner as well as Thorndike is a history of a science that, in its early days, was remarkably heterogeneous. It took place in the lab, but also out of the lab. It concerned instinct as the behavioral background for learning, but also as a research focus in its own right. Some of its practitioners thought it best to presume understanding-free trial and error at work behind animal actions (as Morgan's canon effectively counseled), but others did not. There were attempts to derive lessons for human psychology, but also attempts to understand how animals' natural behaviors fitted animals to far-from-human conditions of life. The species studied were as diverse as the questions asked. It was only gradually, over the next half century, that this eclectic science grew more homogeneous and in Thorndikian directions, becoming ever more closely identified with laboratory studies of learning in the maze-running, bar-pressing rat.[9] To be sure, such studies never took over completely. But they came to dominate the pages of key journals and, more to the point, the imaginations of academically trained psychologists.[10] In a famous 1949 address indicting his discipline's narrowed vision, the Yale experimental psychologist Frank Beach wondered at how his colleagues had so easily come to accept that, as he put it, "in studying the rat they are studying all or nearly all that is important in behavior." "Today," he continued, "the trend has reached a point where the average graduate student who intends to do a thesis problem with animals turns automatically to the white rat as his experimental subject; and all too often his professor is unable to suggest an alternative."[11]

Among the options that dwindled away as experimental comparative psychology stabilized into the study of associative learning in laboratory rats, and as skepticism about higher mental capacities in higher animals spread and hardened, was the primate playback experiment. This chapter tracks its extinction as a possibility from within a science that initially accommodated it comfortably. The chapter begins with a look at the introduction and spread of the puzzle-box paradigm. Wherever they went, Thorndike's puzzle boxes took with them a graphical accompaniment, the "learning curve," and an ideological accompaniment, Morgan's canon—the latter carefully insulated, as we shall see, from the threat that Garner posed in 1892–93. The puzzle box/learning curve/Morgan's canon complex did not sweep all before it, of course, and we will also consider some dissenting responses, notably Hobhouse's. The rest of the chapter tells the tale of the vanished playback by way of an examination of Yerkes's career. Yerkes strove to find a middle way between the Thorndikian and Garnerian extremes—labels that are especially apt in his case, since he knew and admired both men, and his primatological efforts, whether as researcher or patron, bore the stamp of both. Yet even for Yerkes, who did the most in his generation to advance understanding of primate vocal communication, the primate playback experiment ceased to be an option.[12]

The New Puzzle-Box Psychology

The Fortunes of Morgan's Canon

While Garner was in West Africa in 1892–93, looking for evidence of animal language and animal reason, his critic Conwy Lloyd Morgan was carrying out experiments of his own along these lines. The chick experiments Morgan reported in his 1893 paper "The Limits of Animal Intelligence" showed how mere trial-and-error intelligence, "ever on the watch for fortunate variations of activity and happy hits of motor response," could generate effects which, to the casual observer, gave the appearance of reasoned action, derived from "a clear perception of the relationships involved."[13] But, as we have seen, for all the evidence of the sufficiency of associative learning, Morgan nevertheless concluded the paper, in the middle of 1893, with that remarkable caveat. Morgan had staked his reform of comparative psychology on the twinned absences of animal language and animal reason. If gorillas and chimpanzees in the wild turned out to have languages not far removed from savage tongues, the case for that reform would collapse. With his 1893 caveat, Morgan tried to salvage his conceptual distinctions, whatever the fate of his claims about animal reason.

Not long after the publication of the 1893 paper, while Morgan was finishing work on his textbook, *An Introduction to Comparative Psychology*, he took the further step of reformulating his case for reform entirely, so that his new rule now appeared wholly unconnected to his views on animal language and animal reason. The 1894 *Introduction* brought Morgan's "canon of interpretation" (as he now called it) before a much wider audience.[14] After stating the canon, Morgan distinguished three views of evolution, linked to three interpretive "methods." According to Morgan, if evolution always adds higher mental faculties onto lower ones while holding the lower ones constant, then the right interpretive method is the "method of levels," according to which, as Morgan put it, "the dog is just like me, without my higher faculties." If evolution only increases the representation of faculties present in the same ratio in all creatures, then the right method is the "method of uniform reduction," according to which "the dog is just like me, only nowise so highly developed." But if evolution has a free hand both to add genuinely new higher faculties and to tailor the ratio of already-present lower faculties, then the right method is the "method of variation," according to which the dog may not be much like a human at all. "Of the three methods," wrote Morgan, "the method of variation is the least anthropomorphic, and therefore the most difficult." This least anthropomorphic and most difficult method of interpretation was the method promoted in Morgan's canon. The canon was needed, in sum, not because there was (as there had once been) a prima facie case for the absence of animal reason, nor because the simpler explanation was more likely to be the true explanation—Morgan noted that the hypothesis of animal reason often led to more parsimonious explanations of animal behavior—but because evolution had the power to shape animal minds without constraint.[15] Yet Morgan offered no grounds whatsoever for believing that evolution wielded such power (the power to keep animal minds from being remotely like human minds). At the close of the *Introduction*, he admitted his case for the canon had thus left him vulnerable to charges of dogmatism.[16]

Although Morgan now presented his canon without reference to animal language or reason, elsewhere in the *Introduction* he provided an account of the development of language and reason as convoluted as the new canon argument.[17] Readers learned there were three basic levels of mental development and three basic levels of linguistic development. At the bottom level were the brutes, who, mired in the world of sense experience, had little to communicate about beyond their own emotional states and the relation-fringed objects fitfully carried on the waves of consciousness. Morgan praised "Mr. R. L. Garner['s] . . . pioneer work" with the phonograph and monkeys in helping to establish that such "indicative communication" went together with mere awareness of relations. Above the brutes, on the middle level, were those vanished protohumans among whom "descriptive intercommunication" went together with the perception of relations. At the top were modern humans, among whom "explanatory intercommunication" went together with the conception of relations. Only here could reason emerge, for "[t]hat being alone is rational who is able to focus the *therefore*."[18] Once again, and just as Müller had taught, language and reason were intimately bound, and exclusively human; but now the language-reason gap yawned wide enough to swallow anything Garner might discover in the Congo. Even if the apes proved to have powers of description, and so proved to perceive relations, Morgan had made it clear that such proof would not itself count as evidence for ape reason.

Morgan continued for some time after the 1894–96 scandal to cite and defend "Max Müller's dictum that 'the one great barrier between the brute and man is language'" against the claims of Garner and others.[19] In *Animal Behaviour* (1900), Morgan argued that

> [t]he animal "word," if we like so to term it, is an isolated brick; a dozen, or even a couple of hundred such bricks do not constitute a building. Language, properly so called, is the builded structure, be it a palace or only a cottage; hen language, or monkey language, is, at best, so far as we at present have evidence, an unfashioned heap of bricks. It is just because language is the expression of a portion of a scheme of thought that it indicates in the speaker the possession of a rational soul, capable of perceiving and symbolizing the relationships of things as reflected in thought. . . . And though there is no conclusive evidence of its occurrence among animals, yet we have in them the instinctive and intelligent basis on which in due course of evolution it may securely be based.[20]

But while scandal ate away at the credibility of Garner and his claims, Morgan's vision of a reformed comparative psychology enjoyed increasing success and influence. Notwithstanding Morgan's own strong interest in animal instinct and its study, his canon, especially as taken up in the new psychology departments of American universities, helped to shift the focus away from instinct and toward trial-and-error learning as the proper object of inquiry for comparative psychology. At the same time, Morgan's canon contributed much to creating a climate in which anthropomorphic interpretations of animal actions came to be professionally suspect.

Thorndike and the Canonical Study of Animal Learning

What set Morgan's canon decisively on the path to canonization was the allegiance of Edward Lee Thorndike (fig. 6.1). Barely into his twenties, bright and ambitious, disciplined and hardworking, Thorndike had found in psychology a substitute for the abandoned Methodism of his Massachusetts upbringing and had made his way to Harvard, one of the science's centers.[21] It was there, in 1896, while studying under William James, that Thorndike first started doing experiments on instinct and learning in chicks. Neither animals nor association had been of particular interest to him up to that point, though he was surely familiar with the treatments in James's *Principles of Psychology* and Wilhelm Wundt's *Lectures on Human and Animal Psychology* (1894), used as a text in James's classroom. Wherever Wundt was admired, the quantitative, experimental approach to the human mind that he had pioneered in his Leipzig laboratory was emulated. Harvard was no exception. To do

Figure 6.1 Edward Lee Thorndike (1874–1949), date unknown. From Jonçich 1968, facing 22.

well on the Harvard course, one needed to find a topic that lent itself to testing and, preferably, timing.[22] In Morgan's animal work—perhaps brought to Thorndike's attention during Morgan's visit to Harvard early in 1896, when he stayed at James's house—Thorndike seems to have glimpsed his opportunity. True, Morgan himself was not much for counting and timing, but his study of how newborn chicks come gradually to peck at tasty things and not to peck at untasty things lent itself easily to elaboration and quantification. Even Thorndike's great invention, his puzzle boxes, can be traced back to Morgan, in particular a famous discussion, in the *Introduction to Comparative Psychology*, of how his fox terrier Tony had through trial and error learned to open a gate.[23]

Attractive for academic advancement in a Wundtian environment, animal associative learning had its downsides, however. Keeping animals was a time-consuming and often smelly business; Thorndike's animals wore out the patience of more than one host. After travels that included stops at James's basement and Thorndike's parents' house, his menagerie—which soon included cats and dogs as well as chicks—came to rest in the attic of a building in New York City, where Thorndike took up a PhD fellowship at Columbia in the fall of 1897. Freed from having to earn a living, he devoted himself fully to his research: knocking together wooden cat-and-dog-sized boxes, with doors that opened from within by various arrangements of bolts, loops, latches, strings, pulleys, and so on; timing his hungry charges' repeated efforts at opening the doors when placed in the boxes with food visible without (sometimes he left the animals to their own devices, sometimes he tried to show them first how to do it); plotting the results on "time-curves," with successive trials on the horizontal axis and time taken on the vertical one; sorting out his thoughts on his results and their bearing on the scientific picture of the animal mind.[24] As published in June 1898 in the monograph supplement series of the *Psychological Review*, under the title *Animal Intelligence: An Experimental Study of the Associative Processes in Animals*, the dissertation that emerged was equal parts experimental report, theoretical treatise, and methodological manifesto, building to an uncompromising conclusion. Animals, argued Thorndike, learn to do something new only when instinctive motor impulses happen to meet with success, so that thereafter, an association exists between the novel situation and the feeling that brings on the successful movements. That was why, as he found, animals could not learn by watching others or by being put through the motions. Much if not all of the time, they quite literally have no idea what they are doing; they just come to do it. The gradual slopes of his curves betokened, he wrote, "the wearing smooth of a path in the brain, not the decisions of a rational consciousness."[25]

Followers often go further than the followed, and Thorndike here presented himself as taking Morgan's reforms to extremes that even Morgan had flinched from.[26] Throughout, Thorndike hailed the Englishman as "the sanest writer in comparative psychology," "our best comparative psychologist," the "least offender" and most "level headed" when it came to the nonanthropomorphic reckoning of the animal mind, while singling out Morgan's own favored target, Romanes, for heavy abuse.[27] No one before Thorndike, moreover, had taken quite so seriously Morgan's injunction to prefer the least psychologically presumptuous explanation of animal action.[28] Thorndike not only applied the canon scrupulously in interpreting his experiments, but considered how far it could be used to interpret other experiments, notably in a long discussion of whether the animal mind could be shown to contain images or ideas. (Without for a moment indicating that he saw a better alternative, he conceded that, under "the law of parsimony," their existence could never be shown, even if they existed, since explanations supposing their absence would always be possible and always preferred.)[29] For Thorndike, the only trouble with Morgan was his backsliding. He allowed, for instance, for animal imitation and association of ideas.[30] Least forgivably, he held up language as the great difference separating the human from the animal mind. Not so, wrote Thorndike, for man is "no more an animal with language than an elephant [is] a cow with a proboscis." Long before language arose, as a secondary effect of the abundance of eminently associable (hence comparable, hence inference-generating and, ultimately, reason-generating) free ideas, the human mind had acquired a profoundly different character from its ideas-free animal predecessor.[31]

Recognizing homage when he saw it, Morgan was quick to respond, publishing a largely congratulatory review of Thorndike's monograph in *Nature* the next month, then spreading the word further in a number of lectures, articles, and books over the next few years.[32] He was entirely in agreement, he wrote, with Thorndike's major conclusion about the importance of trial and error as the means of animal intelligence, and with more of the subsidiary conclusions than selective quotation from Morgan's works may have suggested. What was more, he regarded Thorndike's experimental approach, together with his keeping of exact records, as without a doubt the way forward. At last, Morgan wrote in his *Nature* review, comparative psychology "is rising to the dignity of a science."[33] That was not, of course, to say that the puzzle-box experiments were beyond criticism. In common with other commentators, Morgan was concerned that starvation and confinement might have distressed the animals, leaving them unable to exercise whatever higher mental powers they might have. (More than once Morgan referred to Thorndike's "victims.")[34] Morgan also noted that since, in some of the experiments, the animals could not actually see the door mechanism in its entirety, their abilities to grasp the mechanical principle before getting the door open were not being fairly tested (fig. 6.2). Still, these were quibbles. On the whole, Morgan acknowledged, and championed, Thorndike's research as advancing his own.[35]

In the meantime, Thorndike had done something just as remarkable: he had found gainful employment as a psychologist. That job description barely existed before Thorndike went looking for a post in the spring of 1898. He

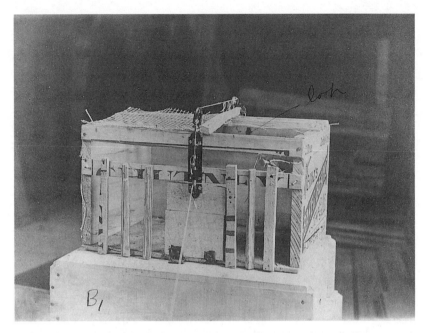

Figure 6.2 One of Thorndike's original puzzle boxes, date unknown. Courtesy of Manuscripts and Archives, Yale University Library (Yerkes Papers, series 569, box 47, folder 920).

did well, teaching educational psychology for a year at Western Reserve University in Cleveland and animal behavior courses at the marine biological wonderland of Woods Hole, on Cape Cod, before landing a plum position back at Teachers College at Columbia, where he would stay for the rest of a long and distinguished career.[36] Naturally enough, his research picked up where it had left off. In *Animal Intelligence*, he identified imitation in primates as a subject of immediate interest in light of his work thus far:

> If it is true, and there has been no disagreement about it, that the primates do imitate acts of such novelty and complexity that only [an] out-and-out kind of imitation can explain the fact, we have located one great advance in mental development. Till the primates we get practically nothing but instincts and individual acquirement through impulsive trial and error. Among the primates, we get also acquisition by imitation, one form of the increase of mental equipment by tradition. The child may learn from the parent quickly without the tiresome process of seeing for himself.[37]

Decades before, T. H. Huxley had emphasized how the ability to learn from others—an ability greatly facilitated by language—had enabled humans to advance far above other species. Thorndike's experiments seemed to confirm that, up to the primates (birds excepted), no animals learned by imitation. But *where* among the primates did imitation begin? Thorndike had in fact already attempted to find out whether monkeys imitate, but the specimen he obtained, as he reported ruefully, "failed in two months to become tame enough to be thus experimented upon." Making his apologies, he promised to try again "as soon as [I] can get subjects fit for experiments." Beginning in the autumn of 1899, now back in New York, he managed it. With, eventually, three more or less tame *Cebus* monkeys from South America to work with, he started experimenting afresh.[38]

The results were surprising. In a 1901 monograph, *The Mental Life of the Monkeys: An Experimental Study*, he argued that, contrary to received wisdom, monkeys in their mental makeup more strongly resembled cats, dogs, and chicks than humans.[39] Once again, experiment had failed to turn up convincing evidence of reasoned understanding, even at the low level required for deliberate imitation.[40] True, the curves generated in these new experiments were different from the earlier curves in a potentially crucial respect: they hardly sloped at all, with downward turns that were abrupt and complete. The monkey curves seemed, as Thorndike wrote, to depict "a process of sudden acquisition" that "rivals in suddenness the selections [of appropriate movements] made by human beings in similar performances."[41] Were these curves not evidence for simian reason, then? No, argued Thorndike, for there were plenty of other possible explanations, and, consistent with Morgan's canon, reason was the explanation of last resort. Perhaps, he suggested, the monkeys just had superior vision or superior motor coordination. Perhaps a difference in the design of the experiments mattered; whereas the chicks, cats, and dogs had needed to escape from their boxes, the monkeys had needed to break into theirs to get their food rewards. Perhaps, rapid though it was, simian learning was still too slow to be the product of reasoning.[42] He concluded that the monkey was a superior learner thanks not to rudimentary reason, but to a capacity for comparatively many, varied, complex and delicately discriminated associations of not-quite ideas, set against a background instinct for undirected action. Only at some point after and above the monkey level of mind did the development of these not-quite ideas and their associations pass a critical threshold, precipitating out true ideas and, with them, inference and reason.[43]

The Dominance of Puzzle-Box Psychology

His monkey-mind experiments marked the end of Thorndike's active involvement with animal learning and its experimental study.[44] After 1901, he turned to projects more directly applicable to human education: putting old ideas about pedagogy to experimental test, promoting the use of the new statistical methods, and so on.[45] Yet the puzzle-box precedent flourished. It set an agenda of learning, laboratories, and loathing of anthropomorphism, which shaped not just comparative psychology but American psychology as a whole. It is, in the words of one historian-psychologist, "awesome to contemplate how great a part of the history of fundamental psychology in the United States during the first half of the twentieth century can be seen to relate . . . to the conclusions that Thorndike based on the informal and crude 'experiments' he reported in 1898."[46] When, in the 1910s, John Watson began to advocate a "behaviorist" approach to psychology, restricting psychological attention to behavior and the laws of stimulus and response that determine it, he did so as a follower of Thorndike, seeking—as Thorndike himself had once done with Morgan's work—to push innovative features to brave new extremes.[47] And when, in the 1930s and 1940s, European zoologists reinvigorated the study of instinctual animal behavior, they did so, as we shall see in chapter 7, in self-conscious opposition to comparative psychology in the Thorndikian, apparatus-and-association manner.

A major vector was Margaret Floy Washburn's 1908 textbook *The Animal Mind*.[48] Going through several editions over several decades, her book taught generations of students to take Thorndike as their guide in obtaining animal-mind facts (prefer experiment to anecdotes) and Morgan as their guide in interpreting those facts (prefer lower- to higher-faculty explanations).[49] On their way into the new lab-experimentalist regime, her readers also learned that "animals have no language in which to describe their experience to us." More was the pity, she added, for "the higher vertebrates could give us much insight into their minds if only they could speak."[50] The possibility that such insight might be there for the taking through assiduous experimental work with the phonograph was no longer even raised. The experiments she described bore much more on the sensorimotor dimensions of animal mind and behavior. As for Morgan's canon, following it, she wrote, was "like tipping a boat in one direction to compensate for the fact that some one is pulling the opposite gunwale," for the "social consciousness of man is very strong, and his tendency to think of other creatures, even of inanimate nature, as sharing his own thoughts and feelings, has shown itself in his past to be almost irresistible." Washburn concluded thus: "Lloyd Morgan's canon offers the best safeguard against this natural inclination, short of abandoning all attempt to study the mental life of the lower animals."[51]

Washburn's attitudes and the popularity of the textbook enshrining them are themselves symptomatic of wider and more diffuse changes. American psychology took professional shape in the late nineteenth and early twentieth century as a service science, providing expert advice on the sorting and educating of people in a new mass society. Puzzle-box and related experiments fit broadly within this educational imperative and the emphasis on "intelligence" that came with it.[52] Just as importantly, they proved—as Thorndike himself had found—excellent for student work in psychology, as they were not too expensive, endlessly adaptable, and exciting for the students. Indeed, it was largely *as* student work that Thorndikian animal experiments and learning curves became a permanent feature of the new American psychology departments (and the phrase "learning curve" a permanent feature of everyday language).[53] The antianthropomorphism the experiments enjoined got an additional boost in the wake of reports of "Clever Hans," the Berlin horse whose celebrated feats of understanding were later revealed to be the result of nothing more than trial-and-error learning.[54] Sensitized as never before to the dangers of looking foolish over rich claims about the animal mind, psychologists made antianthropomorphism a badge of their professionalism.[55] And while questions about evolution, instinct, and heredity were never excluded from learning-focused psychological attention, they were increasingly marginalized. To Washburn's readers, "evolution" was what had brought on a previous generation's anthropomorphic excesses; "instinct," aside from simple physicochemical tropisms, what could sometimes be inhibited through experience. "Heredity" made no appearance at all.[56] As psychology over the succeeding decades became puzzle-box psychology writ large, these tendencies became more extreme and more general.[57]

Even when criticized, the Thorndike-Morgan paradigm won out, in that the most effective criticisms took the form of puzzle-box-style experiments interpreted canonically, but in support of more generous estimates of animal understanding than either Thorndike or Morgan was wont to give. Among the earliest of those provoked to absorbable criticism of this kind was Leonard Hobhouse. A Manchester-based journalist and former Oxford philosophy don, then in his midthirties, he set food-acquiring tasks to a series of animals in and around the autumn of 1900, beginning with his own cat and dog, then with an elephant, otter, rhesus monkey, and chimpanzee at the Belle Vue Zoological Gardens in Manchester. As Hobhouse made plain in the book describing these experiments, *Mind in Evolution* (1901), and in some popular articles ("Diversions of a Psychologist") published the following year, his aim was to see whether animal minds might be shown capable of at least a low level of understanding—below the conceptual kind that language-using humans enjoy, but above the feeling-impulse complexes responsible for learning by trial and error.[58] Trial-and-error explanations, he wrote, had been "applied as a sort of universal solvent by Mr. Thorndike," who had thus created the appearance of an immense gap between the minds of humans and those of all other animals. But, Hobhouse continued, there were reasons for thinking that Thorndike's conclusions were artifacts of his methods, including his failure to engage the animals' attention when testing their ability to learn by watching and his setting of tasks where accidental success was likely.[59]

To correct for poor imitative performance due to lack of attention rather than lack of understanding, Hobhouse designed experiments where food rewards went to the animal watching as well as the animal (often Hobhouse) doing; for, he reasoned, it was quite possible for an animal to perceive and act on relationships that bring it pleasure without, however, being able to perceive and act on relationships that bring pleasure to another. Yet Thorndike's experiments had discriminated only between action-with-reward learning and perception-without-reward learning, leaving out the intermediate possibility of perception-with-reward learning.[60] As for the concern that Thorndike's tasks had made stumbled-upon solutions likely, Hobhouse developed, in addition to experiments involving food shut up in boxes, experiments where the food was put out of reach yet could be obtained by appropriate but unlikely means. In one experiment, he gave the chimp a short stick, then laid out a long stick and a piece of banana at successively longer distances from the chimp's cage, in such a way that the chimp could get the banana only if it first used the short stick to rake in the long stick, then the long stick to rake in the banana. In another, the monkey could reach a too-high tasty reward—on a tabletop, suspended from a basket, and so on—only by pulling over and standing on a nearby box or chair.[61] The results of these and other, similar experiments pointed strongly, in Hobhouse's view, to a capacity in animals for what he called "practical judgment"—roughly, for inference based on perceived relations where these have a direct consequence on the animal.[62] He even suggested that, despite their disadvantageous design, some of Thorndike's own experiments lent themselves to this more generous interpretation of the animal mind. For although Thorndike had described his learning curves as gradually sloping, "a considerable number," observed Hobhouse, "seem decidedly precipitous," in line with the notion that there is in the learning process "a *critical* success, in which the animal does appear to grasp the essential nature of the thing to be done."[63]

Hobhouse's experiments were means to ends quite different from Thorndike's. Concerned to make a career for himself in the new psychology, Thorndike had set out to turn the topic of animal learning into a rigorously quantitative and experimental science. In presenting his achievement, he had wrapped it in a theory of mental evolution, but there was no particular zeal at the beginning to formulate such a theory, much less one positing and explaining radical differences between human and animal minds. For Hobhouse, however, the whole point was to defend a theory of mental evolution, along with a grand intellectual and political project predicated on that theory. He set out his credo in a letter in 1886, during his Oxford days. My hope, he wrote, is

> to ascertain the nature of the mental operations as far as possible, that, working on the Evolution theory, we may see from what & by what steps we have arrived at our present stage, & so be able to [discern] to what & by what further steps we may mount to something higher. If it has been possible for the rudimentary semi consciousness of a zoophyte to develop by successive steps [. . .] into the self-conscious man, with a fully reflective mind that can act upon itself & other minds; and if in man the sphere of mind can be seen enlarging itself as its old faculties are improved while it even acquires some that are wholly new, while it learns to work in concert with other minds, improving both them and itself in the process—then we may confidently [predict] that the growth will not cease, but—no longer the sport of surrounding forces—the mind of man will set its own improvement or development before itself as its great object, & so following yet always modifying the methods will develop itself as it was before developed by external forces into something as immeasurably superior to its present character as the mind of a poet or philosopher of today is to the intelligence of an anthropoid ape.[64]

In substance and texture, it was a Spencerian project, updated for the late Victorian and Edwardian eras, and Hobhouse pursued it in Spencerian fashion, working his way up the sequence of topics of Spencer's synthetic philosophy, from first principles to biology to psychology and on finally to sociology, where he would make his twentieth-century reputation. *Mind in Evolution* belonged to the penultimate stage: a long treatise taking the reader along the scale of life, from zoophytes to zoologists, mostly drawing on the work of others to show that mental progress was the rule, with a future of enlightened collective action the natural culmination. His animal experiments were part and parcel of this attempt to demonstrate the gradual ascent of mind as life's coordinating principle.[65] So too was his campaign, at exactly the same moment, protesting against the atrocities committed by the British in the Boer War.[66] Raising animals in the estimation of psychologists and rousing men out of the barbarisms of war were for Hobhouse expressions of a single, life-shaping, evolutionist vision.

A great deal was at stake, then, in countering Thorndike's conclusions—far more than the professional good health of the science of animal psychology.[67] Yet Hobhouse's experiments, plucked from his wider reformist project, soon entered professional consciousness as interesting variations on Morganian-Thorndikian themes. Morgan himself set the tone in *Nature* with a laudatory review of *Mind in Evolution*, congratulating Hobhouse on an experimental method that seemed "preferable to that of Dr. Thorndike, since the conditions are less cramping to the intelligence," while at the same time declaring Hobhouse's results—his interpretation of them notwithstanding—pleasingly congruent with Thorndike's.[68] In her textbook, Washburn struck a similar note of guarded interest.[69] Soon Hobhouse's experiments acquired a life of their own as models of sanctioned dissent—methods of choice for professional psychologists who wished, without abandoning their discipline, to express dissatisfaction with the trend toward low estimates of the animal mind. These were scientists content to regard artificial problems in food acquisition as windows into the animal mind. The task as they saw it was to find the *right* problems, not to seek alternatives to problem setting.[70] Among the best-known psychologists who made the Hobhousian option their own was Wolfgang Köhler, a German physicist turned Gestalt psychologist whose photographs of chimps reaching for bananas from atop stacked crates have become iconic. According to Köhler, in solving this problem, his chimps demonstrated "insight."[71] Less well remembered than Köhler, but arguably more influential in pursuing and promoting Hobhousian primatological research, was Robert Yerkes.

Yerkes, Puzzle Boxes, and Primate Psychology, 1899–1929

In *Mind in Evolution*, Hobhouse discussed his primate experiments in a separate chapter from the experiments with the other animals, as he believed the former furnished the best evidence not just for practical ideas in the minds of nonhuman animals, but for practical ideas at their most refined—what he called "articulate ideas." An articulate idea, he wrote, is one "in which comparatively distinct elements are held in a comparatively distinct relation." The higher the degree of mental articulation, he explained, the greater the scope for recombining what has been learned in novel and adaptive ways.[72] When, early in 1916, Yerkes wrote to Hobhouse to thank him for sending a copy of the new, revised edition of *Mind in Evolution*—"the most important general discussion of the subject that England has contributed," Yerkes judged—he picked out the notion of articulate ideas as especially valuable. Yerkes continued, "I have found it necessary to refer repeatedly to your work in describing the results of my studies of ideation in monkeys and apes."[73] Those studies were published later that year as *The Mental Life of Monkeys and Apes: A Study of Ideational Behavior*. The resemblance to the title of Thorndike's 1901 study was no coincidence. Yerkes had been Thorndike's first disciple, starting in 1899, the year Thorndike launched his monkey research. In using the phrases "mental life" and "ideational behavior," in studying apes, in conducting Hobhousian experiments that generated sharp-edged learning curves—and led eventually to the reopening of the question of primate language—Yerkes was harking back to the original but, as he saw it, increasingly endangered promise of his science.

Before considering how Yerkes came to apply his profession's (playbackless) toolbox to primate thought and language, it is worth underlining the prestige his investigations commanded among other academic psychologists. The most illuminating example is also the most illustrious. In his book *Thought and Language* (1934), often considered the masterpiece of Soviet psychology, the Moscow-based Lev Vygotsky drew extensively—very nearly exclusively—on the ape studies of Yerkes and Köhler to document the independence in nonhuman animals of thought from language. Determined to show that a materialist, dialectical psychology need not go the route of consciousness-indifferent behaviorism (in the Soviet context, identified most closely with the Pavlovian conditioned-reflex tradition), Vygotsky found in Yerkes and Köhler psychologists as concerned as he was, in his studies with children, to bring objective methods to bear on subjective experience and its development. As he interpreted their work, the world of objects had been shown to engage the ape's intellect but not its speech. He praised Köhler for demonstrating with particular clarity how dependent the chimpanzee is on seeing the whole of a problem at a glance, suggesting the absence in its mind of any stable representation of the problem. Here was evidence of thought in a prelinguistic mode. Most instructive from Yerkes, in Vygotsky's judgment, were the investigations of ape speech carried out in the early 1920s in collaboration with Blanche Learned. While Yerkes tried and failed to teach human words to a chimp, Learned described chimpanzee vocalizations, finding them richly varied phonetically but limited in their use to expressing emotion and influencing, in a self-interested way, the actions of other chimps. Here was evidence of language in a preintellectual mode. For Vygotsky, it was only the growing human child, inheritor of these independent evolutionary legacies, in whom thought and language came to interact, and in so doing to modify each other almost out of recognition, as they transported the child out of biology and into sociohistory.[74]

Experimental Comparative Psychology as Embattled Vocation

To the historian, aware of its immediate and enduring impact on the psychological imagination even of its critics, experimental work in the Thorndikian mode looks to have been a remarkable success. To an informed outsider at the time, such as the journalist Reeve, it looked the same. To the insider, however, the overwhelming experience was of struggle. Yerkes was the consummate insider. Thanks in no small part to his efforts, the experimental study of animal behavior emerged in the first decade of the new century as an identifiable branch of academic psychology in the United States. At Harvard, he introduced animal behavior into the curriculum, soon attracting research students of his own. He experimented and published steadily on a range of problems, from the sensory powers of jellyfish to habit formation in maze-crawling earthworms to the inheritance of habits in a "dancing" variety of mouse. He became general editor for a Macmillan textbook series on animal behavior, inaugurated with an object lesson of his own, *The Dancing Mouse* (1907), followed by Washburn's *The Animal Mind* (1908) and a collection of Thorndike's animal papers (1911). He founded and edited the first specialist research journal, the *Journal of Animal Behavior*, soon the major outlet for Thorndikian work in the new animal psychology laboratories at Chicago, Texas, and Johns Hopkins, as well as Harvard and elsewhere. But it was the vulnerability of the enterprise that preoccupied him, as he strove to win over overzealous partisans and underwhelmed critics alike.[75]

Yerkes's own introduction to comparative psychology came in 1898, around the time that Thorndike published *Animal Intelligence*, and in the place—Harvard—where Thorndike began his experiments. (Yerkes later recalled arriving soon enough after Thorndike's departure for the smell of his chicks still to linger in William James's basement.)[76] Raised on a farm in rural Pennsylvania, educated at a small liberal arts college where he discovered the excitements of laboratory research in physiology, Yerkes went to Harvard in the autumn of 1897 for what he thought would be a year of catch-up studies before beginning graduate training in biology. But the philosopher-psychologist Josiah Royce's classes unexpectedly captivated him, and it was Royce who, toward the end of the academic year, suggested that the young man consider animal psychology as a subject uniting his biological and philosophical-psychology interests.[77] Over the next year, while based in the zoological laboratory, Yerkes started a correspondence with Thorndike. Although just two years older, Thorndike from the first acted as an encouraging teacher toward a gifted student. In February 1899, writing in reply to Yerkes's initial letter, Thorndike—then based in Ohio—said how pleased he was to see someone else "awake to the importance of animal psychology," and advised Yerkes to attend the summer course Thorndike was about to give at Woods Hole.[78] "If there are more than one or two who want to work in comparative psy.," he added in a subsequent letter, "I think it will pay to spend a week or two repeating some of Morgan's experiments and some like mine to get used to methods and only after that start on new work."[79] While at Woods Hole, Yerkes did indeed begin new Morganian-Thorndikian work, on associative learning in labyrinth-escaping turtles. Published in 1901 in the *Popular Science Monthly*, but already publicized in Thorndike's writings, Yerkes's turtle experiment supported Thorndikian conclusions: the turtles learned to escape their labyrinths not through the exercise of reason, nor through any sort of inferential judgment, but, Yerkes wrote, through the preservation of instinctive impulses chancing to meet with success.[80] In 1902, his doctoral research (on habits and instinct in the frog) completed, Yerkes took up a new Harvard post as instructor in comparative psychology.[81]

No good student slavishly adopts a teacher's perspective, and Yerkes rapidly established a program of research that, especially in its focused attention to sensory discrimination in animals, went well beyond Thorndike's precedents. There was, too, little of the self-conscious radical about Yerkes, and he soon lost any sense of obligation toward Thorndike's counterintuitively bleak estimate of animal mental life.[82] Nevertheless, Yerkes's conception of comparative psychology and why it mattered owed much to his time at Woods Hole. For one thing, he saw in Thorndike's approach a degree of methodological scruple characteristic of work done by biologists and other natural scientists but not, yet, by psychologists. The latter tended, Yerkes thought, to have a philosophical appreciation for scientific method, but little by way of practical experience, with the result being poor scientific workmanship. What psychology as a whole needed, in his view, was more of what former biologists like himself were bringing animal psychology.[83] Second, for all that he wanted them to be more like biologists in their standards, Yerkes did not want to see psychologists abandon the study of consciousness—the psychological topic par excellence—for the study of behavior, and that was as true for comparative psychologists as for the rest. "Inner life of animals" was the topic of Thorndike's final lecture in the Woods Hole course that Yerkes took in 1899 (the first of its kind in the United States, according to Yerkes, who kept his notes for the rest of his life); and Thorndike in his 1898–1901 research articles had had, as we have seen, much to say about the quality of the animal's experience as it learned to open its box, and about the evolutionary process that might have produced distinctively human consciousness from animal precedents.[84] Ever after, Yerkes insisted that conclusions about animal behavior were worth forming, and worth forming well, precisely because these were the bases from which sound inferences about animal consciousness and its evolutionary development could be made.[85]

Both of these legacies from the early Thorndike—roughly speaking, the biologism and the psychologism—made for constant professional strife for Yerkes. Among some of the more traditional psychologists, the quasi-biological nature of animal behavior studies, far from recommending them, made them suspect. As Edward Titchener, the Wundtian boss of introspective psychology at Cornell, put it to Yerkes in a letter in 1907, the new animal work did not look up to psychological scratch.[86] The old guard was not impressed and was not hiring; for all that Yerkes's students extended the Thorndikian paradigm in impressive ways, none was able to make comparative psychology a next career step. If they stayed in psychology at all, they tended to become educational psychologists. Comparative psychology was no more popular with university administrators, who saw it as high maintenance—all those animals to keep and feed—and with little to show by way of social application.[87] The powers-that-were at Harvard allowed Yerkes to teach his course and run his lab, but they consistently turned him down for promotion, making it clear that a change of fortune would require a change in research direction toward more human-oriented topics.[88] To a certain extent, they got what they wanted; from 1913, Yerkes divided his days between animal behavior work at Harvard and studies of the patients at the Boston Psychiatric Hospital, where he was soon deeply involved in the practice and theory of human intelligence testing.[89] But he remained animal behavior's professional advocate, and advancement at Harvard continued to elude him.

As the wider academic world disapproved of Yerkes's biologism, his fellow students of animal behavior gave him grief for his psychologism. Was it truly possible to know about the animal mind? A "no" answer was, as we saw in chapter 2, given to dramatic effect by F. Max Müller years before. For Yerkes's generation, the skeptical challenge that mattered was not Müller's but that of a trio of German physiologists. According to Thomas Beer, Albrecht Bethe, and Jakob von Uexküll, writing in 1899, descriptions of animal behavior should make no reference to animal consciousness, since mental happenings in the animal have not been, cannot be, observed. Objectivity demanded a descriptive language purged of subjectivity in even its most subtle guises. Not "eyes," then, with all the illicit visual associations the word brings, but "photo-receptors"; not ears, but "phono-receptors"; and so on.[90] In a 1906 editorial in the *Journal of Comparative Neurology and Psychology* (of which he was coeditor), Yerkes took the Beer, Bethe, and Von Uexküll position—what he called "objectivism"—to task. Yes, he wrote, comparative psychologists should strive wherever possible for the greatest objectivity. But to deny, with Beer and company, that science could pass from truths about animal behavior to truths about animal consciousness was to misunderstand what was needed from science. The brute fact, argued Yerkes, was that all people, even objectivists, make and act upon inferences about the minds of brutes—treating some animals, for instance, as if they suffer pain. In that case, the job of science was to investigate the bases of those inferences and evaluate them, not to cast them out as beneath attention. Rightly conceived, comparative psychology used animal behavior to reckon the animal mind. What is sought, wrote Yerkes, is "the developmental history of consciousness. To say at the outset that there is no such history, or that we can never know anything about it is, to put it mildly, unscientific." In his view, the polar opposites of extreme anthropomorphism and extreme objectivism were alike to be avoided; both disguised ignorance as knowledge.[91]

The objectivism that Yerkes fought off in 1906 is today a historical footnote. The behaviorism that largely took its place is not.[92] Behaviorism had many and diverse sources, including objectivism, but it owed its name and initial fame to a 1913 article in the *Psychological Review*, "Psychology as the Behaviorist Views It," by the journal's editor, the most important American student of animal behavior in the post-Thorndike generation after Yerkes, John Watson.[93] At that time, Watson was best known for research he had completed or started at the University of Chicago in three areas: the behavior of seabirds on the Tortugas, near the Florida Keys; aspects of monkey behavior, including their powers of imitation and color vision; and, most famously, the impact of neurological development and sensory impairment on the learning abilities of box-opening or maze-running white rats.[94] Where Yerkes prized puzzle-box animal psychology as a new field of application for his biological skills, Watson—who came to biological topics in the opposite direction from Yerkes, after an education focused on psychology and philosophy—prized the opportunity to do psychological research that did not much resemble psychological research. As he rose professionally, his irritation with mainstream introspectionists intensified.[95] Now professor of psychology at Hopkins, he wrote dryly in his 1913 manifesto of not wishing, with the introspectionist colleagues who had long looked askance at animal behavior work, to obtain "such proficiency in mental gymnastics that I can immediately lay hold of a state of consciousness and say, 'this, as a whole, consists of gray sensation number 350, of such and such extent, occurring in conjunction with the sensation of cold of a certain intensity; one of pressure of a certain intensity and extent,' and so on *ad infinitum*."[96]

As far as Watson was concerned, there was no science of human consciousness to be had. Nor was there any need for the sort of uncomfortable behavioral-criteria-of-consciousness compromise that Yerkes had struck. ("One can assume either the presence or absence of consciousness anywhere in the phylogenetic scale without affecting the problems of behavior by one jot or one tittle," wrote Watson, "and without influencing in any way the mode of experimental attack upon them.") Psychology as the experimental science of consciousness, now more than fifty years old, was a bust. Psychology as the experimental science of behavior, Watson urged, though just beginning, promised objective knowledge that would also prove of practical use. No, running rats in mazes would not unlock the secrets of producing better-educated, more upstanding citizens—the introspectionists were right about that. But the laws governing stimulus and response in rats and in humans would turn out to be of the same kind, discoverable by the same broad methods. And once such laws were known, behavior could be predicted and controlled. Psychology as the behaviorist viewed it thus offered nothing less than the power to regulate the whole of evolved nature.[97]

Apes, Insight, and the Critique of Behaviorism

Watson's public swipe at Yerkes for overinterpreting should not be overinterpreted. Mutual criticism, in public and private, had figured prominently over the previous decade of what was, for all that, a warm friendship and a productive working relationship. Their extensive correspondence is full of congratulations, consolations, encouragements, news of the (often stalled) progress of research and career, and solicitations of advice.[98] They had recently completed collaborative work on new methods for testing vision in animals, summarized in a celebrated monograph.[99] They were partners on the *Journal of Animal Behavior*, with Yerkes serving as journal editor and Watson as editor of an associated monograph series.[100] Their disagreement about how psychologistic the science of animal behavior should be was deeply felt. But it had early on acquired an ironized, ritualistic element, as if the new science, for its own good and by its own logic, was bound to bring forth just this disagreement, and it had merely fallen to Watson and Yerkes to give expression to the antagonistic extremes. "I am sorry we do not agree about the vivisectional method," wrote Watson in October 1907, in reply to a letter from Yerkes giving advance notice of his mixed review of Watson's most recent rats-and-maze monograph (the rats' senses had been impaired surgically). "It would be a non progressive world however, if all of us agreed." In the same letter, Watson indicated another dimension of the developing debate between them. "I am willing," he wrote, "probably to go further than you in denying a high degree of conscious development to these animals. . . . To my mind it is not up to the behavior man to say anything about consciousness."[101] Two months later, he reported further that, the more he considered his conclusions about rats alongside his recent work with monkeys, "the more I am inclined to go back towards Thorndike's position. I have repeated Hobhous's [*sic*] work hundreds of times on various species of monkeys as tame as kittens and I cannot confirm Hobhous's position." He added that in supporting Thorndikian trial and error as against richer estimates of animal learning abilities, "I may be behind the times but what I tell you is from the depths of my heart."[102]

With Watson over the succeeding years increasingly playing the reductive, Thorndikian provocateur, culminating in his calculatedly outrageous performance in "Psychology as the Behaviorist Views It" (initially a lecture at Thorndike's Columbia), Yerkes took up the role of the Hobhousian dissenter, searching for other-than-associational mental life in animals grappling with artificial problems in food acquisition. Bound, on that precedent, to turn to the apes at some point, Yerkes's ideas about exactly what to do with them derived partly from Hobhouse, but more directly from the recent work of two of his former students. One was Melvin Haggerty, whose research on imitation learning with monkeys and apes at the Primate House at the Bronx Zoological Garden had featured in Reeve's 1909 survey article. Although Haggerty had since drifted into educational psychology, in 1913 he published a piece in *McClure's Magazine*, "Plumbing the Minds of Apes," reporting additional, Hobhousian experiments—variations on the basic task of figuring out how to use a stick with a hook on the end to get food otherwise out of reach—with the zoo's orangutans and a chimpanzee. The orangs had done especially well, one of them using the stick in the right way from the outset. More than trial-and-error learning was at work here, wrote Haggerty, who drew a Hobhousian conclusion: "If we are to speak anthropomorphically, we must say that the animal perceives relations, and that in this perception of relations we have a low order of rationality."[103] The other student providing a useful experimental model, and much else besides, was Gilbert "Dick" Hamilton. A psychiatrist and evolutionist, he had become convinced of the clinical value of experimental study of the reactions of animals, on the view (not uncommon) that mental illness in humans represented a psychological reversion to the prehuman past. In a 1907 paper, he reported the reactions of a dog when placed in a box with four floor pedals, one of which opened the cage's door, the door-opening pedal changing with each trial in an irregular but indicated way. The aim—not much realized—was to call forth powers of mind higher than trial-and-error learning.[104] Hamilton tried again a few years later with more psychiatrically instructive subjects. Having accepted a position as the full-time private psychiatrist to a California millionaire and moved to his estate outside Santa Barbara, Hamilton there established a primatological laboratory/field station, where he soon installed a room-sized version of his box, with four exit doors rather than four pedals. He published the results of experiments with monkeys, humans, and other animals in 1911.[105]

It was thanks to Hamilton, and the largesse of Hamilton's patron, that Yerkes managed to conduct his own anti-Thorndikian experiments with non-human primates. The coming of the First World War had put paid to initial plans for Yerkes to join Köhler on Tenerife, where the latter was already at work on his crate-stacking experiments at the Prussian Academy of Science's anthropoid station. (In the spring of 1914, Köhler sent Yerkes a long letter detailing his work, including a photograph of a stack-perched, banana-reaching chimp; fig. 6.3.)[106] With sabbatical leave from Harvard already arranged for the first half of 1915, Yerkes was relieved when Hamilton wrote offering use of his facilities, all expenses paid. In the event, he offered even more than that, buying an orangutan—Haggerty's star primate—from a San Francisco dealer especially for Yerkes's use, and turning over plans for the redesign of the laboratory to Yerkes. From February to April, Yerkes oversaw the construction of what was effectively a purpose-built laboratory, out among the oak trees of sunny Montecito.[107] On one side was a large room lined with the animals' cages, communicating with a central cage where box stacking and other Hobhousian tests could be run. On the other side were two rooms for other sorts of research and a storage room. And in the middle was Yerkes's own modification of Hamilton's method, the "multiple choice apparatus," though "multiple-choice room" would have been more apt. For several years Yerkes had been working with smaller-scale versions of it, first with patients at the hospital, then with pigs, crows, and other animals. From the back of the room, well away from the tested animals (who, the gentle orang Julius aside, were vicious), the experimenter operated pulleys that opened and shut the entrance and exit doors on nine large boxes in a row. The basic idea was to confront the animal with ten different selections of opened boxes—in the first trial, say, only boxes 2 through 4 open; in the second, only boxes 5 through 9; and so on—with the food reward in each trial always to be found in, say, the leftmost box in the set, or the rightmost box, or the middle box, and so on. Day after day, the boxes were cycled through the same ten permutations, until the animal began consistently to enter the correct box on the first go. If it had continued success even when presented with a "control" sequence—different sets of opened boxes, but with food located according to the same rule—then, Yerkes reasoned, the animal could be regarded as having grasped the idea of leftmost, or rightmost, or whatever.[108]

The next year, Yerkes published his results in *The Mental Life of Mon-*

Figure 6.3 One of Wolfgang Köhler's chimpanzees solving a stacking problem, around 1914. Courtesy of Manuscripts and Archives, Yale University Library (Wolfgang Köhler to R. M. Yerkes, 17 April 1914, Yerkes Papers, series 569, box 57, folder 1090).

keys and Apes. He interpreted the qualitative, Hobhousian experiments and the quantitative, Hamiltonian ones alike as suggesting there was more to the simian mind than associations of perceptions and impulses. Julius's performances were, predictably enough, the outstanding ones. In his success at stacking boxes to get bananas otherwise out of reach, achieved suddenly, permanently, and unexpectedly after over a month of steady failure (though only after being shown how to do it), the orang, in Yerkes's view, showed every sign of having got the idea, and no sign of learning through trial and error.

"No unprejudiced psychologist," wrote Yerkes, "would be likely to interpret the activities as other than imaginal or ideational."[109] Likewise, in his performance in the multiple-choice apparatus—trying out method after method and making no headway whatsoever until one day, and with no discernible run-up, he just cracked it—the orang had shown an ability that transcended trial and error. His actions had yielded a learning curve that was not curved at all, but plunged suddenly down to zero from what had been an erratically fluctuating high horizontal (fig. 6.4). Where Thorndike, confronted with apparently sudden learning, had cast about for alternative, quasi-mechanistic explanations, Yerkes saw evidence of what he, like Köhler, called "insight." Had the same learning curve been obtained with a human subject, wrote Yerkes, it "would undoubtedly be described as an ideational, and possibly even as a rational curve; for its sudden drop from near the maximum to the base line strongly suggests, if it does not actually prove, insight."[110] His experimental records and notes "force me to conclude that . . . the orang utan is capable of

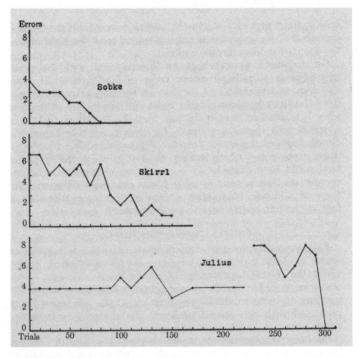

Figure 6.4 Three learning or error curves, published in 1916, and representing the performances of two rhesus monkeys and the "insightful" orangutan Julius in the multiple-choice apparatus. From Yerkes 1916a, 27.

expressing free ideas in considerable number and in using them in ways highly indicative of thought processes, possibly even of a rational order."[111]

Although Watson edited the series in which the monograph appeared, he was no fan, and told Yerkes so. "I have just gone over your book on the apes again," wrote Watson, "and I confess if I had to review it I would say some pretty mean things. You have made statements in there which are based on such flimsy and anthropomorphic evidence that for a while I seriously questioned your scientific spirit." On further reflection, Watson added, he could see that Yerkes was probably being deliberately provocative, in response to what he took to be the other side's excesses. But he had gone too far, and perhaps was already beyond the professional pale. "I have talked now with some half a dozen of the best of the younger men in comparative work," Watson reported, "and I think they will all agree with the sentiments I have expressed above. The papers on the multiple choice stand for a kind of loose type of theoretical interpretation which has been absolutely foreign to the spirit of behavior for the last fifteen years."[112]

Yerkes stood his ground:

> The nearest I have come to the sort of thing which you suggest is the volunteering of certain opinions and the honest and fearless use of a partially subjective terminology. I have used the expression *ideational behavior.* I see how you may take exception to it, how also you may misinterpret that usage. This is simply my reaction against the objective terminology. I know that I am on risky ground, and on the whole I prefer to err rather through strict adherence to the objective than to the careless use of the subjective, and that largely because of the ease and certainty of misunderstanding by non-specialists or carelessly reading specialists. Thus, the newspapers make me say that the orang utan reasons in human fashion. Of course I never have said that, and probably never shall say it.

He added that, had he reviewed "Psychology as the Behaviorist Views It" straight after publication, he would have been no more flattering. "Indeed," he continued, "from the standpoint of information, common sense, and logic, I think that article would entitle me to say almost as unkind things as those of your letter." No such review came, he explained, because he judged it better for comparative psychology that such disagreements be aired privately rather than publicly. By all means, wrote Yerkes, let us be unsparing in our criticisms of the other's work, but let us do so in a way that avoids schism in our science, with Watson's sympathizers segregated from Yerkes's sympathizers. "Whatever else happens," Yerkes closed, "we must hold together, sympathetically and intelligently, subordinating matters personal to the larger interests of scientific work."[113]

The Speech of Apes between Thorndike . . .

Yerkes and Watson did not, in fact, hold together much longer. Over the next few years, each withdrew from the comparative psychological mission, for different reasons. In Watson's case, the withdrawal proved permanent. In 1920, after well-publicized revelations of an affair with a graduate assistant from Baltimore's high society, he lost his job at Johns Hopkins, along with the prospect, for the foreseeable future, of the level of institutional support needed to run an animal laboratory. Needing to make a living, and having already begun the experimental study of behavior in human infants, he now turned his attention exclusively to human behavior, especially those aspects with earning potential. As a salesman for behaviorism, he scored notable successes with two audiences: advertisers looking for surefire ways to stimulate the buying of their products, and parents eager to raise well-behaved children. Although academic psychologists at first kept their distance, within a generation, many had thrown their lot in with behaviorism too, maze-running rats and all. Yerkes, unsurprisingly, remained unbudgingly skeptical. He also, after a brief departure, remained steadfast in his devotion to animal behavior studies, and primate behavior studies in particular. At the start of the war, he had left Harvard for Washington, DC, where he headed up the intelligence testing of recruits to the army, staying on for several years afterward to serve on the National Research Council. Occupied with his governmental work, he was simply too busy in these years to engage in animal research—and had no laboratory to return to in any case. It was only in 1923 that he started again, while on summer vacation, thanks to an unexpected offer of the sale of two chimpanzees (fig. 6.5). Although he could barely afford it, he bought them with his own money, housed them at his family getaway in rural New Hampshire, and, with help from a neighbor, Blanche Learned, picked up where his insight studies at Montecito had left off. Now, however, there was an additional line of investigation, into the speech of apes.[114]

Since the 1890s, there had been a handful of published studies on the subject. For the natural vocal communication of apes, the most notable contribution had been that of Louis Boutan, a French zoologist who, while leading a scientific mission in colonial Indochina between 1904 and 1909, had begun a study of the behavior of the gibbon, including its famously musical vocal expressions. After several years' raising a captive gibbon, Pépée, from near-infancy, first in Indochina and then back in France, Boutan in 1913 published on what he called the gibbon's "pseudo-language." According to Boutan, Pépée's natural vocalizations—recorded on his phonograph, inventoried in his paper, and sounding, he said, exactly like those of the wild gibbons she

Figure 6.5 Yerkes's chimpanzee pupils Chim and Panzee, 1923. The man holding them is probably the Harvard physiologist Walter B. Cannon, a neighbor in New Hampshire. Photograph from Yerkes 1925, facing 244.

had never heard—were mere instinctive accompaniments of feelings, not the meaningful words of a learned language.[115] For the instruction of apes in human language, there were two well-known papers: Lightner Witmer's "A Monkey with a Mind," published in 1909 in *The Psychological Clinic*; and William H. Furness III's "Observations on the Mentality of Chimpanzees and Orang-Utans," published in 1916 in the *Proceedings of the American Philosophical Society*. Both journals were based in Philadelphia, as were Witmer and Furness. A Wundt-trained psychologist at the University of Pennsylvania, Witmer had set up a clinic there in 1907 dedicated to assessing and helping children with learning difficulties. Among the first of its kind, it also sponsored a research journal, edited by Witmer. His article on a mindful monkey told of tests carried out on Peter, a chimpanzee whose stage performance in Boston a few months previously—"skating, riding a bicycle, drinking from a tumbler and eating with a fork, threading a needle, lighting and smoking a cigarette," as Witmer recalled—had impressed Witmer greatly. With the cooperation of Peter's trainers, Witmer subjected the chimp to a range of tests, including some used to tell feeble-minded human children from normal ones: stringing beads, fitting variously shaped blocks into the right places on a "form board," and so on. There was also an attempt to teach Peter, who could already (though reluctantly, and with effort) say "mama," to articulate further sounds of human speech.[116]

Among the many witnesses to these tests was Furness, identified in Witmer's article as "Dr. William H. Furness, 3d, the explorer, author of 'The Home-life of Borneo Head-hunters.'" It was Furness, Witmer wrote, who had first sparked the interest in apes that led him to Peter:

> Sometime during the past winter, in conversation with Dr. Furness, I had maintained that there was no reason why an ape who could be trained to ride a bicycle might not be taught to articulate at least a few of the elements of a language, and I expressed the desire to undertake this experiment the outcome of which, whether successful or not, I thought would be an important contribution to animal psychology. As a result of this conversation Dr. Furness brought for me from Borneo an infant orang-outang, between one and two years of age, which we had begun to train as a scientific experiment. I therefore sought to discover what had been accomplished with a closely related species, the chimpanzee.[117]

In his own article, Furness credited Borneo with giving him the idea for these inquiries as well as the ape needed to pursue them. Was it not likely, he had wondered, that an orang raised in human surroundings would rise to near the mental level of a Borneo headhunter? Humans with terrible deficiencies had acquired language, after all; surely anthropoid apes with senses and faculties intact had a shot. Besides, wrote Furness, it "seems well nigh incredible that in animals otherwise so close to us physically there should not be a rudimentary speech center in the brain which only needed development." The orang Witmer had mentioned was obtained in the winter of 1909 and lived for nearly five years. It was joined in 1911 by a chimpanzee initially acquired for Witmer, but moved out to Furness's house after the clinic setting brought on pneumonia. The chimp was still alive at time of writing.[118]

Witmer and Furness's papers belong to a single, extended, collaborative research effort, inaugurated in an anniversary year that (as we saw with

Reeve's survey) made monkeys, men, evolution, and the experimental psychology of learning an irresistible combination of topics. Peter's stage appearance had itself been part of the festivities. In publicity for the show, it was claimed that Peter had been "born a monkey and made himself a man."[119] Needless to say, Witmer reported, such talk put him on guard that he was about to watch Clever Hans in simian form, mechanically reproducing trained tricks without understanding. But as the investigations proceeded, skepticism melted away. In the end, Witmer agreed that Peter was a truly remarkable creature, a psychological "missing link" between ape and man. His endless displays of cleverness flowed not from blind habit but from an educated being's capacity for pleasure in achievement. Witmer expressed cautious optimism that Peter might even learn to talk, for it had taken no more than five minutes to teach him to articulate a new sound, p; if a languageless child had managed so much so quickly, Witmer wrote, he would predict that in six months that child would speak. But Witmer also noted some discouraging disanalogies between children and Peter. For the former, speech came naturally, thanks to a brain prepared to install it with the right stimuli. For Peter, nothing like that seemed to be the case. Would an ape raised from the first with children and associating with them throughout its life exhibit more natural aptitude? Or would the ape brain impose limits that no environmental stimulation could overcome? At the least, Witmer concluded, Peter's performance "make[s] this educational experiment well worth the expenditure of time and effort."[120] The orang and chimp residing at Furness's place, though absorbing much time and effort, were not subject to anything like the ideal experimental conditions Witmer outlined. Unsurprisingly perhaps, the results were underwhelming: the orang learned "papa" and "cup" before it died, and the chimp thus far had not managed even that much.[121]

Yerkes too, as he reported in his brief 1925 book with Learned, *Chimpanzee Intelligence and Its Vocal Expressions*, found the chimpanzee unrewarding work as a pupil in human speech. The book's title was misleading, in that, for the most part, Yerkes and Learned treated chimpanzee intelligence as having nothing whatever to do with chimpanzee vocal expression. While Yerkes watched for and tested problem-solving ability and other signs of adapting intelligence in Chim and his female partner Panzee, Learned, a musician, listened for and transcribed into musical notation the chimps' spontaneous vocalizations, noting down the circumstances eliciting them (mostly, being around food and being with others). Yerkes was the author of the first, "intelligence" half of the book, documenting Chim's box-stacking successes and other now-familiar evidences of insight; Learned of the second, "vocalization" half, listing one chimp utterance after another—thirty-two in all. Although she called them "words," even identifying a chimp "root word for food" (*gahk*) and what she called a "fruit motive," she did not take the chimps' vocalizations to constitute a language.[122] Nor did Yerkes. "Vocal reactions are frequent and varied in the young chimpanzee," he wrote, "but speech in the human sense is absent." What Learned had transcribed were emotional expressions, not ideational ones. Where chimp speech might connect with chimp intelligence, Yerkes thought, was in the chimp's capacity to imitate human speech. Chim was an irrepressible imitator of what he saw humans do. Could he be enticed to imitate what humans say? If he could, Yerkes never managed it, despite eight months' trying to stamp in an association between getting a banana and making simple sounds such as "ba, ba" or "na, na." He did not, for all that, regard the experiment as a total failure. What it revealed, he judged, was how dependent the acquisition of true speech is on an inborn tendency to vocal imitation—a tendency found in humans, but not, it was now clear, our near evolutionary kin.[123]

The absence of phonographs from these investigations into ape speech was remarked upon by at least one commentator. In his review of the book, an otherwise impressed Conwy Lloyd Morgan wrote of the speech work: "It will doubtless be followed up further as opportunities arise, with the aid of phonograph and kymograph records, which may afford the means requisite for critical evaluation."[124] The apparent indifference to using a phonograph, for documenting ape vocalizations (as Boutan had done) or, more ambitiously, assaying their meaning in playback experiments of the Garnerian kind, is striking. It was not that Yerkes was ignorant of Garner's phonographic work, as we shall see in some detail shortly. Furthermore, as noted in chapter 5, Raymond Ditmars, one of the leading curators at the Bronx Zoo, had been engaged in phonographic inquiries into monkey speech modeled on Garner's own, only two years before Yerkes acquired his chimps through the zoo.[125] To be sure, cylinder phonographs were by then old-fashioned and difficult to find, having been long crowded out of the market by nonrecording disk phonographs. But the well-connected Yerkes could have found one. His indifference has deeper roots, in his long habituation to the professional psychologist's conception of animal mind and its study. His imagination for apparatus and experiment could do nothing with the natural vocalizations of the chimpanzee because a lone chimpanzee confronting a contrived problem in banana acquisition—Yerkes's idea of an experiment—could do nothing useful with its natural vocalizations. For Yerkes, vocalizations were part of the biological background to learning, to be registered dutifully along with height, weight, arm reach, head width (measured, he wrote, using "Hrdlička head calipers") and other traits of no special psychological interest.[126] As a biologist, he recognized a gap in the literature that Learned, with her skills, might fill: hence the half of the book dedicated to vocalizations. As an experimental comparative psychologist, however, he attended solely to that aspect of Chim's vocal behavior that bore on the old Thorndikian issue of imitative learning. True to his disciplinary identity, all of Yerkes's efforts in ape speech set Chim puzzling over a banana-withholding chute or, more often, box.[127]

...and Garner

Phonographless though they were, revealing the utterances of apes to be mere emotional expressions, Yerkes and Learned's investigations nevertheless inspired press coverage that cast them as belated vindications of Garner's work. "More than a generation ago," the *Pittsburgh Gazette-Post* reminded readers,

> a remarkable man, Professor R. L. Garner, conceived the idea that the larger and more intelligent kinds of apes actually had kinds of speech which they never allowed mankind to hear, but which they used among themselves in the thickets of their native jungles. Prepared by years of study of captive apes, Professor Garner visited Africa, listened to the wild apes of the forest, made up what he thought was a list of "words" which he heard these animals use. The scientific professions proved skeptical. Garner's conclusions were attacked by high authority; he was unjustly ridiculed; finally he died, only a few years ago, his theories rejected, but himself still firm in the faith that he had discovered the beginnings of a language which the ape people knew and used. Scientists are beginning to realize now that Garner was right about many of his theories of ape actions. It is now admitted that the animals are much more intelligent than an earlier generation of naturalists gave them credit for. . . . In recent experiments by Professor Yerkes and Miss Blanche W. Learned, carried out on a pair of intelligent young chimpanzees, it was possible to identify a series of no less than thirty-two "words," which these animals used as signals for hunger, or pain, or pleasure, or for some other purpose.[128]

Yerkes had long been familiar with Garner and his claims. At the beginning of Yerkes's career, at least two of his colleagues in the Harvard philosophy department were acquainted with Garner's work, William James and Hugo Münsterberg. As we have seen, James inked a reference to Garner's first "Simian Tongue" article into the margin of his own copy of his *Principles of Psychology*. In 1903, Münsterberg shared the bill with Garner as an after-dinner speaker (Garner being the main attraction that evening).[129] Such was Garner's fame that virtually no one working in or writing about comparative psychology from the 1890s on was unaware of the man and his claims. Watson is a case in point. In March 1907, preparing for his stay in the Tortugas, he was, he wrote to Yerkes, "taking along my two tame Rhesus monkeys, a tame capuchin monk and a baboon—more as comparisons a la Garner than anything else."[130] In his 1913 manifesto for behaviorism, he mentioned in passing that back when biology was immaturely anthropocentric (as, he implied, psychology still was), expeditions were undertaken "to collect material which would establish the position that the rise of the human race was a perfectly natural phenomenon and not an act of special creation." The best known of such expeditions, if not the only such expeditions, were Garner's.[131] In a 1914 textbook on behavior and its study, Watson wrote, in a key chapter establishing that humans alone can form the vocal habits that support language and so thought, of "the misguided efforts of Garner and others to show that the primates have a language."[132] One of those whose efforts aimed to show, to the contrary, that primates do not have a language, Boutan, had devoted a long section of his 1913 paper on gibbon pseudolanguage to quoting and skeptically reinterpreting passages from Garner's *Speech of Monkeys*.[133]

Witmer and Furness can be added to this list. Although neither man acknowledged Garner's precedent as speech teacher to the apes, it is extremely unlikely that they managed to avoid knowing about it. In August 1910, one of their hometown newspapers, the *Philadelphia Evening Bulletin*, reported Garner's arrival back in the States after more than six years at Fernan Vaz. With him was a female chimpanzee, Susie, "one of the most intelligent of his pupils, which he will be glad to submit to the psychological department of the University of Pennsylvania, if the desire is expressed."[134] Garner's long, public advocacy of the use with apes of the psychological tests used with feeble-minded and deaf children made the Witmer clinic a naturally attractive destination.[135] In September 1910, shortly after he published an article in a popular magazine on his recent translational and instructional efforts with Susie and other primates, Garner presented Susie before a packed meeting of the Philadelphia Natural History Society. In print and on the platform, he once again declared his conviction that apes spoke a language suited to express their thoughts.[136] A year and a half later, Garner was again in the Philadelphia papers, now advocating, as the *Bulletin* had it, "the establishment in or near Philadelphia of an independent institution for the study of mental evolution in chimpanzees and gorillas," as "the most important contribution that the time can make to the science of child development." The city's suburbs, the reporter made clear, were already well on their way as ape education centers:

Professor Garner is at Ardmore, finishing a book that is to be a sum-up of his experience since the time when he went to live in a cage in the East [*sic*] African jungles to prove that man is not the only really, truly protoplasmic disturbance in the parade, so to speak. Now, while Professor Garner labors at Ardmore, other important things are doing not far off. Out at Wallingford, in a glass house, among the lilacs on Dr. Horace Howard Furness's place, where gaily chuckling little birds, forever dropping out of the wide skies, are the only disturbers of the peace, Dr. Lightner Witmer, of the University of Pennsylvania Psychological Department, has two educated chimpanzees that bow to him and salute him in highpitched voices as "Mamma" and "Papa."[137]

That was 1912. The following year, Yerkes too began what would become a consuming campaign—eventually successful—on behalf of a scientific station dedicated to exploring the ape mind.[138] The earliest mention of Garner in Yerkes's writings was in *The Mental Life of Monkeys and Apes*, where he referenced all of Garner's books as contributing to the previous scientific literature on ideational behavior in nonhuman primates.[139] He did not comment more substantially on Garner's work until 1925, in a popular book, *Almost Human*, about his recent trip to the primate colony on Cuba run by Madame Abreu:

> Garner . . . , a widely known student of primate speech who devoted the better part of his life to noting, recording, analyzing, and imitating the vocalizations of monkeys and apes, offers in his books much excellent evidence of the existence of vocal language or speech in the monkeys as well as in the great apes. . . . As Garner was not adequately trained for his difficult research and failed to command the scientific resources of his time, his results have not been accepted generally by scientific authorities. It is nevertheless true that many of his observations have been substantially verified, while some have been proved incorrect. Probably his enthusiasm led him to exaggerate the degree of intelligence, and the power of vocal communication, of his subjects. But the writer humbly confesses that the more he learns about the great apes and the lesser primates by direct observation as contrasted with reading, the more facts and valuable suggestions he discovers in Garner's writings.[140]

That passage—from a whole chapter on anthropoid speech—drew an appreciative letter from Garner's son, Harry, who pointed out that his father was on his way to Abreu's colony when he died, after thirty years ("a long training period!") of work on the subject. In his reply, Yerkes wrote warmly of his personal acquaintance with the senior Garner over the years, reiterating the judgment of the value and limitations of Garner's work:

> As you have discovered . . . , I have keen appreciation of your father's objectives, enthusiasm, devotion and achievements, coupled with the conviction that he was seriously handicapped by lack of the sort of scientific training and equipment

which would have made his work easier and infinitely more valuable. I knew your father through one or two meetings and many hours of conference and discussion. I do not reflect unfavorably on him when I say that his scientific training was inadequate for his task. As a matter of fact, I followed his work for many years with genuine interest and always with the hope that he might contribute increasingly important information about the language of the infrahuman primates.[141]

Four years later, however, in his classic, massive review of the scientific literature, *The Great Apes: A Study of Anthropoid Life* (1929), coauthored with his wife Ada, Yerkes was distinctly less generous:

To a surprising number of biologist as well as laymen, speech in monkeys or apes suggests the name Garner, and if [here] we fail to mention his name it might be considered an inexcusable oversight. The fact is that although Garner devoted a considerable part of his life to the intensive study of vocalization and speech in the infrahuman primates, his publications indicate serious lack of scientific competence.[142]

Yerkesian Studies of Primate Language in the 1930s

A half-million dollar grant from the Rockefeller Foundation in 1929 ensured that the Depression years were anything but for Yerkes. By the summer of 1930, the scientific station of his dreams had materialized in concrete and steel among the pines in subtropical Orange Park, Florida, about fifteen miles southwest of Jacksonville. A combination chimpanzee breeding colony and experimental facility, the station represented a magnificent return on Yerkes's recent investments of professional energies, first as a Washington science bureaucrat, cultivating contacts in the world of the big foundations, then, from 1924, as professor of psychology at Yale, where he had established a small, Rockefeller-funded chimpanzee laboratory in a disused barn near campus.[143] From the start of operations in Florida, the Yale Laboratories of Primate Biology (and, more comprehensively, of Comparative Psychobiology) supported work over the broadest range.[144] Two studies in particular, both beginning in 1931, deserve attention here, for they exemplify this diversity as it bears on our themes of the psychological imagination and animal language. Each carried forward a different aspect of the Yerkes-Learned collaboration on primate vocalization.[145] Out of the experimental, instructional efforts of Yerkes came the first in a succession of language-learning studies that, later in the century, provoked the reinvention of the primate playback experiment from within a competitor science, ethology. Out of the descriptive, translational efforts of Learned, meanwhile, came research that, as developed through the 1930s, brought the comparative-psychological tradition as close as it would ever come to serious regard for the natural vocalizations of primates and the research potential of playbacks.

On the experimental, instructional side, the torchbearer was Winthrop Kellogg. With his wife, Luella, and baby son, Donald, Kellogg arrived at Orange Park in the summer of 1931 on a year's funded sabbatical leave from the psychology department at Indiana University. His plan was to raise a baby chimpanzee as a sibling to Donald, with chimp and human treated in exactly the same ways, to see how the chimp fared when exposed to this "humanizing" (Kellogg's phrase) environment. What interested him most of all were not, in fact, humanized chimps like Witmer's Peter, but dehumanized, feral children, like the wolf-raised children of India, then much discussed. It was well known that such children tended never to catch up developmentally when returned to human care and instruction. Did the deficits persist because the children had been born deficient? Or had an impoverished environment in their early years permanently stunted what would otherwise have been normal mental growth? Bringing up a normal chimp as a human was, Kellogg reckoned, the next best thing to the impossible but ideal experiment of bringing up a normal human as a chimp. Thanks to Yerkes's support, and the loan of a baby chimp, Gua, recently imported from the Abreu colony, the experiment started in late June 1931 and ran for nine months. Had speech acquisition been central to Kellogg's concerns, he might well have taken up Yerkes's suggestion, made in light of his failures with Chim, that chimps would probably do better in human language if taught to sign it, like deaf children, rather than speak it. But little Donald was not deaf, and the Kellogg household was a speaking one. As reported in the data-heavy but instantly famous book the Kelloggs' produced afterward, *The Ape and the Child: A Study of Environmental Influence upon Early Behavior* (1933), Gua, though she learned to respond appropriately to lots of words, learned to speak not one.[146]

In the middle of the Kelloggs' remarkable year, in late 1931, another recently PhD-ed psychologist, also sponsored by Yerkes, set off for the first of three extended bouts of field studies in Panama, where the howler monkeys of Barro Colorado Island, in the Canal Zone, gave him materials for an unprecedentedly thorough analysis of natural primate vocalizations.[147] Unlike Kellogg, whose primatological effort was a one-off, Clarence Ray Carpenter came into the Yerkesian fold wanting to turn himself into a comparative psychobiologist. In a letter to Yerkes in February 1931, the Stanford-based Carpenter explained that, though his research up to that point had been on sexual behavior in pigeons (most recently, involving the surgical removal of gonads to observe the effect on mating behavior), he hoped to extend that work now by turning to the primates. Were there, he wondered, any possibilities for doing something with Yerkes at the primate laboratory next year?[148] Yerkes invented one. His outfit had already arranged two naturalistic field studies in Africa, one on the gorilla, the other on the chimpanzee. Neither had gone especially well.[149] While howler monkeys were further from Yerkes's interests than the great apes, the fact that Barro Colorado Island was already known to be a good place to work recommended it mightily. That was the assignment he gave to Carpenter.[150] The result was a benchmark monograph, published in 1934. It covered everything, from feeding habits to population size to social and sexual relations. It also included a table listing nine types of vocalization, indicating for each the "stimulating situation," "subjects," "vocal pattern," "animals responding," "response," and "function of vocalization."[151]

How did someone steeped in pigeon sexual behavior come to take such a deep interest in monkey vocal behavior? A clue lies in the subheading under which Carpenter in his monograph discussed vocalization: "group coordination and control." For Carpenter, the vocalizations of howlers functioned in the first instance not to communicate the emotional state of the vocalizers, much less facts about the observed world, but to stimulate socially appropriate action in other howlers. One kind of vocalization, for instance—a "deep, hoarse cluck" was how Carpenter described it—came from the leading male in a howler clan at the beginning and then throughout movement of the clan from one spot to another. The signal functioned, he judged, to initiate and then to direct the clan's movement. He never heard this sound except when monkeys were collectively on the move or about to be. When the clucking started, movement started. When the clucking stopped, movement stopped.[152] In taking such a view of vocal behavior, Carpenter was merely transferring to a primatological context a perspective long familiar to students of pigeon behavior, who associated it with a classic 1908 paper, "The Voices of Pigeons Regarded as a Means of Social Control," by the American zoologist and animal psychologist Wallace Craig.[153] Marginal though they were to the psychological mainstream as it swelled in the early years of the twentieth century, Craig's interests in animal instinct and the complex behavior it choreographs never died out completely. It was alive, for instance, in the classroom and books of the Torres Straits expeditionist William McDougall, under whom Carpenter began pigeon work while a master's student at Duke.[154] It could also be found, more unexpectedly, in a book that represented behaviorism's most sustained engagement with the problem of the animal origin of human language—a book, moreover, that Carpenter later acknowledged as having had a great impact on him. In *Speech: Its Function and Development* (1927), the Bryn Mawr anthropologist and psychologist Grace de Laguna argued that the way forward was to see human language through Craigian spectacles. "Speech," she wrote, "is the great medium through which human cooperation is brought about. It is the means by which the diverse activities of men are coordinated and correlated with each other for the attainment of common and reciprocal ends." It is, in other words, what pigeons do with their voices, only vastly more complex.[155]

Craig became a hero to the ethologists. It was at the hands of ethologists that the primate playback experiment, as a central feature of publicly celebrated scientific work, would come back into the world. Craig was also, however, a hero to certain psychologists. One of them was Carpenter. In 1937, the same year as Ludwig Koch's zoo playbacks in England, Carpenter conducted a few field playbacks among the gibbons in the jungles of Siam. The fact of their happening is less instructive than their circumscribed nature.[156] Here is the background. With an academic base from the mid-1930s at Columbia, Carpenter was in Siam with a group investigating the anatomy and behavior of the native gibbons.[157] For purposes of studying their vocalizations, he had brought along customized disk-recording equipment and associated paraphernalia (fig. 6.6). His monograph on the expedition presented

Figure 6.6 C. Ray Carpenter in the jungles of Siam with a disk recorder, 1937. Courtesy of the Penn State University Archives, Carpenter Papers (box 60, folder 3), and reproduced by permission.

a detailed typology of gibbon calls, schematized as the howler calls had been, though now with quantitative data on call frequency in the course of a day. Acutely interested as before in the social functions of calls, he dwelt again on the stimulus situations that elicited vocalization and the responses thus stimulated in others.[158] Again following the pattern of his earlier study, he seems to have based his functional characterizations on observations alone.[159] But he also did some field playbacks. In the 212-page monograph he produced about the expedition, there is all of one sentence about them. It is in a passage about how easy it is to stimulate gibbons to vocalize. All a gibbon needs to be set off, Carpenter wrote, is to hear another gibbon, or even "a person roughly imitating" a gibbon. Even a reproduced record of a gibbon will serve, as Carpenter discovered: "[Harold] Coolidge and I had just made our first satisfactory recordings of Type II calls produced by Group 1 and when we adjusted our recorder and played these back, the vocalizations were answered one series after the other."[160] A short article likewise contains one fanfareless sentence: "Several times it was possible to stimulate the wild animals to call by playing back to them the recordings of their own calls which had just been made."[161]

Carpenter's playbacks were so desultory in part, to be sure, because it was hard to do more, under those conditions, with that equipment. But there is also, and more interestingly, the matter of how little was at stake. Consider again De Laguna's perspective on the origin of language. As she saw it, previous thinkers had been too much inside the heads of vocalizing animals and speaking humans. Interpreting the former as expressing emotions, and the latter as communicating ideas, they had stuck themselves with the impossible task of explaining how emotional expressions had evolved into symbolic language. Her behaviorism did not so much solve the problem as dissolve it, by eliminating the mentalistic language in which it was framed. Ask, she recommended, not what vocalizations *are*, but what they *do*. What they do, up and down the scale of complexity, is coordinate social action.[162] Once one sees vocal communication that way, there is no animal-human discontinuity to be explained or—as with the primate playback experiments of the 1890s and 1970s—spectacularly disconfirmed. There is, however, a phylogenetic sequence to be described.[163] Carpenter's recording apparatus was a means to that descriptive end.[164] He was not out to prove a point. Of course, he could well imagine using playbacks experimentally, in a laboratory, to check observationally derived conclusions about what particular vocalizations do.[165] But that gibbon vocalizations served *some* social function or other he already took for granted.[166]

The related question of why, given their paltriness, Carpenter bothered with playbacks at all, is easier to answer. He did them because he could. Having brought with him a playback mechanism—there was no other way for him to hear his own recordings—he was bound to turn it loose on the gibbons eventually. If that savors of technological determinism, then so be it. "Technological determinism" is, like "social Darwinism," a common term of intellectual abuse. It is something coarse that one hopes not to be accused of. Nothing is less surprising than a historian denying that technologies or applications of them were inevitable.[167] On the whole, this book is, on this point, impeccably orthodox, mustering arguments and evidence to show how much needed to be in place for playback experiments with primates to seem worth doing, and worth finding out about, beyond the mere availability of suitable recording devices. When we consider Carpenter's playbacks, however, there is no harm in letting technological determinism do some explaining. Lug all that fancy apparatus into the jungle—and the equipment portraits in his monograph attest to Carpenter's pride in this upgrading of his field naturalist's toolkit—and, at some time or another, you will find yourself reproducing the sounds of the animals and noticing their responses. It is, in such a circumstance, an obvious thing to do. And it was never more so than in the era when Garner's work was still within living memory.[168] No doubt there were other be-phonographed ethnologists and musicologists who did the same, in otherwise quiet moments on the reservations and in the backcountry villages. When recording in the field was still novel, those who did it were naturally curious about the fidelity of their recordings. Were they good enough to "fool" live animals? Carpenter's were, and his playbacks went no further. They testified not to the quality of his ideas about primate communication, but to the quality of his gear. They were not experiments except in the most generalized sense, nor did Carpenter describe them as experiments. There is nothing extraordinary about playbacks in themselves, given the existence of recording technology capable of performing them. What is extraordinary is that the experimental playback ever became the pivot of a major scientific research effort into primate communication, and that the results were of interest to a wider community. We turn now to consider how that happened a second time.

THREE

7

MR. MARLER'S SPECTROGRAPH

◆ ◆ ◆

IN LATE MARCH 1949, two young men from London arrived on the island of Pico, in the Azores. They were botanists by training, and Pico was their field site. Over the next weeks, they hiked back and forth over the island's northern slopes, through thick evergreens, orchardlike woods, scrub, and heath. They noted the kinds and distribution of plants, took soil samples, and speculated on succession: the ordinary business of botanists with an interest in ecology. But the two also noted the kinds and distribution of birds, transcribed some songs, and speculated on the evolutionary pressures that had altered the songs and calls of Pico species compared with forms of the species elsewhere. This was work far off the botanical agenda. It was nevertheless what they had come to Pico to do, why this small island in the middle of the Atlantic had been chosen in the first place. If the botanical side of the expedition was not exactly a cover story, it was close enough, especially for one of the pair, Peter Marler (fig. 7.1). The idea for the trip was his, and so, a quarter century later, would be the idea for the primate playback experiment in its reinvented form. Between the Pico birds and the Amboseli monkeys there stretched an astonishingly productive career, dedicated to the structure, function, and evolution of animal vocalizations, and bringing together some of the most innovative trends in midcentury English biology in a singular way.[1]

Figure 7.1 Peter Marler (1928–) in California, 1958. Courtesy of Christopher Marler and reproduced by permission.

He was born in 1928, the son of a toolmaker in Slough, near London. Despite the industrial setting and the disturbances of war—the poet Betjeman's "friendly" bombs did fall on Slough, though not on the Marler home—his early life was typical of boy naturalists in that accommodating era.[2] There were talented science teachers, at a grammar school generous in sponsoring field trips. There were home chemical experiments, parentally endured. Above all, there was birdwatching, and the collecting of flowers and birds' eggs, on family walks into the countryside.[3] It was more the sight than the sound of birds, their sheer visual exuberance, that first delighted him. Marler speaks still, in his midseventies, of plumage as having "the charm of miniatures," and of his preference—not all that common among bird enthusiasts—for small rather than large birds. His appreciation for the visual found another youthful outlet in paintings, of seascapes and human figures, but also more imaginative, sub-Blakean scenes. There was never any question of a career outside the sciences, however. He found chemistry uncomfortably mathematical, though, and could not imagine earning a living from ornithology. On the advice of a teacher, Marler enrolled at University College London to study botany. The postwar agricultural boom beckoned.

London was convenient; he could go to university while continuing to live cheaply at home. It was also, in contrast with Oxford and Cambridge, a traditional destination for students from factory families like Marler's. For a scientific education, London was, in any case, second to none. At UCL, Marler attended lectures from some of the most innovative figures in biology, including the anatomist J. Z. Young and the geneticist J. B. S. Haldane. Back home, meanwhile, there were naturalists' excursions and meetings, endless rounds of them, including those of the Slough Natural History Society (which Marler helped establish). In both his formal and his informal studies, the main tasks that absorbed him were the classification of species and the gathering of data on species communities. It was empirical science at its most unapologetic. A co-authored paper from the time, reporting fieldwork conducted in the far north-western corner of Scotland in the summer of 1948, consists of around fifteen densely printed pages of short statements detailing the surroundings where a particular kind of plant was found.[4] But Marler was good at it—so much so that his UCL botany professor, W. H. Pearsall, invited him to stay on for a PhD in plant ecology. Marler began to prepare himself for research on plant succession at Esthwaite Water, in the Lake District of northern England.

The trainee professional's immersion in botany had not quite extinguished the amateur's passion for birds, however. When Marler heard about the student ornithological conferences starting to be held at the Edward Grey Institute for Field Ornithology at Oxford, he decided he had to go, attending his first one shortly after Christmas 1948. That was where he encountered David Lack, and Lack's latest ideas.[5] Lack was director of the institute and also of these student gatherings. For Marler, the event prompted a massive mid-Atlantic detour: London to Lancashire via Pico. It also set him in a new scientific direction, away from plant ecology and toward animal ecology, more precisely toward the question of how and why the songs of birds differ in different places. This was, to give it a name, the geographical ecology of birdsong.[6] Although, as we shall see, there were overlaps, it was not quite ethology, the then-new science of animal behavior that Marler in the 1950s would make his own. This chapter dwells on his bird-focused research of this period and the intellectual, institutional, and technical developments that made that research possible, taking Marler, within a few years, to the edge of the debate over animal language. En route we will see what comparative psychology's great rival, ethology, did for Marler, but also what he did for it, thanks not least to Lack.

Peter Marler's Early Studies of Chaffinch Songs and Calls

Lackian Imperatives

When Marler met him, Lack was not yet forty, but already famous for incisive, pioneering studies of bird behavior and ecology, notably a book on the robin and, more recently, *Darwin's Finches* (1947), based on an expedition to the Galápagos Islands before the war. Along with the peppered moth, introduced to the wider world by Lack's Oxford zoological colleague Bernard Kettlewell, the Galápagos finches rapidly fixed themselves in midcentury textbooks as, in Lack's phrase, "a living case study in evolution."[7] In his Darwinian view, natural selection had acted on the famously varied beaks of the finches to adapt them to different kinds of food, according to the locally prevailing conditions of competition on the islands. Species competing with closely related species in the same habitat had diverged, each becoming more uniformly specialized. Species freed from such competition, however, had become more variable and less restricted in their habitats. Liberated from the confining pressure of competition, some finch populations had even become adapted to habitats elsewhere tenanted by other species. In the absence of woodpeckers and warblers on the Galápagos, there were now, as Darwin himself had observed, woodpeckerlike and warblerlike finches.[8]

In his book on the finches, Lack elegantly set out these ideas about geography, ecology, variation, and competition. He did not relate them much to the songs and calls of the birds, concentrating instead on their anatomies and habitats.[9] In a subsequent paper, however, completed in 1948 and published the following year, differences in the vocalizations of island and mainland birds joined habitat differences as central foci. The island birds in question were not those of the Galápagos but those of Tenerife, in the Canary Islands, which Lack had visited with a colleague, the population ecologist Mick Southern, in March and April 1948. For both voices and habitats, the Lackian interpretation seemed to hold: where competition was weaker, variation was greater, and, in their phrase, "specific distinctiveness" thereby reduced. As to voices, they ventured two generalizations. First, among the Tenerife birds, "the songs of several species are shorter and simpler, though not infrequently louder and coarser, than those of their British counterparts." They named several species that had degraded vocally under the more relaxed regime of the island, including the goldcrest and the chaffinch. Second, the Tenerife birds sometimes sing or call in ways similar not to forms of the same species in Britain, but to other species, absent from the island. On Tenerife, they noted, the "Robin sings more varied phrases, often recalling those of Song Thrush or Nightingale, while the Blue Tit includes calls reminiscent of almost every species of British tit." An explanatory gloss followed: "In the absence of these other species from Tenerife, there is much less selective value in having a uniform song or call."[10]

Lack and Southern emphasized the need for further, similarly comparative studies of the land birds of other Atlantic islands, including the Azores.[11]

At the Christmas conference, Marler took the suggestion to heart. Only a few months later, he was camped above a village on Pico, inspecting its flora and fauna. He had secured funding from UCL and also a collaborator, fellow botanist Derrick Boatman. True to their professional aspirations, they conducted a thorough survey of the plant communities on the island's north slopes.[12] But, for Marler at least, the birds were the main attraction. In 1951, "Observations on the Birds of Pico, Azores," coauthored with Boatman and completed, between bouts of drilling mud cores in Esthwaite Water, in spring 1950, appeared in the premier British ornithological journal, the *Ibis*. It marks more than Marler's debut as a professional student of birdsong. It records a major shift in intellectual ambition.[13] Thanks to Lack's example, the door to theory and explanation now opened wide.[14] Far from merely recording what he had seen and heard, Marler was now reasoning about it all, in the Lackian manner:

> It would appear from the variation in song and call-notes which occur in the Canary Islands that when a bird is adjacent to closely related species, the conditions are such that its song or call-notes may not overlap with those of its neighbours. Teleologically speaking, it must not permit a situation to arise in which confusion between species, leading perhaps to hybridization, may occur: or, to put it another way, any individual strain which uses notes resulting in confusion and hybridization with another species will be eliminated by reasons of sterility, even if any progeny resulted from the cross. When related species are absent, however, these selective factors are withdrawn, and the vocabulary of the individual bird may expand. The Tenerife Blue Tit is a good example of this . . . and on Pico the Goldcrest is another.[15]

According to Marler and Boatman, several of the Pico species held broader ecological niches than usual. The goldcrest, for instance, though in Britain restricted to pine forests, on Pico could be found in roughly equal numbers over the whole of the northern slopes. Just as pronounced, they reported, had been the expansion into unoccupied acoustic niches. As among the Tenerife birds, the songs and calls of the Pico species were sometimes degraded compared with mainland forms, and sometimes reminiscent of other species altogether. Marler and Boatman exhibited the Pico chaffinch as a case in point: its song in general was simpler than that of the British chaffinch, while some of its calls were sometimes far more characteristic of the bullfinch, absent from Pico. Again following Lack's precedent, they attempted to build back from these facts about variation into the evolutionary past.[16] Since some of the calls of the Pico goldcrest (genus *Regulus*) sounded like calls of the absent titmouse (genus *Parus*), they argued—somewhat convolutedly—"it would seem to imply either that this phenomenon may occur between species not necessarily closely related, which is very unlikely, or that *Regulus*, as has been suggested in the past on morphological grounds, does in fact bear some close connection with the Paridae."[17] In other words, the light pressure of competition on Pico had apparently released the song equivalent of an atavism in the goldcrest, revealing it as sharing a common ancestor with the titmouse.[18]

Where Marler and Boatman pushed decisively beyond their model was in their quantitative treatment of chaffinch song. Lack and Southern had written merely that, compared with the British chaffinch, the Tenerife chaffinch's song was "much less" elaborate and "less musical" in tone, though "similar" in length. The younger men furnished much more. They had transcribed the twenty-five songs heard, using a quasi-musical notation to indicate the number and relative pitch of notes. Dividing each song into two parts, an opening "trill" of descending cadences and a concluding "phrase," they found, as they explained in their paper, three different kinds of trill and four different kinds of phrase among the Pico chaffinches. Most often, the phrase consisted of just one note, or was absent altogether. By contrast, the most common phrases found among chaffinches heard in Britain contained between two and six notes, of varying pitch. Thus on the remote island as against the mainland, they concluded, "there is a trend towards simplicity."[19] Their one caveat was that, in Britain itself, chaffinch song seemed so variable that all generalization about the British form was hazardous, at least until better data were available.

Marler now set himself the task of providing that data. Over the next two years, between February and June—the busy season for chaffinch breeding, hence for male singing—wherever he went, he collected chaffinch songs. In 1949, back from Pico, he transcribed in the song-filled valleys around Esthwaite Water, moonlighting from his PhD research (duly completed). The following year, he continued in the southern Scottish Highlands, while scouting out conservation sites for the Nature Conservancy, a newly founded and, in more ways than one, remarkably benevolent government agency. A break in June 1950 in northwestern France netted him chaffinch songs from the Seine Valley. Passing through Slough to see family and friends, he snuck in trips to the countryside, to fill his notebooks with the songs of the middle Thames form. Combined with the Pico data, the set swelled to include over five hundred songs from five regions, yielding, on Marler's analysis, fourteen kinds of trill and forty-five kinds of phrase. He publicized this hard-won expanded classification in his next paper, his first as sole author, "Variation in the Song of the Chaffinch *Fringilla coelebs*" (1952).[20] Its arguments form the bridge spanning his prior, ecological interests in birdsong and his emerging ethological ones. We turn now to its intellectual scaffolding.

Nature, Nurture, and Geography

There were two connected theses in this second ornithological paper, one stating a new fact, the other explaining it. According to Marler, there is no such thing as a regional variety of chaffinch song—"the" southern Scottish Highlands song or "the" north Lancashire song. An area big enough to count as a region will contain more than one song, understood as a particular combination of trill type and phrase type. What is characteristic of a region is rather a particular statistical distribution of song variants. On Pico, for instance, thirty-five percent of all trills heard belonged, in Marler's new scheme, to the second type, seventeen percent to the third type, and forty-eight percent to the seventh type. Furthermore, as a general rule, no variant was exclusive to a particular region, although the six-note phrase came close to being the signature of the Thames Valley chaffinch. But if there was greater diversity than was sometimes held, it was nevertheless, for Marler, an ordered diversity. For one thing, a comparison of theoretically possible phrases and empirically attested ones showed a regular mapping: the more phrase types that, for a given number of notes, theory indicated as possible, the more phrase types that one actually found, up to a certain point at least. He applied standard statistical tests to check that the regional distributions he detected were real, and not mere artifacts of the limited sampling of a uniform population subject to chance perturbations. Comparing his data with an older study on geographical variation in chaffinch song, based on fieldwork in Moscow and the western Urals, Marler found that, in both data sets, the same kinds of trills bulked the largest (though not the same kinds of phrase). Finally, Marler's own observations in Scotland—home of nearly half the chaffinches whose songs he had transcribed—suggested that song types, as he wrote, "are not dispersed at random but as dialects which tend to be restricted to certain topographical areas."[21]

Why should this be the pattern? What accounted for the existence and maintenance of stable dialects at the subregional level? The paper's second major argument looked to the development of song in individual chaffinches for the answer. The mosaiclike distribution of dialects arises, Marler reasoned, because the song of the chaffinch is in part inherited and in part learned. This is by no means true of all bird species. For the grasshopper warbler, for instance, the song appears to be wholly innate. For the Baltimore oriole, by contrast, the song is wholly learned. But there is a third category of song birds, including the canary, the song sparrow, and the chaffinch, which, wrote Marler, "inherit some characteristics of the song, although acquiring others by learning."[22] Strikingly, he now gave a developmental twist to the older, evolutionary interpretation of the simpler song of the Pico chaffinches: "The song may be said to be more 'primitive' than elsewhere in that it is the least removed from the innate song."[23] As the young chaffinch matures, its innate song is modified and elaborated until it sounds like the song variant that happens to prevail in that locality, whatever the juvenile's genetic similarity to or difference from its elders. The song variant acquired then endures as a local dialect, just as if it were genetically specified within that population, because, Marler continued, chaffinches tend to breed where they learn their songs. In sum, then, "the geographical variation and development of dialects in the song of the Chaffinch are phenotypic variations, arising and persisting because of the two processes of learning to vocalize from associates and of retaining a preference to breed in certain localities."[24]

Marler was no stranger to the possibility that apparently innate traits could have environmental causes. He had, after all, learned his genetics under Haldane. A larger-than-life figure, whose research in several branches of the science had attained classic status, the Weldon Professor of Biometry addressed the UCL undergraduates with matchless authority.[25] One of his oft-repeated lessons was that the interactions of nature and nurture were multiple, and there was no way of telling in advance how a particular trait had arisen. For Haldane, this was a point of logic; indeed, he introduced his taxonomy of nature-nurture interactions in a philosophical journal.[26] But there was more than logic at stake. As a man of the left throughout his scientific career, and a Marxist from the 1930s, he constantly urged vigilance against the class and race prejudices that vitiated so much of eugenics.[27] In his 1941 book *New Paths in Genetics*, for instance, he asserted that, until the understanding of nature and nurture in humans was much more advanced, proposals to promote or discourage breeding "will generally be expressions of their authors' political rather than biological opinions."[28] While far from thinking that environment was all—he defined genetics as the science dealing with "innate differences," and saw eugenics, purged of prejudicial assumptions, as an altogether rational, laudable enterprise—he always emphasized the need for alertness to the role of the environment.[29] During Marler's UCL years, this environmentalism had come to special, controversial prominence. Almost uniquely among Western geneticists, Haldane had refused to condemn the Soviet agriculturalist T. D. Lysenko, whose doctrines, including the Lamarckian inheritance of acquired characters, became official policy in the Soviet Union from 1948. Doubtful in the past about Lamarckian inheritance—often given a socially conservative spin in Britain at that time—Haldane now began to consider whether it might, after all, be induced in nonhuman organisms. His defenses of Lysenko were never robust. But he used them to emphasize again the claims of the environment on the geneticist's attention.[30]

Haldane thus imbued his better students, at least, with a skeptical attitude toward claims about the innateness of individual differences, even apparently inheritable ones. But even without his example, Marler could hardly have avoided the possibility that stable differences in environments, not genes, accounted for chaffinch song dialects. That was the view of the Soviet population geneticist and field ornithologist A. N. Promptov, whose 1930 study had provided such useful comparative data.[31] Promptov's conclusions had been highlighted in several prominent anglophone discussions of his research (published in German). In 1940, the Cambridge zoologist W. H. Thorpe, writing in a volume that Julian Huxley edited on the "new systematics," gave a succinct summary:

> A very curious example of biological differentiation in different geographical areas is provided by the work of Promptoff (1930). This worker showed that chaffinches (*Fringilla coelebs* Linn.) in southern Russia can be divided up, solely on the basis of variation in song, into well-defined populations each confined to a given area. These differences appear not to be wholly hereditary but to some extent at any rate to be handed on from parent to offspring by force of example, the young learning their song from that of their parents and of other adult individuals in the same neighbourhood. That isolation of this sort can persist is probably due to the fact that, although migration takes place, there is a strong tendency to return to the same restricted locality for the breeding season which is, of course, the song period.[32]

Two more famous works of the midcentury evolutionary synthesis also included discussions of Promptov's research: Ernst Mayr's *Systematics and the Origin of Species* (1942) and Julian Huxley's *Evolution: The Modern Synthesis* (1942).[33] More so than Thorpe, Mayr and Huxley raised the possibility that the mosaic of chaffinch song diversity might reflect genetic diversity, and could therefore have evolutionary as well as developmental dimensions. For his part, Mayr acknowledged how difficult it was to judge whether behavioral differences between geographically bounded races of a species arose from genetic differences. He also noted that from time to time, new habits and preferences did spread through populations, as successive individuals imitated a successful innovator. "Such new habits," however, he continued, "are usually lost as quickly as they are acquired, unless they add measurably to the survival

value of the species." Although Mayr placed the Russian chaffinch-song findings among those pointing toward nurture over nature in determining a bird's
songs or calls, in his concluding summary, he tipped the balance slightly the
other way: "It is therefore likely that genetic factors enter into the formation
of song races in birds, even though conditioning may play a major role."[34]
Huxley, meanwhile, suggested that nature and nurture could be working together among the chaffinch populations, with learned songs gradually acquiring a genetic basis, though a process then known as "organic selection." He
nevertheless emphasized the importance of learning for keeping song variants
stable across time and space. It was only thanks to the learned basis of birdsong, he wrote, that "the different geographical groups will tend to maintain
their song-differences in spite of a considerable amount of exchange of populations through the wanderings of young birds."[35]

At the start of his 1952 paper, and then again at the end, Marler came
out swinging against the idea that chaffinch populations with distinctive songs
were new species in the making, genetically distinct from one another.[36] If,
as we have seen, neither Promptov nor his commentators—all cited in this
connection—had quite endorsed that idea, it was nevertheless true that recent
experimental and theoretical work, also cited, had put the environmental contribution to song learning under the spotlight as never before. In Denmark,
Holger Poulsen had found that chaffinches experimentally isolated from exposure to adult song during a critical period—the early spring of their first
years—failed to develop normal song, and could not acquire it thereafter.
Notwithstanding the impoverished upbringing of these birds, however, their
songs showed the characteristic structure of a longish trill followed by a terminal phrase. Furthermore, of the two song parts, the phrase seemed more
dependent for its final form on learning—a result congruent, in Marler's view,
with his own data, showing much wider geographical variation in the phrase
than in the trill. As for the restricted period when song could be learned, and
the apparent preference of chaffinches for learning the songs of their own species, Marler paraphrased Thorpe, who, in a more recent paper commenting
on Poulsen's experiments, had suggested that song acquisition in the chaffinch
"may be an example of 'imprint' learning."[37]

"Imprint" was ethological jargon, and in Britain Thorpe was among the
most energetic converts to the new science. By the time Marler submitted the
paper on chaffinch-song variation, he was working alongside Thorpe as the
first research fellow—soon the first PhD student—in ethology at Cambridge.
Some good luck and excellent contacts in the upper levels of the Nature Conservancy brought Marler and Thorpe together. Among the agency's directors
was E. M. Nicholson, an ornithologist of considerable standing and, it hap-

pened, strong interests in birdsong and its study. Nicholson knew of the intellectual transformations in Marler. They were, indeed, hard to miss; far from hiding his slackening interest in plants, Marler had made a formal proposal to stop doing conservation work altogether, in order to begin full-time research on birdsong. What Nicholson also knew—and Marler did not—was that, at that very moment, Thorpe was settling into a new field station for ornithology in a village outside Cambridge, and that one of the station's main projects was the experimental study of song learning in the chaffinch. With the Conservancy's blessing and, crucially, their funding, Marler thus bid farewell to botany. In late spring 1951, he moved from Edinburgh to Cambridge, ready to start again as an ethologist.[38]

The Making of a New Science of Animal Behavior

Ethology and Its Founders

A maturing science comes to be known through its textbooks. In that same year, Niko Tinbergen published the first ethological primer, *The Study of Instinct*. Its frontispiece became iconic: a small fish up-ended before its own reflection in a mirror. Here was a male three-spined stickleback in the iron grip of instinct. As Tinbergen explained, when sexually active, a male on home territory assumes fighting form at the sight of another male of the species. The threat posture is one element in an automatic and undeviating pattern of behavior. Since the behavior pattern is innate, a complete inventory of such patterns—an "ethogram" for the species—serves to characterize it as surely as a description of its morphology. The first task for the ethologist is the building up of such inventories, through prolonged observation of animal species in their natural setting, or something close to it. After all the innate behaviors of a species have been reliably identified, the tasks of explanation begin. According to Tinbergen, a full explanation should span four domains. There is causation, with some causes, such as hormonal changes, being internal to the organism, and other causes, such as changes in its visual field, being external, with the two kinds of stimuli often occurring in combination (a complicated business, taking up more than half the book). There is development, including the role of the environment in the shaping of innate behavior. There is function—how an innate behavior pattern adapts an organism for survival in its typical surroundings. And there is evolution, glimpsed especially through comparisons with similar patterns in related species. At whatever point in their investigations, Tinbergen urged, ethologists should steer toward experiments, and away from the subjective life of animals.[39]

If this program can be said to have a beginning, it is 1936, the year that Tinbergen, then not quite thirty, first met the slightly older Konrad Lorenz, at a conference on instinct in Leiden. Already in correspondence, each was a fitting counterpart to the other: the quietly charismatic Dutchman an expert field naturalist, skilled in the experimental dissection of instinct; the loudly charismatic Austrian a devoted raiser of birds and a creative theoretician of instinct and its environmental molding. In retrospect, there seems little that was truly unprecedented in either man's approach. The study of instinct had been a venerable part of natural history, and the label "ethology" had attached to it fitfully since the turn of the century. Among those promoting the term was one of Lorenz's mentors, the Berlin ornithologist Oskar Heinroth, who, long before Lorenz, had drawn attention to the usefulness of comparative studies of instinct for phylogeny reconstruction, and to the phenomenon of instinct-guided learning that Lorenz, following Heinroth, called "imprinting." (The phenomenon gave ethology another iconic image: goslings walking behind Lorenz, having fixed on him, during a critical period in their development, as their mother.) Another mentor, the Hamburg zoologist Jakob von Uexküll, had introduced the key Lorenzian notions of the "umwelt," the world as the animal experiences it and acts upon it, and the "releaser," a pattern of stimuli that unlocks an adaptive, instinctive motor response. Lorenz's developing model of instinctive energies as fluids in central reservoirs echoed similar ideas in the work of William McDougall, Wallace Craig (a correspondent from the mid-1930s), and, more recently, the Berlin physiologist Erich von Holst. Tinbergen too owed much to the older generation. His field studies of bird instinct resembled those of one of his instructors at the University of Leiden, Jan Verwey, whose own heroes included Heinroth, Julian Huxley, and the great British amateurs in field ornithology, such as Edmond Selous. Tinbergen also learned from A. F. J. Portielje, director of the Amsterdam zoo. An admirer of Heinroth and Von Uexküll, Portielje had done experiments using cardboard cutouts to discern precisely what stimuli released attacking behavior in captive bitterns. In the mid-1930s, Tinbergen did exactly this sort of thing with three-spined sticklebacks.[40]

What was new in ethology, then, was neither its ideas nor its methods. What was new was the will to make of these an autonomous, flourishing branch of biology.[41] At the 1936 conference, Lorenz and Tinbergen presented a united, deliberately iconoclastic front for their kinds of investigations. Soon they were collaborating in research, with Tinbergen coming to stay at Lorenz's grand home-laboratory at Altenberg, outside Vienna, in the spring of 1937. Together they analyzed the egg-rolling behavior of hybrid greylag geese and,

in a series of elegant Portielje-esque experiments, the visual stimuli that released instinctive escape responses in geese, turkeys, and ducks. On his return
to Holland, where he had a post in the Leiden zoology department, Tinbergen
published a reinterpretation of his stickleback results, fitted out with the new,
Lorenzian vocabulary. In 1938, he traveled to the United States, promoting
the new science in public lectures. Meanwhile, his Leiden program in animal
behavior, established in the mid-1930s and involving both laboratory work
(sticklebacks) and fieldwork (herring gulls and digger wasps), continued to
thrive, attracting a number of stellar students. For his part, Lorenz had proved
less successful at finding a university base. He had, however, effectively taken
control of the editorial board of the new journal *Zeitschrift für Tierpsychologie*, turning it into the ethological house organ. He also continued to publish
innovative research, including, in 1941, a long paper reconstructing the evolutionary genealogy of one of the duck families, the Anatinae, through comparison of instinctive repertoires.[42]

The struggle to institutionalize ethology had two aspects, constructive
and destructive. It was of course important to show by example how exciting and productive was the ethological attitude, to secure a place for it in the
teaching curriculum, and to dedicate a research journal to publicizing its findings. The difficulty was that, in the 1930s and 1940s, the science of animal
mind and behavior was not up for grabs. As we have seen, almost everything
the ethologists had to offer was being done in a dispersed way within other
sciences, including zoology, physiology, and psychology (witness the name of
the ethological journal). One well-established science in particular, comparative psychology, had impeccable credentials as an objective science of animal
behavior, free of the taint of vitalism or appeals to the subjective, inner state of
the animal. It was, on balance, the ethologists who rather invited such suspicions, with their talk of entering the animal's mental world. If ethology was to
succeed, comparative psychology had to be shown up as deficient in ways that
ethology was not. Soon there was a stock indictment. Psychologists, it was
said, brought to the study of animal behavior a distorting interest in understanding human behavior. They concentrated so singlemindedly on learning
in animals not because learning was thought so much more important than
instinct in the animals' natural lives, but because learning obviously mattered
a great deal to human social life. With animals thus conceived as stand-ins for
human learners, experimentalists contrived situations in which only the most
generic facts about learning might be revealed—and ended up concluding that
the only behavioral facts worth knowing were generic ones about learning. By
contrast, ethologists made animal behavior an object of study in its own right.

Ethologists asked about learning, but also about instinct, and how, separately and in combination, these served the animals, as members of their species and in their natural settings. Indeed, this interest in the animals in themselves had led ethologists to create a truly comparative psychology, based, like comparative anatomy, on the study of groups of closely related species throughout the tree of life, rather than a small number of species scattered in the rough vicinity of humankind.

At the heart of the new science's sense of itself there was thus a strong oppositional stance. To be for ethology was to be against comparative psychology, or at least against the human-oriented perspective that, according to the ethologists, kept psychologists from investigating the behaviors that mattered most to the animals.[43] Tinbergen and Lorenz put the point vividly, as did Thorpe and—most important for our purposes—Marler. The locus classicus is Tinbergen's 1942 paper "An Objectivistic Study of the Innate Behaviour of Animals," in many ways a precursor of his textbook a decade later. Tinbergen arraigned the behaviorists on three counts of anthropocentrism. Although declaring their psychology "comparative," they looked almost exclusively at mammals, and in the main at a small number of highly domesticated species, in stark contrast with the ethologists (as Tinbergen persuasively demonstrated by comparing the 1939 volumes of the *Journal of Comparative Psychology* and the *Zeitschrift für Tierpsychologie*). Moreover, what the behaviorists really wanted to find in the few species they examined was evidence of, in Tinbergen's phrase, "the prehuman in the animals," which was why they ignored innate behavior, under the mistaken impression that it was too lowly to throw much light on the human mind. But from the ethological point of view, he continued, that was precisely backward. The goal of understanding the human mind was better pursued first "by studying the animals for their own sake, and after that, tracing the animal in Man." Finally, the behaviorists had become so entranced by their experimental apparatus that their boxes and mazes had themselves become the objects of inquiry, rather than a means for illuminating important problems about nature. "Comparative Ethology," he concluded, "consciously returns to a sounder mode of approach, in which the problem again is primary, the method secondary."[44]

Speaking at Cambridge in 1949, on "The Comparative Method in Studying Innate Behavior Patterns"—published the following year—Lorenz too lambasted the behaviorists. Along with the Pavlovian mechanists of old, he complained, they "conducted only such experiments as were beforehand destined to confirm the theory" of animal behavior as environmentally determined:

This is about the worst fault a working hypothesis can have. With exceedingly few exceptions, the experiments of mechanists confined themselves to letting some sort of stimulation impinge upon the organism and then to record its answering reaction to this stimulation. This kind of experiment could not but create and confirm the opinion that the function of the central nervous system was restricted to receiving and reflecting external stimuli. No mechanist ever thought it worth while to observe what healthy animals do when left to themselves. So the central nervous system, poor thing, never got the opportunity to show that it could do more than answer to stimulation.[45]

Lorenz also reiterated Tinbergen's jibe at the patchy comparative base of the psychologists' studies, so useless for phylogenetic purposes: "I must confess that I strongly resent it . . . when an American journal masquerades under the title of 'comparative' psychology, although, to the best of my knowledge, no really comparative paper ever has been published in it."[46]

Thorpe and the Postwar Reconstitution of Ethology

These two statements, from 1942 and 1949, respectively mark the end of the first era of ethological enthusiasm and the beginning of a second one. In between came the worst of the war and its aftermath. Lorenz and Tinbergen both spent years in prison camps, Tinbergen under the Nazis, Lorenz under the Russians. Their science lost more than just momentum in the war years. While Tinbergen had opposed the Nazis, Lorenz had been one. Indeed, Lorenz had thrived under Hitler, winning prestige and posts in part by showing how well ethology supported Nazi policies on racial purity. His efforts turned on a key doctrine, in place well before the *Anschluss*: the uniformity of wild instincts. According to Lorenz, unlike learned behaviors, which are hugely variable, instinctual behaviors are uniform, and never more so than in wild animals. The claim did triple duty, justifying the ethologist's preference for observing wild animals under natural conditions, furnishing a criterion for telling instinctive actions from learned ones, and underwriting the treatment of instinct as a species-typical character, useful to the systematist.[47] But from 1938, Lorenz added a racialist twist. He began to argue that healthy organisms harbor instincts for mating only with healthy individuals of their own kind. Hybridization was thus a symptom and source of biological decay. Another source was domestication, which shielded the weaker animals from the full force of natural selection, and even induced damaging mutations. In Lorenz's view, as "innate releasing mechanisms" erode, an increasing range of stimuli trigger instinctive behavior patterns, including reproductive ones.

The animal thus loses its natural affinity for what is best for its kind. He drew an explicit parallel with humans, contrasting—in common with Nazi propagandists—vigorous, decent country folk with morally and biologically degenerate urbanites.[48]

The Nazification of ethology could well have left it, like eugenics—another science of biological innateness—too ideologically suspect for resuscitation. In the postwar United States, such was indeed ethology's fate.[49] But in Britain, the strength of attraction was great enough to overwhelm any reservations. The attraction had several dimensions. In the scientific study of animal behavior, the British had never embraced American, maze-running behaviorism, but neither had a coherent, widely adopted alternative developed out of the native contributions of Darwin, Romanes, Morgan, and Hobhouse.[50] The ethologists offered just such an alternative, even acknowledging the British field ornithologists of an earlier generation among the science's forebears. At Oxford, furthermore, the now-strong tradition in animal ecology and population biology had precipitated out a new Department of Zoological Field Studies, which included Lack's Edward Grey Institute. Tinbergen was a natural for such a department, and, several years after the war, gladly accepted the offer of a post there (and thus the chance to leave behind smashed, starving, embittered Europe).[51] And at Cambridge, there was W. H. Thorpe, swiftly establishing himself as a patron par excellence. When Tinbergen, still in Leiden, had mooted the possibility of a new ethological journal in 1946, Thorpe signed on as a coeditor, and also suggested its eventual title, *Behaviour*.[52] The two soon started planning a conference to introduce ethology to the anglophone scientific world. On learning of Lorenz's return—he had been presumed dead—Thorpe went on a pilgrimage to Altenberg, arranged for some research funding courtesy of the English man of letters J. B. Priestley (who had royalties tied up in Austria), and invited Lorenz to the conference, to be held in Cambridge in July 1949. It was here that the reunion of Lorenz and Tinbergen took place, at Thorpe's house. When Thorpe inaugurated the new Ornithological Field Station at Madingley the following year, he had Lorenz—still jobless—in mind to be its curator.[53]

All of this extraordinary activity on ethology's behalf occurred when Thorpe was in midcareer as an agricultural entomologist. He had come to share the ethologists' concerns with instinct, learning, and evolution—and to discover Lorenz's impressive work on imprinting—in the course of research into insect parasites, and whether these could be conditioned to parasitize other host organisms besides their usual hosts.[54] It was a question with large practical and theoretical consequences. One application in prospect was the efficient control of agricultural pests, and during the war, Thorpe, a soon-to-be Quaker and conscientious objector, made his contribution through research efforts along these lines. But it was the theoretical side that attracted him most, especially the possibility that host preferences could be passed from one generation of parasites to the next—could even become defining features of geographically distinct groups—without those preferences being genetically determined. Early in his career, he had suggested that evidence for the stability of new preferences could best be explained as instances of Lamarckian inheritance. But throughout the 1930s, he became increasingly doubtful. Other, non-Lamarckian explanations seemed to work equally well. In the same 1940 paper in which he summarized Promptov's research on geographical variation in chaffinch song, as an illustration of how populations can be biologically differentiated through mechanisms invisible to traditional, anatomy-fixated taxonomists, Thorpe described his findings on host conditioning in a moth parasite. The parasite concerned had an instinctive preference for meal-moth larvae. But if reared not on the usual meal moth but on a wax moth, the parasite was later attracted to wax moths, on which its own offspring would then be reared, acquiring the same noninstinctive preference, and so on. Such conditioning traditions, wrote Thorpe, "provide a non-hereditary ecological barrier which may serve as the first stage in evolutionary divergence, tending to aid the establishment of a new variety in exactly the same way as do geographical barriers."[55]

This was imprinting in all but name, and indeed, within a few years, having read Lorenz, Thorpe was referring to "locality imprinting."[56] Soon he adopted more than just the ethologists' vocabulary. He too laid in to the shortcomings of the psychologists' experiments. In the 1951 paper from which Marler plucked the word "imprinting," Thorpe repeated the now standard accusations: "It is not . . . always realized how artificial and isolated the classical conditioned reflex is and how completely passive and otherwise unresponsive the animal must be before it can be demonstrated"; "When using puzzle boxes and mazes it is extremely easy to set a problem which, while seemingly simple to the experimenter, is in fact far too difficult for the animal"; and so on.[57] The title of the paper was "The Learning Abilities of Birds"—Thorpe had also switched from insects to the ethologists' favorite organism. The change was not completely unprepared, for Thorpe was an amateur birdwatcher of long-standing, with occasional publications on birds to his credit. Almost immediately after the war, he had begun campaigning for an ornithological field site. By 1950, he had it: four fence-enclosed acres of fields and woods in Madingley, with a small building to house aviaries, and another just off site to serve as an office. Mindful, no doubt, of Promptov's work, and latterly of Poulsen's, Thorpe latched onto song learning in the chaffinch—a denizen of

Madingley woods—as an especially promising focus for a study of imprinting.[58] Lorenz in the end stayed in Germany, landing a prize job in Germany at a Max Planck station in Buldern. So the position, and with it responsibility for day-to-day management of the site, went to Robert Hinde, a young field ornithologist as devoted to Tinbergen's vision of ethology as Thorpe was to Lorenz's.[59]

Hinde, Marler, and the Madingley Field Station

Thorpe and Hinde had first met a few years earlier at Cambridge, where Hinde, fresh from war service in the air force, was studying the natural sciences, including zoology.[60] A keen amateur ornithologist, Hinde was one of the first generation of Cambridge undergraduates to receive their first sustained exposure to the new science of behavior from Thorpe's lectures. Hinde in turn came to Thorpe's attention—and the attention of other ornithologists throughout the country—with the discovery, near Cambridge, of a number of warblers belonging to a species never before seen in Britain. Thorpe came out to have a look. One of the few ornithologists of note who did not come out was David Lack, who was nevertheless impressed enough to offer Hinde a position as the first postgraduate student at the Edward Grey Institute. In 1948, Hinde left Cambridge for Oxford, intent on carrying out a behavioral study of birds. Lack's *Life of the Robin* had of course been splendidly attentive to behavior, with discussions of Lorenz and Tinbergen's ideas and even a report of an experiment very much in the ethological spirit, involving a stuffed robin. Observing how one robin continued to attack the specimen after its head had been ripped off, Lack attempted "to see just how much of a stuffed robin was needed for a wild robin to treat it as an intruder and attack it":

> So the headless specimen was further reduced. First its tail was removed. It was still attacked. Its legs had already been replaced by wires to facilitate fastening it to suitable perches, and as a further step its wings were detached. Many individual robins still attacked it. Finally the whole of the body and back were removed, so that the specimen was reduced simply to the red breast feathers with some white feathers below, these being stitched on to a supporting wire. Half the specimens to which this bundle of breast feathers was shown displayed at it with typical threat posturing.[61]

The lesson Lack drew was no less ethological:

> We tend to assume that the world that a robin sees is much like the world which we see. Suitable experiments show how false this impression is. A headless, wing-

less, tailless, legless and bodiless bundle of red breast feathers appears as a rival to be attacked. . . . The world of a robin is so strange and remote from our experience that into it we can scarcely penetrate, except to see dimly how different it must be from our own. Yet it is a world well adapted to everyday life, and leads under normal conditions to actions which appear rational and which therefore deceive us into assuming that the mind which inspires them is not unlike the human mind.[62]

But, as Hinde soon discovered, Lack had changed direction. With questions about speciation and behavior increasingly well catered for, he had decided that the resources of the Grey Institute should be concentrated on comparatively neglected questions about populations. "The object of this Institute," he declared, "is to find out why birds are as numerous as they are."[63] As the institute's pioneering student, Hinde thus found himself encouraged to start a PhD not on instinctive displays, but on feeding ecologies. He declined, and in the end Lack agreed to let him do a field study of behavior in the institute's house birds, the tits. Although this initial conflict in no way ruptured the relationship—Hinde later credited Lack with teaching him much, not least about the ecological viewpoint—Hinde's allegiance was to Lack's past, not his present. The 1949 Cambridge conference was thus all the more important for Hinde, who got to meet Lorenz and Tinbergen, and to renew his relationship with Thorpe. When Tinbergen arrived in Oxford later that year to take up his new post, Hinde effectively adopted him.

By the time Hinde started at the field station in Madingley, he had thus become a loyal disciple of ethology, and Tinbergen's version of it especially. Marler's case was interestingly different. To be sure, Marler had absorbed aspects of ethology, reading about Lorenz, reading him directly, and even hearing Tinbergen give an inspiring lecture at the student ornithology conference where Marler presented his Azores findings.[64] But where Hinde had come to ethology more or less in spite of Lack, Marler had followed a Lackian route into the new science. This singular path into ethology accounts, as we shall see, for what is most distinctive about his first ethological contributions.

Those were not, it turned out, on birdsong and its development. Shortly after arriving in Cambridge, in late spring of 1951, Marler discovered, to his chagrin, that Thorpe regarded birdsong as his exclusive research subject. He wanted Marler to assist with experiments, but not to conduct his own.[65] The intellectual territoriality did not end at the field station's boundaries. Thorpe had earlier sought, and received, Lack's assurances that Oxford would steer clear of song learning and the facilities needed for its study, leaving the Cambridge group to develop research without the threat of competition.[66] There was, of course, nothing unusual about such gentlemen's agreements in British science at that time. The best known is exactly contemporary with the ones Thorpe extracted, and involved Cambridge: the ring-fencing of the solving of DNA's three-dimensional structure, reserved for King's College London. The young American geneticist James Watson, who arrived in Cambridge shortly after Marler and famously ignored the DNA prohibition, later put such agreements down to the "combination of England's cosiness—all the important people, if not related by marriage, seemed to know one another—plus the English sense of fair play."[67] The ethology community was cozy to the point of claustrophobia, and would remain so for some time.[68] Marler did as he was told, and kept off birdsong. Other aspects of Cambridge life put him in his place no less firmly. Thorpe, though "Bill" to his friends, was "Dr. Thorpe" to Marler, and remained so for years. Marler, meanwhile, was merely "Mr." Marler, since his London PhD was unacknowledged in Cambridge. By the end of 1951, he duly reregistered as a PhD student in zoology.[69]

But for all that, Cambridge was immensely exciting. Its tiny budget notwithstanding, the Madingley station had unrivalled resources for the scientific study of vocalizations in birds. The station building housed separate rooms for breeding chaffinches, for raising and observing them, and, within a few years, for isolating individual birds from the rest, under more or less soundproof conditions. There was recording equipment, including a disk-cutting machine, a magnetic-tape machine, a microphone, and, for reducing extraneous noise on the recordings, a parabolic reflector.[70] Although forbidden from making chaffinch song a research topic in its own right, Marler, as Thorpe's assistant, spent a great deal of time recording song, in the captive chaffinches and the wild ones. Meanwhile, Marler began a field study of chaffinch behavior—effectively, the assembling of the chaffinch ethogram—as his PhD thesis topic. Song may have been off-limits, but the rest of the bird's vocal repertoire was not, and Marler was all the more attentive. And when it became too cold for fieldwork to be fruitful, he turned to aviary studies, including a collaboration with Hinde on courtship in captive finches. Hinde was a combination of teacher, colleague, and rival—just five years older than Marler, and no further along professionally, but better acquainted with ethological theory and practice, and possessed of a scorching critical intelligence. Outside Madingley, Thorpe's imperium extended to a lab in the zoology building and to rooms at Jesus College, where he was a fellow; and in both places, Marler found stimulating scientific company. The college—where Marler too was a member, and later a research fellow—became especially important as the site of a joint seminar on behavior that Thorpe ran with the neurolinguist Oliver Zangwill. Through all this activity and much else besides, including participa-

tion in soon-legendary ethological congresses, Marler's identity as an ethologist began to set.[71]

The Transforming Encounter with
American Telecommunications

The balance sheet for the war's impact on ethology has two sides. There were,
as we have seen, losses of momentum and, at Lorenz's hands, of innocence.
But there were also gains. In ethology, as in other natural sciences, from particle physics to plant physiology, scientists reaped the benefits of new ideas, apparatus, and techniques developed in the course of wartime research. Again as
with other sciences, some of the links were personal. David Lack, for instance,
worked on the radar project during the war, and subsequently pioneered the
application of radar to studies of bird migration.[72] Other links were more indirect but more consequential. At Cambridge, the ethology of birdsong and
calls advanced with the aid of two analytic tools imported from the war effort
in the United States. One was a piece of apparatus, the sound spectrograph—
the acoustic equivalent of a prism, capable of resolving complex sounds into
their components and displaying these visually. The other was a set of ideas,
information theory. Both had been invented, wholly (the spectrograph) or in
part (information theory), at Bell Telephone Laboratories, one of the greatest
industrial-military research laboratories of the era, indeed of all time.[73]

The Origins and Early Impact of the Sound Spectrograph

Work on the sound spectrograph began in early 1941, before the entry of the
United States into the war.[74] It was not, of course, the first device to visualize
sound. A half-century earlier, Garner had used the soon-forgotten phoneidoscope. Successor technologies included several that transcribed the curves on
the surfaces of phonograph and gramophone recordings.[75] By the 1940s, the
cathode-ray oscilloscope was the display device of choice.[76] But with complex
sounds, it sometimes left much to be desired. An oscillographic portrait of a
spoken word showed a horizontal line punctuated with fuzzy bulges of various sizes. As the Bell radio engineer Ralph K. Potter, who headed the spectrograph project, later explained, in oscillograms, "two of the basic dimensions
of sound, those of frequency and intensity, are indiscernible[,] much as letters would be in a printed word if they were piled one upon the other instead
of being arranged side by side in a row."[77] A word visualized on the sound
spectrograph gave much more information, appearing, on etched paper or a

luminous screen, as a sequence of vertical, patchily shaded bands, read from left to right. The word's phonemic structure was thus made visible. The higher a shaded region, the higher the pitch or frequency of that part of the word; the darker a region, the greater the loudness or intensity.[78] Potter himself had laid the foundations for such superior resolution in a 1930 paper, describing a new method for measuring distortion in a radiotelephone circuit (used, for instance, in transatlantic calls, when two wire systems were connected through a shortwave link).[79] The immediate aim of the 1941 initiative was something much loftier: to help the deaf, by giving them the means to read incoming telephone calls, and to improve their own speech, by seeing what their words sounded like and what they should sound like. Concern for the welfare of the deaf stood in a long tradition at Bell, reaching back to its founder, who was, like his father, a teacher of the deaf.[80] With the arrival of the war, however, another aim took over. As the most powerful means yet invented for picturing complex sounds, the sound spectrograph promised unparalleled exactness in the detection of submarines and other enemy vessels, all of which had distinctive acoustic signatures in the water. The project was upgraded to a war project, and it was as a weapon of war that the sound spectrograph rapidly progressed.[81]

From ploughshare to sword and back again: even before the end of the war, a training program for reading spectrographically rendered human speech was under way at Bell.[82] In November 1945, the sound spectrograph and its peacetime uses at last went public, first in the pages of *Science*, then in *Time* and *Life* magazines.[83] In 1947, Potter and his colleagues summarized their work in a sort of textbook, under the title *Visible Speech*—a self-conscious homage to Alexander Melville Bell.[84] Soon, a New Jersey firm with close ties to Bell, the Kay Electric Company, began manufacturing and selling versions of the sound spectrograph, now called the Sona-Graph.[85] But despite the good publicity, swift commercialization, and noble intentions of its inventors, the device was never widely taken up among the deaf, who found it too complicated to use as a speech translator. The main constituency turned out to be scientists, including biologists interested in vocal communication in birds. In the *Science* paper, Potter had included spectrograms of the songs of a cardinal, a robin, a mockingbird, a brown thrasher, and a screech owl. His group had analyzed recordings of these songs, he wrote, in order to test the machine's capacity for handling rapid shifts of tone in an otherwise simple acoustic setting. "But these sound patterns are obviously revealing," he continued. "With such patterns as these it will be possible to analyze, compare, and classify the songs of birds, and, of even more importance, it will be possible to write about such studies with meaningful sound pictures that should enable others to understand the results." In *Visible Speech*, he expanded enthusiastically:

> If detailed analysis of song patterns is possible, there would seem to be a wide new field of study open to the ornithologist. Perhaps bird books and periodicals of the future will be filled with song pictures, and serious readers may become well enough acquainted with this sound language to read visible patterns of bird music in the way a musician reads a musical score.

This came to pass, as did another prediction-cum-exhortation:

> Little attention has been given thus far to the patterns of animal sounds, although here also would seem to be opportunity for extensive study. Unlike the songs of birds, that are usually made up of clear, high notes, with colorful variety in rapid excursions of pitch, the sounds of animals (at least the larger ones) would be expected to contain many overtones and conspicuous resonance shaping. Certainly, these animal sounds would present a type of resonance modulation and pitch inflection different from that of the human voice, although there may be interesting comparisons found between human sounds and those produced by the anthropoid apes. Possibly a study of these could contribute a new chapter to the story of evolution.[86]

In the early 1950s, this new means for analyzing animal sounds was joined by powerful new means for recording them. The result was a period of technological exuberance in the scientific study of animal communication, or "bioacoustics," as the field rapidly came to be known.[87] Ornithologists were especially quick to seize on the advances.[88] Along with the sound spectrograph, a 1953 review in a British ornithological journal described the portable magnetic tape recorder, the parabolic reflector, new types of microphone, and a battery-powered pack for radio transmission of recordings from the field.[89] Even in such glittering company, the sound spectrograph stood out. Although available for purchase, it was forbiddingly expensive, and few people had access to one or had even seen it in action, including the review authors. To make use of a sound spectrograph, it helped to have personal connections of one form or another to Bell Labs, as did all of the first group of spectrogram-flourishing ornithologists. There was Donald Borror at Ohio State University, who had learned of the spectrograph project during the war, while working in navy intelligence. There was Nicholas Collias at the University of Wisconsin, who teamed up with a member of that project, the linguist Martin Joos, now at the same university.[90] And there was the ever-resourceful Thorpe, who arranged a visit to Bell Labs while on a lecture trip to the United States in 1951–52. Back in England, he managed to gain access to one of the only sound spectrographs in the country, a supposedly top-secret model at the Ad-

Figure 7.2 The Sona-Graph that Peter Marler and W. H. Thorpe used in their studies of birdsong at Cambridge in the 1950s. Held in the Whipple Museum of the History of Science, University of Cambridge, inventory no. Wh.3658. Photograph by Boris Jardine and reproduced by permission of the Whipple Museum.

miralty Research Laboratory in Teddington. Once a month, for around two years, he traveled there from Cambridge with the latest batch of recordings, ready for analysis. A grant from the Rockefeller Foundation at last enabled him to purchase a sound spectrograph of his own and install it in his Cambridge lab (fig. 7.2).[91]

The Spectrographic Chaffinch Studies of Thorpe and Marler

Thorpe's first scientific paper on the birdsong research acknowledged the debt to technology unabashedly: "The Process of Song-Learning in the Chaffinch as Studied by Means of the Sound Spectrograph," published in *Nature* in March 1954. In fact the spectrograph's role was rather more indirect than the title implies. The spectrograms accompanying the paper are snapshots not of song learning in progress, but of the mature songs of captive chaffinches raised under different kinds of conditions. Some of the birds—the control group—were left alone, so that their songs defined "normal song." The other birds endured various sorts of isolation: separation from parents after five months; separation from parents after a few days, then raising by hand; and so on. Their spectrograms showed something of what Thorpe called "the interlacement of instinct and learning" in song acquisition in the chaffinch. Especially revealing, in his view, was that the parent-raised birds produced songs conforming much more closely to normal song than those of the birds raised by hand. Even when the birds were too young to imitate what they heard, they must have been learning. But even birds deprived of such learning sang something like a degraded version of chaffinch song. "The experiments with the hand-reared birds," he concluded, "suggest that there is an inborn basis to the song but that it is extremely generalized . . . amounting to little more than the ability to produce a song of about the normal length (2.3 sec.) and consisting of a *crescendo* series concluded by a single note of relatively high 'pitch.' All further refinements have to be learnt."[92] The sound spectrograph had thus made impressively precise a lesson already familiar in the scientific literature. Where the device truly came into its own was in demonstrating that mature song was itself less fixed than previously thought, undergoing small changes in duration and frequency range from the first year to the second—changes, wrote Thorpe, "so minute as to be practically imperceptible to the naked ear."[93]

This paper put the ethological community on notice to the opening of a new and, in Thorpe's estimation, superior avenue into the study of imprinting and other aspects of innate behavior. For, in his view, where "vocalizations were formerly the most difficult of all releasers to investigate precisely, they have now become far more readily amenable to analysis than are many patterns of visual or olfactory stimulation."[94] Using the spectrograph at this time was not quite as straightforward as Thorpe implied, however. Despite all the advances in technology, the raw material of birdsong recordings remained problematic, with distortion from variable tape speeds an especially persistent worry. And good as it was, the early sound spectrograph was, as Marler has recalled, "temperamental to operate, and obtaining reproducible results was an art."[95] But once that art had been acquired, the device proved an indispensable ally. To Marler's delight, Thorpe gave him free reign over the Cambridge spectrograph, and over the vast library of bird recordings that Thorpe had managed to secure from the BBC.[96]

Marler's own first Cambridge publication featured spectrograms prominently. Like Thorpe's paper in the same journal the year before, "Characteristics of Some Animal Calls," appearing in *Nature* in July 1955, was an instant classic.[97] It argued that the vocalizations of chaffinches, along with those of other birds and other animals, have the structures they do—the acoustic properties of frequency range and so on—because of the functions they serve. Marler distinguished two broad classes of function. Some vocalizations, such as the song of the wooing male chaffinch, need to make the singer or caller easy to locate in space. Other vocalizations, such as the alarm call that chaffinches give on seeing a hawk in flight, aid survival to the extent that they frustrate easy location. These contrasting functions ought, Marler reckoned, to be served by sounds with contrasting physical characters. Taking into account the physics of sound transmission and the physiological psychology of sound perception, one expects to find that the location-facilitating vocalizations cover a wide range of frequencies and intensities, with an internal structure of several clearly separated component sounds. The location-confusing vocalizations, by contrast, are predicted to be confined to a narrow band of fairly high frequencies and to be continuous in character—in spectrographic terms, high horizontal smears rather than chunks, checks, and spikes. The published spectrograms bore out these predictions remarkably well (fig. 7.3).[98] They also showed that different bird species had converged on alarm calls with roughly equivalent physical structures. Neither by ear nor by spectrograph was it easy to tell the flying-hawk alarms of chaffinches from those of blackbirds—two species not at all closely related.

In certain respects, the paper extended Marler's earlier work on variation. Following in Lack's footsteps (almost literally), he had looked for, and

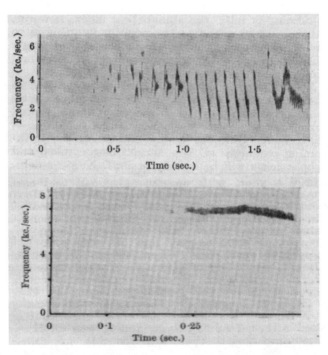

Figure 7.3 The acoustically contrasting spectrograms of the song (*above*) and alarm call (*below*) of the male chaffinch. From Marler 1955a, 6, 7.

found, correlations between, on one side, the degree of variability in the songs and calls of bird species in a particular place, and on the other, the degree of competitive pressure from closely related species in that place. Where competition was higher, the evolutionary premium on specific distinctiveness was greater, and the variability thus lower, restricted to whatever range ensured that members of a species could recognize each other reliably enough to propagate their kind. In other words, for Marler and for Lack, the range of variation was *adaptive*. There were functional reasons why the birds in a place sang and called with as much variety as they did. Now Marler embedded this earlier conclusion in favor of functionality within a more comprehensive analysis.[99] Broadly speaking, he now argued that songs and calls were adaptive not just in relative terms—according, that is, to the particular vocalizations of the other species with which a species happens to cohabit—but in absolute terms, fitted to the general properties of all terrestrial environments in which birds find themselves, and the nervous systems of all animals that birds typically encounter in those environments. Indeed, specific distinctiveness could itself be seen now as one of several characteristics contributing to easy or difficult location. For all the variability in chaffinch song, it still always sounds like chaffinch song, because a male chaffinch that sings too much like other birds will likely be passed over by females who do not recognize him as a potential mate. The flying-hawk alarms, by contrast, have virtually no specific distinctiveness, since the alarms of all species have to meet roughly the same narrow physical specifications in order to frustrate easy location by predators. The paper's concluding sentence stated a confident generalization: "Thus it appears that many of the sounds used in bird and insect 'language' have not been chosen arbitrarily, but are directly adapted in structure to the function they have to perform."[100]

At the same time, however, the paper represents a shift into intellectual territories encountered only after his move to Cambridge. The germ of the paper was, in time-honored fashion, a series of humble observations, made in the course of listening and looking in Madingley Wood. Marler noticed that while it was sometimes easy for him to locate a vocalizing chaffinch, at other times it was almost impossible. Some alarm calls, indeed, seemed to be "ventriloquial"—coming from anywhere but the location of the vocalizer. And when he did manage to find the bird responsible for one of these calls, it often turned out not to be a chaffinch at all, but a great tit or some other kind of bird. Curious to learn more about sound localization (of which he knew nothing), he sought ought two Cambridge figures known to be working in the area. One was John Pringle, a specialist on cicadas, whose songs also seemed to be designed to frustrate easy location of singing individuals. The other was

Donald Broadbent, an expert in the psychology and physiology of hearing, based in the Applied Psychology Research Unit. Between them, these men— and especially Broadbent, who was keenly interested in behavioral questions, and well-disposed toward ethology, becoming a participant in Thorpe's seminar—introduced Marler to the classic scientific literature on sound localization and to some of the most innovative current research.[101] By the summer of 1954, when he presented some of his work in London, he had worked out the argument he would publish the following year in *Nature*. Strikingly, the abstract for the London talk refers as much to structure and function in bird vocalizations as to structure and *information*. "Some relationship," he wrote, "can be found between the structure of a note and its information content." Chaffinch alarm calls "appear to give a minimum of clues as to location of a caller in space," whereas their songs "give a wealth of location clues which connects with the abundance of intra-specific, social information that song conveys."[102]

The Origins and Early Impact of Information Theory

To the scientific imagination of the 1950s, few words were more enchanting than "information." The immediate sources of the fascination were two rather different books: Norbert Wiener's *Cybernetics, or Control and Communication in the Animal and the Machine* (1948), and Claude Shannon and Warren Weaver's *Mathematical Theory of Communication* (1949).[103] Wiener was a mathematician extraordinaire based at MIT. Shannon had studied mathematics and engineering at Wiener's MIT before starting work at Bell Labs in 1941. Weaver, an important patron of both men, was director of the Natural Sciences Division of the Rockefeller Foundation. He came to prefer the spare rigor of Shannon's work to the ostentatiously learned splendors of Wiener's, and such has been posterity's verdict too.[104] Their differences in style notwithstanding, both men took a statistical approach to the analysis of communication. According to the "Wiener-Shannon" theory, as it was sometimes known, the amount of information in a transmitted message depends on the probability of that message being selected from a finite set of possible messages. The less probable a message, the more informative it is, as measured (in Shannon's formulation) in "bits," short for "binary digits"—famously, either 1s or 0s, standing for mechanical or electrical switches turned on or off, or statements being either true or false. Systems of communication as a whole could now be described quantitatively, including the optimal transmission rates or "capacities" of the channels linking sources of information with their destinations.[105] Although Shannon and Wiener converged on these ideas independently, they drew on a common fund of ideas, some of them developed at Bell Labs in the 1920s.[106] Moreover, for both men, the theoretical impetus derived from the practical demands of the Second World War. Wiener had sought general principles for the design of weapons systems with self-correcting feedback mechanisms. Shannon had been part of a cryptography team at Bell, and his effort to generalize his student work on the logic of switching advanced in dialogue with his attempts to devise fiendishly hard-to-crack codes.[107]

Like wartime Bell Labs' other contribution to ethology, the sound spectrograph, information theory started life as a means for representing what gets sent down telephone lines and other communication channels. Unlike the spectrograph, however, information theory found its intended audience, and then some. "Since [Shannon's] work has been published, interest has surged like a forest fire," observed Colin Cherry, reader in telecommunications at Imperial College London, lecturing in London in 1953. "So great is the anxiety today concerning parochialism in science, that any hint of a general all-embracing theory is seized upon eagerly. Thus, Communication Theory has been grabbed like a straw, by linguists, by psychologists, by physiologists, by sociologists, by someone from every 'ology.'"[108] The intellectual tinder included more than fear of overspecialization. Among information theory's earliest converts were scientists familiar from their war service with military signaling and its technical challenges. To these men, the prospect of a more comprehensive understanding of communication systems had obvious appeal. Military experience also bred a taste for working on problems requiring expertise from different branches of the traditional sciences. Again, information theory, especially as Wiener presented it, seemed to offer just such opportunities.[109] More generally, there was growing awareness that in the postwar world, with its unprecedentedly large organizations and rapid communications, the need for scientific principles to guide management of information flow and storage was pressing.[110] From the outset, the new theory enjoyed institutional goodwill. In the United States, the Josiah Macy Foundation sponsored a series of annual conferences on cybernetics, bringing together senior figures from a range of physical, biological, and social sciences.[111] In the United Kingdom, where several large symposia were held at the Royal Institution and elsewhere, there was a new Communication Research Centre, founded at Marler's alma mater, University College London. The center's inaugural lecture series in 1953 featured, in addition to Colin Cherry, the philosopher A. J. Ayer, the classicist T. B. L. Webster, the art historian R. Wittkower, the phonetician D. B. Fry, the English scholars Ifor Evans and Randolph Quirk, the National Coal Board member Sir Geoffrey Vickers, and the UCL biologists J. Z. Young and J. B. S. Haldane.[112]

Haldane's advocacy was especially notable in bringing information theory to the attention of British biologists, and vice versa. Wiener was an old friend, and had corresponded during the war about his new ideas at "the edge of some biological work," as Wiener put it in a 1942 letter to Haldane.[113] Five years later, Wiener was Haldane's guest for a three-week visit to Britain. It is easy to imagine them enjoying each other's company, with their shared passions for universal learning, progressive politics, and the mathematics of stochastic processes. The same faith in scientific modernity that prompted Haldane to embrace eugenics and Marxism now enfolded cybernetics. In the introduction to *Cybernetics*, Wiener wrote of finding interest in the new science on that trip "about as great and well informed in England as in the United States," mentioning Haldane as one of several English professors who appreciated how urgently its perspectives were needed.[114] In the main part of the book, Wiener praised Haldane's tentative proposals for understanding self-replicating genes and viruses as quasi-cybernetic systems.[115] The book explored several other connections with biology, including analogies between nervous systems and computing machines, and the role of communicated information within organisms and their societies.[116] Haldane responded with enthusiasm. "I am gradually learning to think in terms of messages and noise," he wrote to Wiener in November 1948. "A mutation seems to be a bit of noise which gets incorporated into a message. If I could see heredity in terms of message and noise I could get somewhere." Over the next few years, Haldane sent Wiener periodic updates of progress in working out cybernetic ideas about evolution and embryology.[117] In his lectures and writings, however, Haldane tended to stay closer to the territory mapped in *Cybernetics*. Most of his 1953 lecture in the UCL series "Communication in Biology"—all the lectures were collected and published in book form in 1955—dwelt on communication between cells and between members of the same species. But where Wiener the biological outsider had gestured vaguely to recent work in these areas, Haldane dealt in specifics, from physiology and from the new science of signaling in animals, ethology.[118]

What most impressed Haldane about the ethologists' discoveries was the elaborately *ritualized* nature of the signals they described, and the parallels he saw between these and rituals in human life.[119] In Haldane's view, descriptive language in humans was a more or less recent development out of the vocalizations that, in ancestral humans, as in some nonhuman animals now (notably birds), accompanied ritual action. He termed these vocalizations as "evocative," serving less to communicate facts about the world than to bring others to action, or to emotional states that in turn lead to action. Reviewing

Lorenz and Tinbergen's work in his UCL lecture, Haldane noted one unpleasant legacy from instinctive signaling in the ancient, prehuman past:

> Fifty million years ago our ancestors probably communicated largely by a system of releasers and innate releasing mechanisms, mostly operating through the sense of smell. Then they took an arboreal life and largely lost the sense of smell and the quasi-mechanical behaviour associated with it. This was a prerequisite for such freedom of the will as we possess, and probably for the development of language. But even now smells can affect us so powerfully as to overcome our wills, for example causing vomiting, as a musical discord or an ugly pattern will not.[120]

According to Haldane, our animal past also explains the persistence of rituals even in the modern world. New military recruits are drilled as they are in part so that they come to feel warlike, ready for combat on behalf of the group. The imitation of action generates the corresponding emotion, as William James taught. In Haldane's view, moreover, much of human linguistic production has this older, emotional, evocative, collective, forward-looking aspect. But not all languages are equal on this score, as he pointed out in another 1953 paper. Never one to miss a chance to undermine house prejudices, or to flaunt his awesome breadth of knowledge, Haldane declared Indo-European languages to be more animal-like than Semitic languages. The reason he gave was that the simplest verb forms in the former are often the imperative singular forms ("do it"), while in the latter the simplest are the past tense forms of the third-person singular ("he did it"). Objectivity was thus within easier reach in the Semitic languages. Might this be, he wondered, "why Moses, Jesus, Mohammed, and Spinoza achieved, on certain matters, an appalling clarity of thought"?[121]

The animal signaling research that Haldane most admired, inspiring his most sustained reflections on animal communication and information theory, was Karl von Frisch's work on the dance language of the bees. Like Lorenz and Tinbergen, Von Frisch was a European zoologist whose elegant experimental studies of perception and communication in a nonhuman species were beginning to attract attention in the anglophone world in the 1950s. The achievements of these three men would continue to be discussed together, culminating, in 1973, in the awarding of a joint Nobel Prize. But Von Frisch was considerably older than the other two, and for all the similarities in aims and methods, he did not much share in their programmatic evangelism.[122] Whatever Von Frisch's ethological credentials, or those of his student Martin Lindauer, Haldane thought the world of their discoveries about bee communication, ranking them, as he told his UCL audience, "quite as high as those of Champollion, Rawlinson, or Ventris."[123] In brief, Von Frisch had found that bees returning to the hive from a food source perform ritual movements—"dances"—informing other bees about the distance, direction, and amount of the food. More recently, Lindauer had reported observing bees dancing their way to consensus about the site of a new nest.[124] In 1954, Haldane and his wife Helen Spurway published a long and remarkable paper on this work, "A Statistical Analysis of Communication in 'Apis mellifera' and a Comparison with Communication in Other Animals." It ranged widely, offering, among other things, calculations of the accuracy with which the message recipients in Von Frisch's experiments flew toward the food; speculations about the messenger's abdominal "waggling" as an indicator of distance; a reiteration of the view that the bee dance, in common with other animal rituals, is not descriptive of facts but evocative of actions; and a detailed conjecture about the evolutionary history of the bee dance.[125] But most innovative of all was a section entitled "Ethological Cybernetics." Wiener's cybernetic perspective, wrote Haldane and Spurway, "has been applied to communication between men and between machines, but not yet to communication between animals." Using Shannon and Weaver's measure of information, they proceeded to calculate the information transmitted in the bee dance as to direction. Their conclusion: between two and three cybernetic units of information—a little less informative than the English message "fly north north-west."[126]

Marler on the Information in Bird Vocalizations

Was it from Haldane that Marler picked up on information theory, when he was a student at UCL or later on? Certainly other biologists in Haldane's orbit caught the enthusiasm.[127] In 1950, the UCL geneticist Hans Kalmus published "A Cybernetical Aspect of Genetics," a brief but incisive article examining the genetic system as a site of messages, memory, feedback, and control.[128] In the same year, J. Z. Young referred extensively to information theory in a series of prestigious BBC radio lectures on the brain, published in book form in 1951. Young's major theme was communication and its biological underpinnings, as revealed, in his words, through "a combination of the techniques of evolutionary biologists and communication engineers." At the time he was engaged in research on the neurological basis of behavior, using the octopus as his model, and he compared the brain of an attacking or retreating octopus to the control system of a guided missile. In both octopus and missile, he suggested, information about the environment flows in through receptors, calculations are made, and a correct response selected.[129] In New York in 1952, at the Macy conference on cybernetics, he presented recent experimental work on memory and learning (discussion lit on how many bits of information the octopus had learned).[130] In London the next year, in the lecture series launching the Communication Research Centre at UCL, he again recommended information theory as a source of insight for the biologist.[131] A 1954 essay, which Marler knew and cited, explored the case for reviving old analogies between memory and heredity—the recording of the past in, respectively, the individual nervous system and the species germplasm—using the new language of information, storage, and coding.[132] At Haldane's UCL, then, and within its wide sphere of influence, information theory became all but unavoidable. Even so, another inspiration for Marler's delving in could well have been his Cambridge expert on sound localization, Donald Broadbent. Along with Colin Cherry, Broadbent was looking at how the human nervous system selectively filters and processes incoming speech streams, as if it were a communication channel of limited capacity.[133] At a 1955 London symposium on information theory, organized by Cherry, Broadbent issued "a plea for the qualitative use of information theory in psychology," as a remedy for well-known deficiencies in behaviorism (as Broadbent illustrated with an information-theoretic reading of Pavlov's dog).[134] In *Perception and Communication* (1958), his influential summary of his research to that point, Broadbent included one of the first information-processing flow charts in modern psychology.[135]

From whatever source it derived, information theory had a pronounced effect on Marler's conception of his studies of birdsongs and calls. As he presented it in his abstract for the July 1954 meeting of the Association for the Study of Animal Behaviour, his field study of the chaffinch vocabulary now embodied "some ideas of elementary communication theory" in its theoretical approach. With a view to teasing out information content, he was attempting to correlate vocalizations with all the associated changes in the vocalizer's inner and outer environments. So far as acts of vocalizing always went along with a characteristic suite of changes, a song or call, he wrote, "may be regarded as containing potential information about these environmental factors." Of course, one of the elementary lessons of communication theory was that the information transmitted from the source was different from, and inevitably greater than, the information received at the destination. How much of the potential information in the vocalizations was being received? Observation of responses revealed different answers in each case. When, for instance, a male intrudes on another's territory and hears his rival's song, the intruder "may act upon six of the nine items of potential information that have been listed for this particular vocalisation." At the end of the abstract, Marler ventured a fourfold taxonomy of transmitted information among the

chaffinches: "identificatory" information, to do with the vocalizer's identity
(as a chaffinch, as a male chaffinch, as *this* male chaffinch, and so on); "loca-
tive" information, giving or withholding clues to location in space (according
to physical structure and biological function); "emotive" information, concerning the motivational state of the vocalizer; and "environmental" information, reporting facts about external conditions and variations in them.[136]

The meeting took place in London, so Haldane might well have attended.
He almost certainly kept up with the association's *Proceedings*, since he was
then becoming more directly involved in ethological debate. In June, Haldane
and Spurway had participated in a conference in Paris on instinct, along with
Von Frisch, Lorenz, and the American ornithologist Daniel Lehrman, recently
notorious for an excoriating public rebuke of Lorenzian instinct theory. Lehrman had set a comparative-psychological cat among the ethological pigeons,
and his critique, along with the political suspicions still hovering over Lorenz,
and personal animus arising from (it seems clear) Lorenz's love affair with
Spurway, awakened Haldane's own critical faculties. So his antennae were
even more closely attuned to developments in ethology when Marler presented his research.[137] Impressed, Haldane pushed his former student to publish his structure-function discoveries as soon as possible.[138] The great man
also featured Marler's findings prominently, with all due credit, in a series
of lectures later that year, "Animal Communication and the Origin of Language," given at the Royal Institution. Unsurprisingly, Haldane emphasized
those aspects that ran counter to Lorenzian doctrine. Where Lorenz insisted
that birdsongs and calls were everywhere the same, Marler had demonstrated
enormous variation. And where Lorenz assumed that animal signals were
arbitrary in their structure, Marler now showed that this was not always the
case. Haldane even paid Marler the compliment of reproducing his sound
recordings of chaffinches for the Royal Institution audience.[139] He and Marler also became friendly socially, along with their wives (Marler had recently
married his hometown sweetheart, Judith Gallen). As Marler later recalled:

> We came to know him and his wife, Helen Spurway, and to admire especially their
> capacity for vintage cider, at The Mill in Cambridge, and for beer at various Lon-
> don pubs. Spurway [was] well equipped vocally, and became even more loqua-
> cious as evenings wore on. We had to leave one pub hastily, in the face of verbal
> abuse, when she threatened to drown out the resonant voice of Winston Churchill
> coming through the radio on the bar. Much of the time with them I was content
> just to sit and listen to Haldane's displays of erudition.[140]

Marler summarized his studies of chaffinch vocalizations and the information these convey in a new paper, "The Voice of the Chaffinch and Its

Function as a Language," completed in autumn 1955 and published in 1956, again in the *Ibis*.[141] There had, of course, been earlier inventories of the calls and songs of bird species, with attention both to their sounds and their functions. The American ornithologist Margaret Morse Nice, for instance, in her famous 1943 monograph on the song sparrow, had listed twenty-one "chief vocalizations" of the bird.[142] In 1945, the Russian expert on chaffinch song, A. N. Promptov, together with his colleague E. V. Lukina, had published a paper on the twenty calls at the command of great tits (Haldane was fond of discussing their results).[143] More recently, Tinbergen, in his popular book *The Herring Gull's World* (1953), had catalogued that bird's most common calls.[144] But the description of sounds in these works was onomatopoeic and qualitative, and the contexts in which the calls were given lightly sketched. Marler's paper furnished a wealth of detail on both counts. Nearly all of the fourteen vocalizations he described came with an accompanying spectrogram, and sometimes several. Through comparison of spectrograms of these and other vocal phenomena, including the developing alarm calls of immature birds and the occasional "intermediate" calls of mature birds (mixing together properties of other calls), he at one point conjured a sound-picture of the evolutionary ancestor of the chaffinch, with about half the vocabulary of its descendants. Noting that this hypothetical bird resembled members of another modern finch genus, the carduelines, he further suggested that the carduelines and the Fringillidae had "a much closer relationship than would be inferred from recent taxonomic studies."[145]

What was truly remarkable about the paper, however, was its treatment of call function. Despite Haldane's pathbreaking efforts, and despite a common language of "signal codes," ethologists in the Lorenz-Tinbergen tradition had so far given the new information theory a wide berth.[146] The barriers to combining these perspectives on animal communication were indeed considerable. *The Herring Gull's World* illustrates the conventional ethological view. No less than Marler, Tinbergen was sure the calls of the herring gull serve biological ends. He insisted that most if not all of them "have a communicative function." But he rather gave the impression that to describe a functional call as informative would have made as much sense as describing a key turning a lock that way. An unlocked door is set in motion, and that, as Tinbergen (following Lorenz) saw it, is the effect of a call on an animal. As for the calls themselves, they were mere "outlets," as Tinbergen put it, in an echo of Darwin's argument in the *Expression of the Emotions*; "through them an animal can 'get rid' of impulses if there is no other way open." An excited animal vents excess nervous energy vocally, and, thanks to natural selection, the vocalization triggers an adaptive motor routine in conspecifics. The whole exchange is mechanical through and through.[147] Lorenz had conveyed a similar message in his own popular work, *King Solomon's Ring* (1952). "Animals," wrote Lorenz, "do not possess a language in the true sense of the word," because their calls are the vocal equivalents of reflexes: inborn, invariant, automatic. Animal calls express emotions, and nothing more.[148]

Marler now set forth a quite different way of thinking about communicative function in animals.[149] If a call elicits a biologically appropriate response, then, he argued, it follows that some information has been transferred. A characterization of the call's function can therefore consist of a listing of the items of information—propositions about the caller and/or the world—contained in the call. The assembly of such a list, he suggested, implied nothing about the minds of callers or hearers. Whether the information in a call is intentionally transmitted or not, whether it is consciously responded to or not, what mattered, in his view, was that "the recipient responds *as though* certain items of information have been received."[150] Marler proceeded to unpack the propositions transmitted in each of the chaffinch vocalizations. In one of the alarm calls, for instance, he inferred two items of information: that the caller "has perceived strong danger," and that the "danger is from above." (Recall that one conclusion of Marler's earlier work was that alarm calls suppress clues to location and, since other species have often converged on the same calls, identity.) In the song, as given by a male to an unmated female, Marler inferred six items: that the caller "is a Chaffinch," that he "is a male," that he "is at a particular place," that he "is within his territory," that he "is ready to make reproductive attack on other Chaffinches," and that he "is ready to mate."[151] Again, Marler distinguished four types of information, using nearly the same taxonomy as before (the "emotive" calls were now "social" ones).[152] Unlike Lorenz and Tinbergen, who saw mainly differences between human language and animal communication, Marler concluded that "the voice of the Chaffinch can surely be said to serve as language."[153]

How so? Yes, he conceded, the chaffinch vocabulary was small and rigid compared with human language. Moreover, chaffinch calls, and chaffinch responses to chaffinch calls, were inborn, quite unlike the human case. But some calls, like song, were subject to dialect variation, and chaffinches could learn to respond to other species' calls. Perhaps, Marler suggested, the learned or instinctive basis of vocalizations and responses was, for purposes of understanding communication, not all that important. Whether birds acquire their calls and responses in their own lifetimes, or whether these calls and responses are inherited, as legacies from other, earlier members of the lineage, the information transferred between caller and hearer is the same. As for the nature of the information transferred, Marler followed Haldane in distinguishing evocative from descriptive communication. Chaffinch calls are evocative; even when informing a hearer about "danger from above"—a state of affairs external to the caller, so no mere expression of an emotion—they evoke actions, or moods appropriate to them. Humans alone, it seemed, described. But, as Haldane had emphasized, human communication too was full of evocative elements, and was probably even more so in the past. "The early Greeks," wrote Marler, drawing on recent scholarship on European thought, and no doubt conversations with Haldane, "often did not distinguish clearly between the emotions evoked by a word, and its descriptive implications, between the particular and the general . . . , between intellectual activity and emotional activity."[154] Marler would return to the complicated relations between the emotional and nonemotional aspects of language, animal and human, repeatedly over the next twenty years. But the ancient human world would not long detain him. Soon his inquiries were taking him further back still, beyond the Greeks, to the animal past of human language, and to Garner's old subject: the vocalizations of wild apes and monkeys.

8

SIMIAN SEMANTICS

✦ ✦ ✦

"IT WAS WITH real fanaticism that Richard L. Garner, an American who was very well-known when I was a student, set about the task of investigating simian speech." So wrote the Leipzig-based Georg Schwidetzky, in a book published in English in 1932 as *Do You Speak Chimpanzee?* and based on lectures and radio talks he had given around Germany, summarizing research begun in 1918. "The aim of my work," he wrote, "has been to discover, through systematic investigation of the speech of animals (in particular apes, monkeys and prosimians), the first beginnings of speech and the languages of prehistoric and fossil men."[1] In the main, Schwidetzky had attempted to identify vestiges of prehuman speech in human speech, through close, phonetically precise comparisons between human vocal sounds and animal vocal sounds, as studied in the zoological gardens of Europe. Reviewing earlier studies, he devoted several pages to Garner: his background, his introduction of the phonograph, his Congo expedition, and his subsequent fall from grace. He reckoned Garner a pioneer in the translation of ape speech and the instruction of apes in human speech. As for Garner's "trifling" deceptions concerning the expedition, especially in light of his further African sojourns "to the detriment of both health and fortune," they had "to be judged leniently."[2]

Schwidetzky's attitude was robustly empirical and evolutionist. He promoted animal activity as furnishing facts useful to taxonomy and the reconstruction of phylogenies. He insisted that the first step was to learn what animals say, and only then to begin comparisons with humans. He was impressed by the power of experiment to disentangle instinctive actions from learned ones. (One of Garner's most important contributions, wrote Schwidetzky, was to provoke Boutan's superb work along these lines.) He was even an institution builder of sorts, having started a German Society for the Study of Animal and Primeval Languages. So far, so ethological. But there was a major and, it would turn out, unforgivable breach of the developing ethological code. In line with a tradition of theorizing that went back to Linnaeus, Schwidetzky argued that a great deal of evolutionary change, right up to the emergence of humans, was due to hybridization. Monkeys belonging to separate groups had crossbred to produce apes, and apes from separate groups had crossbred to produce humans; or, as Schwidetzky rather pungently put it, "anthropoid apes are bastard forms of lower monkeys, and men are the bastards of bastards."[3]

For Lorenz, such talk was anathema. When his mentor Heinroth wrote to him in December 1939 and again in January 1940, expressing anxiety over the "dangerous" Schwidetzky and his opinions, Lorenz—eager for opportunities to demonstrate the congruence of Nazism and ethology—leapt to action. He contacted Walter Greite, a senior figure in Nazi biology, alerting him to Schwidetzky's views and the need to stamp them out. As Lorenz reported back to Heinroth:

> Regarding the poor madman [Schwidetzky]: just as a sex killer is, so is a poor madman, who personally can do nothing about his deficit mutations. Since he however is enormously harmful for the people as a whole, one slaughters him justly! I have sent the whole Schwidetzky package to Greite, who is just now very busy with the authoritarian suppression of trash in biological literature. The manner and way Schwidetzky derives *us* from hybrids out of hybrids (what an idea he has of the fertility of cross-breeds! Try once to cross the baboon and the gibbon) is nothing short of propaganda for crossing human races, racial shame on the large scale. It will be forbidden immediately, as well as the whole Society for Primitive Language Research.[4]

Forbidden led soon enough to forgotten.[5] More than twenty years would pass before ethological attention returned to the natural vocalizations of nonhuman primates. By then the struggle to give the new science a firm footing had been won. In 1959, the year the scientific world celebrated the centenary of Darwin's *Origin* and his "ape theory" of human origins, the ethological imperium stretched across Britain and continental Europe. The United States too was beginning to come within the fold, thanks in part to the arrival there of ever larger numbers of emissaries, among them Marler, who left Cambridge for Berkeley in 1957. The expansionist mood of the times was as much intellectual as institutional. The charming accounts of natural history that had introduced a wider audience to ethology in the 1950s gave way in the 1960s to deliberately provocative studies of human nature, such as Lorenz's *On Aggression* and the former Tinbergen student Desmond Morris's *The Naked Ape*. Other ethologists traveled less noisily and more cautiously to similar destinations. For Hinde, the path to humans and other primates led through collaborations with psychologists, anthropologists, and other workers in what the twentieth century liked to call "the behavioral sciences." Such openness to cross-disciplinary exchange explains in part why two of the most famous modern primatologists, Jane Goodall and Dian Fossey, were Hinde's supervisees, though—like Robert Seyfarth and Dorothy Cheney a little later—neither came from zoological backgrounds.

Marler's trajectory to apes and monkeys was different again. The United States was to prove an encouraging place for an ethologist with a passion for understanding vocal communication in animals and its relation to human language. At Berkeley and then, from 1966, at the Rockefeller University, the legacies of Marler's ethological training intersected with new trends in American anthropology and old ones in American psychology. This chapter tells how those intersections came about and how their consequences made the primate playback experiment a live option once more.

The Legacies of Marler's Cambridge Years

Six years under Thorpe's benign reign, with no responsibilities but the care and feeding of the aviary chaffinches, had enabled Marler to achieve a great deal. When he left Cambridge, there were a number of fine observational and experimental studies to his name, including a monograph-length inventory of chaffinch behavior (Marler's dissertation),[6] four papers on aggression and social relations in chaffinches,[7] and one on their defense of territory and personal space.[8] There were wide-ranging synthetic papers, on the Lackian topic of specific distinctiveness in bird signaling[9] and, for Thorpe and Zangwill's seminar group, the Broadbentian topic of stimuli filtering.[10] There was even a note on the evolution of finch colors in *The Middle-Thames Naturalist*, the annual report of his old natural history society.[11] But it was the papers on the structure and function of calls and (more hesitantly) songs—the mid-1950s "Bell Labs" papers, making use of those two gifts of war, the sound spectrograph and information theory—that had made Marler's reputation. The former botanist was now an ethological insider, with an insider's contacts. Tinbergen had examined his dissertation. Lorenz had sent a letter of recommendation for the Berkeley job, as had Thorpe and Haldane.[12]

The ascent had been smooth, but not frictionless. Marler had bridled at the chauvinism of Cambridge, where Judith, as a woman, was excluded from college meals, and where men like himself, from working-class backgrounds, were forever reminded of their outsider status. Every conversation at the Madingley station made audible the social gap separating Marler from Thorpe and Hinde, both products of English public schools. But even without the class differences, bosom friendship was never in the offing. Thorpe was just too guarded, Hinde too combative, the system too hierarchical. There were, to be sure, moments of warmth, or something like it, as when Thorpe, bearing the good news of Marler's college fellowship, extended him the privilege of using Thorpe's first name. But there were also tensions, and occasional eruptions. Marler recalls an especially severe dressing down over dealings with the press. Thorpe had tried to keep publicity about the station to a minimum, mindful of growing concerns about the welfare of animals in scientific experiments, and of the station's fragile finances. When a newspaper reporter came around one morning, asking questions, Thorpe and Hinde duly turned him away. At lunch in a local pub later that day, however, the reporter tried again with Marler, who, unaware of house policy, chatted freely. When Thorpe found out, he was furious with Marler. The short item in the paper proved innocuous. But the chill in relations now set even deeper.[13]

Ethology was bigger than Cambridge, however, and Marler had found the science an exceptionally comfortable fit. He had come to identify strongly with Lorenz and Tinbergen's vision. He used their language of "releasers," conducted experiments modeled on theirs (notably in his aggression studies), and, as we shall see, conceived his research within the old contrast-pair of good ethology, bad comparative psychology.[14] But he had also arrived at several conclusions that sat awkwardly alongside the founders' teachings. The divergences from Lorenz were most stark. Where Lorenz talked down the similarities between animal vocalizations and human language, Marler talked them up. Where Lorenz disdained variation in animal signals as a sign of degeneracy, arising when animals were buffered from natural selection, Marler saw signal variation quite differently—and in Lackian fashion—as natural selection's plaything, expanding or contracting according to local adaptive needs.[15] And where Lorenz insisted that animal signals could be assumed as arbitrary in their form or structure, Marler raised the possibility that signal structure could be functional. This last dissent cut the deepest, as Marler hinted in his 1955 *Nature* paper, and made plain in his paper on specific distinctiveness, published in *Behaviour* in 1957. According to Lorenz, evolutionary relationships could be inferred from instinctive displays only if the forms of those displays were arbitrary—that is, only if animals display as they do because their ancestors displayed that way, not because the environment demands it. Independent, adaptive convergence on the same form was ruled out.[16] Now Marler was ruling it back in.[17]

Near the end of Marler's time in Cambridge, two new research possibilities swam into his ken. One was the playback experiment. The bioacoustics boom of the late 1940s and 1950s had brought a turn in the experiment's fortunes. As with zoological use of the sound spectrograph, the ornithologists had led the way.[18] One of the first spectrographic studies, Collias and Joos's 1953 paper on chicken vocalizations, made use of playbacks to distinguish between sounds that attracted and sounds that repelled.[19] Marler was of course familiar with their work. While he never conducted playbacks in a sustained way at Cambridge, he did use them from time to time. Some of his playback results even came to public notice, thanks to Haldane. In the printed version of the latter's 1955 Royal Institution lectures, he reported experiments by Marler showing that unmated chaffinches, in Haldane's words, "respond in the same way to reproductions of the song" as they do to a singing male chaffinch. These experiments had been conducted in the Madingley aviary. Although Marler referred to them in his 1956 summative monograph and elsewhere, they were never published in their own right, as he came to regard them as a failure, due to technical limitations.[20] These no doubt put him off using playbacks in his field studies of territoriality—an application that made the name of a young Toronto researcher and visitor to Madingley, J. Bruce Falls.[21] Much more active than Marler in exploring the potential of playbacks were Hinde and Thorpe. Not long before Marler left Cambridge for the post at Berkeley, in 1957, Hinde started looking at whether birds respond preferentially to playbacks of their own songs. Around the same time, Thorpe introduced playbacks into his song-learning studies, via a specially modified tape recorder called the Song Tutor, used to find out which songs birds can learn and when they can be learned.[22]

Growing alongside Marler's appreciation of the possibilities in playbacks was an interest in primate communication.[23] He had made some initial forays into the literature while writing up his aggression experiments, consulting some of the classic studies on primate social life, including the British physiologist Solly Zuckerman's 1932 book *The Social Life of Monkeys and Apes* and a 1942 paper by Carpenter on the same topic.[24] More focused reading around communication in primates followed, thanks to an invitation from a botanist friend, Peter Bell. He was putting together a volume to commemorate the centenary of the *Origin of Species*, and asked Marler for a chapter on the science of animal communication. Marler responded with a clear and comprehensive survey, taking in chemical, visual, and auditory communication; the evolution of birdsong; the origins of communication (he found a number of overlaps between ethological ideas and Darwin's explanations of expression); and the parallels between animal and human language. As much the work of a man hitting his intellectual stride as of someone anxiously filling gaps before students found them, the chapter showed Marler operating at a new level of generality. His examples came from all over the tree of life: birds, of course; but also slime molds, grasshoppers, bees, fiddler crabs, lizards, minnows, hamsters, cats, zebras, wolves—and monkeys and apes. There were discussions of Boutan's and Carpenter's findings on vocabulary size in gibbons and howler monkeys, respectively; the limited capacity for mimicry in chimpanzees as reported in the Yerkes' ape compendium; and the Russian psychologist Nadia Kohts's study of facial expression in the chimpanzee.[25] The most recent of these contributions was more than twenty years old. None came from within the ethological tradition or its sanctioned precedents.

The discussion of animal and human language revisited familiar themes, including evocative versus descriptive language and the idea that signals communicate information whether or not they are voluntary. As we have seen, both themes were, for Marler, Haldanian in origin. In Haldane's hands, they had tended to underscore a sharp difference between animal signals and human utterances. Bees might dance two to three bits of information about the direction of the food source, but, for Haldane, the dance still only ever prepared other bees for action of a certain kind, by priming their emotional states. But Marler now took the opposite view. Given the information transmitted in the dance language, he wrote, it "must be regarded as descriptive, in the same sense in which we apply this to our own language." The deeper lessons he drew were that, once one began reckoning the information content in messages, the old distinctions between descriptive and evocative, or between unconscious and conscious, did not seem all that fundamental. While conceding that much animal communication was evocative, and that it was "a tremendous step" from the alarm-call message "'here is something to flee from'" to the message "'hawk,'" the change was not, Marler argued, "essentially a change in the information carried in the signal, but rather in the way in which the information is responded to, a change in word usage rather than word structure." And while conceding that much animal communication was instinctive, he insisted, as he had before, that animals "respond as though they have received certain information," and that this was true whether or not that information was consciously communicated or understood.[26]

Animal and human languages could thus be seen as different chiefly in the complexity of information transmitted. The more complex a signal, the more complex the information it was capable of transmitting. Here was a generalization that held across the human-animal divide: an alarm call conveyed simpler information than a song, and a song conveyed simpler information than a sentence. Furthermore, in Marler's view, there were adaptive reasons why, bees excepted, animals had such small vocabularies, with different kinds of information packed into each signal:

> If we consider the function their language has to perform in the natural state, there is a certain consistency. The lives of most animals, and particularly birds, are generally dominated by a limited number of factors which overwhelm all others in importance. This is as true of their social life as of their interaction with the environment. For such activities as helping others to avoid danger (especially the mate and young), competing for food, keeping in contact with members of the same species and, above all, reproducing with them, it is probably most efficient to have a relatively limited vocabulary, which is instinctively used and understood the first time it is needed. Apart from certain exceptions such as bird song, where learning may have an important part to play, it appears that the more subtle advantages of elaborate communication by learned signals need a highly complex society for their value to be felt.[27]

So it was no surprise that bees, with their highly complex societies, were the exception. Complex societies shield the individual from the most severe threats to survival, while creating conditions favorable to the growth and differentiation of vocabulary. As more signals become available, a higher degree of information parceling becomes possible. In the case of humans, Marler suggested (following Haldane and others), the major transition probably occurred in the upper Paleolithic, when tool use flourished. But not even the later acquisitions of grammar and so on altered human language's status as a system of information exchange, adapted to the needs of the organisms that use it. Language was not different in kind from the other such systems; rather, Marler concluded, "its distinctiveness results from a particular combination of attributes which, considered alone, are not unique to man."[28] Darwin had been right all along.

A New Analysis of Animal Communication

Absorbing the Lessons of Cherry and Hockett

The end of a paper or book often delivers the ideas close to an author's heart but not especially to his or her (or anybody's) competence. Marler's comparisons of humans and animals were suggestive, but they betrayed little sign of an acquaintance with linguistics or the more technical side of communication theory. He raised his game considerably over the next few years, largely in response to two developments. One was his reading of Colin Cherry's 1957 classic *On Human Communication*. Given Marler's interests, the circles he moved in, and the sheer excellence of Cherry's introductory survey, it was bound to fall into Marler's hands at some point. For the twenty-first-century reader, the final chapter, "On Cognition and Recognition," is the immediate standout. Here Cherry sketched psychology in the image of information theory, or, as it would soon be known, cognitive science, dealing with the ways the mind filters, classifies, manipulates, and stores information. For a midcentury ethologist of animal vocalizations, however, there was much else besides to profit from. A lucid account of the principles of spectral analysis taught Marler a lot about what spectrographs could and could not do. But it was the book's far-ranging and interlinked explorations of logic, language, philosophy, and information theory that opened new vistas. Marler acknowledged his debt in his next attempt at grand theory, a paper called "The Logical Analysis of Animal Communication," published in 1961. "As a student of animal behavior who has been grappling with problems of animal communication," he wrote, "the writer has been struck by the relevance to zoology of many of the ideas expressed in Cherry's book. This paper tries to apply some of them to animal communication and to show that they open up new avenues to the understanding of the kind of evolutionary problems with which many zoologists are concerned."[29] We will return to this paper below.

The other sharpening stimulus at that time was Marler's encounter with the work of an American anthropological linguist, Charles Hockett. In the late 1950s, Hockett achieved something remarkable: the rapprochement of the Boasian tradition with Darwinian animal studies. The intellectual breadth that this required was a hallmark of the Hockett style from early days. He was born in 1916, the son of a professor at Ohio State University, where the (very) young man started in 1932. The publication of Leonard Bloomfield's masterpiece *Language* early the next year occurred just in time to become the text in a course Hockett was taking on the scientific study of language. (The Boas-inspired Bloomfield had taught at Ohio State in the 1920s.) Thanks to that exposure, Hockett found himself drawn, as he later put it, to "anthropology wrapped around linguistics." A fellowship in hand, he started graduate studies in anthropology in 1936 at Yale, where another great Boasian, Edward Sapir, presided. Boas himself was still active at the time, and Hockett along with other anthropologists in the region met with the old man at regular meetings. In the now-traditional manner, Hockett wrote a doctoral dissertation based on fieldwork with a native Amerindian people, the Potawatami. More fieldwork followed his graduation in 1939, as well as some interaction with Bloomfield. Drafted shortly after America's entry into the Second World War, Hockett spent most of the next four years providing linguistic services of one kind or another for the military. Along the way he picked up Chinese and Fijian. When the war ended, he got a job at Cornell, where, besides teaching Chinese for beginners, producing technical papers in theoretical and descriptive linguistics, and doing much else besides, he started in earnest on the great questions about language and humankind.[30]

He put the case for the need to address these in a 1948 article in the *American Scientist*, "Biophysics, Linguistics, and the Unity of Science." Here he pledged himself to one of the great philosophical movements of the day: the "unity of science" movement, based, like Bloomfield, in Chicago, and dedicated to reconstructing all the sciences in a common, physicalist language, referring only to what could be observed. Ranging widely over the philosophy of science and the history of science as well as linguistics, Hockett here called for a new discipline that would stand to the social sciences as biophysics stood to biology. In Hockett's view, the successes of biophysics—he had in mind the Scottish biologist D'Arcy Thompson's mathematico-mechanical studies of growth and form—had vanquished the arguments of those great disunifiers of an earlier era, the vitalists, and their claims that biological phenomena could never be brought within the domain of the physical sciences. Now, argued Hockett, the time had come to take on a still influential group of disunifiers: those social scientists who talked of "mind" and "human nature," and treated biology as having no explanatory purchase whatsoever on human social phenomena. As this unifying science of the future did not yet have a name, Hockett gave it one, on the model of "biophysics": "sociobiology."[31]

Coming from someone trained in Boasianism, this proposal was, to put it mildly, unexpected. Hockett's urging on linguists a central role in sociobiology amounted to an especially audacious reversal, for, as we have seen, it was precisely on the assertion of human language's autonomy from biology that Boas, Kroeber, and the others had erected their science.[32] But times had changed. If Boasianism had been a modernist success story, so had the "unity of science" movement. One of its supporters was Bloomfield, and his precedent, including his vigorously stimulus-response, behaviorist approach to language, must have emboldened Hockett.[33] More surprising was the encouragement Hockett received from Alfred Kroeber. The two men were fellows together in 1955–56 at the Center for Advanced Studies in the Behavioral Sciences in Palo Alto, California. Like so many people, Kroeber had become fascinated by Von Frisch's discovery of the bee language, and had started again to think about humans, animals, and language.[34] In this supportive environment, Hockett began to educate himself in the biological literature on language and communication. He read zoologists, ethologists, psychologists, paleontologists.[35] Among the latter was George Gaylord Simpson, one of the architects of the "modern synthesis," itself a unificatory enterprise of the first order, integrating Darwinian natural selection with Mendelian population genetics. Hockett reviewed Simpson's *Meaning of Evolution* for a linguistics journal—a further bid to bring linguists into the sociobiological fold.[36]

Hockett's ruminations on human language and animal communication found their first public expression in the final chapter (where else?) of his *Course in Modern Linguistics*, published in 1958 and soon a standard introductory textbook in the field.[37] The chapter bore T. H. Huxley's old title, "Man's Place in Nature."[38] After locating humankind within the grand taxonomic scheme, Hockett introduced four classic examples of communicating animals: dancing bees, courting sticklebacks, feeding herring gulls, and calling gibbons. "The fact that we class the four examples given above, along with human language, as communicative behavior," he continued, "implies something of our general definition of communication: communicative behavior is those acts by which one organism triggers another."[39] "Triggers" was of course redolent of Lorenzian-Tinbergian ethology. But it was just as redolent of Bloomfieldian linguistics, with its close links to behaviorism.[40] Ever the unifier, where others saw conflict, Hockett had found consilience. His irenic definition of communication now in place, he proceeded to the question of what, precisely, made language different. His answer had seven parts—the seven properties found in all human languages. They were "duality" (when a single utterance can be parsed in two ways, into elementary units of sound and elementary units of meaning); "productivity" (the ability to utter and understand new things); "arbitrariness" (the condition of an utterance that does not resemble what it refers to—as distinct from "iconicity"); "interchangeability" (when there are no restrictions on who can give and who can receive messages); "specialization" (how far communicative signals are used for nothing except communication); "displacement" (the intervening of time and space between a signal's being received and being responded to); and "cultural transmission" (the teaching and learning of communicative habits).[41]

While these could be found in various combinations in animals, only humans, in Hockett's view, enjoyed them all. He summarized the similarities and differences on a grid, with the seven properties arrayed vertically on the left and the five compared species horizontally on the top. Bees came closest to the human total, with four "yes"s, one "slight," and one question-marked "no."[42] The gibbons ran a none-too-close second, but because, like humans, the gibbons were surviving members of the superfamily Hominoidea, they were, Hockett argued, much more instructive than the bees for reconstructing the evolutionary origins of language.[43] Noting that the gibbon call system showed specialization, interchangeability, and glimmers of cultural transmission, Hockett attributed these properties to the communication system of the ancestors common to humans and gibbons. Of the four remaining properties—duality, productivity, arbitrariness, and displacement—he picked out productivity as most likely to have evolved next, both because all of the others could have emerged in its wake (he gave some reasons why), and because it might have begun in a fairly humble way, through the blending of pre-existing calls. As the system grew more productive and flexible, older properties such as cultural transmission would have been strengthened, with dividends for survival, since, as Hockett observed, learning "via already understood symbols may obviate the considerable dangers involved in learning via direct participation." Underpinning these changes was the emergence of *Homo*'s large brain—though Hockett counseled caution as to whether language was primarily cause or effect of this enlargement.[44]

Hockett here showed he had learned well from the synthetic evolutionists. There was nothing remotely Lamarckian about his suggestion that the use of language could have enlarged the brain. He had in mind the possibility that, as he put it, "a new functional development, such as productive language, might be the key factor favoring a genetic selection for larger brains." In other words, language's benefits would have been such that, wherever it found a foothold, individuals who chanced to be born with genes fostering larger, better functioning brains would have enjoyed a selective edge, because they would have been better able to exploit language to survive and reproduce. In thus pressing home their natural advantage, these individuals would in their turn have developed and extended language, thereby increasing its benefits to those who could use it, and placing an even higher premium on genes making for larger brains—genes that the descendants of these language experts would have inherited.[45]

A momentous corollary of this synthetic-Darwinian treatment of the origin of language was the explanation Hockett gave for the old Boasian, Bloomfieldian doctrine that all human languages are equal. According to Hockett, all existing languages had to be equal, since they had all survived the prunings of natural selection. Whatever differences there may have been in the past between inferior and superior languages must have long ago disappeared:

> At the time of the earliest foreshadowings of productivity, there may have been striking differences in the genetically determined abilities of various groups of

Hominoidea to acquire culturally transmitted communicative habits, and correspondingly striking differences in the "languages" and quasi-languages found among the groups. But the workings of natural selection, on both the genetic and the cultural level, eliminated the inefficient strains and the inefficient culturally transmitted habits in relatively short order, so that what we know now as human language is, and for many millennia has been, about equally efficient for all human communities.[46]

He noted how this scenario made intelligible the gulf now stretching between humankind and its nearest relatives. A familiar lesson from Darwinian biogeography was that better-adapted varieties tend to drive to extinction the similar but less well-adapted varieties vying for the same ecological niche. When this lesson was brought to bear on human origins, the inescapable inference was that our ancestors, as the first to evolve a more languagelike communication system, had outsurvived and outreproduced related hominoid lineages; and these, unable to keep up, had eventually disappeared. Man's place in nature is so lonesome, Hockett concluded, because natural selection had favored language so ruthlessly.

In the case of the adaptation or adaptations which changed prehumans into humans—first, and quite early, the genetic changes which were permissive for cultural transmission, and later those which were permissive for language—the differential advantage was enormous, for a wide variety of ecological niches: no half-way genetic adaptation among kindred strains had a chance.[47]

"The Logical Analysis of Animal Communication"

As a fresh rethinking of an elusive subject, it was a bravura performance. There was more to come. Even before the book came out, Hockett's seven properties of language acquired the name under which they would become best known—"design features"—in a lecture he presented at an anthropological meeting in Chicago in December 1957.[48] The following August he introduced more substantial revisions at a symposium on animal sounds and communication in Bloomington, Indiana.[49] His title was "Logical Considerations in the Study of Animal Communication." The design features of language were now upgraded, from mere empirical generalizations about known human languages to, very nearly, the set of necessary and sufficient conditions for a thing's being a language. In principle, Hockett stated, languages might thus be found among some hitherto unknown animal communities, and found absent among some hitherto unknown human ones.[50] The design features themselves, meanwhile, were proliferating. By the time the paper was published, there were thirteen in all. The six new ones were "the vocal-auditory channel" (language is spoken and heard); "broadcast transmission and directional reception" (speech sounds radiate outward every which way, yet listeners can usually locate the speaker); "rapid fading" (speech is transitory, leaving no record of itself); "total feedback" (speakers can hear everything they say—all emitted signals are monitored); "semanticity" (sounds function as symbols—there are "associative ties with things and situations, or types of things and situations, in the environment"); and "discreteness" (when signals are clearly demarcated, and therefore not "continuous").[51]

Of these, the feature that drew the most surprising commentary was semanticity. Here Hockett dispelled any suspicions that, his objective stance notwithstanding, he was out to define language beyond the reach of the animals, and keep it for humans only.[52] Hockett made it clear he was in no doubt that some known animal communication systems exhibited semanticity. The bee dance had it. So did gibbon calls. He took to task those who had failed to give animal systems such as the gibbon's due semantic credit:

> Some anthropological theorists have tended to imply, perhaps unintentionally, that only human communicative systems are semantic. . . . Under our definition this is clearly not so. A hungry gibbon reacts to the sight or smell of food by approaching the food, by emitting the food call, and, presumably, by salivation and other familiar anticipatory behavior. A hungry gibbon reacts to the sound of the food call by this same concatenation of behavior: motion in the direction of the source of the call, repetition of the call, and doubtless the other food-anticipating reactions . . . [implying] that there is some sort of associative tie between food and the food call, whereby either food or the food call elicits a pattern of reactions different from that elicited by, say, danger or the danger call. This is all the evidence we need to class gibbon calls as a semantic system. . . . A word is, as we say, a symbol for something. A gibbon call is a symbol for something.[53]

The whole must have made for uncomfortable listening for Marler, taking part in the symposium with a paper on the more circumscribed topic of birdsong and mate selection.[54] Here was someone else who regarded ethological and protoethological studies as crucial to understanding animal communication, how it differed from human language, and how human language might have evolved. Hockett's revised grid showed, moreover, that each of his design features of language could be found, to varying degrees, in at least one animal communication system—exactly Marler's conclusion.[55] But there were other aspects of Hockett's approach that had to be less appealing. For one thing, Hockett's remarks about how easy it was in general for animals to locate the source of a sound showed he had not read Marler's work in the area.[56] For another, there were some rather caustic asides about methods Marler relied upon heavily. Hockett pointed out, for instance, that, for all that oscillograms and spectrograms were improvements on the old verbal renderings of songs and calls, they "cannot show which features of the sound are communicatively significant for the species or for the particular animal community." In a similar vein, he noted that an "ornithologist's classification of the observed and recorded songs of a particular species or variety of birds may reflect a functional discreteness for the birds, but it may also—though this seems unlikely—be an artifact of the sampling and of our human tendency to pigeonhole rather than to scale."[57] But what seems to have annoyed Marler most was that Hockett had made man the measure of all things linguistic. First came the design features that made language language, then came the assaying of animal communication systems for the presence of these features. Hockett made no apologies. Occasionally, he argued, it makes sense to use "our own species as a point of departure"; the comparative study of communication was just such an occasion.[58]

Marler found Hockett himself good company, and the two were soon in correspondence.[59] In 1960, Hockett published a version of his Bloomington paper in *Scientific American*, thus bringing his design-feature analysis before a much wider audience.[60] Marler's response, "The Logical Analysis of Animal Communication," came out the next year in the less accessible but lively pages of the new *Journal of Theoretical Biology*. At the outset Marler flew the disciplinary flag, contrasting the "anthropocentric approach" to animal communication of comparative psychologists and linguists with the much more satisfactory "objective approach" the ethologists were pioneering. The standard accusations followed: comparative psychologists, Marler complained, had sought not to understand animals, but to discern what made human language unique; over and over again, "this prejudice has influenced the questions that are asked, and therefore the answers obtained." Hence their vain struggle—Hockett was an honorable exception here—to formulate behavioral criteria for purposive communication, in hopes, Marler surmised, of showing that humans met them and animals did not. To escape such bias, one had to turn, Marler advised, to the "'ethological' school" of Lorenz, Tinbergen, and Thorpe. "Instead of approaching animal communication with anthropocentric preconceptions, they set out to describe the natural behavior in objective terms, seeking to derive conclusions about the evolutionary basis of behavior."[61] What followed, then, would be about animal communication in its own right.

For Marler watchers, the thrust of the paper was familiar: when we ask about the information in animal signals, we find clues to why natural selection has shaped different signals in different ways. Familiar as well were most of the animal examples, from singing chaffinches to grinning chimps. But the co-ordinating ideas were completely new. There was, first of all, a three-way distinction, learned from Cherry, between "syntactics" (the study of the formal aspects of signs), "semantics" (the study of their meaning), and "pragmatics" (the study of their usage, their functional contexts). Cherry had presented these as three levels of a science of "semiotic," illustrating them as nested spheres, with syntactics—the most abstract—at the center, embedded first in semantics and then, surrounding the whole, pragmatics. Semiotic in the anglophone world soon acquired a plural name and a largely European identity, indeed a raffish one, calling to mind "readings" of magazine advertisements or plates of food. But Marler's discussion, following up hints in Cherry's, was all about the theory of signs as developed by the nineteenth-century American philosopher (and friend of Chauncey Wright) Charles Sanders Peirce, and more recently by the Chicago philosopher Charles Morris. From Morris's pragmatics, Marler took another set of borrowed distinctions, glossing them in informational terms. Of the classes of signs Morris distinguished, the most important for zoological purposes, Marler wrote, were the "designators," which conveyed information about the signaler or the environment. A version of Marler's old taxonomy of the different kinds of information transmitted in animal signals—species-specific, sexual, individual, motivational, and environmental—now reappeared as subclasses of the designators.[62]

Plaited in with the new semiotic strands was a more sophisticated understanding of information theory. Wiener, Shannon, and Weaver were now points of reference. Marler also now made clear, in a way he never had before—and, for interesting reasons, in a way he would shortly cease to do—that when he talked about information, he was not talking about meaning. Animal signals informed, but that did not mean that they meant, that they stood for things outside themselves. This gap between meaning and information was one of the more counterintuitive aspects of information theory, as Weaver had been at pains to emphasize. Urging readers to see the bright side of this limitation, he had compared an information transfer system thus conceived to "a very proper and discreet girl accepting your telegram," who "pays no attention to the meaning, whether it be sad, or joyous, or embarrassing."[63] Cherry had imparted the same lesson to his readers, declaring information theory for syntactics, not semantics.[64] Marler now did likewise. Reviewing the items of information conveyed to a female chaffinch in the male's song, he added: "This does not imply that the song has any meaning for her, only that it performs selective actions upon her, appropriate to a certain input of information."[65] The successfully transmitted song, in other words, selects from within the female's preexisting repertoire of responses to the world the one that leads her to approach the male.[66] In a similar spirit, Marler argued that syntactics (form) and pragmatics (function) carved up the study of animal signals between them, leaving no room for semantics. "Semantics," he declared, "are of doubtful value in animal studies."[67] It was under the banner of pragmatics, then, that he raised the question of what, exactly, animal signals transmitting environmental information signify. For the most part, warning signals, it seemed to him, communicated the degree of danger, but not its nature, not the sort of agent making things dangerous. But he allowed that animals might be found with calls conveying more precise environmental information—perhaps the European willow warbler.[68]

Primate Communication in Focus

Marler, Washburn, and the New Physical Anthropology

At Berkeley, the bird that increasingly absorbed Marler's research energies was not the willow warbler but the white-crowned sparrow. On the basis of published reports, Thorpe had earlier described it as California's answer to the chaffinch, with a song that varied region to region according to local, learned traditions.[69] Marler soon found it was even better than that. Working with a graduate student, Miwako Tamura, to plot the song map of the white-crowned sparrows in wider Berkeley, he noticed that, unlike chaffinch songs, the songs of white-crowned sparrows were highly uniform within a region. A prediction about the song a white-crowned sparrow would acquire under normal circumstances could thus be much more precise than a comparable prediction about a chaffinch, making the California songster an even better candidate for the song-learning laboratory.[70] The matériel of such a lab was soon in place—spectrograph, soundproofed booths, devices for playback of recorded song, and the rest.[71] The white-crowned sparrow soon proved its usefulness, and the Berkeley birdsong lab flourished, attracting excellent students and launching some eminent careers.[72] Two of the early stars were Fernando Nottebohm and Masakazu "Mark" Konishi, who examined the role of feedback in song learning—the importance of a bird's hearing its own developing song—through the surgical deafening of birds at various stages in the learning process. Both went on to pioneer the neuroethology of birdsong.[73] Meanwhile, the white-crowned sparrow work started on a journey to undergraduate fame, thanks in part to prominent positioning in a textbook that Marler and a colleague published on animal behavior in 1966. The conservatism of textbook writers has ensured that even today, tyro biologists learning about imprinting pass their eyes over spectrograms of white-crowned sparrow songs.[74] In fact, the bird's time as the white rat of "vocal learning" studies (as they came to be known) proved short-lived, as it proved hard to breed in captivity. Successor species such as canaries and zebra finches have had longer working lives. But at Berkeley, the white-crowned sparrow gave Marler superb access to the questions that drew him to ethology in the first place.

Until the early 1960s, all of Marler's professional work as an observer and experimentalist of animal communication—as distinct from his more theoretical and synthetic writings—dealt exclusively with birds, indeed local birds. When the time came to extend his firsthand knowledge in a new direction, it would have been natural enough to turn to the sorts of species studied by his nonbirdsong students: the insects, fish, or small mammals of California. That he turned instead to the monkeys and apes of Africa was due largely to the influence of the physical anthropologist Sherwood Washburn and one of his graduate students, Irven DeVore.

Washburn was a phenomenon, almost single-handedly leading his discipline into what, from the early 1950s, he had been calling "the new physical anthropology" (fig. 8.1). The old physical anthropology for Washburn meant the kind he had learned at Harvard in the 1930s: typological, race-obsessed, and devoted above all to measuring in exhaustive detail the hard parts of human and hominid bodies. He had sensed early on how estranged this project was from the best of modern scientific thinking. The new physical anthropology as he promoted it dealt instead in populations and the functional demands that coordinated bones, tissues, and behavior into adaptive wholes. It avoided even raising questions of "racial quality." Changes in practice went with these changes in orientation. Alongside measurement, according to Washburn, there needed to be experiments, of the sort he had done during the war years at Columbia medical school, cutting and reconnecting muscles in the heads of young rats in order to understand the forces shaping hominid skulls. There also needed to be observations of social behavior in living hunter-gatherers and nonhuman primates. Evolutionary functionalism was the intellectual glue holding these diverse activities together. The dynamic processes available to scientists now—the growth of muscle and bone in a rat's head, the grooming habits of a troop of monkeys—could be regarded as surrogates for the processes at work in the past because, in Washburn's view, Darwinian natural selection adapts organic form to function, whenever and wherever it operates. "The task of the anthropologist," affirmed Washburn, in a paper delivered before G. G. Simpson, Ernst Mayr, Theodosius Dobzhansky, and other proponents of the new synthetic evolutionary biology, gathered at a famous 1950 symposium on human origins, "is to fit knowledge of the primates into the framework of modern evolutionary theory."[75]

Figure 8.1 Sherwood Washburn (1911–2002), 1963. Courtesy of the Bancroft Library, University of California at Berkeley (Washburn Papers, series 7, carton 6, folder 31), and reproduced by permission.

The infusion of functional thinking and experimental tinkering into anthropological anatomy was the central reform, since anatomical studies were the main business of physical anthropology, and nonfunctional, nonexperimental approaches had long dominated. As Washburn wrote in 1952:

> Traditional physical anthropology was based on the study of skulls. Measurements were devised to describe certain features of the bones, and, when the technique was extended to the living, the measurements were kept as close to those taken on the skeleton as possible. From a comparative and classificatory point of

view, this was reasonable, and for a while it yielded useful results, but it brought the limitations of death to the study of the living. Whereas the new physical anthropology aims to enrich the study of the past by study of the present, to understand bone in terms of function and life, the old tried to reduce the living to a series of measurements designed to describe bones.[76]

Physical anthropology as an exclusively descriptive science was coming to an end. Washburn hailed the power of a functional, experimental approach to give meaning to the measurements that otherwise accumulated so pointlessly. To the traditional anthropometrist, a nose was just an opportunity for anatomical measurement, along this, that, and the other dimension, with patterns then sought in the results. But considered functionally, a nose was part of a face adapted to the demands of chewing—demands revealed through experimentation and other nonbiometrical studies:

> Far from being an independent structure which can be described by itself, the nose is an integrated part of the face, and variations in its form can be interpreted only as a part of the functioning face. The form of the nose is the result of a variety of factors. Just how many and how they are interrelated can be discovered only by research, but it seems clear that the most important ones, as far as gross form is concerned, are the teeth and forces of mastication. But these are not included in the traditional descriptions of the nose, nor will looking at skulls or measuring them give this kind of information.[77]

The usefulness of behavioral studies, in addition to experimental ones, was something he had glimpsed as a graduate student, when he had accompanied the Asiatic Primate Expedition to Siam, charged with assisting the anatomist Adolph Schultz with dissecting dead gibbons and the psychologist Carpenter with observing live ones.[78] It was not until the mid-1950s, however, and some chance observations while in East Africa of free-ranging baboons—more conspicuously intelligent and complexly social than gibbons, and much more easily observed—that Washburn began to incorporate behavioral studies in a serious way into his disciplinary reforms.[79] By now, he commanded the institutional resources to do it very effectively. A professor in the stellar anthropology department at the University of Chicago, with positions in all the major professional organizations and journal editorial boards, he was well placed to extract large sums from granting bodies, including the deep-pocketed Wenner-Gren Foundation, set up recently to advance anthropological understanding. The first beneficiary of Washburn's well-funded behavioral turn was Irven DeVore, a Chicago graduate student whose background in social rather than biological anthropology was, in Washburn's view, just what a decent observer of baboons needed. At first reluctant, DeVore quickly understood that this was not an invitation open for him to refuse, and within a few weeks, Washburn's intelligence and enthusiasm had won the younger man over completely.[80] In June 1959, a report on their joint field observations of baboons was presented to a conference memorialized two years later in *Social Life of Early Man* (1961), edited by Washburn. Washburn and DeVore's paper in the volume included what became a kind of emblem of the Washburnian approach: a chart comparing baboons and "preagricultural humans" along eleven variables, including group size, diet, play, sexual behavior, dominance, and use of sounds and gestures.[81]

By 1961, Washburn had moved to Berkeley, taking the new physical anthropology with him, DeVore and all. Soon Marler found himself teaching, examining, and generally getting to know a growing number of anthropologists interested in animal behavior. Ethological zoology and Washburnian anthropology were not, to be sure, wholly congruent. As we shall see, the different weightings they assigned to anatomical experiments in part led, a few years later, to Marler's change of mind about the semanticity of animal calls. Still, there was overlap enough for fruitful collaboration. Along with the comparative psychologist Frank Beach, Marler and Washburn created Berkeley's first animal-behavior station. It became home to a primate colony where anthropologists trained before heading out to the field. Marler thought Washburn himself enormously impressive—a kind of latter-day Cuvier, seemingly able to conjure vanished forms out of single bones.[82] For his part, Washburn encouraged Marler's primatological interests, asking him, for instance, to review a new book on field studies of the mountain gorilla by a young American zoologist, George Schaller, for *Science*.[83] But the senior man was far too busy for the kinds of informal discussion that can transform curiosity into commitment. It was his students, and especially the gregarious DeVore, who did the most to tempt Marler toward primate behavior.[84] No doubt they were largely responsible for his agreeing, at rather short notice, to survey recent advances in the study of primate communication when the invitation came around New Year 1963. As he set to work educating himself, primates were still no more than a side interest. But by the end, they had become rivals with the birds for his attention.[85]

The invitation was from the Primate Project, a year-long symposium organized in part by Washburn, and taking place at the nearby Center for Advanced Studies in the Behavioral Sciences at Stanford (host to Hockett some years before) in 1962–63. As DeVore explained in his editor's preface to the culminating volume, *Primate Behavior: Field Studies of Monkeys and Apes* (1965), the aim "was to bring together a group of people who had just completed long-term field studies of monkeys and apes, and by providing them with sufficient opportunities for discussion over many months, to arrive at mutual understandings concerning the description and interpretation of primate behavior."[86] Just why, at that moment, did so many people find themselves studying monkeys and apes in the field, or engaged in comparably serious studies of captive or semicaptive primates? We have looked at the considerations that led an American physical anthropologist to channel his students that way. But rather different considerations led others—and the bodies funding them—in the same direction.[87] In England, for instance, there was now a colony of rhesus monkeys under observation at Madingley, thanks to Robert Hinde's collaborations with the psychologist John Bowlby and their mutual interest in using primates to model the effects of parent-child separation in humans. As Hinde's expertise with monkeys developed, he attracted the attention of the Anglo-Kenyan hominid paleontologist Louis Leakey, who, independently of Washburn, had decided that living apes offer a window onto the prehuman past. Leakey had offered the job of ape watcher to his then-secretary, Jane Goodall, got her started observing chimpanzees at the Gombe Stream Chimpanzee Reserve in Tanganyika, and then arranged a crash course for her in behavioral zoology with Hinde.[88] Distinctive yet again was Stuart Altmann's path to the primates. As a Harvard graduate student in biology in the mid-1950s, he had set out to revive and update the prewar Carpenter tradition, observing monkey societies in the field but theorizing about them as cybernetic systems. The project was regarded as so idiosyncratic that the only faculty member prepared to supervise was a young professor of ant behavior, Edward O. Wilson.[89]

In the early 1960s, then, the primatologists—as they were not yet called—were a diverse group, with much to talk about. Marler was not of their number. But he had a matchless knowledge of animal signaling and a demonstrated ability to synthesize disparate materials. When the Primate Project fieldworkers turned to him, about midway through their year together, to distill some generalizations from their data on communication, he found the pickings extremely rich. A table assembled for his published paper, comparing the sizes of the vocal repertoires of various species, drew almost wholly on recent work. For the gorilla, there was Schaller's field study of mountain gorillas in the Congo. For the chimpanzee, there was, in addition to Goodall's work (represented elsewhere in the volume, though not here), the field study of London-based Vernon and Frances Reynolds in the Budongo Forest in Uganda. For the baboon, there were the field studies of DeVore (in Kenya) and the Bristol zoologist Ron Hall (in southern Africa). For the rhesus, there were Altmann's field studies on Cayo Santiago, and the more recent work of Hinde and Thelma Rowell with the Madingley rhesus colony. For the Japanese macaque, there was data from Hiroki Mizuhara, a fellow at the Center for Advanced Studies in the Behavioral Sciences and representative of a thriving Japanese primatological tradition. For the langur, there was the field study of a former Washburn student, Phyllis Jay, in north India. For the howler monkey, there was another of Altmann's field studies, as well as Carpenter's. And for the brown lemur, there was the work of Richard Andrew, a graduate (like Rowell) of Cambridge ethology, now based at Yale and absorbed in investigating vocalizing primates in his lab and at the Bronx Zoo.[90]

What did Marler make of it all? For the most part, he was struck by how different primate communication systems—and those of the higher primates especially—were from the bird systems he knew so well. On his sifting of the data, three major contrasts emerged. One was that, where the vocal or visual signals of birds tended to be complete unto themselves, the signals of monkeys and apes tended to be composite. Interacting at close range, primates seemed to express themselves with face and body and voice all at once. By and large, a signal in one sensory modality could thus be understood only in the communicative round.[91] A second, related difference was that, where bird vocalizations were often highly discrete in structure, and could be quite melodious, primate vocalizations tended to fall within graded, rather unlovely continua. Graded signaling made life much more difficult for the investigator, for, as Hockett had warned in his "Logical Considerations" paper, it was all too easy to identify as two or more signals what were in reality varieties of a single signal. Rowell, indeed, had reported tumbling into just this trap in her rhesus studies, realizing only gradually that what she had picked out as nine separate sounds in fact made up a unitary system.[92] The same properties that put human students of primate communication at a disadvantage were, however, advantageous to the primates themselves, according to Marler. Composite, continuous, close-range signals had much richer potential than their opposites, in his view, to communicate "information about the slightest changes in the nature and intensity of moods in the signaling animal."[93] Furthermore, he argued, the high proportion of such signals among some primates made sense on selectional grounds. Rhesus monkeys, baboons, gorillas, and chimpanzees rarely cohabited with creatures whose signals might become confused with their own, so evolutionary pressure for signal distinctiveness tended to be low. At the same time, these animals had such complex social lives, soaking up so much communicative energy, that there was a premium on the use of signals capable of conveying graded information—hence the abundance of graded, composite signals.[94]

The near-monopoly of social matters among signaling primates marked the third major contrast that Marler found with birds. Where birds in their vocalizations could sometimes convey fairly precise information about their environments, primates seemed to converse about their environments hardly at all, and what they did say was unimpressively vague:

> By far the greatest part of the whole system of communication seems to be devoted to the organization of social behavior of the group, to dominance and subordination, the maintenance of peace and cohesion of the group, reproduction, and care for the young. Interindividual relationships are complex enough in monkeys and apes to require a communication system of this high order of complexity. But there is little application of the communication system to events outside the group, beyond the existence of signals signifying potential danger. There are no calls associated with, say, water or food, such as are known in some birds. . . . Environmental information, present or past, figures very little in the communication systems of these animals, and a major revolution in information content is still required before the development of a variety of signals signifying certain objects in the environment and a system of grammar to discourse about them can be visualized.[95]

Musing, in closing, on this designative deficiency among the nonhuman primates, Marler ranged widely over the ontogeny and phylogeny of language. He noted that human children use graded vocalizations before learning the local, discretely organized language. All around us, in other words, graded communication systems mature into discrete ones—and in primates no less. What, asked Marler, could have motivated a parallel change in the past: the shift in a single hominid lineage from a graded, genetically inherited communication system to a discrete, culturally inherited one? Natural selection was the obvious short answer. To go further, one needed to know under what conditions, precisely, it had become advantageous to acquire language through vocal learning. The answer to that question, Marler concluded, with a Washburnian flourish, can be reached only with further study of communicating primates in nature.[96]

The Primatological Field Studies of Marler and Struhsaker

Unprecedented in quantity and quality, the new field data, in Marler's view, nevertheless left much to be desired. The fieldworkers' main interest was social life, and as Jane Lancaster, a Primate Project contributor, later noted, they had tended to treat communication as "a means of entrance into the workings of the social system, . . . not an object of study in itself."[97] At the start of his survey, Marler sketched what he considered a better way. Investigations should proceed, he advised, in two steps: first, the assembly of a trustworthy inventory of signals and the responses these typically evoke; second, the test-ing of generalizations with field experiments, of the sort J. Bruce Falls was pio-neering with birds.[98] In his most recent paper, Falls had shown that playback of artificially modified recordings of song—altering pitch, length, sequencing of elements, and so on—could be used to identify the song properties clue-ing birds to whether a songster belongs to their species.[99] Here was a kind of acoustic updating of what Lack had done with a stuffed robin and Tinbergen with cardboard hawks. Falls's new method also provided a brilliant riposte to Hockett's complaint that bioacoustic studies ignored the question of what animals attend to in their own sounds. Synthesis and playback of birdsongs were more easily accomplished, Marler conceded, than the same with primate vocalizations, given the often graded and composite character of the latter. Nevertheless, field experimentation would, in his estimate, be a major task for the primate communication students of the future.[100]

How hard it was even to reach that stage was something Marler would shortly understand more vividly. In 1964–65—the year student unrest boiled over in Berkeley—he took sabbatical leave in Uganda in order to study the vocalizations of its richly varied apes and, especially, monkeys. Nothing in his field experiences with birds in Europe and America had prepared him for the trials of studying forest monkeys in Africa. They proved very wary of humans, so that observation and recording often had to take place from the road. Worse, they lived high up in the dense canopy, making them hard to see and even harder to track. In the end, Marler concentrated his efforts on the vocal communication of two kinds of colobus and two kinds of guenon. He recalls doing a few playbacks; none, however, made it into the published reports, which stuck instead to description, spectrographic analysis, and the probing of structure-function relationships along familiar lines.[101] A couple of years later, the Marlers returned to Africa, for a summer with the chimpan-zees at Gombe Stream in Tanzania (as it now was) as the guests of Jane van Lawick–Goodall (as she now was). Over the years, Goodall had developed a system of baiting—little cement cubicles loaded with bananas and topped with lids opened from a distance—which brought lots of chimps right up to the camp. For six weeks, as Goodall's husband, the acclaimed wildlife pho-tographer Baron Hugo van Lawick, filmed the chimps in full expressive flight, Marler recorded their vocalizations and other sounds on tape (fig. 8.2), while Goodall or one of her assistants described what was observed into a Dicta-phone. "The vocalization study is going *EXCELLENTLY*," Goodall wrote to Leakey that August. Besides the film—a unique, and uniquely useful, docu-ment—the summer project yielded the first evidence that individual primates

Figure 8.2 Marler's recording of chimpanzee vocalizations at Gombe Stream is interrupted, 1967. Courtesy of Peter Marler and reproduced by permission.

have individualized calls. There were not, however, experimental playbacks, since Goodall forbade these, fearful that the wrong recording played to the wrong animal might disturb the social structure catastrophically.[102]

By now, Marler was professor at the new Institute for Research in Animal Behavior, a joint venture of the New York Zoological Society—sponsor of the Bronx Zoo—and the Rockefeller University, also in New York City. A kind of research mecca for the sciences, the Rockefeller gave its professors freedom from undergraduate teaching and what amounted to personal departments. Working alongside Marler as junior faculty were two recent PhDs from his Berkeley laboratory, representing his two main research interests: Fernando Nottebohm and Thomas Struhsaker.[103] Although Nottebohm collected some recordings of monkey vocalizations in the course of a field study of parrot vocal behavior in Trinidad, his forte was, and would remain, laboratory studies of birdsong.[104] Struhsaker, however, had become a bona-fide field primatologist, indeed one of the few from Berkeley who had arrived via zoology rather than anthropology. Beginning in late spring 1963, after six months' searching for the right site and the right subjects, he had spent around a year at the Masai-Amboseli Game Reserve in Kenya, studying the behavior and ecology of another guenon, the vervet monkey (*Cercopithecus aethiops*). The circumstances proved ideal. With prodigiously varied fauna and Mount Kilimanjaro as a backdrop, Amboseli was picture-book Africa—just the sort of setting that had attracted Struhsaker to primate research (his first published paper was on flight in bats). The site also offered crucial practical advantages, notably its short grass, making for easy tracking and observation of the often ground-dwelling monkeys. Just as important, the vervets turned out to tolerate human company well (fig. 8.3). There was even scientific expertise on hand, in the form of Stuart Altmann and his wife, Jeanne, a mathematician. They had been in search of a field site when they met Struhsaker in Nairobi, about a month after he had settled at Amboseli. The rapport was instant. Struhsaker invited the Altmanns to join him at Amboseli, which they promptly did, soon launching what became a long-term project on baboon social behavior. For the next year, the three dined and discussed together nearly every evening. Effectively Stuart Altmann became Struhsaker's field supervisor, and when the time came to write up the vervet data, he did it back at Altmann's academic base, the University of Alberta.[105]

Vervets, Struhsaker found, do a lot of gesturing and vocalizing. Given the volume and variety of these activities, a serious interest in vervet communicative behavior was always on the cards, even without Marler and Altmann as teachers. As part of Struhsaker's comprehensive study of vervet behavior and ecology, he carried out a detailed observational and spectrographic survey of the vervets' vocalizations. In December 1964, at a conference Altmann organized in Montreal on primate social communication, Struhsaker presented his findings. On his count, there were twenty-one different "stimulus situations" evoking vocalizations in vervets, and upward of thirty-six different vocalizations. The phrase "stimulus situation" had been Carpenter's, of course, and the centerpiece of Struhsaker's paper was a table clearly modeled on Carpenter's vocalization tables. Stretching over four pages in its published form,

Figure 8.3 Thomas Struhsaker with vervets in Amboseli, Kenya, around 1963. Photograph by Stuart Altmann. Courtesy of Thomas Struhsaker and reproduced by permission.

the vervet version listed, for each stimulus situation, the sounds elicited, the age and sex of the animals making the sounds, the probable "message" communicated (where Carpenter's tables had "probable function"), and the responses observed in surrounding vervets and other animals.[106] Situations 14 through 19 dealt with nonvervet threats in the environment, including predators.

In light of the later fame of vervet alarm calls for snake, eagle, and leopard, a couple of points are worth noting. First, as the numbering scheme might suggest, the alarm calls were given no special prominence in the paper. There was no flagging up of the alarming situations as somehow especially important, for questions of animal semantics or anything else.[107] Second, with the exception of what Struhsaker called the "snake chutter," the famous trio of predator calls is not easy to locate in the six alarming situations.[108] Two of the situations involve eagles—numbers 18, "initial perception of major avian predator," eliciting a "rraup," and 19, "proximity of major predator (mammalian and avian)," eliciting both a "threat-alarm-bark" and a "chirp." Among the occasionally proximate mammalian predators the vervets met were leopards. Like eagles, leopards elicited "chirps," as did cheetahs, serval cats, and lions. Furthermore, according to Struhsaker, vervets "*uh!*"-ed at spotted hyenas and Masai tribesmen, "*nyow!*"-ed at baboons and bush cats, and chuttered in non-snake-chuttery ways at human observers like himself.

Moreover, the "chirp" and other vocalizations functioned, in his view, not just to alarm the preyed-upon, but to threaten the predators.[109]

It was Marler, not Struhsaker, who picked out just three alarm calls from this rich deposit and described them as tantalizingly semantic, representational, symbolic:

> Field study of the vocalizations of the vervet monkey reveals at least six sounds that seem to be provoked by the presence of predators [citing Struhsaker]. The call given correlates with the identity of the predator, so that a snake elicits a different call than does an eagle or a leopard. The behavior of the predator is also significant. Flying and perched eagles elicit different calls, for example. There is some evidence that the signals represent different environmental situations to respondents. The initial responses that they elicit are in some cases different. The "snake chutter" of the vervets elicits approach and examination of the snake from a distance. When they hear a "chirp" call, given in response to a leopard, the monkeys run to trees and climb to the topmost branches. In response to a "rraup" call, given for an eagle, they run from open areas into thickets and descend from treetops. Direct perception of these three types of predator may elicit the same three patterns of response. Although the relationships are not specific enough for us to think of the three signals as names for the three types of predator mentioned, in principle, these signals begin to approach the phenomenon of object naming.[110]

The occasion was a review paper, "Animal Communication Signals," published in *Science* in August 1967. It had started life two years earlier as Marler's contribution to a Wenner-Gren symposium on animal communication. Once again, his brief was to tour the research horizon from an ethologist's perspective. Fresh from his year in the Ugandan forests and mindful, no doubt, of the psychologists and linguists present, Marler announced at the start that any nonanthropocentric study of animal communication had to deal in animal responses to animal signaling. The rest of the paper advised attention to such matters as how responses orient animals spatially, the different kinds of behavior stimulated by signaling, correlations between signal variation and response variation, the way the different parts of a signal function (and how useful Fallsian experimental playbacks with artificial signals can be here), the range of sensory modalities in play in animal communication, and the range of situations stimulating particular signals. It was under this last heading that Marler introduced the vervet alarm calls to the wider world.

In his 1961 "Logical Analysis" paper, he had dismissed semantics as irrelevant for understanding animal communication. Revisiting the subject now, he showed plenty of the old skepticism. He noted with disapproval, for instance, the way that semantic considerations forced distinctions with little biological or logical warrant, as between alarm signals and sexual signals.

Both kinds of signal, he pointed out, serve the Darwinian ends of survival and reproduction. Sexual signals, moreover, could be construed as representing or symbolizing inner states. Yet only alarm calls, supposedly representing or symbolizing things in the external world, receive the honorific "semantic." Restricting attention to such apparently externally referring signals, there were, in Marler's view, a number of interpretive difficulties. Consider that an animal responding to a signal enters what is arguably a new stimulus situation. Is the ensuing behavior a response to this new situation, or to the former, signal-evoking one? Marler illustrated by way of Hockett's assertion that a call eliciting feeding behavior in gibbons symbolizes food. Marler distinguished two possibilities. It might be that, on hearing the call, a gibbon begins to salivate and to respond in other ways preparatory to feeding. But it might also be that the call merely stimulates a hungry animal to approach, with feeding behavior arising only after the respondent sees or smells food. "In the latter case," wrote Marler, "it is less easy to decide whether the food call really represents food to the respondent, for other stimuli will also elicit approach even in a hungry animal." Even the impressively semantic vervet calls had their share of complications. He noted that the calls in themselves do not always suffice to determine the pattern of response. Vervets in a thicket when they hear the call for "eagle" tend to crouch down, whereas vervets that hear the call when out in the open tend to run for the nearest thicket. "In such cases," he concluded, "the role of the context must be taken into account in a consideration of whether a semantic system is involved."[111]

In what sense, then, were *chutter*, *rraup*, and *chirp* the vervets' names for, respectively, snake, eagle, and leopard? Marler offered no clear verdict, and his lingering discomfort with the question was manifest. Why raise it, then? Undoubtedly one reason was sheer surprise at what Struhsaker had found. As we have seen, Marler came away from the Primate Project persuaded that the vocalizations of monkeys and apes convey hardly any information about their environments. The vervet calls seemed to fill precisely the gap Marler had described, and on that account alone, they must have struck him as deserving wider notice. Less obvious, however, are his reasons for publicizing them under the banner of semanticity, symbolization, object naming—all concepts he had previously kept at arm's length. We turn now to the other developments behind Marler's semantic turn.

Neuroscience, Emotions, and Vocalizing Primates

Like all good mysteries, this one has a red herring: the work of W. John Smith. Six years younger than Marler, Smith was a Canadian-born, Harvard-trained zoologist with interests closely paralleling Marler's own. In a 1963 paper, "Vocal Communication of Information in Birds," and a 1965 paper, "Message, Meaning, and Context in Ethology," Smith had argued that animal signals in themselves convey only general messages, which acquire specific meanings in context.[112] With Marler in attendance, Smith had presented his case at the 1965 symposium on animal communication.[113] When Marler wrote of the need to take context into account in deciding the semanticity of the vervet predator calls, he cited Smith. More than ten years later, when Dorothy Cheney and Robert Seyfarth returned from doing playbacks with vervets in Amboseli, they presented their results, to funding agencies and then the world at large, as overturning a Smithean consensus.[114] But Smith's thesis could hardly have provoked Marler all that much. For one thing, Smith had credited Marler as a pioneer in bringing signal context to the ethological fore (and thanked him for helpful discussion).[115] For another, as we have seen, Marler regarded primate signals as overwhelmingly composite in nature. At most, Smith had promoted to a rule what Marler was prepared to accept as an often-correct generalization, especially as applied to apes and monkeys. Indeed, what Marler conceded about a vervet's location determining its response to a "rraup" call was all Smith required for the call's meaning to count as contextual. The vervet calls were thus Smithean whether or not—to pose the most famous contextual question later settled by the playback experiments—the calls evoke predator-specific responses (looking up at the eagle call, down at the snake call, and so on), or merely prompt general looking around, with adaptive responses following the sighting of a predator.[116]

The one clear disagreement between Marler and Smith turned on a point neither original to Smith nor, at first, put with much vehemence. In Smith's view, animal signals in themselves referred not to the world, but to the inner state of the signaler—to, as he put it in his 1965 article, "some aspect(s) of the state of the central nervous system (CNS) of that individual," including a state of readiness to act in a certain way. He did not present this argument as especially revisionist or skeptical or placing a barrier between animals and environmentally referring humans. On the contrary, he gave the impression that his argument was orthodox communication theory and applied equally to humans and nonhumans.[117] He did not comment in print on the vervet alarm calls until his 1977 synthetic book *The Behavior of Communicating: An Ethological Approach*. There, true to form, he presented the calls as conveying escape messages, albeit unusually narrow ones. *Rraup* referred, for Smith, not to eagles, but to a readiness to escape into the thickets; *chirp* referred not to leopards, but to a readiness to stay in cover or go find some up a tree. Looking next at birds, he argued that claims about alarm calls that "effectively name a single class of predator" have always evaporated on closer inspection, as lots of different things have invariably been found to elicit the call. "The conservative interpretation at present," Smith concluded, "is that alarm displays have no referents external to the communicators."[118] Here a gauntlet was laid down.[119] But even in 1977, the theme of environmental reference in animal signaling occupied a small part of Smith's attention. Back in 1965, it barely registered at all.

The anthropologists, however, were another matter, as Marler soon learned. His symposium talk discussing vervet calls took place in Austria in June 1965. A few months later, he was again in Austria, again for a Wenner-Gren symposium, this time on primate social behavior. Another participant was one of Washburn's PhD students at Berkeley, Jane Lancaster. As noted above, she had been part of the Primate Project. Now she presented an interim report from her dissertation research on—to use the title of her published paper (also of her dissertation)—"primate communication systems and the emergence of human language." Her thesis did not comport well at all with the possibility that vervets refer in their calls to snakes, eagles, and leopards. For Lancaster, humans alone had the ability to name objects in the environment. Indeed, she saw the evolution of this ability as the true beginning of language, since language, in her view, was first and foremost a system of names, organized in traditional ways. The communication systems of the other primate species, she wrote, "have little relationship with human language, but much with the ways human beings express emotions through gesture, facial expression, and tone of voice." In support, she appealed to three classes of evidence: communication in monkeys and apes, so overwhelmingly social in character; the acquisition of language in children, whose first words, she reported, tend to name objects, and who do not express emotions with words until quite late (not because children recapitulated the evolutionary past, but because they developed in ways shaped by that past); and third, anatomical and physiological studies of the brains of monkeys, apes, and humans. According to Lancaster, object naming had been shown to have a separate anatomical basis from emotional expression—and a uniquely human one.[120]

She took her neuroscience straight from the Boston neurologist Norman Geschwind's long, two-part review of the scientific literature, "Disconnexion Syndromes in Animals and Man," published that same year. The disconnections concerned were between different parts of the brain, as in aphasia, Geschwind's specialty. In outline, the Geschwindian picture—a composite of old and new facts, principles, conjectures, theories—was as follows. In monkey, ape, and human brains, different regions of the cortex have different functions. Some of the regions receive sensations; one region receives visual sensations, another auditory sensations, and so on. Each sensory region connects with an adjacent "association area," which in turn connects with other parts of the brain, enabling associations between sensations and movements, or between sensations and emotions. In nonhuman animals especially, there are abundant connections with the limbic system, lying below the cortex, and widely associated with the emotions. In no animals are there abundant connections between the sensory association areas themselves; so, by that route, there can be only weak associations between one kind of sensation and another kind. In human brains, however, there was a further association area, found in the inferior parietal lobule, in a region containing a structure called the "angular gyrus," barely present or absent from monkey and ape brains. Connected to each of the different sensory association areas, this structure enabled humans easily to form "intermodal" or "cross-modal" sensory associations, such as those required for naming (as when the sight of a snake comes to be associated with the sound "snake"). On this picture, the limbic system and the angular gyrus make for sharp contrasts. The limbic system is animal, ancient, impassioned, inflexible, submerged. The angular gyrus is human, recent, dispassionate, flexible, surface. Language emerged with the evolution of the angular gyrus and the intermodal associations it made possible.[121]

The consequences for interpreting vervet calls were straightforward. However much they look like names for snakes, eagles, and leopards, this cannot be the case, since vervets lack the cerebral equipment needed for such precise environmental reference. As Lancaster put it:

> The rare examples of nonhuman primates' communicating some information about the environment in their high-intensity alarm calls are not relevant to the evolution of a system of object-naming. The similarity between these abilities is only a minor, superficial one, and the underlying mechanisms are entirely different. . . . [I]t is apparent that the ability of the vervet monkey to refer to the environment, a design feature which is superficially similar to man's object-naming but which is in fact based on quite different underlying neurological mechanisms, cannot be suggested as representing a possible step toward language. Object-naming is unique to man because the anatomical basis of the ability is also unique to man.[122]

She did not here cast doubt on whether different kinds of predator evoke acoustically distinct calls from vervets, or whether those calls elicit adaptive responses in surrounding vervets. She took all that to be true. What she doubted was that the vervets were representative. For her, they were a special case, intelligible but not to be generalized from. They were, on her presentation, rather like apes that had learned a few words after extensive training by humans. What was striking about apes, Lancaster argued, was not that a little language could be drummed into them, but that it could be done only with extraordinary difficulty. Similarly, the vervet case should teach us not that animals can name objects in their environments, but that, when species have come under sufficient pressure from a number of quite distinct dangers, natural selection may have differentiated their alarm calls in ways that mimic environmental reference. But there was all the difference between as-if naming and the real thing.[123]

On the face of it, this slamming down of a language barrier between humans and animals was the last thing to be expected from the new, baboons-and-bushmen physical anthropology. But in fact Lancaster's argument pressed a couple of Washburnian buttons. It fit well, as she showed, with one of Washburn's major themes, the importance of tool use in the making of modern humans. Like tool use, object naming would have greatly expanded the power of our hominid forebears to exploit the environment in ways that gave them a selective advantage. Even better, she had reached this conclusion in fine Washburnian fashion, combining the results of field studies of primate behavior and classy research in functional anatomy.[124] And the research came no classier than Geschwind's, whose 1965 paper was bringing an end to decades of disenchantment with localization studies. There had, to be sure, been much progress since the days of Ferrier—the sensory association areas were identified in the 1890s and 1900s, the limbic system in the 1930s and 1940s—but it was felt that no useful general principles linking brain and behavior had emerged.[125] This was what Geschwind's paper supplied. It was as much backward-looking as forward-looking—an attempt to restore a classical but neglected neurological heritage. To anyone familiar with the 1860s, indeed, its conclusions about language have a strongly déjà vu quality: the angular gyrus as the hippocampus minor come again; the distinction between neocortical object naming and limbic emotional expression an updating of the Jacksonian doctrine distinguishing higher, propositional, humans-only language from lower, emotional, animal expression.[126]

There was more to the new comparative neuroscience of language than old ideas, of course. Some innovative experimental work was on view at the Montreal meeting where Struhsaker talked of vervet alarm calls. Bryan Robinson, from the Yerkes Regional Primate Research Center in Atlanta, announced the development of a "tele-stimulation" unit that would enable a new mixing of lab and field: while monkeys with electrodes embedded in their brains, and solar-powered stimulation and transmission units on their heads, interacted with their social group, a researcher hidden out of sight would, by remote control, selectively stimulate different parts of different individuals' brains.[127] Detlev Ploog, at the Max Planck Institute of Psychiatry in Munich, reported work moving in a similar direction, integrating studies of brains, vocal and other behaviors, and social dynamics—in his case, with laboratory colonies of squirrel monkeys.[128] But it was Geschwind's synthesis that gave such investigations moment, showing how they bore on the big questions about mind, brain, language, and evolution.[129] When W. H. Thorpe, at a November 1965 meeting on language and the brain, argued for parallels between birdsong and human language, he got a Geschwindian rejoinder from Horace Magoun, a pioneering student of the cerebral bases of emotional expression. According to Magoun, animal experiments and anatomical studies all "seem to oppose the view that man's capacity for speech evolved from the abilities for emotional vocalization present in lower animals":

> On the contrary, man's communication by symbols, both vocal and written, appears to represent an entirely novel functional increment related to acquisition of associational cortex. . . . One can conclude that there are two unrelated central neural mechanisms for vocal expression in vertebrates: one for nonverbal affective communication, widely present in the animal brain stem, and a second for verbal communication, present only in the lateral neocortex of the brain of man.[130]

Two years later, Robinson, citing Geschwind on the human brain—and Marler on how primates vocalize mainly "in situations bearing some emotional valence"—echoed the point:

> it appears that human speech and primate vocalization depend on two different neural systems. The one is neocortical; the other, limbic. This suggests that human speech did not develop "out of" primate vocalization, but arose from new tissue which permitted it the necessary detachment from immediate, emotional situations. The neurological evidence suggests that human language arose in parallel with primate vocalization, surpassed it, and relegated it to a subordinate role.[131]

Psychology, Symbols, and Language-Trained Apes

Provoked by neuroscientifically inclined anthropology to reconsider simian semantics, Marler soon found that shift reinforced by developments in comparative psychology. When Marler and Lancaster spoke in Austria, the best-known attempt to teach the names of objects to an ape was that of Keith and Catherine Hayes.[132] In 1947, shortly after arriving at the Yerkes labs in Florida, they had begun an experiment to see how far an infant chimpanzee raised in a human family would develop the abilities—especially linguistic—that children acquire. Neither Hayes was an obvious candidate for such a study: Keith was, in Cathy's words, a "'rat psychologist' at heart," fresh from the doctoral program in psychology at Stanford, while Cathy was a journalist. But they were fascinated by the central psychological problem of nature versus nurture, and by the prospects a home-raised chimpanzee would offer for illuminating the human case. Raise a chimpanzee as a child, they argued, and whatever differences arise from children can be put down to differences in genetic endowment. They were aware, of course, that the experiment had been done before. But they were intent on avoiding what they saw as the shortcomings of the earlier attempts. Certainly the Yerkes labs, with experts including Karl Lashley and Henry Nissen on hand, offered as advantageous a backdrop as could be hoped for. Brimming with confidence, the couple took home a few-days-old chimpanzee named Viki. They loved, stimulated, and spoiled her, and she advanced in lots of ways. But she remained mostly silent. Having decided that their adopted daughter was never going to pick up speech the way human children do, they launched an intensive program of speech therapy, lasting the rest of Viki's life (she died of a sudden illness in 1954). It yielded only four, hoarsely whispered, coarsely referential words: "mama," "papa," "cup," and "up."[133]

For the Hayeses, as for Lancaster later on, the main lesson, as Cathy Hayes wrote in her charming memoir *The Ape in Our House* (1951), "lies not in the fact that [Viki] has learned a few words, but rather in her great difficulty in doing so, and in keeping them straight afterward."[134] Here was a vital clue to what made humans human. The source of Viki's difficulty, according to the Hayeses, was not her vocal equipment. In her (infrequent) spontaneous vocalizations, they reported, Viki uttered most of the sounds found in spoken English. Nor was the problem a failure of general intelligence. In their view, a battery of problem-solving tasks set to Viki had shown that, in every way apart from language, she was an intellectual match for human children of her age. The diagnosis they favored initially was neurological: what held Viki back from language was the absence of the part of the brain that in humans enables speech.[135] Indeed, when, in the spring of 1950, they took Viki to a speech clinic at the University of Michigan, they were told that her condition most resembled that of human aphasics.[136] Later on, however, they backed off from this explanation, attributing Viki's failure instead to a low drive for communicative play. Unlike children, Viki did not babble—something her adoptive parents fretted over from the start. In a 1954 paper, "The Cultural Capacity of Chimpanzee," the Hayeses suggested that selection for babbling was what ultimately separated the hominid lineage from the rest of the primates. Babbling infants grow into speaking, cultured humans, they argued, because babbling leads to vocal mastery, and that mastery is the entry point both for speech and, through language, for culture.[137]

That same year, the Yerkes labs hosted another couple who would go on to make their names doing and writing about experiments with apes and human language. Like Keith Hayes, David Premack arrived as a newly minted PhD in experimental psychology—in his case, from Minnesota.[138] His introduction to the natural communicative life of chimpanzees occurred in the course of his first night, spent on the grounds of the labs. As his wife, Ann, later recalled, they awoke "to the sound of shrieking. The shrieks increased in volume, surrounded us, echoing and reechoing until even our intestines were vibrating. Sixty animals shook the bars of their cages in unison. After several minutes, although the thunderous noise subsided, my terror remained."[139] Taking up a post at the University of Missouri shortly thereafter, David Premack started an experimental program looking in several ways like the Hayeses' approach. The plan was for infant chimpanzees to be raised in a lab-based simulacrum of a human home, with loving "mother surrogate" minders, visual surroundings that would arouse curiosity, and exposure to lots of language. But Premack was only partly motivated by the nature-nurture concerns of the Hayeses, and his choice as to *what* language to train the chimps in reflects this difference. Unlike the Hayeses, Premack saw himself as creating a model system for the experimental study of language. Almost by definition, such a system had to be simpler than what it modeled in order to be useful scientifically. Furthermore, the Hayeses had already shown how futile it was to expect a chimpanzee to acquire spoken English. So Premack dedicated himself to inventing an artificial language, tailored to the ape's physical and mental limitations, but nevertheless representative in its essentials of the natural languages that interested him. What he came up with was a language based, like spoken English, on sound elements that could be combined to create meaningful words. But there the obvious parallels ended, for this was a language accessed by manipulating not lips and lungs, but hands and a mechanical device—a joystick, capable of cueing a complex range of sounds, combining different frequencies, amplitudes, levels of white noise, and so on.[140]

When Premack described his efforts at a 1965 symposium on psycholinguistics, there was not much to report by way of concrete results. The words and rules of the artificial language were still to be invented. More awkwardly, not even the humans in charge of the experiment had managed to conquer the joystick.[141] In light of such slow progress, it is little surprise that Premack's move the following year to the University of California at Santa Barbara brought a change of tack. The use of an artificial language remained, as did some of the (now adult) chimps, but much else was scrapped or simplified. No longer did the experimental design call for the animals to live in a stimulating, loving, homelike environment. Their homes now looked very much like traditional animal psychology labs, cages and all. Where the emphasis previously had been on letting the apes absorb language through constant exposure and their own activity, they now sat through training sessions, subjected to the same reinforcement methods—rewards for correct responses, none for incorrect ones—deployed to coax complex behaviors from rats or pigeons. The new language was composed not of meaningless elements, but of meaningful words: object names such as "apple" and "banana"; relational terms such as "same" and "different"; even logical operators such as "if/then." The most striking thing about the new vocabulary, however, was its silence: just colored pieces of plastic, of various sizes and shapes. Premack's star pupil, Sarah, eventually proved capable of responding correctly when presented with long, motley sequences of these tokens (see fig. 8.4). She also showed, on Premack's interpretation, that she grasped the symbolic nature of the object tokens, for when asked (in the token language) to describe the features of, say, an apple, she answered the same way whether presented with a real apple or with the blue triangle that served as the token "apple." Such performances, for Premack, showed that the token called forth in Sarah's mind a mental representation of the associated—but absent—object. For Sarah as for her human trainers, the word symbolized the thing.[142]

Before Sarah's successes became widely known, however, another language-using chimpanzee, named Washoe, stole the spotlight. Her education had taken place a few hundred miles away from Sarah's, in Reno, Nevada, beginning in 1966, the same year that Premack revamped his program. Again, the starting point was a lesson learned from Viki's failures—in Washoe's case, by Allen and Beatrice Gardner, based in the psychology department at the University of Nevada at Reno (in Washoe County). Allen Gardner was a Bronx-born experimental psychologist with the standard expertise in the study of learning in rats. Trixie Gardner was an Austrian-born ethologist who had trained under Tinbergen at Oxford, where she had investigated the instincts of jumping spiders. The Gardners were living proof that psychology and ethology could get along, though as individuals they conformed, in the eyes of students, to disciplinary stereotype: he hard-nosed, she soft-hearted.[143] The idea for "teaching sign language to a chimpanzee"—the title of their joint paper in *Science* in August 1969—sprang from watching one of the sound films the Hayeses had made of Viki speaking. They noticed that Viki had been much more naturally expressive with her hands than with her mouth, and wondered whether a gestural language, such as the deaf use, might have worked better than a spoken language.[144] In short order, the Gardners acquired a wild-born young chimp and began raising her at their suburban home as if she were the deaf child in a deaf extended family (who liked to spend most of their time

Figure 8.4 David Premack's chimpanzee pupil Sarah on the cover of *Scientific American*, 1972.

in the yard, the garage, and a trailer). Much of the day-to-day educating was done by graduate students schooled, like the Gardners, in American Sign Language. All day long, Washoe's minders played with her, signed to each other and to her, and tried to teach her new signs and their meanings.[145]

The big questions about nature, nurture, language, and humanness do not seem to have been much in the Gardners' sights. For them, the challenge was a rather narrowly technical one: could a cross-fostering experiment with an ape be designed in such a way that an ape acquires human language? The success of Project Washoe, as their enterprise came to be known, testified for them not to the crumbling of the wall between man and beast, or to a new understanding of what language is and how it emerges, but to the need to accommodate the natural propensities of the animal subject in a learning experiment. Their 1969 *Science* paper drove home the point vividly:

> Psychologists who work extensively with the instrumental conditioning of animals become sensitive to the need to use responses that are suited to the species they wish to study. Lever-pressing in rats is not an arbitrary response invented by [the behaviorist B. F.] Skinner to confound the mentalists; it is a type of response commonly made by rats when they are first placed in a Skinner box. The exquisite control of instrumental behavior by schedules of reward is achieved only if the original responses are well chosen. We chose a language based on gestures because we reasoned that gestures for the chimpanzee should be analogous to bar-pressing for rats, key-pecking for pigeons, and babbling for humans.[146]

In keeping with this pragmatic ideal, the Gardners' team deployed a range of training methods, unconcerned by whether they corresponded to how humans learn language. The one that proved most effective was guidance: manipulating Washoe's hands into the correct configuration for a sign, showing her what the sign meant (by presenting the associated object or action), then rewarding her for making and using the sign correctly (in the case of a desired object, by giving her the object). By such means, Washoe by twenty-two months had acquired, on the Gardners' count, over thirty signs, including signs for "flower," "cat," and "dog," and could understand many more. They reported that she had little trouble generalizing—using a sign introduced in connection with a particular thing to mean all things of that class. And as might have been expected, as her vocabulary grew, her words seemed to become more precise in their meanings.[147]

All of this added up to a scientific sensation. The image of an ape conversing in sign with an appealingly hippyish young man or woman became an icon of 1970s popular culture.[148] When most people thought of apes and language, they thought of Washoe and her successors. What for the Gardners was a triumph of clever psychological methods became for the rest of the world a reassuring reminder—like the moon landing that same summer—that science in the nuclear age could still be benign, indeed ennobling. No one watching Washoe on television, or seeing her photograph in countless newspaper reports and magazine articles, was put in mind of conditioned rats. She seemed a world away from the behaviorism satirized so bleakly in *Brave New World* or, more recently, *A Clockwork Orange*. Scientists had given Washoe the gift of language, had empowered her, set her free. Simian semantics set throats tightening. "It is unlikely," wrote the anthropological linguist Jane Hill in the mid-1970s, "that any of us will in our lifetimes see again a scientific breakthrough as profound in its implications as the moment when Washoe, the baby chimpanzee, raised her hand and signed 'come-gimme' to a comprehending human."[149]

9

PLAYBACKS IN AMBOSELI

◆ ◆ ◆

INVESTED WITH HIGH HOPES and lofty sentiments, the ape language projects flourished. After the successes of Washoe and Sarah, there were new projects and new pupils: Lana the chimp, Chantek the orangutan, Koko the gorilla, and others. Media coverage remained copious and respectful, with upbeat documentaries, major magazine articles, popular books, even novels.[1] All kept the marvelous talking apes in view, so that when, at the end of the 1970s, the backlash finally hit, it struck before a large popular audience. "Are Those Apes Really Talking?" was the deadly question *Time* magazine posed in March 1980. A "raging academic storm" was in the offing, readers learned, over whether the instructed apes had acquired genuine linguistic competence, or whether—like Clever Hans so long before—they were merely simulating competence, well enough at least to fool people eager to find evidence for it. An emerging champion of the ungenerous view was Herbert Terrace, a psychologist at Columbia University and sponsor of yet another signing-chimp project, Project Nim. "Nim Chimpsky" was Terrace's ironic salute to Noam Chomsky, the most influential linguist of the era and a thinker notoriously dismissive about animal language. A photo showed Terrace with Nim outdoors, playing piggyback.

As Terrace told it, he had set out to show, contra Chomsky, that a nonhuman animal could learn the basics of human language, including the feature that Chomsky had singled out as definingly human: the ordering of words into sentences according to rules, or syntax. But Nim had never even approached the goal. At best, Terrace and his students concluded, Nim had shown himself a skillful mimic, capable of performing the actions that got him the rewards he craved, but without doing so in a consistently creative or comprehending way. Worse, when Terrace examined films of other instructors at work, he found nothing to indicate that the other apes were any different from Nim. So the apes had made monkeys of the humans. According to the *Time* reporter, the Gardners had taken the news very badly indeed. There had even been talk of a lawsuit. Other, less combative project psychologists had been similarly thrown on the defensive. Meanwhile, a linguist, Thomas Sebeok, was taking up Terrace's cause in a new anthology on the projects. The article gave the last word to a gloating Chomsky: "It's about as likely that an ape will prove to have a language ability as that there is an island somewhere with a species of flightless birds waiting for human beings to teach them to fly."[2]

For the next year or so, allegations chased counterallegations, to the delight of editors everywhere. On one side, critics condemned the project psychologists for wishful thinking and lax methodology. At a widely publicized conference Sebeok organized in New York City, Terrace spoke on the same program with the Amazing Randi and other experts in detecting deception.[3] On the other side, the project psychologists protested Terrace's misrepresentation of their work, while casting doubt on his portrayal as reluctant skeptic—a coinvestigator had long been hostile to the projects—and on the quality of Nim's education at the hands of an ever-changing roster of teachers in a cramped, windowless cell. ("The one thing Terrace actually proved," Washoe's main teacher Roger Fouts later wrote, "was that a socially deprived chimpanzee locked in a prisonlike environment will *not* learn American Sign Language.")[4] Between the already raised profile of the projects and the sometimes jaw-dropping bitterness of these disputes, interest in the question of primate language and its scientific study was stoked still further—so much so that when, in November 1980, the journal *Science* published a paper purporting to show that monkeys in nature communicate symbolically, the *New York Times* put the story on its front page. Under a headline announcing the discovery of a "rudimentary 'language'" among African monkeys, the story described the first major attempt since Garner's day to use playback experiments as a means for translating the simian tongue.[5]

The experiments had been done in 1977–78 by Robert Seyfarth, Dorothy Cheney, and Peter Marler, all based at the Rockefeller University. They had aimed to test whether the alarm calls of the vervet monkeys of Amboseli supply information about different predators, as Thomas Struhsaker's observations had suggested, or whether the calls merely express the emotional state of the caller. The Rockefeller group's conclusion in favor of alarm-call symbolism or semanticity came to be celebrated as revealing a new dimension in animal-human continuity. At the same time, the method they used—the playback of recordings of the natural vocalizations of the vervets and observation of the responses evoked—struck many as a brilliant alternative to the psychologists' increasingly discredited efforts. This chapter tells how the vervet alarm playbacks came to be done and how the results came to be interpreted in such a consequential way. As we shall see, the experiments' transit from vague ambition to research reality to uplifting exemplar of "cognitive ethology" was a complex affair, with the ape language projects providing crucial counterpoint long before 1980. Moreover, for all that the experiments were a natural outgrowth of the Marlerian program, combining his interests in primate vocalization, field studies, playback experiments, and animal semantics (including the semantics of vervet alarm calls), at first, the idea of seeking out humanlike attributes in the vocalizations of nonhuman primates had little general appeal. We do well at the outset to recall the far from congenial climate of opinion prevailing when Cheney, Seyfarth, and Marler started their experiments.

The Ends and Means of Vervet Playback Experiments

Animals, Language, and the Anticontinuity Consensus, 1965–75

At the end of the 1950s, Marler had extended science's blessing to Darwin's claim, in the *Descent*, that "the faculty of articulate speech" does not "in itself offer any insuperable objection to the belief that man has been developed from some lower form."[6] A decade later, the same line, quoted on the back cover of a popular introduction to linguistics, came with the tag: "Perhaps not: but it makes a mighty big difference."[7] Despite the efforts of Marler, Charles Hockett, the ape language psychologists, and others, the wider intellectual world between 1960 and 1970 grew doubtful about animal communication as a source of insight into human language and its origins. Behind the new skepticism were developments in a range of sciences. One of them, as we have seen, was neuroscience. The Darwin-baiting *New Horizons in Linguistics* summarized the latest findings thus, by way of explaining why the alarm calls of vervets are not names:

> Even the alarm calls of the vervet monkey which seem, superficially, to be "naming" the type of predator are more plausibly regarded as expressing no more than

the relative intensity of the fearful and aggressive emotions aroused by the various predators. . . . An emotive interpretation of primate calls is also suggested by the observation that vocalizations can be obtained by electrically stimulating sites in the limbic system of monkeys. The brain area is known to be crucially implicated in the mediation of emotional behaviour [citing Robinson's study]. . . . In order for vervet monkey calls to qualify as "names," a particular call would have to be associated with a particular object irrespective of the emotion aroused by the perception of that object. In the human case, the expression "leopard" is appropriately used in referring to members of the class of leopards. Whether one is attracted or repelled by leopards is irrelevant.[8]

Among linguists, the shift against animal communication was due much more directly to the ideas of one of their own: Chomsky. Based at MIT, barely in his forties, Chomsky had given the scientific study of language a cachet unknown since F. Max Müller's heyday.[9] What was being hailed as the "Chomskyan revolution" amounted—though this has been little noted—to a comprehensive repudiation of Hockettian notions. There was, first of all, Chomsky's main theoretical contribution: an analysis of the rules needed to generate all and only the grammatical sentences in English, or, as it came to be known, a "transformational generative grammar." Introducing his grammar in his classic book *Syntactic Structures* (1957), Chomsky first explained why Hockett's statistical, information-theoretic attempt to do the same failed. (The famous slogan of Chomskyan linguistics, "colorless green ideas sleep furiously," is a deeply improbable sentence that we nevertheless judge grammatical, and hence, according to Chomsky, a sentence that no statistical grammar can account for.)[10] Next consider Chomsky's views on language acquisition, first aired in an annihilating 1959 review of the book *Verbal Behavior*, by the Harvard behaviorist psychologist B. F. Skinner. Against the behaviorists, Chomsky denied that children acquire language wholly through stimulus-response conditioning and analogy. Rather, he suggested, the fundamentals of language grow in the minds of children as their brains grow. The rules underlying grammar are part of the built-in structure of the human mind; indeed, they have to be, for, he argued, there is no way children could infer those rules so rapidly and so unfailingly from the haphazard snatches of language actually heard. In fact, Skinner had not addressed language acquisition at all. But Bloomfield and Hockett had, along just the behaviorist lines attributed to Skinner. Attacking Skinner was a high-profile way of rubbishing an idea dear to Hockett.[11] Rounding out the core Chomskyan teachings on grammar and its acquisition was, from the mid-1960s on, a dismissal of animal communication systems as even remotely related to human language, conceptually or evolutionarily. Here Chomsky snipped the most distinctive strand in the Hockettian bow. In *Language and Mind* (1968), Chomsky wrote that, compared with the bounded communication systems of animals, "human language, it appears, is based on entirely different principles. This, I think, is an important point, often overlooked by those who approach human language as a natural, biological phenomenon; in particular, it seems rather pointless to speculate about the evolution of human language from simpler systems." The origin of language for Chomsky warranted talk not of evolution, but of emergence.[12]

As much for this rethinking of language as for his opposition to the Vietnam War, Chomsky by the early 1970s had become unavoidable—the subject of everything from a Modern Masters primer to Woody Allen gags.[13] Such was Chomsky's authority that his view of language's place in nature echoed in the most unlikely places, including E. O. Wilson's ultra-Darwinian *Sociobiology: The New Synthesis* (1975).[14] But even if one managed to ignore the buzz around Chomsky, there were still, as Wilson noted, at least two other lines of scientific research tending in the same skeptical direction. Philip Lieberman, of the famed speech-centered Haskins Laboratories, had conducted a detailed study of the vocalizations and vocal anatomies of monkeys, chimpanzees, and gorillas, and concluded that nonhuman primates lack the equipment—specifically, the means for controlling the larynx and for modifying the sound-shaping region above it—required for speech as complex as human speech. In outline, his conclusion was not, of course, a new one, but the methods behind it were impressively up-to-date, as were some speculations (later developed using computer modeling) on speech in fossil hominids.[15] Better known still than Lieberman's work, of course, were the ape language projects. For Wilson, however large and varied Sarah's vocabulary, neither she nor any of the other language-trained chimpanzees showed anything like the creative verbal drive that characterized humans.[16] To be sure, there were plenty of onlookers who drew an opposite message. In another influential book from 1975, *Animal Liberation*, the philosopher Peter Singer noted how the teaching of language to apes was playing havoc with a criterion that, since Descartes—Chomsky's hero, by no coincidence—had been used to separate man from the beasts.[17] But even the project leaders took it for granted that, unlike humans, the apes used their voices merely to express emotions. The projects had been successful, it was said, precisely because ways of circumventing the dependence of language on a vocal channel had been found. Yes, the animals could be *trained* to communicate symbolically, but left to their own devices, they did not—and certainly not with their voices.[18]

Here was the backdrop against which Marler set two new postdocs following up Thomas Struhsaker's work on vervet vocalizations. Marler had tracked the changes in neuroscience and linguistics just described, but he was much closer to what was happening in psychology.[19] To put it in psychological terms, the vervet playback experiments were Marler's response to the stimulus of the project scientists, in particular Premack. More so than the Gardners, Premack played to disciplinary type, and thus helped reignite an old disciplinary rivalry.

Marler's Dialogue with the Ape Language Psychologists

The Gardners had been acquaintances for some time—Trixie, like Marler, was a product of 1950s English ethology—and he remembers talking with them as they planned the chimpanzee project. He thought their idea of giving sign language a try had great potential, for two reasons. First, Struhsaker's observations in the field had shown that some primate signals were tied so specifically to a particular circumstance or object referent that, to Marler's mind, they were most implausibly considered expressions of emotions such as fear and aggression. Teaching a chimp to communicate in sign would thus amount to enriching and exploiting a symbolic capacity already manifest in a much more limited and focused form in natural behavior. Second, as Jane Goodall's work in particular was revealing, chimps in the wild use their hands for signaling in a very rich way, so a symbolic language built of gestures stood an especially good chance of take-up.[20] Far from being dismissive of Project Washoe, Marler was enthusiastic, and would remain so. Shortly after the project started, he dropped by to see for himself how things were going. "On a visit with Washoe in Reno," he later wrote, "I recall her eagerness to be admitted to the locked garage where her toys were kept. She turned to Trixie, raised her hands, and signed 'gimme key open door,' greatly excited when her request was granted."[21] Before the Gardners' *Science* paper had come out, Marler incorporated their findings into the latest of his survey articles on animal communication, juxtaposing photographs of wild chimps in full gestural flight with a list of Washoe's growing vocabulary. He noted her ability to string signs together in meaningful ways, confessing how hard it was to avoid "the impression that a primordial syntax is emerging." The accuracy of this impression—syntax was the fortress on the border Chomsky drew between animals and humans—would, however, "only be determined by further results from this most interesting project."[22]

At around the same time, he was completing a preliminary account of his own chimpanzee research, from his summer at Gombe Stream. His shift toward semantic concerns since the mid-1960s can be gauged by setting his concluding remarks in this report, published in 1969, beside some broadly similar ones at the end of his Primate Project survey, published in 1965. Then Marler had asked about the evolutionary transition from the graded, inherited vocalizations still found in chimpanzees to the discrete, learned vocalizations making up human speech. Now, he characterized the transition rather differently, as from a system of graded, emotional vocalizations to one of discrete, symbolic ones. He cited with approval the Japanese macaque expert Jun'ichiro Itani's suggestion that, in Marler's summary, "one of the requirements for the origin of speech is the release of the sounds from strict determination by an external situation and by a single overwhelming emotion," and that the "grunting sounds of a relaxed macaque"—in other words, vocalizations produced when the monkeys are not highly aroused—"might be looked to if one were to postulate an origin for speech sounds in such animals." Among chimpanzee vocalizations, Marler pointed to the "rough" grunts associated with feeding as good candidates for an Itani-like scenario. Here was a call that varied along a number of parameters; that was given in response to a range of external stimuli, from salt licks to bananas; and that sometimes involved low-intensity emotions. "We could imagine," Marler speculated, "that different food items might come to be correlated with different phases in the array of [acoustic] possibilities. Thus there would be an elementary step towards the evolution of different words for different objects in the environment."[23]

Another strut in the new emotions-versus-symbols analytic framework took shape through Marler's interactions with Premack. Marler had read with interest in the mid-1960s about the joystick-language research program, but, he recalls, he first got to know Premack only a few years later, during an official "site visit" to Premack's token-language lab. At the time, Marler was a member of the experimental psychology section of a major granting agency, and such on-the-spot inspections were a routine part of the job. He came away hugely impressed, not least with Premack's engagement with the interpretive problems surrounding primate communication.[24] "Of all of the ape researchers," Marler later judged, "Premack has the deepest understanding of the underlying linguistic and philosophical issues."[25] But some members of the committee were skeptical. In common with the other ape language programs of the era, after all, Project Sarah was hugely expensive, scientifically unorthodox, and based at a non-elite institution—all factors guaranteed to heighten suspicion. Marler and other supporters won the case, however, and funding continued unabated, as did Marler's professional relationship with Premack. The flow of ideas between them soon proved, for Marler, disappointingly one-way.[26] Premack had no experience of primates in the wild, and—mindful perhaps of that first, body-quaking night at Orange Park—he seemed not at all curious as to whether they might already be communicating symbolically, using their own evolved, voice-based system. On the contrary, he was quite sure, from early days, that they did no such thing. In summarizing his pre-Sarah research, for instance, he had argued that "the assumption that speech did not evolve out of the [primate] call system is apparently more compatible with the psychological and neurological evidence than the opposite assumption." In a note, he had even thrown doubt on whether the phrase "call system," implying some kind of determinate reference for different vocalizations, was justified, observing that "the cries that our caged animals make appear to depend upon intensity of stimulus no less than upon kind; different stimuli of equal intensity may prove to produce the same cry."[27]

At a 1972 symposium in Ireland, Premack went further, stating categorically that untrained animals never communicate symbolically, but merely express emotions. As he put it in the published paper—later reprinted in the *Handbook of Psychobiology* (1975)—"Man has both affective and symbolic communication. . . . All other species, except when tutored by man [here he cited the Gardners and himself], have only the affective form." He went on to argue, however, that, under certain conditions, a wholly affective communication system can be highly informative about the world—a privilege conventionally reserved for symbolic communication only. So long as the animals in a group have the same likes and dislikes, ordered in the same way (so preferring apples to peaches, say, and fearing tigers more than snakes), then, said Premack, displays of positive or negative affect, of varying degrees of intensity, will correlate reliably with external events. Animal A will hear B over there sounding very pleased and will guess, rightly, that there are apples where A is located. According to Premack, symbolic communication will emerge only once this background of "concordant preferences" disappears—a hypothesis that lent itself admirably to testing in a combined lab-field study. First, take chimpanzees into the laboratory and induce, through the usual conditioning methods, systematically discordant preferences. Second, put the animals together in an interesting environment, and see whether their inability to glean environmental information from each others' affective displays leads to the invention of symbols.[28]

For Premack, then, there were good reasons why a capacity for symbolic communication—demarcated sharply from affective communication—should remain latent among wild primates. These reasons were at the same time, of course, a rationale for the psychologist's traditional preference for lab work and learning studies.[29] What was left unclear, however, was why the capacity for symbolic communication had come into being in the first place, available for development under the appropriate conditions (a lab full of humans with tokens and food, a field or forest full of chimps with unnaturally discordant preferences), but otherwise undeveloped. Not long after, Marler raised the prospect of future field researchers finding the capacity not so latent after all:

> The accomplishments of Washoe and Sarah raise a curious dilemma. If chimpanzees can achieve competence with language when provided with an appropriate vehicle, why has not this been demonstrated in nature? One possibility is that our knowledge of processes of natural communication in animals is still in such infancy that we cannot say whether language-like processes are present or absent. We have surely only begun to skim the surface of many and varied patterns of social exchange that occur in wild animals.[30]

Alongside the discussion of the ape language projects here were treatments of two related topics: the vervet predator calls as instances of animal semantics, and the problems inherent in categorizing all animal signals as "affective." Marler at one point acknowledged both how hard it was to make sense of the vervet calls as emotional, and how hard it was to make sense of them as anything but:

> The calls do not intergrade, and one is hardly inclined to ascribe the differences to varying levels of arousal. In fact, the animals tend to be highly aroused in each case. To explain the vervet monkey's complex alarm-signaling behavior, several different physiological states must be postulated, more specific in nature than the cluster of emotional behaviors denoted as "fearful." Conforming with another common attribute of emotional behavior, however, the vervet alarm calls are each associated with a complex of other behaviors, and rarely occur in isolation from them. Insofar as monkeys are not known to engage in relaxed discourse about past events or those in the distant future, there will be few if any occasions when a signal will be disassociated from the signaler's other overt responses to a situation.[31]

The vervet alarm calls, the ape language projects, and the problematic affective/symbolic dichotomy traveled together in Marler's papers of the mid- to late 1970s. He presented his most sustained reflections on the trio in October 1976, at the Yerkes Regional Primate Research Center.[32] The occasion was a conference honoring the centenary of the birth of Robert Yerkes. Since the tenures of the Hayeses and the Premacks, the Yerkes complex had changed location, to Emory University in Atlanta, as well as name. Once again, moreover, it was host to a well-known ape language experiment. Led by the psychologist Duane Rumbaugh, the "LANA Project" was training chimps to communicate in an artificial language via a computer system that—an advantage over gestural signs and plastic tokens—kept records as it was used, enabling a much more accurate charting of the learning process.[33] With a nod to the local achievement, and its continuity with Yerkes's own efforts, Marler took up the question of his title, "Primate Vocalization: Affective or Symbolic?" His answer was that at least some primate vocalizations were both. It was a mistake, he argued, to regard affective communication as something wholly distinct from, and inferior to, symbolic communication. Rather, pure examples of each marked out extremes on a continuum. At the affective extreme, there were indivisible clusters of signals, emotions, and actions, all arising in response (and only in response) to events that were often quite heterogeneous. At the symbolic extreme, there were narrowly referential signals that could be used and encountered at any time and without emotions being roused. In between were, for instance, the vervet alarm calls. These should be understood, suggested Marler, "as representing highly specific classes of dangers, each favoring a particular escape strategy. In this sense they may be viewed as symbolizing such dangers. If they also carry the hallmarks of affective behavior, with simultaneous signs of arousal, perhaps we could think of these as supplementing and enriching the symbolic function rather than excluding it." As he had done from his days as a geographical ecologist of birdsong, Marler here exhibited his great confidence in the power of natural selection to shape gradually and adaptively. In his view, a signal sat at a particular point on the affect-to-symbol continuum according to the adaptive needs of the signalers. Often in nature, overwhelming affective response was optimal. At least once, however, in one hominid lineage, circumstances favored the severe curtailing of such response. Pure symbols thus emerged at the endpoint of a piecemeal paring away of affect, under the complex social conditions that alone made this shift advantageous.[34]

Seyfarth and Cheney as Students of Primate Social Relationships

That same month, October 1976, two young primate ethologists, Robert Seyfarth and Dorothy Cheney, took up appointments as postdoctoral fellows in Marler's Rockefeller lab.[35] With Marler's assistance, they would go on, between spring 1977 and summer 1978, to conduct the field playback experiments later celebrated as vindicating the semantic nature of the vervet alarm calls. Between Marler's prior, public engagement with the issue of vervet alarm semanticity, and the high impact of the papers reporting the experiments, a straightforward ordering of events suggests itself.

1. Marler had a well-defined project that needed doing—vervet alarm playbacks.
2. Two primatologists in need of a job came knocking.
3. Marler put the primatologists and the project together.
4. The rest is history.

As a first approximation to what happened, this is not terrible. Certainly it is close to how Marler came to see things. In 1979, he wrote to the head of the Rockefeller, Joshua Lederberg, of how, when his new recruits first joined, "I was bemoaning our ignorance about semantics of animal communication signals, and I had become convinced that it was time to make another try at playback studies with primates in the field. The vervet monkey seemed the natural candidate in view of Struhsaker's previous work. Dorothy and Robert took the project with enthusiasm, and have now made it their own."[36] Unpublished documents from 1975 to 1978 give a more complex picture, however. As we shall see, when the couple first ventured into Amboseli, they regarded the experimental settling of alarm semanticity as at most a side project. It was, moreover, one that had appeared in their field plans rather late, long after they had begun to prepare for a playback-based investigation in Amboseli. Their primary concern was to study social relationships between vervet and vervet, not semantic relationships between vervet calls and the nonvervet world. Indeed, as late as November 1977, when the program of alarm playbacks was far advanced, Seyfarth and Cheney reckoned that their experimental results *weakened* any claim for semanticity.

To understand Seyfarth and Cheney's intellectual trajectory in the Rockefeller years, we need first to see that, along with broad primatological expertise, they brought with them a number of quite specific ambitions.[37] The interest in primate behavior was initially Seyfarth's. When he was a freshman at Harvard in 1966, with no idea what he wanted to do for the rest of his life, he hedged his bets with a diverse set of introductory courses. Among them was "Man's Place in Nature," dealing with the fossil record of human ancestry and—courtesy of the recently arrived Irven DeVore—the new field studies of living primates and hunter-gatherers. The class was fun and fascinating and not all that hard. Biological anthropology duly became Seyfarth's major. He thus entered a lively but, at the time, undersubscribed department, where undergraduates mixed easily with graduate students and faculty, and the talk ranged from the hominid past to the Vietnam present. The science and politics of the day came together most eye-openingly in an independent study course that Seyfarth took with Robert Trivers, then a graduate student, and a nonanthropologist friend of DeVore's. Trivers was deep into a profound rethinking of Darwinism and behavior, along lines quite different from the reconstructive efforts of DeVore. Following David Lack and others, Trivers had become persuaded that animals, when acting adaptively, never do what they do for the benefit of their species. Rather, natural selection, like sexual selection, operates always *within* species, on individuals, whose reproductive interests may well not coincide. In the Triversian view, even in apparent cooperation—the raising of young, the giving of alarm calls—there often lurked competitive conflict. Working out the consequences of this insight, with the help of mathematical modeling, was the task ahead.[38]

This was sociobiology in all but name. When published in the early 1970s, and subsequently publicized as core components of E. O. Wilson's synthesis, Trivers's arguments electrified a generation.[39] For Seyfarth, who watched those arguments take shape—he recalls Trivers's delight on receiving an encouraging letter from his own scientific hero, the English theorist W. D. Hamilton—Trivers dramatized how exciting the scientific life could be. There were other, less high-minded reasons for Seyfarth's sticking with science through to graduate school, however. He had not found anything else he especially wanted to do; watching apes in Africa still sounded terrific; and, if it all went badly, well, there were always friends' fathers and, distasteful though the thought was to a baby boomer, the corporate jobs in their gift. So he started applying, taking to heart some advice given to Trivers: for the sake of rigor and breadth, a scientific interest in primate behavior was better pursued within biology than within anthropology. Among biology departments, Madingley soon emerged as especially attractive, again for mixed reasons. Compared with a year at law school or medical school in America, a year in England was cheap, leaving plenty of parental funds available for European travel. Another bonus was that Seyfarth's girlfriend Dorothy, a political science student at Wellesley (fig. 9.1), was due to spend the year at the London School of Economics, just down the rail line from Cambridge. And while primatological training was on offer elsewhere in Britain, notably at Bristol, Madingley had Robert Hinde. Despite working mainly with captive monkeys, his reputation as a supervisor of field studies of primates was unsurpassed. Indeed, Madingley under Hinde was widely, and correctly, seen as a way station to the most glamorous field site of them all, Gombe Stream, where Hinde's now-famous former student Jane Goodall presided.

So Madingley it was. Arriving there in the fall of 1970, Seyfarth soon discovered that the English idea of graduate school was rather different from the American one. There was nothing much resembling instruction at all—no lectures or seminars or qualifying exams. Even the crucial tasks of designing a field project and managing its logistics were, like the rest of the education, wholly up to the student, who was expected to pick things up from research seminars and tea-room conversations. It was all amiably stimulating, and, by June, it yielded Seyfarth a plum project, studying the baboons at Gombe Stream. With everything in place but the visa to Tanzania, he went back with Dorothy to the States, got a temporary job in the Harvard admissions office, and waited. The months rolled by, but the messages from Goodall remained

Figure 9.1 Robert Seyfarth (1950–) and Dorothy Cheney (1948–) at Cambridge, around 1975. Courtesy of Robert Seyfarth and Dorothy Cheney and reproduced by permission.

upbeat. Any minute now . . . In December 1971, Robert and Dorothy got married.[40] By the following June, there was still no visa. Goodall at last signaled that the problem was not going to be resolved any time soon. Now Hinde, who had agreed to supervise, intervened. He had recently been in contact with a young South African field primatologist, Graham Saymann, who knew the baboon sites in South Africa, and might be able to suggest a good place there. After some correspondence, and some soul-searching about working in the midst of apartheid, the Seyfarths—the jobless Dorothy was coming too—decided to go. They got permission, flew out, and around September 1972, after a little looking around, settled on tiny Mountain Zebra National Park, in Cradock, Cape Province.[41]

At first, things did not go at all well. The park was new, and the days were not long gone when farmers had shot at the baboons there. It thus proved impossible to get near the animals. In the jargon, the baboons never "habituated." Instead, they had to be tracked from a long way off, through lousy, headache-inducing binoculars. Bad as the conditions for observation were, the observers soon realized they had little idea how to turn raw sightings into usable data. With no guidance coming from Madingley, they began writing letters to people whose papers they had admired, asking for tips. One who replied was Thelma Rowell. She advised that, one way or another, they had to learn to tell the baboons apart, so that records of individuals' behavior could be made. With the help of local farmers, the Seyfarths thus set about tagging the baboons. In many places in Africa, this could never have been done without embarking on slow, expensive hagglings with government officials for permission. But to rural Afrikaner game wardens and farmers, trapping baboons was entertainment, and in no time, the job was done. A second helpful reply came from Jeanne Altmann, who sent a draft of a new paper, essentially a how-to guide for observers of social behavior keen to guard against statistical biases creeping into their data. Published the next year (1974), Altmann's paper has since been hailed as a landmark in the rigorization of primate field observations. For the Seyfarths, the upshot was a new, systematic operating procedure, centering on "focal-animal sampling." In an order randomized across several variables, they observed each baboon for a half-hour stretch, noting all its behaviors on a checklist. Once all the animals had been thus observed, the cycle started anew.[42]

Thanks to Rowell and Altmann, data on the baboons was at last accumulating on secure foundations.[43] The empirical side of their work now flourishing, the Seyfarths began to develop the theoretical side. Here the key figure was Hinde. Since his early studies of mother-infant separation, he had been much interested in social bonds. Of late, he conceived them within a three-tiered scheme. At the bottom level were "interactions," the doings of individuals with each other—particular bouts of playing, fighting, grooming, and so on. In Hinde's view, interactions were what observers of primate social life observe, and, strictly speaking, all they ever observe. Above the interactions were "relationships." These were generalizations, inferred from the interactions. Relationships had predictive and explanatory power. If A and B were said to have a close relationship, based on their past interactions, then one could guess how A would respond to any number of future events involving B. Above the level of the relationships, an even more comprehensive generalization could be inferred—the "social structure" of the group. Such a bottom-up approach to social life was unfashionable in the social sciences of the time. It did, however, have striking affinities with the work Trivers was doing—work the Seyfarths were drawing on as they puzzled over the checklists strewn across the floor of their farmhouse.[44] No sooner, it seems, had Hinde begun casting about for field data that might exemplify his analysis than the Seyfarths wrote him a long missive from South Africa outlining their own different but congenial ideas. Hinde's encouraging response did for the Seyfarths what Hamilton's had done for Trivers, letting them know, when they needed most to hear it, that they were being taken seriously. The Cambridge-Cradock post now opened to intellectual traffic.

What had started as something of a lark had transformed itself, and transformed them. In the spring of 1974, after eighteen months with the Mountain Zebra baboons, the Seyfarths left for Madingley and the business of their doctoral theses. In a rough and ready way, they divided their data between them, Robert taking the data on the adult animals and Dorothy the data on the immatures.[45] In Dorothy's case, of course, there was a complication. She had never actually been admitted to Cambridge, indeed had no formal scientific training of any kind. While still in South Africa, she had raised the possibility with Hinde of doing a PhD under him. Long noncommittal, he now took her on, with a proviso: she was to put the thesis off for a year, and spend the time instead writing catch-up essays for him on the biology and psychology of behavior. This she did, so impressively that Hinde—the author of two textbooks on these subjects—found himself instructed.[46] Meanwhile, the integration of Triversian and Hindean perspectives continued. The question of why adult females partner up as they do for grooming proved an especially fruitful one. In the baboons the Seyfarths had studied, and in lots of other primate groups, first-ranking A and second-ranking B groomed together; third-ranking C and fourth-ranking D groomed together; and so on, down the hierarchy. The usual explanation gestured toward was kinship—that kin naturally cooperate with kin. Yet the pattern appeared even where kinship was known *not* to be a factor. Á la Trivers, the Seyfarths turned instead to competition and advantage-seeking. Suppose, they reasoned, that grooming functions mainly in the building and maintenance of social relationships. Suppose further that the higher an animal's rank, the more desirable she is as a grooming partner, since a close relationship with her will confer all sorts of advantages—she can be counted on to form alliances, to allow feeding next to her, and so on. The pattern of partnerships should therefore arise as individuals compete for the high-ranking grooming partners. A, at the top, will of course have her pick, and thus choose B. B will get A, so B will be satisfied. Third-ranking C will look around, find A and B taken, so settle with the next best thing, fourth-ranking D. It was social structure from the bottom-up, by way of a simple, selfish rule.[47]

When Hinde wrote to Marler on the Seyfarths' behalf in July 1975, to ask whether he might help these two students of "quite exceptional promise" find a base for future field research, they had a successful, multifaceted project up and running.[48] Between 1976 and 1978, it yielded a number of papers, including a 1977 *Journal of Theoretical Biology* paper on the model of social grooming, extended by way of some innovative computer simulations. (At Cambridge, they had taken a class in the user-friendly computer language of the day, Fortran, and developed a program that calculated the outcomes when abstract entities variously related and ranked followed the structural rule.)[49] Other major themes were male-female relations among adults,[50] rank and its acquisition,[51] and the different patterns of interaction that males and females experience while growing up, leading females to be socially integrated and males peripheralized, priming some of the latter for transfer into other groups.[52] Here was a lifetime's scientific itinerary, complete unto itself, and with no overt connection to the question of animal semantics. It is little surprise, then, that in the Seyfarths' initial dealings with Marler, both parties worked to graft aspects of his program onto theirs, rather than the reverse.

Plans Rejected and Revised, 1975–77

The first plan agreed on had nothing to do with vocal communication or with vervets—despite its origins with Tom Struhsaker. He was now in Uganda, engaged in a long-term study of its colobus and other forest monkeys. Of all Marler's primatological associates, Struhsaker had kept the interest in vocalizations most firmly in check, subordinated to the tasks of describing behavior, sociology, ecology, phylogeny, and the links among them.[53] No longer on the Rockefeller faculty, he was still part of its extended teaching family when, in the summer of 1975, Marler wrote him about the Seyfarths. Impressed with their credentials, Struhsaker suggested a project—"perhaps they could do a nice interspecific comparison of infant sociology as it relates to our log of ecological and other sociological data."[54] Marler passed the suggestion along to Seyfarth, encouraging him to put in for a postdoctoral fellowship through the Rockefeller. The Ugandan animals, Marler added, were "all now superbly habituated, which would fit well with your past experiences."[55] In a subsequent letter, Marler, having read some of Seyfarth's thesis (submitted in October), reaffirmed his confidence that Struhsaker's monkeys were "well enough habituated to permit the kind of fine-grain analysis of interactions that you have broached with the baboons. There are many fascinating possibilities there, such as for example a comparison of developing mother-infant and sibling-sibling relations in the two species of colobus [red and black-and-white]."[56]

But the Seyfarths—Marler had made it plain that Dorothy was included in these plans—were not at all keen. They had, as they wrote to Marler, two concerns about the Ugandan study: the likelihood of disruption "in a country which is politically so unstable" (this was the Amin era), and doubts that the forest monkeys were "really well-enough habituated to permit individual recognition at a distance, even in the large red colobus groups, and even despite possible breaks in field observation." But maybe, the Seyfarths continued, with Rockefeller backing, they could establish a new research base for the study of baboons, in attractively stable Botswana? The authorities there had already signaled their support for such an endeavor.[57] But Marler was unbudging. Fellowship applications, he replied, fare better when the project furnishes students with new educational opportunities, not more of the same. Yes, the politics were a concern, but, he advised, these should be assessed separately from the scientific merits of a project. As for the latter, again, the colobus were, he assured, "relatively easy to approach and identify"; and in light of the Seyfarths' express interest in comparative work, the unusual opportunity afforded among the Ugandan colobus—two closely related species, living together, but with contrasting social systems—seemed unbeatable.[58] At this point the Seyfarths backed down.[59] Without much enthusiasm, but with no serious alternative prospects in sight, they threw themselves into learning about colobus monkeys.

In December 1975, Robert and Dorothy flew back to the States for the holidays. Around the middle of the month, they paid their first visit to the Marler lab, on the day of its Christmas party. Some years before, the lab had moved from New York City to the town of Millbrook, where Marler, now overall director of animal behavior studies at the Rockefeller, presided over a dedicated field research center (fig. 9.2).[60] To outsiders, "Millbrook" could be an intimidating place. But the mood that day was relaxed and convivial. Marler chatted a while with his new students, when, as they recall, quite out of the blue, he ran a new idea by them. What would they say, he asked, to a project examining vocal communication among Struhsaker's Amboseli vervets, making use of experimental playbacks? The Seyfarths jumped at the new offer. As they remember it, the proposal had obvious appeal along several dimensions. For one thing, their work up to that point, in common with almost all primatological field research, had been strictly observational. The idea of learning to do experiments—a specialty of the Millbook approach—was novel and exciting.[61] They also liked the idea of adding expertise with vocalizations to their repertoire. (In South Africa, they had usually been too far away to hear any but the loudest baboon vocalizations.) But most attractive of all, it seems, was the fact that the Amboseli vervets—as highly social, savanna-dwelling monkeys, in a country that was stable and scientist-friendly—would make excellent surrogates for the Botswana baboons they had wanted to study. Indeed, when Robert had asked about a Botswana base, he had named vervets as well as baboons as the subject species. Thus was the vervet project born. Robert spent the rest of the party cooped up with pad, pencil, and a pile of books and papers, learning about vervets and vocalizations.

There was, in fact, already a model close to hand of a field investigation

Figure 9.2 Peter Marler surrounded by the chambers used for rearing songbirds in acoustic isolation at Millbrook, early 1970s. From Moser 1974, 163.

melding seriously good monkey sociology with seriously good monkey bioacoustics. The role of primatological expert in Marler's lab had passed from Struhsaker to Steve Green, whose paper "Variation of Vocal Pattern with Social Situation in the Japanese Monkey (*Macaca fuscata*): A Field Study" had just been published.[62] It reported the findings of fourteen months spent observing and recording Japanese macaques on their native ground in the late 1960s, when Green had been a Rockefeller graduate student. As he explained, for purposes of studying primate social life, the Japanese situation could hardly be bettered. Thanks to the quasi-ethnographical orientation of Japanese primatologists, a number of troops had been under close, continuous observation for a long while, so that answers to questions about who was who, who was related to whom and how, who ranked above whom and by how much, what kind of history had shaped individuals and groups—all the hard-won background facts needed to make sense of social behavior—were firmly in place. As a result, Green was able to characterize in great detail the social contexts of vocalizing.[63] On his return from the field, after running spectrographic analyses, he was thus able to spot a remarkable relationship between the social interactions eliciting a particular graded call, the "coo," and variations in the acoustic structure of the call. While coos spanned a considerable frequency spectrum, and all of the interactions they accompanied were friendly ones ("affinitive" rather than "agonistic"), the monkeys appeared to partition the spectrum into seven subcalls, according to the precise nature of an interaction. In general, Green concluded, the more actively an animal was soliciting contact, and the higher the rank of the individual with whom contact was solicited, the higher the pitch of the coo type used—most likely because, as Darwin had long ago pointed out, arousal causes the muscles in the larynx to tense up, producing vocalizations of a higher pitch.[64] The discovery seemed to confirm Marler's suggestion, in his Primate Project survey, that the graded calls of primates were optimal for conveying nuances of motivation.[65]

What of experimental playbacks? Green had tried to use them, but without much success—though here too Marler's insights had been borne out. For some time, Marler had emphasized the distinction between calls given from far away, functioning without aid from visual signals, and calls given at close range, when visual signals can accompany vocal ones. In Green's Japanese study, as well as in his earlier study of colobus monkeys in Uganda, he had found that playbacks of the informationally self-contained long-range calls managed to evoke typical responses. With the close-range, composite calls, however, the difficulties for the playback experimentalist were considerable:

> For sounds used at close range . . . once the Japanese monkeys came within sight of the broadcast apparatus, their attention turned to scanning visually and otherwise searching for the "hidden monkey," including lifting and looking under the speaker apparatus. After the monkeys auditorially localized and approached the source of playback vocalizations, an unusual response, searching, was then elicited from them. This response appears to be governed more by the unusual context of vocalization occurring without a monkey visible than by the nature of the sound. Field techniques more sophisticated than simply a playback of recordings will be required to elucidate the function of the audible portion of composite signals for those parts of vocal repertoires employed principally in close-range intra-troop communication.[66]

Perhaps it was Green's attempts at close-range playbacks that Marler had in mind when he recalled thinking, around the time the Seyfarths showed up, that "it was time to make another try at playback studies with primates in the field." Certainly the potential for playbacks of long-range calls had already been amply demonstrated. 1975 saw the publication not only of Green's paper on the macaque coos, but of a report in *Nature* by Peter Waser, a Millbrook graduate student, on playback experiments with gray-cheeked mangabeys in the Kibale Forest in Uganda. The aim was to test whether, as previously suspected, the loud calls known as "whoopgobbles" served to keep neighboring mangabey groups at certain distances from each other. The experiments took two people: one, out of sight from the test group, and at a known distance, to play a recording of a whoopgobble from a member of a nontest group; the other, to observe the response of the test group to the played-back recording. Throughout, steps were taken to ensure that the monkeys never saw the play-back apparatus and never got habituated (that is, stopped responding) to the broadcast whoopgobbles. Though primitive compared with what was now being done with birds, the experiments were judged a success, twice over. Not only did they support the view that the call functioned to regulate intergroup spacing, but, wrote Waser, they showed that "playback techniques can be used successfully with free ranging primates."[67] Field playbacks thus were on the collective primatological mind at Millbrook.[68] For Marler, of course, vervet vocalizations and field playbacks had been long-running, though hitherto isolated, preoccupations. Barely a week before meeting the Seyfarths, indeed, he had given a number of lectures at Indiana University on "zoosemiotics," one of them a precursor of his October 1976 paper on vervet alarm calls and the affective/symbolic dichotomy.[69]

Whatever the exact impetus, the Seyfarths left Millbrook excited about the new project. Not long after, they posted back a first shot at a brief pro-posal, emphasizing, they wrote, "the possibilities of applying research on vo-cal communication to individual social relationships and to the social struc-ture which those relationships produce"—or, in other words, of annexing the playback technique to the Seyfarthian program.[70] In the full fellowship appli-cation, incorporating comments from Marler and Steve Green and submitted to the National Institute of Mental Health (NIMH), a three-stage plan was sketched: first, observational study of social relationships and social structure in the vervet groups; second, recording of vocalizations, from a large number of individuals in multiple social and ecological contexts; third, experimen-tal playbacks of some of the recorded vocalizations, in four separate trials. One series of experiments would involve playing calls to the same group in open country and in forested areas, to see whether the different visual contexts—unimpeded and impeded, respectively—made a difference to response. Another, Waserian set would look at how vervet groups regulate their spacing and maintain their territories. Another would take up the question of the social consequences of intergroup male transfer, through playback of male calls before and after transfer. A final set would consider vocalization during development, and the possibility that, as the baboon research had suggested, mothers respond preferentially to the cries of their own infants.[71]

No mention was made of the vervet predator calls. The absence is instructive. Although alarm calls are paradigmatically "altruistic," and in that sense social (for, it is supposed, calling benefits others while putting the caller at risk), alarm calls are otherwise little connected, in the first instance, with social life and its upkeep.[72] Recall that Marler had found the vervet alarm calls so intriguing because, unlike most other primate vocalizations, the calls seemed to be outward-looking, bearing not on the group but on its environment, in an unusually focused way. For the Seyfarths, however, it was precisely the inward-looking vocalizations that mattered. It was not Struhsaker on vervet alarm calls, but Green on macaque coos—where acoustic structure and social situation had been mapped so closely—whom the Seyfarths aspired most to emulate. As they wrote in Robert's NIMH application, "Green's study suggests that by examining vocalizations in the context of other social behavior it may be possible to formulate principles concerning the ways in which vocal communication is used to regulate social interaction among non-human primates."[73] Such regulation, Green had made plain, depended on vocal signals expressing affective states. Yes, he allowed, those signals could be described as "meaningful," in the sense that signal variations and signal responses were related systematically. But the macaque coos were not even candidates for names, of objects or anything else. They were social lubricants, customized to the occasion. To use the old distinction, a study along Green's lines engaged animal pragmatics, not animal semantics.[74]

Throughout the first half of 1976, the Seyfarths were back at Cambridge, doing academic odd jobs and working up papers (and, in Dorothy's case, completing her thesis). In Dorothy's applications that spring for small awards—both granted—from the National Geographic Society and the Wenner-Gren Foundation, she proposed research plans in the same social vein Robert had tapped.[75] As the months crept onward, the need to know whether the Amboseli expedition was really going ahead or not became gradually more pressing. By late June, however, there was still no news about Robert's fellowship. At this point, some time-honored academic realpolitik—Robert got an offer of a temporary lectureship at University College London, and, with the clock ticking, let the Rockefeller make a counteroffer—led to a salaried research position at Millbrook for Robert. The timing was exquisite; three days after Marler's office cabled the good news, NIMH came back with bad news—Robert's application had been turned down.[76] Pockets went so deep at the Rockefeller that it did not really matter, however. What is more, in the course of helping the Seyfarths secure funding, Marler had become steadily more impressed with their potential. When the three met that August in Cambridge for the Sixth International Primatological Congress, detailed planning began in earnest.[77]

Based at Millbrook from that fall until early the next spring, Robert and Dorothy in fact had little chance to take in the intellectual ambience. In mid-September, Dorothy's father, a career diplomat stationed in the Philippines, was killed in a plane crash outside Manila. For the next few months, the needs of her family overtook most other things in their lives.[78] The moments they could grab for science were largely devoted to finalizing arrangements for the Amboseli trip and finalizing papers on their baboon research. Nevertheless, the grant applications they filed that December—Dorothy to the National Science Foundation, Robert to NIMH again, both of them successful—reflect a much more thorough absorption of Marlerian themes. Most strikingly, for the first time, the couple now evinced an interest in whether vervets communicate symbolically, and whether experimental playbacks of their alarm calls could help decide, in the Seyfarths' words, "whether the arousal associated with much animal signaling should be viewed as conflicting with, or supplementing, any additional symbolic function." They listed four questions for playback elucidation. First, will animals in diverse conditions consistently respond to different alarm calls as if these represent specific external dangers? Second, when there is uniformity of response, is each animal responding independently to the call, or is something else involved? Third, assuming the calls do evoke consistent responses under diverse conditions, just how atypical can conditions be? Fourth, does it matter *who* gives an alarm call—are there differences in response according to the caller's age, sex, and track record?[79]

It was this last line of inquiry that appealed most, it seems. To show that the alarm calls inform vervets about predators would, after all, merely confirm what Struhsaker had already claimed. But to show that vervets respond in systematically different ways, depending on who gives an alarm call (say, a high-ranking versus a low-ranking adult male), would reveal something new and interesting about the subject of primary interest, vervet social life. "To date," wrote the Seyfarths, "no one has attempted to measure how much information [the vervet alarm calls] actually contain."[80] Someone in pursuit of the fourth question would be well-positioned to make just that missing measurement—a point Robert and Dorothy reiterated in a talk to colleagues about their upcoming research trip.[81] Still, the alarm calls at this point were far from occupying center stage. The main ambition remained the study of the vocalizations accompanying social interactions. Far more extensive and detailed than the alarm-call plans were the plans for determining how the presence or absence of this or that nonalarm call affects the course of the interaction it precedes. The playback experiments described a year before—testing mothers' responses to recorded infant cries and so on—remained in the project under this rubric. The playback-based alarm study was explicitly a backup plan, prepared, Robert wrote, to allow for "flexibility" in the field, should events overtake the main, nonalarm study.[82] And so, for a time, events did.

In Amboseli, 1977–78

"We're finally settled in Amboseli—at least for the next 10 days or so—and are beginning to watch vervets regularly," Robert and Dorothy wrote to Peter on 20 March. "The last two weeks have been taken up primarily with buying a car, fridge, and other equipment, plus scurrying from one government office to another."[83] As with Garner so long ago, roughing it in isolation was never to be part of the experience. Home in Amboseli was a *banda*, a thatched hut with a shower and a communal kitchen, run by the park council. Neighbors were plentiful, including other foreign scientists, notably the Altmanns and their students, as well as local Masai park wardens, rangers, and assistants. For permits and supplies, a trip to Nairobi was needed. A four-and-a-half-hour drive from the park, the city brought a welcome change of pace. It was Kenya's scientific as well as political capital, bustling with anthropologists, paleontologists, botanists, geologists, and zoologists. After exhausting journeys or long periods in the field, these were people looking to enjoy themselves. Their party mood, and the first-world comforts on offer at third-world prices, made Nairobi a treat. The Seyfarths got right in the spirit. It had long been part of the plan that Marler would join them in the summer to help with the initial playback trials. The couple wrote in mid-June that they had booked rooms for the Marler family, due to arrive in early July, in a Nairobi hotel that was the new "watering spot for itinerant biologists." "All in Amboseli," they added, "look forward to your arrival, and we should have great fun." As hosts, Robert and Dorothy would "endeavor to purchase a goat during your stay for a large communal roast."[84]

Compared with the Mountain Zebra baboons, the Amboseli vervets were turning out to be dream subjects. They were easy to find, easy to observe, and had quickly habituated to the new scientists in their midst. In no time, the

Seyfarths had learned and named all of the individuals in each of the three troops under observation and, on the model of their South African research, had begun the regular sampling of social behavior. They had also embarked in earnest on their new career as bioacousticians, using microphones and tape recorders from the Rockefeller to start building up a library of recorded alarm calls and other vocalizations. It was only now, during these first three months in the field, that the couple began to glimpse for themselves the fascination of the vervet alarm calls. In part this was a matter of hearing and seeing alarmed vervets in action. Nothing witnessed among the baboons or read in advance had prepared them for the sheer swift spectacle of contests pitting the monkeys' early-warning systems against pouncing leopards, swooping eagles, and literal snakes in the grass. After a short while, it became easy to tell simply by listening to the vervets what kind of predator had entered the neighborhood.[85] It was all very impressive—though, as a defense against predation, the system was imperfect, and its failures underscored how vulnerable the vervets truly were, how dependent for survival on the predator calls. "In the early weeks of the present study," Marler and the Seyfarths later wrote, "a group of vervets was observed giving alarm calls to a python with a large lump in its stomach on the morning following the overnight disappearance of an adult female."[86] Being in the field helped bring the alarm phenomena to life for the Seyfarths as never before. It also provided them with the time and leisure, during the sun-scorched afternoons, at last to read and talk in a serious way about animal communication. Until then, they had simply been too busy, and Marler's preoccupations just too distant from their own to be fully grasped. Dorothy recalled in interview the moment in Amboseli when, on her umpteenth reading of Marler's grant proposals and his paper on primate vocalization and the affective/symbolic question, it all suddenly made sense. Now, unexpectedly, the throwaway backup study on the alarm calls started to look like a crucial experiment, capable of deciding a crucial issue. The Seyfarths were hooked.

By the time Marler arrived (fig. 9.3), bearing a high-quality sound movie camera courtesy of the National Geographic Society, all was ready. The basic technique for the playback experiments had been worked out back at the Rockefeller before the Seyfarths had left. In its settled form, it was roughly as follows.[87] When a group of vervets was observed moving in a certain direction, someone would walk up ahead and hide a loudspeaker behind a bush or tree. (Recall that Steve Green, who had had such trouble with intragroup playbacks, had not concealed his playback apparatus.) Connected to the loudspeaker was an amplifier, and connected to that was a tape recorder, containing a tape of a prerecorded alarm call, dubbed from the original recording to isolate the call from potentially distracting clicks and pops.[88] With someone at

Figure 9.3 Dorothy Cheney, Peter Marler, and Robert Seyfarth together in Amboseli, August 1977. Courtesy of Peter Marler and reproduced by permission.

the tape recorder ready to play the call, a second person took up position behind the camera. When the vervets hove into view, then, provided a number of conditions were met—there were no predators nearby, the vervet group had a representative enough mix of males and females and adult and juveniles, they stayed close enough together to be filmed properly, the call they were about to hear had not come from a member of the group—the experiment began. First, the group was filmed for ten seconds, so that behavior after the alarm call could be compared with behavior before the call, to rule out the possibility that the alarm call had not influenced behavior at all. Then the recording was played, with filming kept up for a full minute afterward. Throughout, a third person (or, when it was just Robert and Dorothy, the camera person doubling up) narrated what was observed into a tape recorder. As soon as possible, a map was drawn up, indicating the positions of the speaker and camera, and identifying and locating all the participants, vervet and human. Later that evening, the verbal narrative record was transcribed.

From the initial attempt, it became clear that the technique was a success. The played-back call evoked neither imitative calling, nor searching for the hidden monkey, nor any other disappointingly aberrant behavior, but a general state of alarm and a flurry of evasive actions.[89] Soon, with Marler's son, Christopher, helping out at the tape recorder, the number of trials began to rise. The pace was steady, but slow. To keep the monkeys responsive, it was clear, the experimentalists had to avoid overdoing it. Too many predator calls and not enough predators would soon teach the monkeys to ignore the playbacks. To keep such data as they acquired rigorous, moreover, required strict observance of the rules laid down about when an experiment could go ahead. One result was a lot of abandoned experiments—more, in the end, than completed ones.[90] But the counterpart to the high attrition rate was a data set meticulously expunged of bias. The attention lavished on experimental design can be appreciated in the handling, for instance, of call loudness or amplitude. In general, they found, leopard calls were louder than eagle calls, and eagle calls were louder than snake calls. Wanting to smooth out this natural variation in loudness, they first attempted to do so by dubbing their playback tapes at a single, standard level of intensity. But this process, they discovered, made the snake alarms sound distorted. So instead the team standardized loudness of calls through speaker placement, making sure that leopard calls were played from farther away than eagle calls, and eagle calls played from farther away than snake calls. And so it went, throughout the summer and beyond, all the while keeping watch against certain kinds of predator calls being played more than others, or the calls of certain kinds of caller being used more than others, or certain individuals being at the receiving end of playbacks more than others.[91]

For all the energy they absorbed, the alarm playbacks were relatively rare occurrences. Usually there were no more than three attempts each week, one with one group on Tuesday, another with the second group on Wednesday, and another with the third group on Thursday—the spacing generous enough, it was hoped, to ensure the vervets would have some genuine predator encounters before more fake ones were introduced the following week. In the meantime, there was the sampling regime to be kept up, and voluminous data to be archived. There were also the intrigues of park life to be savored, as the Seyfarths related in a letter to Marler in mid-September, a few weeks after his departure:

> Life at Amboseli continues to meander elastically. There was a farewell party for one of the Assistant Wardens last Saturday at "The Ol Tukai Social Center," the existence of which came as a complete surprise to most. In addition to the de rigueur speeches and hard drinking, there was a fist fight that broke down according first to tribal lines and second to the rival lodges in the Park. The "new" generator is now completely out of service again since Lodo, the mechanic . . . threw a hammer into the engine in a drunken frenzy.[92]

When Marler wrote back a week later, however, it was the work that was uppermost in his thoughts. "I am still glowing with enthusiasm from the success of what we accomplished," he wrote. Indeed, his mind had started to turn to a question they had raised with him, of finding someone willing to keep the daily behavior sampling going after their departure.[93] The matter would soak up lots more correspondence over the next months.[94]

In the meantime, the experimental procedure continued to develop. At first, the question of whether the vervet alarm calls symbolize different kinds of predator had been treated as equivalent to the question of whether the characteristic escape behaviors arise in response to the alarm calls alone, in the absence of visible predators. If, the reasoning went, eagle calls were found to elicit the same behavior in vervets as eagles themselves elicited—behavior, furthermore, that was adaptive for escaping eagles (running into the bushes) but not for escaping leopards and snakes (who could well be in or around the bushes)—then eagle calls could be regarded as representing eagles to vervets, in the same way that blue triangles represented apples to Sarah the chimp. The attraction of playback experiments was that they enabled the field experimentalist to present alarm calls in isolation from the more complex signals in which they were usually embedded, and thus, potentially, to separate symbol from symbolized, label from labeled, name from object, on a par with what the laboratory experimentalist did as a matter of course (when, say, hiding all the apples before posing questions to Sarah about blue triangles). As the alarm playbacks progressed, however, a new interpretive problem began to press. It was noticed that leopard calls tend to consist of repeated units or "syllables," making for relatively longer calls, whereas eagle and snake calls only rarely cluster in this way.[95] How, then, could one be sure, given longer leopard calls and shorter nonleopard calls, whether escaping vervets were responding to the length of a call or to its acoustic structure?

Mere playback of the calls kept both possibilities open. Worse still—for anyone seeking to determine alarm-call semanticity, that is—a longer call could well correlate with a higher level of arousal. It was entirely plausible that the more frightened a vervet was, the more frequently it repeated the alarm in any bout of calling. Maybe, to vervets, leopards are simply the most frightening predators around. That would certainly be consistent with the observation, already noted, that leopard calls are the loudest of the alarm calls. A Premackian possibility thus took shape: perhaps the vervet alarm calls *did* express nothing more than level of fear, as registered—not perfectly but tolerably close—in varying call loudness and length. If vervets consistently fear leopards more than eagles and eagles more than snakes, and if the three levels of fear associated with these three predators consistently elicit alarm calls at, respectively, the high, medium, and low end of the amplitude-and-syllables spectrum, then, as Premack had argued, calls could inform in specific ways about the environment without, however, being symbolic.[96] If that was true of the vervet alarm calls, then the fact that the acoustic structures of the three calls were so unmistakably different was a mere curiosity of vervet physiology. In principle, or so the Amboseli team now reasoned, on the affective interpretation, a snake call made progressively louder and more repetitive should thus elicit eagle-escape behavior and, at the upper limit, leopard-escape behavior, while a leopard call made progressively more quiet and less repetitive should elicit eagle-escape behavior and then, at the lower limit, snake-escape behavior.

It is not clear from the surviving documents exactly when considerations like these began to emerge.[97] For Marler, who had time and again given the functionality of acoustic structure the benefit of the doubt, the notion that the vervet calls' structures were functionless quirks was intuitively repugnant, and it is easy to imagine him being alert for ways of staving off that interpretation if at all possible.[98] There is, however, another, less theory-driven possibility: once the differences in alarm-call length had been noticed, a simple striving for standardization of conditions motivated the switch to multisyllable calls. Whatever the rationale behind the change, on 11 October, the Seyfarths wrote to Peter that "we're now well into the second 'block' of 24 trials, using alarms all of which have been standardized to a length of 5 syllables"—not, it should be noted, by artificially stitching together multiple snake calls and multiple eagle calls, but by restricting the recordings used in playback (fig. 9.4) to ones from those unusual occasions where vervets gave multisyllable snake calls and eagle calls.[99] Now both loudness and length variations were being controlled for. As the couple explained, however, the playback experiments as a whole had encountered difficulties, and the new procedure had brought interpretive complications all its own:

> Experimentation is proceeding a bit slower than when you were here, for two reasons. First, gale force winds come up by 2 p.m. each day, and often as early as 11. These make any playback inaudible, not to mention the discomfort as far as we're concerned. We've also had to get the Nagras [the tape recorders] cleaned regularly since you left. Second, we're being extra careful about having a fair bit of time pass between any two experiments on the same group, since the 5-syllable alarms give the animals much more time to locate the speaker and habituate to the experimental apparatus. We've purchased some camoflouge [sic] material with which the speakers are covered on most trials.[100]

If the experimental alarm project was now much more than a subsidiary study, it was nevertheless but one project among several. A progress report sent to the Wenner-Gren Foundation in mid-November described, in addition to the alarm trials, two other areas of ongoing investigation: the social development of male and female monkeys, and the social function of vocal communication.[101] Both, of course, had been imagined at length in the Seyfarths' prefield plans, and the report made plain how dominant the social side of the vervet studies remained in their own conception of the work. Social success, they wrote, following a new but already influential argument,

Figure 9.4 Dorothy Cheney recording in Amboseli in 1978. Courtesy of Robert Seyfarth and Dorothy Cheney and reproduced by permission.

was most plausibly what primate intelligence had evolved for. The spur to bigger, better-functioning brains was the demanding social environment in which primates lived, not their ecological environments.[102] "As a corollary to this hypothesis," they continued, "we assume that language originally evolved as the result of natural selection operating on the vocalizations used by individuals in their attempts to regulate their social interactions." It was thus the study of *these* vocalizations in vervets—the vocalizations occurring when vervets try to cope with each other, not with their nonvervet environments—that, in the Seyfarths' view, promised most thoroughly to illuminate the origin of human language. For all their absorption in the vervet alarm work, then, they still saw it as somewhat beside the point, belonging to the asocial dimension of vervet life. Although promoted from being a mere backup study, the alarm-call playback experiments were thus still very much subordinate. As portrayed in the report, they seemed at most a useful means to a much more exciting end—a conceptual and practical warm-up for the more difficult, more consequential experiments ahead on the subtle, graded, intragroup calls that, preliminary findings indicated, served vervets in something like the way the coo subtypes served Japanese macaques. It was not just that alarm calls, like the intragroup vocalizations, exhibited the "categorical assignment of particular sounds to particular contexts," in the Seyfarths' phrase. It was that, for both sorts of communication, the calls and the responses of immature animals seemed, they wrote, to be "generalized," only gradually acquiring specificity as the animals grew older.[103]

The couple's letter to Marler on 20 November expanded on these themes, and then some. It is worth quoting from at length:

> Our apologies for not having written in greater detail about the research. Both experiments and observation seem to be proceeding well, though the wind interfered with both for a while. On the experimental side, we have been using only the multi-syllable alarm calls we discussed just before you left: leopard, eagle, and snake calls all of which attempt to match the 5-syllable length of one of the male leopard alarms used in our earlier trials. Results are tending to support our earlier tests, although in a number of cases the responses may be stronger when compared with the same trial using a single syllable alarm call.
>
> By this stage in our experimental work there seems little doubt that responses to alarm calls are context-dependent, thus perhaps weakening any claim for the calls' semanticity. Numerous observations and recordings, however, strongly suggest that vervets do in fact "label" their predators with discrete vocalizations. We therefore would very much like to get some experimental data to back up our observations on this latter point. Consequently we have now begun a search for (a) recordings of leopards, lions, hyenas, and martial eagles, and (b) stuffed versions

of the same predators as well as pythons to present to the monkeys. The former should be easy to come by, the latter *much* more difficult, and of course time may limit our ability to perform as many trials as we'd like. Our eventual aim (results pending!) will be to provide additional evidence showing that, although responses to alarms may be context-dependent, the relations between alarms and predators nevertheless are in many ways context-independent. In the meantime, we are managing to conduct playback trials with just the two of us: one filming and speaking commentary into a portable cassette recorder (the camera's microphone is placed as far away as possible), with the other placing speakers and performing the Nagra-Amp-Watch tasks so ably carried out by Chris.

As far as observation and recording are concerned, a number of interesting areas are beginning to crop up. To begin with, you already know that our alarm playback experiments seem to be indicating both that immatures show more generalized responses to predators when compared with adults (for example, by giving more than one different alarm to the same predator), and that immatures' responses to playbacks are also more extreme than those older animals. Similarly, we believe we can show differences in intragroup vocalizations by immatures when compared with adults, and it should be possible to relate these differences to the changing social relationships of immatures as they mature. [Here they suggest reading the Wenner-Gren report for more.] Secondly, as the individual-distinctiveness of vocalizations in all three groups has become increasingly apparent to us, we have become increasingly frustrated at our inability to give convincing proof that the animals make use of individual differences in their social interactions. Consequently we are attempting some additional experiments in which screams of immatures are played back to a group of adult females, one of whom is the immature's mother. We're using a procedure similar to that used in the alarm experiments, with filming before and after the playback. As a further control we will take screams from an immature in one group and play them to females in another. Of course, this series of tests will require a number of trials, since observational data indicate that at best mothers respond to about 35% of their offsprings' screams. We've recorded a number of screams during observation, however, so multiple trials should be possible without duplicating any one call.

Observation and recording are also suggesting that there are at least 4 or 5 different classes of vocalizations which may resemble the macaque coos, in that subtly different calls within a class are used in different social contexts. This possibility, of course, cannot really be investigated until we get back to the lab; however, from what we can see at the moment such an inquiry into the social function of a graded series of calls could potentially complement our work on more discrete alarms and screams.

Although not directly related to the vocal work, we are continuing to observe relations between groups and male transfer, with slowly increasing success. We now observe 15–20 encounters each month between groups in which we know all the members, and another 10–15 between groups in which all the males are known. Thus we are beginning to build up data on male behavior in the "original" troop preceding emigration, and on male behavior in the adopted troop after transfer. This should eventually lead us to some idea of how the transfer of males affects subsequent interactions between groups, particularly in those cases where the male knows all the individuals involved. These observations really only scratch the surface of a very interesting problem, and the real results should accrue at an increasingly rapid rate over time.[104]

The surviving correspondence from the remainder of this first field trip in Amboseli—the Seyfarths left in early June 1978, handing over to Phyllis Lee, a Madingley graduate student—shows the steady fulfilling of this program.[105] Indeed, with one exception, the conclusions and ambitions of this private letter map straightforwardly onto the published papers that followed. In "The Ontogeny of Vervet Monkey Alarm Calling Behavior: A Preliminary Report" (1980), the Seyfarths exhibited the data showing that as vervets mature, their alarm calls and alarm responses become less generalized.[106] In "Vocal Recognition in Free-Ranging Vervet Monkeys" (1980), they recounted experiments begun in December 1977 to test the responses of adult females to recordings of infants' screams—the "three females" experiments, as the trials came to be known, demonstrating that while the mother of the screamer looked in the direction of the scream, the other females looked at the mother.[107] The male-transfer observations saw publication under the title "Intergroup Encounters among Free-Ranging Vervet Monkeys" (1981).[108] Due to pressure of other work, the playback study of a graded, intragroup vervet call had to wait until the second visit to Amboseli, in the first half of 1980, but this study eventually came to light as "How Vervet Monkeys Perceive Their Grunts: Field Playback Experiments" (1982).[109] Other, as it were interstitial, papers appeared as well, on the grooming behaviors of adult female vervets (1980)[110] and the selective forces shaping vervet alarm calling (1981).[111] The latter contained the sole mention in print of the stuffed-predator experiments. On the whole, these proved unsatisfying, in part because the facsimile animals left much to be desired—the leopard had a weak chin and newspaper poking out of its anus, while the python (fig. 9.5) came with a "life-like" wriggle that, unfortunately, suggested not a live python, but a cobra or mamba—and in part because the results largely confirmed what already seemed well enough confirmed. Still, the leopard furnished its owners with impressively quantitative data showing that the leopard alarms

Figure 9.5 The stuffed python in front of the Seyfarths' *banda*, 1978. With Dorothy is their stuffed-predator supplier, the anthropologist Andrew Hill, then based at the National Museums of Kenya. Courtesy of Phyllis Lee and reproduced by permission.

of high-ranking adults last longer than the alarms of their lower-ranking groupmates.[112]

The exception to this private-public paralleling is, however, remarkable: *"By this stage in our experimental work there seems little doubt that responses to alarm calls are context-dependent, thus perhaps weakening any claim for the calls' semanticity."* In the published work, of course, just the reverse is asserted: when it comes to eliciting distinctive escape responses, the different alarm calls are sufficient unto themselves, independent of context, thus strengthening a claim for call semanticity. The titles of the joint 1980 papers with Marler on the alarm playbacks, in *Animal Behaviour* — "Vervet Monkey Alarm Calls: Semantic Communication in a Free-Ranging Primate"—and in *Science*—"Monkey Responses to Three Different Alarm Calls: Evidence of Predator Classification and Semantic Communication"—put the claims for semanticity right out front. Indeed, the failure to find among the monkeys "rudimentary semantic signals," in the phrase picked up on by the *New York Times*, would scarcely have been newsworthy.[113] So what was it about context that, at a quite advanced stage, had sown doubts about semanticity? And why, eventually, did the Seyfarths change their minds?

The Private and Public Lives of the Amboseli Field Data

A Change of Mind about Alarm-Call Semanticity

It will help first to clarify what the Seyfarths most likely did *not* doubt.[114] Consider again the later papers on the alarm work. There the claim for semanticity rests mainly on two sets of evidence: evidence from the experimental varying of the acoustic properties of the calls; and evidence from the experimental varying of their contextual settings.[115] Here is a summary from the 1980 *Animal Behaviour* paper: "Playback confirmed observations and showed that (1) alarm length, amplitude and alarmist's age/sex class had little effect on response quality; and (2) context was not a systematic determinant of response."[116] The letter of 20 November 1977 shows that something like conclusion 1 was already firmly in place. Updating their supervisor on the progress of experiments with the multisyllabic alarms, the Seyfarths reported that "results are tending to support our earlier tests"—in other words, the long versions of snake and eagle calls had, on the whole, so far evoked snake-fleeing and eagle-fleeing responses, respectively. The switching from short to long versions of the calls had not, that is, converted all of the vervets' responses on playback into leopard-fleeing responses, as a purely affective reading of the calls might have predicted.[117] Perhaps, they noted, the longer calls had evoked stronger snake-fleeing and eagle-fleeing responses than their shorter counterparts. If so, then the switch did have an effect, but it was quantitative, not qualitative, and that was wholly consistent with the Marlerian notion that affect and symbolism work together—that while, say, all snake chutters convey the information that a snake is present, a longer chutter conveys greater fear than a shorter chutter. Along similar lines, as we have seen, the experimentalists had done their best to even out the natural differences in call volume, making all the played-back calls about as quiet as a snake chutter—yet, by and large, the vervets had still responded differently to each of the three calls.[118]

So one of the two main evidential supports for claiming call semanticity was likely already strong when the couple expressed doubt about the strength of the other support. The change of mind we are seeking to reconstruct, though important, was local, not global. It centered on just one of two main classes of evidence. Robert and Dorothy did not mean there is *no* evidence for call semanticity. It was already clear that, consistent with the semanticity hypothesis, the experimental flattening of variations in loudness and length preserved the differences in evoked response. Another thing they almost certainly did not mean was that the alarm calls were context-dependent, and therefore at best weakly semantic, because the calls failed to evoke *exactly* the same response from the vervets no matter where they were: up a tree, or down on the ground, or wherever. Long before the Seyfarths ventured into Amboseli, they knew perfectly well that, in this sense, context matters hugely to determining response. An impeccably symbolic leopard call, heard by vervets with interpretive abilities and survival instincts in good working order, would of course evoke different responses from vervets on the ground—who would need to flee up a tree and fast—than from vervets who were already safely up a tree. No one would expect to see such animals behaving literally the same way, point for point. Indeed, no one who demanded strict identity of responses across contexts before conceding that alarm calls were semantic—as arguably W. John Smith did—would need to bother with playbacks at all, since vervets so obviously fail to meet the standard.[119]

We can also set aside a third, related possibility: at the time of the 20 November letter, the Seyfarths had failed to detect the context-independent, adaptive responses that they later discovered, thus securing conclusion 2, that "context is not a systematic determinant of response." In the later papers, the golden example of a context-independent response to an alarm call is the looking-up of monkeys that have heard an eagle call. It is, indeed, the sole instance cited in the most sustained discussion of context, responses, and semanticity:

> In our experiments, context was not a systematic determinant of the responses of vervets to alarm calls. Different alarms evoked different responses in the same context, and responses to some alarms remained constant despite contextual variation. For example, monkeys looked up when they heard eagle alarm calls, regardless of whether they were on the ground or in the trees. Given the variable role of context in determining responses to alarm calls, the most parsimonious explanation would appear to be that, for all those within earshot, each alarm represented a certain class of danger. . . . Individual monkeys then responded according to the nature and degree of their vulnerability to that danger at the time.[120]

It is easy to imagine the looking-up response as so subtle that its discovery awaited the post-Amboseli analysis of the playback films. But the documentary evidence to the contrary is straightforward. The report to the Wenner-Gren Foundation is dated 16 November 1977. In it, the couple wrote that the monkeys "generally . . . look up when they hear a vervet 'eagle' call."[121]

What, then, did the Seyfarths have in mind, if not global skepticism, Smithean literalism, or the absence of cross-contextual responses? In part, most likely, they were mindful of something well attested in the later papers: vervets, on hearing an alarm call, do not always act as if they have acquired knowledge of the presence of the associated predator. Sometimes they act that way, of course. But sometimes they do nothing much at all. At other times they do the "wrong" thing—looking up on hearing snake calls, looking down or running up a tree on hearing eagle calls, running into bushes on hearing leopard calls.[122] Very often, the papers reported, the first thing vervets do is look in the direction of the caller/speaker and scan in other directions "as if they were searching for additional cues, both from the alarmist and elsewhere."[123] Such behavior is hardly what one would expect if the calls were informationally sufficient, in no need of contextual completion.[124] Yet so common is visual cue-searching that Marler, Cheney, and Seyfarth tended to portray it as a kind of generalized complement to the more specific escape responses. "Alarm-call playbacks," they wrote in their *Science* paper, "produced two kinds of response. Subjects in all age classes and of both sexes looked toward the speaker and scanned their surroundings more in the 10 seconds after a playback than before. . . . In addition, each alarm type elicited a distinct set of responses."[125]

Here, then, is a first clue: measured against the semanticity hypothesis, the data on playback responses were full of anomalies. A second clue lies in the grant application Robert submitted a few months before entering the field. Recall that this document set forth the fullest plan up to that point for the experimental use of playback to illuminate the affective versus symbolic nature of the vervet alarm calls. Of the four questions specified, the first and third dealt with context dependency. In full, they read:

(1) Do recordings of different alarm calls consistently evoke different responses when presented to the animals under various conditions (see also below)? In other words, can such calls really be viewed as "representing" different types of external danger?

. . . .

(3) Are responses to alarm calls environment-specific? Records will be kept of the environment in which each call occurs, and playback experiments will be attempted in "atypical" environments, for example using leopard alarm calls in open country.[126]

The idea of distinguishing between typical and atypical environments, and the importance of doing so for purposes of semanticity testing, is not something that survives in the later papers.[127] But the point is not difficult to grasp. Suppose scientists perform leopard-call playbacks only when vervets are in circumstances where typically they do encounter leopards, and thus typically do hear leopard calls. Even if the playbacks uniformly elicit as-if-escaping-a-leopard responses, that result can be explained with or without supposing call semanticity. It may be that the vervets responded to the fact that a bark or a chirp, corresponding to a leopard, was heard, rather than a rraup (eagle) or a chutter (snake). Or it may simply be that vervets in those circumstances respond to any alarm call—of whatever acoustic structure—as if it had been given to a leopard, since it is typically leopards, not eagles or snakes, that threaten vervets in those circumstances. Thus, potentially, there is much interest in finding out how vervets respond to calls played in circumstances not strongly associated with hearing those particular calls and/or seeing the corresponding predators. Arguably, if vervets respond to a leopard call as if it means "leopard" even when the vervets are not, as it were, expecting to hear a leopard call, because they are in open country or somewhere else where leopards are seldom encountered, then a claim for call semanticity will be much strengthened.[128]

Turning again to the later papers, we find a markedly different conception of what it is for a call's meaning to be dependent on context. The question raised is not whether atypical contexts flummox the vervets, but whether context is "a systematic determinant of the responses of vervets to alarm calls"—in other words, whether, say, a vervet on the ground will run up a tree no matter whether a bark or a rraup or a chutter is heard, and will never react in the same way to the same call when on the ground as when up a tree. Assessed along these lines, the vervet responses look impressively context-independent, for, again, "[d]ifferent alarms evoked different responses in the same context, and responses to some alarms remained constant despite contextual variation." En route to this conclusion, moreover, the *Animal Behaviour* paper offers a number of ingenious explanations for anomalous responses to playbacks. Vervets failing to respond at all, for instance, were often adult males and infants. Since adult males are less vulnerable than other sorts of vervets to predation, the males' lack of response, far from showing that they had failed to understand a call's meaning, showed just the opposite.[129] As for infants, they are extremely vulnerable, and if they could respond to alarms, it would be very much in their interests to do so. But alas, as was clear from early on, alarm-call behavior is not wholly instinctive, but grows through learning. Younger animals can thus be expected to make—and they do make—a lot of mistakes, sometimes failing to give alarm calls to the right animals, and sometimes failing to respond to alarm calls in the right ways, if at all. On inspection of the data, indeed, it was the junior, still-learning animals who accounted for many of the maladaptive responses.[130] There was even a plausible explanation offered for why juveniles in trees tend to look down when they hear eagle calls (it was because juveniles generally run out of trees in spurts, running then stopping then looking down then running some more, whether or not they are responding to alarm calls).[131] But the exculpatory masterstroke—reserved for the paper on ontogeny—was the explanation for the general looking-for-cues that so faithfully accompanies the adaptive escape responses. Cue searching was not a symptom of informational deficiency in the calls, but a strategy of the young for learning what the calls mean—which explained why infants and juveniles were more likely than other classes of vervet to look at adults and copy their responses.[132]

In sum, at a certain point, the Seyfarths stopped thinking about semanticity the way they did in Robert's 1976 grant application, as something to be assessed by comparing responses in typical versus atypical contexts, and started thinking about it in the manner of the papers completed in 1979, as something to be assessed according to whether context alone suffices to determine responses. More speculatively, it seems that this change went along with a change in the way anomalous responses were classified and explained, from an analysis that emphasized differences of context, with anomalies arising more frequently in atypical contexts, to one that emphasized differences between classes of individual, with various vulnerabilities, competencies, and so on. Chronologically, the Seyfarths were closer to their 1976 selves than to their 1979 selves when they remarked on the apparent context dependency of alarm calls. If we suppose that, conceptually as well, they were closer at that time to their 1976 selves, then a plausible interpretation of the remark lies to hand. "By this stage in our experimental work," they wrote, "there seems little doubt that responses to alarm calls are context-dependent, thus perhaps weakening any claim for the calls' semanticity." In other words, outside a restricted range of typical circumstances, played-back alarm calls failed on the whole to elicit adaptive, predator-specific responses. The more atypical the circumstances, the more likely the production of anomalous responses, suggesting that it is the call in context, not acoustic structure alone, that informs vervets about what kind of predator needs to be escaped. Any claim for the calls' semanticity was, accordingly, weakened.

It is not possible to say with confidence when the crucial change or changes of mind occurred.[133] The Seyfarths in interview recalled no dramatic about-faces on semanticity, in the field or out of it. Perhaps the reorientation was so gradual that in the end it was not noticed and cannot now be recollected.[134] Whenever and wherever the change to a more bullish viewpoint on call semanticity first took hold, and whatever its initial promptings, there are several reasons to think that the year and a half spent at Millbrook (from around June 1978 to December 1979) provided, at a minimum, ample reinforcement. For one thing, it was at Millbrook that the Seyfarths first really saw what had happened in the alarm playback trials. During the experiments themselves, the couple were often preoccupied with the tricky business of performance—making sure that all of the apparatus was working, that none of the numerous precautionary rules were being violated, and so on. But even an observer free of duties would have been hard-pressed to track all the potentially diverse responses in a group of vervets in full, frenzied alarm mode. Now the filmed trials were subjected to painstaking, frame-by-frame scrutiny, with quantitative criteria used for deciding, for instance, how far a vervet head had to have tilted upward to count as "looking up."[135] As the Seyfarths got to grips with the full range of their data, they no doubt became aware of hitherto invisible patterns of response, as well as new possibilities for analyzing them. For example, despite efforts in the field at corrective speaker placement, natural differences in call amplitude had persisted in the alarm trials. Now, however, it was possible to identify playback trials where the amplitude levels had nevertheless been very nearly equivalent, because a louder-than-usual snake call, a quieter-than-usual leopard call, and an averagely loud eagle call had been used. With amplitude variation now smoothed away even more comprehensively, the distinctions between the snake, eagle, and leopard responses, however, still held good.[136]

Marler was delighted with such excellent results. What intrigued him most of all in the Amboseli work, however, was the data on the development of vervet alarm calling. Strikingly, for all his long years as a student of vocal ontogeny in birds, and for all that the Seyfarths had arrived at Millbrook with interests and expertise in primate ontogeny, the study of vervet vocal ontogeny was not, as we have seen, a major or even minor objective for the Amboseli trip. Like the alarm-call study more generally, the developmental substudy had bootstrapped itself up the research agenda only once the Seyfarths were in the field. They had taken to it, however, with their customary energy. Their years at Madingley, where they had worked alongside experts in children's cognitive development, had primed them to appreciate the interest of the phenomenon of overgeneralized speech in children. Now at Millbrook, with Marler as their enthusiastic guide, the couple started making extensive forays into the psychology of categorization.[137] The psychologist Richard Herrnstein's classic 1964 paper (with Donald Loveland) "Complex Visual Concept in the Pigeon";[138] new anthologies of recent papers in cognitive science;[139] and, extensively, the books and papers coming out of the ape language projects, especially Premack's work, so rich in insights on how to test for understanding of hierarchical relationships in a symbol-wielding primate[140]—all of these and more started reshaping the Seyfarths' intellectual world. At one point, as they recall, it had occurred to them to represent their data on the gradual narrowing of the alarm-eliciting stimuli as the vervets mature—from, say, lots of different kinds of mammal eliciting leopard calls to just leopards doing so—by way of some classificatory trees, on the model of the trees found in the pages of Jeremy Anglin's book *Word, Object, and Conceptual Development* (1977). For the Seyfarths, the diagrams made for a handy visual summary of their findings, as well as a deliberately provocative suggestion that, as they recently recalled, "the same mechanisms might underlie the development of words as labels in children and in monkeys."[141]

This turn toward the scientific literature on cognition and categorization affected not just the way the couple thought about their developmental data, but the way they thought about their projects as a whole. Ultimately they reclassified all of their Amboseli experiments, and even some of their pre-Amboseli work, as studies in classification.[142] The three-females experiment, for instance, which had been conceived as a playback-based twist on ideas about mother-offspring interactions and rank acquisition, now became a test showing that adult female vervets classify juveniles in complex ways—not just as "my own" or "somebody else's," but also as "hers over here" or "hers over there."[143] In a similar spirit, the *Science* paper on the alarm research, with its subtitle "Evidence of Predator Classification and Semantic Communication," opened with some reflections on how the small sizes of natural signal repertoires meant that a study of natural semantic communication—in contrast with the ape language projects—had to consider the "extent to which nonhuman species divide objects into groups for the apparent purposes of communicating about them."[144] Such thoughts, of course, had come at the end of the inquiry described, not at its beginning.

Finally, we should notice that the Seyfarths were doing all of this at the time and in the place of the birth of cognitive ethology. The phrase "cognitive ethology" was Donald Griffin's, introduced in his slender 1976 volume *The Question of Animal Awareness: Evolutionary Continuity of Mental Experience*. Best known for his discovery of echolocation in bats, Griffin was the founder and former head of animal behavior studies at the Rockefeller, where he was still based and still active.[145] A longtime admirer of Karl von Frisch's work on the symbolic dance language of the honeybees, Griffin had spent the 1960s and early 1970s defending that work against critics who doubted the bees' dances truly bore the precise information claimed.[146] The experience seems to have pushed Griffin to a more fundamental examination of the rights and wrongs of anthropomorphism in the science of the animal mind.[147] The result was *The Question of Animal Awareness*, where, in the name of evolution, he called for casting off old taboos about exploring the mental awareness of animals. For Griffin, the experimental study of animal communication was the promising route forward into the animal mind. He concluded: "The future extension and refinement of two-way communication between ethologists and the animals they study offer the prospect of developing in due course a truly experimental science of cognitive ethology."[148] For the most part, reviewers were unimpressed, some because they thought Griffin was wrong, others because they felt he was knocking at an open door.[149] Seyfarth and Cheney recall siding very much with the latter. Certainly they did not see themselves as signing on to a program of Griffin's making. Marler, however, was an old talking buddy of Griffin's, and very sympathetic. There is more than a touch of Marler in Griffin's book,[150] and more than a touch of Griffin in Marler, Cheney, and Seyfarth's argument for, in their words, a "view of alarm calls as a form of semantic signalling, probably involving the formation of internal perceptual concepts, or symbols."[151] Furthermore, however the Seyfarths categorized themselves, the rest of the world claimed them for cognitive ethology. "The Seyfarths now propose new research on several aspects of what one might characterize as the 'cognitive ethology' of vervet monkeys," Marler wrote in a letter to a granting body in October 1979—the first of many similar labelings to follow.[152]

Behind the label, however, there lies a more subtle but perhaps more important phenomenon. The Rockefeller was more than just the place where Griffin and Marler happened to be. It was a place where the highest esteem went to scientists who, through a combination of rigor and imagination, demonstrated parallels between human and animal communication. In the early 1970s, Fernando Nottebohm had done it, discovering that the bird brain is lateralized for song.[153] In the mid-1970s, James Gould had done it, showing, with a series of brilliant experiments (or so most onlookers concluded), that honeybees do indeed receive information about food sources from the bee dances.[154] In the late 1970s, Steve Zoloth had done it, as part of a team effort revealing the first evidence of perceptual specialization for own-species sounds in a nonhuman animal, the Japanese macaque.[155] These were the achievements that kept Millbrook in the pages of *Science*. They also formed the backdrop against which Cheney and Seyfarth laboriously re-examined and rethought their Amboseli field data. At Millbrook, no one scored points for failing to find human-animal parallels. Cheney and Seyfarth found them. And the points scored were beyond anyone's expectations.

The Meanings of the Vervet Alarm Calls

"Now that the Science paper is out we will no doubt get a bit of publicity," Marler wrote in a memo to the Seyfarths in mid-November 1980.[156] They were back at Millbrook, having spent most of the year in Amboseli, pushing forward with new research and flush with grant awards. They were now excited not about the alarm-call playbacks—ancient history to them by this point—but their more recent experiments and observations, above all their playback tests of meaning in the vervets' subtly varied grunts. Given hundreds of times a day, grunts are part of the weave of vervet social life. They were thus a far better match than alarm calls for the Seyfarths' major interests in the organization of primate societies and, now, of social knowledge. Moreover, because, to human observers, all grunts sound nearly the same, and evoke nothing like the easily detected responses that alarm calls or juvenile screams elicit, the challenge of doing experimental playbacks was that much greater. But the experiments were a success, revealing the call, on the Seyfarths' analysis, to contain four subtypes, distinct both in acoustic structures and eliciting situations.[157]

Here was work that was truly the Seyfarths' own. It took cues from Green and Struhsaker, but it went well beyond their precedents, in technique and in interpretation, while advancing a distinctive research program that, as we have seen, predated Millbrook days. Indeed, for all that the couple would become indissolubly linked with the vervet alarm study, they were much more proud of their work on grunts. Aside from the reasons just mentioned, there was always the feeling that Struhsaker had shown it all anyway. To those already persuaded that the vervet alarm calls name their predators, the playback experiments offered pleasing confirmation of old news. And to those not so persuaded, because convinced that, like all other primate vocalizations, the vervet predator calls had to be limbic-emotional in nature and so could not possibly name or symbolize, there was little in the new evidence to bring them around.[158] Certainly there was no contrary neuroscientific data—a liability borne home to the Seyfarths during talks at Berkeley in 1979, when Sherwood Washburn made it abundantly clear that he was not impressed.[159] With all signs, then, that the response of the scientific community would be at best lukewarm, the alarm work seemed unlikely to break out into wider consciousness.[160] There was even some early evidence that it would not. When a friend from Madingley days, Jeremy Cherfas, now a science writer, did a story on the Seyfarths' vervet research for the *New Scientist*, it made the magazine's cover in June 1980, but otherwise made no impression on the media world.

A few months later, things were different. Now, with the ape language controversies intensifying, and the alarm-call papers at last migrating into print, journalistic antennae started to twitch. Throughout late November and early December, led by the *Sunday Times* of London and the *New York Times*, newspapers carried stories on the vervet alarm calls and their scientists, under headlines such as "Clues of a Monkey 'Language'" (*Newsday*) and "Primitive

Monkey Language Revealed" (the Toronto *Globe and Mail*).[161] "A new blow has just been struck in the ongoing battle over just how unique we human beings are in the evolutionary scheme of things," ran a leader in the *Wall Street Journal*, with the title "Descent of Man." Summarizing the vervet predator lexicon, the writer added some heavy-handed whimsy of the sort with which the press of an earlier day greeted Garner's work on the simian tongue: "O fortunate vervet—you who know how to tell your dangers apart and know what to do about them. Maybe someday we will recover your skills and learn the difference between 'world running out of oil' and 'overregulated energy industry,' between 'Soviet expansionism' and 'bloated U.S. military establishment.' Until then we'll just have to keep watching in envy."[162]

A constant theme was the contrast between the ape language projects and the vervet playback trials. "Although chimpanzees and gorillas can be taught in the laboratory to represent different objects by manual sign and plastic shapes," the London *Sunday Times* told its readers, "the monkey 'words' are believed to be the first evidence of this kind of ability in the wild."[163] An April 1981 segment on the *Today* program, on the American television network NBC, opened with a shot of Terrace's Nim, accompanied by the following voiceover from science correspondent Robert Bazell: "This is Nim Chimpsky, one of several primates who were taught to speak American sign, the language of the deaf. The experiments have been very interesting, but when monkeys and apes are living in their natural environment, they don't communicate with one another in languages which they've learned from scientists, and the question is, do they have a language or a way of speaking?" At this point, viewers learned of the Rockefeller group's work, illustrated with film clips of vervet monkeys standing up and looking around in response to a played-back snake call, fleeing into the bushes at an eagle call, and climbing into the trees at a leopard call. Bazell continued: "The scientists, who work under the direction of Dr. Peter Marler, say we are getting our first look at the language of animals."

SEYFARTH: We're trying to tackle the monkeys on their own terms, to go into their habitat, learn their systems of signals.
CHENEY: Calls, that humans in the past have assumed simply manifest an emotional state, such as excitement or fear or hunger, really can stand for something.
BAZELL: The next stage in this research is to study the grunts, which the monkeys utter as they are just sitting around chatting. It turns out that these, too, have very specific meanings, although they are much more difficult to decipher. [To the presenter, Jane Pauley] Wouldn't it be great if we could really understand an animal language?

PAULEY: Yes. . . . I always thought it a little arrogant for scientists to teach monkeys our language, when you think it would be easier for scientists to learn theirs, right?

BAZELL: And that's what's happening now.

PAULEY: Good.[164]

The following year, Seyfarth published a popular article on the vervet work in his own right, for *International Wildlife* magazine. He began by setting two scenes side by side: a chimp and a psychologist chatting in sign while out for a walk in Manhattan; and another scientist "more than 5,000 miles away, . . . in an East African national park," watching monkeys quietly feed when one of them gives a grunt, in response to which the others look toward the horizon. After discussing the apparently endless debates surrounding the work of the psychologists on primate communication, he wrote:

> In the midst of this hoopla, a newer group of scientists, in which I include myself, has emerged. We argue that animals in the ape language projects have been brought into a human world, taught a human sort of communication and been tested in human terms. Given the fact that few Americans, for example, would perform to the best of their ability if quizzed, say, in Nepalese, the design of the ape language projects may have led researchers to underestimate the natural abilities of their subjects. Why not, then, try to enter the monkeys' and apes' habitats, learn their systems of communication and evaluate their abilities on their own terms? . . . [Our] results gave us convincing proof that monkeys in the wild could do at least some of the things that had previously been done only by apes in the laboratory. It seemed clear that if we were patient enough, and designed our experiments correctly, we could get the animals to tell us how they communicate with each other, and we could do this without teaching, in an environment where the monkeys were free to choose their own method of communication and topics of conversation.[165]

We have been here before. The setting of biologically inappropriate tasks to animals whose abilities thus remain hidden; the presumption that natural behavior must prove uninteresting; the general failure to study "the animals for their own sake" (Tinbergen)—these were the allegations that the first generation of ethologists had flung at their psychologist competitors, and that had remained disciplinarily defining ever since. As critique, there was a grain of truth to them, no more. We can imagine the project psychologists in their defense answering that they had used sign language with the apes precisely to accommodate their natural abilities. No matter. The point was to make vivid the need for the ethological reform of psychology. We saw that, in a complex way, Marler had conceived the vervet alarm playbacks as an ethological riposte to the ape language projects in their most extreme, Premackian form. Now, thanks to an accident of timing—publication at a moment of high media interest in the difficulties surrounding the projects—the vervet playbacks could play their assigned role to the hilt. In contrast with the disintegrating ape language projects, the vervet playbacks now looked not just interesting and clever, but virtuous. Unlike the laboratory projects, which had bent the wills of the animals to the experimenters' interests, the field playbacks had let the animals be free. Where the project scientists had brought animals into the scientists' world in order to teach, the playback scientists had taken themselves to the animals' world in order to learn. The projects thus stood for failure born of arrogance; the playbacks, for success born of humility.[166]

But the echoes reach back even further than the 1930s, all the way to the 1890s, when Garner invented the primate playback, at a time of intellectual crisis over the origins of language. The time has come for a general reckoning of the resemblances between the invention and reinvention of the primate playback experiment. How far do they extend? And what, if anything, do they reveal about the nature of the historical processes that produced them?

Conclusion

The history of the sciences is the history of certain questions coming to be asked and, sometimes, answered. A major task for historians is to reconstruct the conditions that gave life to questions, making them real, compelling, irresistible, for a community of inquirers—and perhaps the wider world—at a particular time and place. A connected task is to understand how once-live questions lost their vitality. To say this is hardly novel. Two of the most historically minded philosophers of the sciences in our time, Nicholas Jardine and Ian Hacking, have raised just this question about questions.[1] And yet, surprisingly few historical works address it squarely. A metaquestion lurks here, about the conditions under which historians take seriously the questions scientists pose. But that is not the business of this book. It has used the long run of scientific studies of nonhuman primate communication to show how unobvious was an apparently simple idea: to find out what primate vocalizations mean, one should record them, play the recordings back to the animals, and watch what happens. The question of what would be revealed through playbacks with nonhuman primates only *became* an obvious one to pursue, in a focused and persistent way, under highly particular circumstances. The curious twist in this tale, of course, is that this happened twice, in the latter parts of different centuries.

1890 and 1980 in Retrospect

When we consider the experiment's nineteenth-century invention, it looks at first blush as if the mere existence of the improved phonograph was sufficient to prompt its use in the experimental study of animal language. Garner invented the primate playback experiment almost as soon as it became technically possible. He did so, however, not as a phonograph buff looking for new ways of applying the talking machine, but as an evolutionist searching for a better means of translating the utterances of monkeys. The key to Garner's innovation lies, as we have seen, with his engagement in the simmering debate over language as fatal to the evolutionary theory. His own, decidedly non-elite understanding of that theory also matters; a more subtly Darwinian thinker might have been less confident that, in an evolved world, simians must be found speaking a tongue slightly more primitive than the tongue of the most savage humans. We should note too Garner's animus toward the academy. In traveling with cage and (aspirationally) phonograph to the French Congo, he presented himself as the true scientist, willing to take the intellectual and physical risks before which the so-called professionals trembled.

For the twentieth-century reinvention of the experiment, of course, none of the above holds. The basic hardware had been around since Garner's day, improving steadily, and even, from time to time, used for recording and playing back the vocalizations of primates. Evolutionary theory had become imperturbably secure and, after the Second World War, dissociated from old ideas of racial hierarchy. Furthermore, there were now professional scientists, ethologists, who made a habit of sometimes demanding field studies of the natural behavior patterns of animals. What brought the primate playback experiment back into existence, above all, was the flaring up of a disciplinary antagonism around an ethologist with a unique cluster of interests: in the learning of birdsong, studied in the lab with experimental playbacks; in primate vocalizations, studied in the field; and in the characterization of animal signals as bearers of information or meaning. No one but an ethologist could have made a career in these areas. And no one was quite as prone as an ethologist to criticize psychologists for studying animal behavior on human terms rather than animal terms. The psychological projects in ape language tutoring that flourished from the late-1960s to the late-1970s found their comeuppance in the vervet alarm-call experiments that Marler assigned to Cheney and Seyfarth.[2]

So twice, following two quite different paths, history generated the primate playback experiment. To what should we attribute such robustness?[3] Maybe nothing. Maybe things just turned out that way, and there is nothing more general to say or explain. Such explanatory minimalism, or maybe nihilism, recommends itself on several counts. It is cleanly intelligible. It requires no extra work on the part of the historian. And it conforms to recent trends in science historiography, where the signature tone is relentlessly contingentist. All scientific debates are discussed as if they could have gone any which way at any given moment, but for the accidents of human character and position. The conclusions agreed upon as to what is true, and the courses of action taken to find out what is true, are presented as in the end inseparable from the vagaries of social history. The implication, of course, is that different social matrices would have brought about different conclusions and different courses of action. Nothing in science is inevitable. The fact that the primate playback experiment came around twice, independently, does not show that the experiment was bound to happen, given a certain level of technical sophistication and enough people committed to learning about the natural vocalizations of animals. It shows only, once again, that science is a social construction, if sometimes with weird contours.

Certainly the history of the primate playback experiment is rich in contingency. Most conspicuous are the events conspiring to drive the experiment to the margins for nearly a century, despite huge positive publicity.[4] A tantalizing possibility is that, if Garner had returned from the Congo with phonographic evidence of languagelike abilities in the apes, there might never have arisen a comparative psychology for ethology to define itself against. Conwy Lloyd Morgan had announced himself, in 1893, quite prepared to drop his arguments for a presumption of trial-and-error learning in animals if presented with such evidence. He was not, and comparative psychology grew up in the shadow of his stern rule of inference, Morgan's canon. Then there is the complex counterpoint of developments in the ape language projects and in Marler's immediate sphere. It is easy enough to imagine the Struhsaker alarm-call observations remaining forever on Marler's "to do" list without Premack's provocation, or to imagine the results of the alarm-call experiments getting little attention outside a small circle of ethologists and primatologists. And so on. It is not difficult to multiply counterfactuals in a similar spirit. Yet what distinguishes the history recounted here are less the elements that lend themselves to a contingentist reading than the elements that do not.

Consider a striking contrast in rationale between the 1890 and 1980 instantiations of the experiment. Something absolutely fundamental to Garner's self-conception had no place whatsoever in the later work: namely, that playback experiments needed doing because, on the evolutionary theory, the world ought to contain animals with humanlike language and humans with animal-like language. By the time of the vervet playbacks, the consensus opinion, virtually unchallenged and perhaps, now, unchallengeable, was that all human languages are equal in their complexity. None are more animal-like than any other. So while the Rockefeller team believed in gradual evolution no less firmly than Garner did, they in no way saw themselves as demonstrating the existence of a predicted hierarchy in nature. For them, if anything, evolutionary considerations took an antihierarchical form. Marler wrote of how amazing it would be if the capacities for symbolic communication in the ape mind, on which the successes of the ape language psychologists depended, went unused in the course of the animals' natural, autonomous lives, but emerged only under a human-imposed tutoring regime.[5] If this does not amount to a straightforward reversal of Garner's rationale (whatever such a thing would be), it is both clearly evolutionist, indeed adaptationist, and free of even a whiff of natural hierarchy upheld. And while the scientific racism of the later nineteenth century—what we might call its evolutionary politics—did not disappear in the twentieth century, it did become officially unwelcome, and increasingly so, not least among professional biologists in the United States.[6]

The primate playback experiment appears, then, to have been invented with one evolutionary politics in mind and reinvented with an opposite one in mind. It first arose against the backdrop of an unarguably hierarchical evolutionary theory, and arose a second time against the backdrop of an arguably antihierarchical evolutionary theory. Two quite different social histories, in other words, produced roughly the same conclusions about what is true and about how to find it out. In this important respect, the experiment seems independent of a certain kind of social-scientific determinant. That does not mean, of course, that Garner's context and the Rockefeller team's context had nothing relevant in common—that the conditions that prompted Garner's question were totally unlike the ones that prompted Marler's. Around 1890 and 1980, the animal origins of language had become hugely controversial. At both times, there was a feeling that haughty scientists were underestimating the natural abilities of nonhuman primates. And at both times, the conducting of a field experiment brought with it an unbeatable combination of moral and epistemic authority: the authority of experiment over observation or speculation, and the authority of nature over artifice.[7]

Can we identify further commonalities? As noted, the pattern of circa 1890 appearance/circa 1920 disappearance/circa 1980 reappearance can be related only weakly to patterns of technological change. Patterns of theoretical change, however, are another matter. Here is one of several crude periodizations that have been offered for the long run of evolutionary theory. From 1860 to 1890, saltationism was little more than a minority interest. From 1890 to 1930, it was much more than a minority interest. From 1930 onward, it went back to being a minority interest. The chronological match for the primate playback case at least raises the possibility that the experiment's prospects go up as saltationism's popularity goes down.[8]

The Vitality and Virtue of Saltationism

Those with saltationist (or catastrophist) sympathies take the view that evolution proceeds at least sometimes not by small, gradually accumulating increments but by large and sudden leaps between sustained periods of stability. Certainly it is striking how common were such sympathies among the scientists involved in the changes that, on the analysis ventured in this book, eliminated the primate playback from the newly professionalized, early-twentieth-century disciplines of cultural anthropology and comparative psychology. We have explored this theme most fully in connection with Horatio Hale. He bears further reflection, for he reminds us that, from T. H. Huxley's writings through to Galton's and beyond, there was plenty of saltationist help to hand in the post-*Origin* era to translate into evolutionary terms an essentially pre-*Origin*, old-style ethnological view of humans as more or less degraded descendants of a fully human first pair. Consider the following passages from Hale's 1886 Buffalo address, briefly summarized earlier, but deserving full quotation for the confidence they convey in presenting a wholly up-to-date and compellingly evidenced argument:

That the "speaking man" of our era is a descendant of the "speechless man" of the River-drift period cannot be doubted. We have not to deal with the origin of a new species, but simply with that of a variety. There can be no question but that this variety arose in the usual way, by what is termed the process of heterogenesis or, in other words, the law by which the offspring differs from the parents. As every child has two parents, it cannot resemble both, and, in point of fact, it never exactly resembles either of them. Ordinarily, this unlikeness is restricted within certain defined and rather narrow limits; but occasionally, as when dwarfs or giants are born to parents of ordinary stature, it is very great. Among the lower animals, when such offsprings [*sic*] propagate their like, a new variety or breed arises, which sometimes differs very widely from the original stock,—as occurred, for example, in the Ancon or otter breed of sheep, which thus originated in New England, and in the hornless cattle which have overspread several provinces of Paraguay. *That in some family of the primitive speechless race two or more children should have been born with the faculty and organs of speech is in itself a fact not specially remarkable.* Much greater differences between parents and offspring frequently appear. Among these, for example, is one so common as to have received in physiology the scientific name of polydactylism, a term applied to the case of children born with more than the normal number of fingers. M. de Quatrefages mentions that in the family of Zerah Colburn, the celebrated calculator, four generations possessed this peculiarity, which commenced with Zerah's grandfather. In the fourth generation four children out of eight still had the supernumerary fingers, although in each generation the many-fingered parent had married a person having normal hands. Plainly, he adds, if this Colburn family had been dealt with like the Ancon breed of sheep, a six-fingered variety of the human race would have been formed; and this, it may be added, would have been a far greater variation than was the production of a speaking race descending from a speechless pair. The appearance of a sixth finger requires new bones, muscles and tendons, with additional nerves leading ultimately to the brain. There is good reason to believe that the first endowment of speech demanded far less change than this. All the anthropoid apes can utter cries of some sort, and some of them can make a variety of sounds. Professor Hartmann expressly informs us that the larynx in these animals resembles in the main that of man. We cannot doubt that our primitive ancestor, the *Homo alalus*, in spite of his name, could utter many sounds, and possessed the usual vocal organs. Professor Huxley has dwelt with much force on the slight anatomical difference which might exist between the speechless and the speaking man. A change of the minutest kind, he tells us, in the structure of one of the nerves which communicate with the vocal chords, or in the structure of the part in which it originates, or in the supply of blood to that part, or in one of the muscles to which it is distributed, might render all of us dumb. And he adds . . . : "A race of dumb men, deprived of all communication with those who could speak, would be little indeed removed from the brutes. The moral and intellectual difference between them and ourselves would be practically infinite, though the naturalist should not be able to find a single shadow even of specific structural difference."[9]

Then a little further on:

And here it is proper to remark on the mistake, or the confusion of processes, which has led some esteemed writers to suppose that the first speaking men, originating from parents of weak mental capacity, must have partaken of that intellectual feebleness. Elaborate works have been written on this subject, in which the whole argument has been based on the supposition that the earliest of speaking men were inferior to their successors, not merely in accumulated knowledge,—which was a matter of course,—but in mental power, which is a very different affair. The lowest tribes of our time—the Australians, Hottentots, Fuegians, and other savages— have been assumed to be fair representatives of what our earliest ancestors must have been when they were first endowed with the faculty of speech. This supposition is contrary both to reason and to the known facts. It confuses two processes, which are totally unlike in their working and in their results. The changes caused by climate and the other external influences which are commonly known as the "environment" are gradual. The changes which arise from heterogenesis are sudden, and are at once complete. In the cases of polydactylism, we do not find that a mere germ or stump of a finger first appears, and gradually becomes longer and stronger in succeeding generations. The perfect finger appears at once. So in the lower animals: the Ancon or otter breed is known to have sprung from a single sheep, born with abnormally short legs, which became no shorter in its descendants. The hornless cattle of Paraguay are known to be all descended from a single animal, which was born without horns. There is no reason for supposing that the earliest speaking men may not have been endowed with the highest intellectual faculties of the human race. There is every reason to believe that they were so endowed.[10]

It is unsurprising that the saltationism driving this account of the origin of language has drawn little comment from historians, since Hale himself, although influential in his day (and especially on this topic), has become a rather neglected figure. More surprising is the silence on the saltationism of Franz Boas, the ur-figure in American cultural anthropology and the subject of a large and sophisticated historiography. If his commentators have mostly failed to pick out this aspect of his biological thought, the blame lies in part, no doubt, with the notion that Boas's ideas—indeed everyone's ideas about biology—must fall on one side or the other of a "typological/population" dichotomy: within, that is, an old-fashioned Platonic vision of biological reality as composed of types, which living organisms instantiate to greater or lesser degrees; or a modern Darwinian vision in which biological variation *is* the reality.[11] Routinely placed in the population camp, Boas is credited with striking the fatal blow against typological thinking about human races, insisting—in what some of his admirers register as neo-Darwinian tones—that human racial types are not absolutely fixed.[12] In fact Boas's position was more complicated. It is true that he regarded apparently fixed features of human races as subject to modification in new environments. In 1911, the year that he published *The Mind of Primitive Man*, he also published a soon-famous anthropometric survey purporting to document exactly this kind of change among America's new immigrants.[13] Moreover, in *Primitive Man* Boas contended that there was more variation within the races than between them.[14] But although that contention was later glossed as showing that races have no biological reality, Boas himself did not draw that lesson.[15] He took both variation and types to be real, stressing that racial types were not fixed, but also that they could change only so much—in other words, that there were limits to how far they could vary.[16] He put the case for typological stability plainly, and subversively, in his contribution to the Darwin celebrations of 1909, a lecture called "The Relation of Darwin to Anthropology." Never published, it was delivered as part of the same series at Columbia University where Boas's Columbia colleague John Dewey presented his rather better remembered views in "The Relation of Darwin to Philosophy."[17] Up to the present, Boas suggested, the evidence on the human races did not remotely support a picture of Darwinian gradual evolution:

The anatomical study of man, and archaeological investigation, have . . . failed to furnish any material that could be utilized to prove or disprove Darwin's opinions relating to the manner in which man has developed from lower forms, and in which the different types of man originated from the same ancestral forms. In fact, so far as our knowledge of the distribution of human types during the last ten thousand years goes, the races of man, as well as his numerous sub-types, appear stable; and we are not in a position to show a single case of the transition of one type into another, however closely they are related. Neither can we give any proof that the civilization which had existed in Europe and in China at least for several thousand years has had any cumulative effect upon the organization of the human body. This stability of the recent human types seems particularly remarkable when we consider that, from a biological point of view, man must be considered in the same way we consider domesticated animals, the extreme degree of variability of which has been so elaborately discussed by Darwin.[18]

Typological thinking about race has a reputation for being racist. In Boas's case, the opposite was true. In *The Mind of Primitive Man*, he made the same point made in his 1909 lecture in order to bolster his case against primitive men having fixedly primitive minds. It was, he argued, precisely because the human racial types are so old and stable that evidence showing different languages to be equivalently complex can be taken as evidence that the minds behind the languages are equivalently powerful.[19]

For Boas, it seems, the human races came into being much as we find them today, each equally far removed in body, mind, and language from any animal antecedents. Like his quasi-mentor Hale, Boas here brought up to date arguments for human biological stasis and equality that stemmed from a pre-Darwinian anthropological tradition—in Boas's case, the German tradition of Virchow, Adolf Bastian, and others.[20] Also like Hale, Galton's saltationism was probably a crucial inspiration and resource. For all his bitter opposition to eugenics, Boas was an admirer of Galton's mathematical anthropometry. In light of these enthusiasms, and the prior predisposition toward saltationism in Boas's thinking, it may well be that what have been described as the "inconsistencies and obscurities" in Boas's works on race in his era—on the one hand, his antiessentialism; on the other, his persistence in talking of stable "types"—is better interpreted as a consistent application of Galtonianism (the language of "types," including "the stability of types," were favored Galtonian terms).[21]

Among the psychologists as well, we find similar saltationist opinions. I have already emphasized the combined, consiliently saltationist influences of Huxley and Müller on Morgan's formulation of his canon, and how Edward Thorndike made that canon, and the human-animal mental divide it presupposed, foundational for the new experimental psychology. But consider as well Robert Yerkes, Thorndike's disciple, writing in his key 1905 paper "Animal Psychology and Criteria of the Psychic": "It is furthest from my purpose to argue . . . that there are no crises in organic development; on the contrary, I should admit that it is practically certain that sudden changes do occur, that

Nature does make leaps, that gradual development in one direction suddenly makes possible some apparently new process of change."[22] Or consider John Watson, in his equally important *Behavior* of 1914:

> How, then, shall we account for the notion that man is something and has something which the brute is not and has not? The feeling that a cataclysmal difference exists has been strong through all the centuries, and is as firmly fixed in the popular mind today as ever. Among scientific men the conviction that the gap is not so wide as was formerly supposed is growing; yet we find scientifically-minded men still searching for the "missing link" [here citing Witmer]. The feeling that a gap exists is shown all through the behavior work from studies on amoeba to those on the anthropoid ape. . . . If we examine man and animal . . . , we find that there is a difference in structure which at once accounts for the popular and the scientific feeling that a break exists between man and animal. We refer to the lack of well-developed speech mechanisms in animals and to the consequent lack of *language habits*. So far no language habits have ever been found in animals, nor has anyone succeeded in developing such habits in them. . . . It is pretty clear that the mutant man, when thrown off from the primate stock, sprang forth with a vocal apparatus different from that of the parent stock, and possessing abundant richness in reflexes, even far surpassing that found in the bird.[23]

The saltationism so popular in the decades around 1900 left an inquiry like Garner's looking, if not quite motiveless—Hale was encouraging, after all—then certainly much less urgent; for if evolution makes leaps, then it might well have made one in creating humans, leaving in its wake a mental and linguistic gap between humans and apes. On such a reading, at any rate, the fact that saltationism's rise coincided with the primate playback's fall starts to make sense.

Taking fuller account of the era's saltationism might also help us explain that remarkable shift in evolutionary politics mentioned above. To be sure, the sources of scientific antiracism in the nineteenth and twentieth centuries were multiple. Hale, Boas, and many of those who followed them came to hold antiracist attitudes independently of their saltationism. After the Second World War, revelations about Nazi abuses of biology further engendered widespread eagerness for a theory of evolution with no predictions of racial hierarchy. But scientists rarely justify their theoretical allegiances on programmatic, political grounds. At a minimum, most require enough by way of uncontentious observations, plausible conjectures, and at least arguable assumptions to explain to themselves and to others why they hold the views they do. A possibility worth exploring in future is that the Boasian example taught key thinkers in the next generation that modern biology and antiracism could work together. To put the point another way: it may well be that the theory of evolution that many people feel comfortable with today—a gradualist theory without any ineluctable racist commitments—became possible only after the era of saltationist enthusiasm.[24]

A Post-1980 Postscript

If the downturn in the fortunes of the primate playback experiment after Garner's heyday can be attributed in part to an upsurge in saltationist ideas, then we should expect to find later saltationist crestings likewise bearing the experiment toward a trough. In a modest way, later history vindicates this prediction. Recall that Chomsky kept the saltationist option alive throughout the 1960s and 1970s as a respectable stance on the animal origins of human language. In Chomsky's view, language sprang whole and complete out of the animal communication systems with which it otherwise had nothing interesting in common. But though he argued this point often and passionately, he never did so at length, tending to dismiss any attempt to inquire further as fruitless speculation.[25] His picture of language became comprehensively embedded in a saltationist picture of the origin of modern humans only with the publication in 1990 of the linguist Derek Bickerton's remarkable book *Language and Species*. There Bickerton proposed that at some moment between 140 and 290 thousand years ago, a *Homo erectus* female in Africa gave birth to a female child with the first brain wired for syntax, due to a genetic mutation that linked together brain areas active in her *erectus* forebears when they communicated in an asyntactic "protolanguage."[26] Bickerton defended the "catastrophic" (his term) element in this scenario by appeal to a general theory of evolution then at its peak of popularity, the theory of "punctuated equilibria," developed in the 1970s by the paleontologists Niles Eldredge and Stephen Jay Gould, and positing long periods of stasis punctuated by brief periods of rapid change. The Gould-Eldredge theory had returned saltationism, if not quite to the mainstream, then to a major sidestream, where the likes of Bickerton might find it and use it to make sense of data suggesting that discontinuity in nature was real.[27]

Which data? First up for Bickerton were the vervet alarm-call playbacks. It is tempting, he wrote, to suppose that our word "python" is the descendant of a gradual process of semantic narrowing, word accretion, and grammatical elaboration that began among our distant ancestors with calls very like the vervet calls for python. But there are, he went on, a number of differences between the word and the call that cast doubt on any such continuist scenario. Consider, for instance, the range of meanings we can give to our word "python" via inflections. Yes, there is "python!" as in "Get away fast! There's a python!" But there is also the dismissive, not-at-all-alarmed "python" ("Yeah, right, there's a python nearby—tell me another one"); the inquisitive "python?" ("Pardon? Did you say 'python'?"); and the neutral, list-reading "python" ("Large snakes include the boa constrictor, the python, . . ."). What is more, we, unlike vervets, can speak of pythons even when there are none around, and when there is no immediate gain to be had through such speech. We can also, unlike vervets, talk about things having no obvious bearing on our survival—even things that do not exist.[28] Bickerton's point can best be grasped in relation to the continuum Marler proposed between wholly affective calls and wholly symbolic ones. For Bickerton, even if nature could drain out the affect from the vervet call for python, the result would not be identical to the word "python," because the word functions as a counter in a system of thought—of representation—that humans alone enjoy, and that enables humans alone to talk of pythons and so much else in other-than-alarmed ways. Language for Bickerton is the climactic achievement in the long evolutionary history of progress in the nervous systems that in animals represent the external world, the better to coordinate adaptive action. When *erectus* groups in Africa began labeling their concepts and communicating with each other via the labels and other items in a growing lexicon, protolanguage was born— and with it, a new if still modest ability to step back from the world as given and imagine it otherwise, remodel it mentally, plan its tailoring to needs. And when a brain-altering (hence also skull-altering and, Bickerton suggests, vocal tract–altering) mutation in one of those groups gave rise to a new species, *Homo sapiens*, whose members could not help but organize protolanguage syntactically, language was born—and with it mental powers that made them the dominant animals.[29]

How exactly all of this happened, and why, are questions Bickerton dwelt on at some length. When he catalogued the sciences likely to clarify the story in future, he included neurology, genetics, evolutionary biology, paleoanthropology, and linguistics—not, pointedly, playback studies of animal communication.[30] Perhaps, he allowed at one point, a pre-*erectus* call for a predator survived to become the vocal template for the word for that predator in the protolanguage. But if so, that fact would nevertheless be utterly without significance for understanding the forces that drove word formation in the first place. Start reconstructing the peculiar vulnerabilities and talents of early hominids via the archaeological, paleontological, and geological records, Bickerton argued, and you rapidly come to appreciate that, for instance, words for prey would have been far more useful for purposes of collective planning, and so far more likely to have sprung up early.[31] Nor is it just by reasoning on such indirect evidence that we can appreciate, in Bickerton's phrase, "the gulf between language and animal communication."[32] For proto-language, he argues, is with us still, if we know where to look. It is to be found in the asyntactic speech of human under-twos (whose still-developing brains are presyntactic), of "wild children" who grew up without normal human interaction (and so missed crucial experiences during the early critical period of language development), and the asyntactic signings of scientist-taught apes (who thus demonstrate the prehuman mind's readiness for protolanguage, requiring only the right circumstances to bring it forth). Most intriguingly and famously, Bickerton argued that protolanguage is to be found in the pidgins that exist wherever adults from different language communities are made to live together, with no single shared language imposed upon them. When children growing up with pidgin pass the two-year threshold, and begin organizing it syntactically, a true language, called a creole, is born. Here—if Bickerton's reading of the documents is correct—is evidence of the shift from protolanguage to language in a single generation, with no intermediate steps. On a scientific principle dear to Darwin, Bickerton ventured that what happens in the present may well have happened in the past.[33]

Once again, then, saltationist convictions drove a wedge between the urge to find out how language began and the urge to go out into the world and listen and experiment among vocalizing nonhuman animals. After 1980 as much as before it, saltationism has traveled as part of a package with skepticism about the value of the primate playback experiment. Nevertheless, the connection should not be exaggerated. For one thing, there have been plenty of sources of criticism of the vervet playback results besides saltationism. Animal learning psychology, with its tradition of associationist explanation, has remained potent. A quasi-Pavlovian possibility that increasingly nagged at Cheney and Seyfarth after 1980 went roughly as follows. Suppose that—as they believed they had established—vervets really do attend to the acoustic structure of their calls and not just to call volume and call length. A structure-responding animal may understand that, say, rraup *means* "eagle." But the observed facts are also consistent with another, less impressive interpretation: that the structure-responding animal has merely come to associate the rraup sound with eagles or, even more mechanically, with actions appropriate to escaping eagles. To discriminate experimentally between these meaningful and meaning-free scenarios, Cheney and Seyfarth borrowed a technique from child development psychologists, the habituation/dishabituation (or "hab/dishab") experiment. Having played a certain intergroup call to a particular vervet so frequently that the animal paid no attention ("habituated") to the call, they switched to playing a broadly synonymous call—that is, a call with a very different acoustic structure, but used in similar situations to similar effect. The question was whether the animal would respond to the switch, on the view that, if there was a response, the animal was attending to acoustic structure (which had changed), and if there was no response, the animal was attending to meaning (which had not changed). Their nonresponse results were taken to vindicate the semanticity of the calls—though inevitably these experiments too have been criticized.[34]

Even those carrying no brief for either saltationism or behaviorism could give the vervet playbacks a wide berth in theorizing about the origins of language. In 1990, the same year that *Language and Species* came out—as did Cheney and Seyfarth's *How Monkeys See the World*—another Chomskyan linguist, Steven Pinker, published (with psychologist Paul Bloom) an important paper in the journal *Behavioral and Brain Sciences*, "Natural Language and Natural Selection," aiming to show that neo-Darwinian theory, gradualist to its core, was sufficient to explain how grammar came to be a universal human endowment. Pinker and Bloom set themselves three textbook tasks: first, to demonstrate that there is genetic variation for grammatical ability; second, to undermine arguments suggesting that grammar had to be all-or-nothing, that there could be no functional intermediates; and third, to identify advantages for survival and reproduction accruing from greater grammatical ability. A further task that they explicitly denied having to take up was that of uncovering language's "phyletic continuity." It was just a mistake, they contended, to suppose that anything produced through natural selection must have antecedents discoverable today. If the genetic variation that led to the first grammar-supporting brain occurred among the australopithecines, then the search for language's roots among apes and monkeys—not our direct ancestors anyway—is pointless. "We must," wrote Pinker and Bloom, "be prepared for the possible bad news that there just aren't any living creatures with homologues of human language, and let the chimp signing debate come down as it will." The vervet playback experiments were not even mentioned. As in Bickerton's book, animal communication studies were conspicuous by their absence from the roll call of sciences to be consulted for future insights.[35]

Other Chomskyan linguists have followed Pinker's lead in trying to bring together gradualist Darwinism and Chomskyan linguistics. Among them is the former archsaltationist Bickerton, recently active conjecturing intermediate stages between protolanguage and language.[36] Another of the new synthesizers is—even more unexpectedly—Chomsky himself, in alliance with Cheney and Seyfarth's most accomplished former student, Marc Hauser. As Hauser tells the story, around 2000 he attended a lecture Chomsky gave on language and the brain, the message of which was the reverse of Pinker and Bloom's.

In Chomsky's view, the endless hypothesizing about how natural selection might have brought language into being had proved scientifically sterile, generating plausible arguments but no new empirical research programs. Far better for those interested in the origin of language, he continued, to concern themselves with its phylogeny and with the comparative animal studies that throw light on it, in particular by identifying the language-enabling capacities that humans share and do not share with other species. The shining example Chomsky held up as pointing the right way forward was Hauser's monumental volume *The Evolution of Communication* (1996).[37] Thus began a collaboration that resulted, with additional help from Hauser's then-postdoc W. Tecumseh Fitch, in a long coauthored review paper in 2002 in *Science*, "The Faculty of Language: What Is It, Who Has It, and How Did It Evolve?" Here the question of gradual versus saltational evolution was raised as one of several up for grabs as researchers trace the origins of the language faculty's component systems, both broadly construed (the nonlinguistic systems on which language depends—the sensorimotor and conceptual-intentional systems) and narrowly construed (the computational system whose operations Chomskyan linguists seek to uncover). For Hauser, Chomsky, and Fitch, the primate playback experiments took their place within a range of studies concerned with the phylogeny of the conceptual-intentional system, from research on whether chimpanzees attribute minds to each other to whether vervets and other monkeys modify their vocal productions according to audience composition. The authors' attitude toward "the classic studies of vervet monkey alarm call work"—since extended, they report, to macaques, Diana monkeys, meerkats, prairie dogs, and chickens—comes across as robustly positive. The verdict on the vervet calls as wordlike, however, is rather Bickertonian: "many of the elementary properties of words—including those that enter into referentiality—have only weak analogs or homologs in natural animal communication systems."[38]

Beginnings and Endings

The experimental program launched in Amboseli in 1977 thrives still. So, despite the dip around 1980, does the semirival program of the ape language psychologists. Any current reckoning of the science of the evolutionary origins of language will need to take account of impressive innovations in both programs, and much else besides.[39] This book, however, has had other aspirations, and it is fitting that its main narrative finishes where it does, around 1980. First of all, the book has sought to explain how the primate playback experiment, as publicly important science, came to be invented when it was, how it came to disappear when it did, and how it came to be reinvented when it was. Another self-limiting feature of the book has been the interest in using the primate playback story as a window onto the emergence of key twentieth-century behavioral sciences. For all the evidence that has accumulated since 1980 on the question of the animal roots of human language, the period has not seen the emergence of new sciences comparable in scope or strength of disciplinary identity to, say, Boasian cultural anthropology or Thorndikian comparative psychology. Equally constraining has been the style of explanation attempted here, with no skipping of historical periods and no skimping on documentary evidence. It is an open question whether the story of post-1980 developments can be told to this standard. Consider that where Peter Marler has kept a crowded lifetime's correspondence, including—to this study's immense benefit—his correspondence with Dorothy Cheney and Robert Seyfarth, they have a file-cabinet folder's worth of mementos.

A letter found in the Yerkes papers at Yale exemplifies this point about changing habits of archiving. It also makes vivid the connections across time that bind the events chronicled in this book into a single story. The letter was sent to Seyfarth in March 1982, although he no longer has a copy. Its author wrote to say how much she had enjoyed the article in *International Wildlife*:

> Your analysis of the vervets' language of grunts, and testing of your understanding of it by playing back tapes to the monkeys, called back to my mind an old question of mine, as to how much the vocalizing of monkeys and apes has been analysed musically. I ask because, as Robert Yerkes' daughter, I was on hand in the summer of 1923 when my father's first two chimpanzees arrived at the farm in Franklin, N. H., and he and our musician neighbor Blanche Learned made the observations that led to their joint book, *Chimpanzee Intelligence and its Vocal Expressions*.

Mrs. Brand Blanshard, of New Haven, Connecticut, went on to explain that her curiosity was personal, not professional, as she was not a scientist but a retired editor. But being Yerkes's daughter she had kept up contacts with the world of primatology, including Seyfarth's corner of it:

> Incidentally, on the plane from New York to Atlanta in 1976 my seat partner was revising a manuscript and I was reading during the first part of the trip. But when I glimpsed a review of a book by Donald Griffin in his newspaper and asked to see it, we discovered that we were bound on the same mission, to speak at the Yerkes Centennial Symposium. He was Peter Marler.[40]

With a little imagination, we can see the whole of the primate playback story nestled in on those plane seats, as the daughter of Garner's keenest admirer among the twentieth-century psychologists engaged in conversation with the man who, in a short while, and in keeping with his discipline's antipsychological traditions, would reinvent what Garner had invented, again to wide acclaim.

In a famous fable about reinvention, Borges told of Pierre Menard, a French man of letters in the early twentieth century who set out to write *Don Quixote* word for word as Cervantes had published it in the early seventeenth century, but without copying Cervantes's text and without remotely trying to live as Cervantes had lived. A Menard-composed *Quixote*, Borges suggested, would be a completely different work, the change of context bringing about a wholly different significance.[41] The present study can be read as the most comprehensive attempt to date to test Borges's moral for the history of science. We have seen what the experiment-guided exploration of the meaning of primate vocalizations meant for Garner, his contemporaries, and his successors, with these meanings recuperated through reconstruction of contexts from Darwin to Chomsky (and Chimpsky), down to the vervet research of Marler, Cheney, and Seyfarth. The primate playback that in Garner's hands revealed evolution's hierarchies undermined them in Marler's, Cheney's, and Seyfarth's hands. To contemplate that difference, and the changes that created it, is to glimpse something of the possibilities for scientific reasoning as applied to the problems of human evolution. "Light will be thrown on the origin of man and his history," Darwin prophesied in the *Origin*.[42] So it has. But what Darwin did not foresee, quite, was how that thrown light would come to acquire a history as richly and instructively intricate, as shot through with contingency and inevitability, as the history it disclosed. T. H. Huxley, at the end of *Man's Place in Nature*, hymning the glories of speech, came closer:

> Our reverence for the nobility of manhood will not be lessened by the knowledge, that Man is, in substance and in structure, one with the brutes; for, he alone possesses the marvellous endowment of intelligible and rational speech, whereby, in the secular period of his existence, he has slowly accumulated and organized the experience which is almost wholly lost with the cessation of every individual life in other animals; so that now he stands raised upon it as on a mountaintop, far above the level of his humble fellows, and transfigured from his grosser nature by reflecting, here and there, a ray from the infinite source of truth.[43]

Notes

Introduction

1. Schmeck 1980.
2. Dennett 1983. On this paper's impact, see Seyfarth and Cheney 2002.
3. See, e.g., Bickerton 1990, 10–16; Diamond 1991, 125–49, esp. 127–37; Pinker 1994, 351–52; Dunbar 1996, 21–22, 46–51, 140–42; Deacon 1997, 54–68; Anderson 2004; and Mithen 2005. Jared Diamond in particular (126–27) emphasizes how crucial the vervet playback experiments have been in modern theorizing about language origins.
4. "Can Monkeys Talk? A Queer Scientific Experiment at the Smithsonian Institution," *St. Louis Daily Globe-Democrat*, 21 September 1890. I am grateful to Pamela Henson of the Smithsonian Institution Archives for sending me this article.
5. For a recent attempt to revive talk of prematurity in science, see Hook 2002.
6. On his return to the States after that first expedition, Garner too made the front page of the *New York Times* ("Thinks Well of Gorillas: Prof. Richard L. Garner Tells of His Life in a Cage," *New York Times*, 26 March 1894). He became a common point of reference internationally. In Sweden, for instance, the opening of a new monkey house prompted a reporter to comment: "now and then, a cry was uttered, that perhaps could be translated by Professor Garner, but which we assume was meant to call Malle, the capuchin monkey, to order." Chevalier, "Skansens Nya Aphus," *Svenska Dagbladet*, 22 September 1901. And in Britain, during the First World War, in a pamphlet for children about a dog-hero of the trenches (*Terrier VC: A Black and Tan Hero Who Won the Victoria Cross*, n.d., 13), Julia Lowndes Tracy wrote: "Do you remember how, some years ago, a famous scientist and animal-student lived for quite a time in an iron cage in the depths of an African forest in order to study the language of monkeys and apes? Well, some day we shall see some great scientific person taking the trouble to study the conversation of dogs." I am grateful to Sofia Åkerberg for the first reference—I quote her translation from the Swedish—and to Jonathan Burt for the second.
7. See, e.g., the 1924 newspaper article reproduced in Haraway 1989, 20.

8. All unreferenced citations and quotations in this introduction can be found in the chapters that follow.

9. Until now there have been no sustained historical studies of Garner or of Marler, Cheney, and Seyfarth. For overviews of the long history of speculation on the origin of language, see Stam 1976, Aarsleff 1982, and Gessinger and Rahden 1989. The bibliography assembled in Hewes 1975 appears in condensed and chronological form in Hewes 1996, 572–82. There is little specialist historiography for the origin-of-language debate in the twentieth century. For the British debate in the later nineteenth century, however, the pickings are richer. The most comprehensive studies are Ferguson 2006, on animal language as dealt with in Victorian science and popular fiction, and Knoll 1987, which, however, deals more or less exclusively with the views on language and mind of Darwin, Müller, and Romanes. An indispensable collection of source material (including Garner's "Simian Tongue" articles) is Harris 1996. On the related history of analogies between species change and language change, see Alter 1999.

10. Cf. Numbers and Stephens 1998, an examination of attitudes about evolution in the South of Garner's day, based on the writings of Southern seminarians and university professors of science. Garner is not mentioned, though he was by far the most famous Southern evolutionist of the era.

11. The phrase derives from Lakoff and Johnson 1980.

12. Dessalles and Ghadakpour 2000. But see Kohn 1995 on signs of racism re-emergent in science today.

13. For views of the present state of debate on anthropomorphism, see Mitchell, Thompson, and Miles 1997, and several of the contributions in Daston and Mitman 2005. For contrary assessments of the value of Morgan's canon from two distinguished philosophers, see Sober 1998 and Fodor 1999. While Elliott Sober here seeks to put Morgan's canon on firmer foundations, Jerry Fodor recommends rejecting the canon altogether—as does, more recently, Simon Fitzpatrick (forthcoming).

Chapter One

1. The best general history of the "gorilla war" is Rushing 1990, although the concise treatment in Gould 1998 is superb. On Du Chaillu, see McCook 1996. On the "hippocampus" controversy, see Gross 1998, Wilson 1996, and Cosans 1994. On Owen's role, see Rupke 1994, 259–322. On Huxley's role, see Desmond 1994, chap. 16. The *Punch* verse of 18 May 1861 is reprinted with commentary in Burkhardt et al. 1985–2006, 9:426–28.

2. Müller 1861, 340, emphasis in original. The lecture—reported in the *Illustrated London News*, 15 June 1861, 565—is reprinted in Harris 1996, 7–41. On the crowds at Müller's 1861 lectures, see Müller 1898, 169.

3. The Huxley-Wilberforce encounter at Oxford in the summer of 1860 became celebrated only subsequently as the defining moment of linkage. See Gould 1991.

On apes and humans in the pre-Darwinian transmutation theories of Buffon, La-marck, and Chambers, see Rushing 1990, 14–24.

4. See, e.g., Ruse 1999a, Bowler 2003, and Oldroyd 1983. The exception here is El-legård 1990, 319–20.

5. See, e.g., Amigoni 1995; Knoll 1987, 1986; D'Agostino 1983; and Dowling 1982.

6. Quoted from an obituary in the *Outlook*, Papers of Friedrich Max Müller, "Vol-ume of Press Cuttings, Reviews, Letters etc.," MS.Eng.b.2034, Bodleian Library, Oxford. It is the spirit of the quotation, not the letter, that matters. Apart from Du Chaillu and Grimm's law, the references are to works published well after 1861. The reference to an arboreal ancestor is to Darwin's *Descent* (1871, 2:389), while the reference to protoplasm is probably to Huxley's 1868 lecture "On the Physical Basis of Life." See Huxley 1869.

7. "Tyndall, Huxley, and Max Müller . . . for half a century represented to the world at large the oracles of their respective fields of knowledge." Hopkins 1900, 395.

8. Müller 1861, 340.

9. On the cultural dimensions of the debates over geological method, see Radick 2003, 159.

10. For a concise overview of the geological background, see Ruse 1999a, 36–56.

11. On romanticism and the sciences generally, see Cunningham and Jardine 1990.

12. On the European discovery of Sanskrit and its consequences, see Rocher 1995. For a concise overview of the history of philology from the Göttingen seminar in the eighteenth century to Müller's battles with Darwinism in the nineteenth cen-tury, see Smith 1997, 378–85. More detailed histories of comparative philology include Amsterdamska 1987 and Pedersen 1962. Wells 1987 discusses the debate on the origin of language during this period.

13. "For a time, everybody who wished to learn Sanskrit had to come to England. Bopp, Schlegel, Lassen, Rosen, Burnouf, all spent some time in this country, copy-ing manuscripts at the East-India House." Müller 1861, 158. For biographical details on Müller, see Radick 2004b and Chaudhuri 1974—the latter drawing heavily on Müller 1901 and 1902. For an assessment of Müller's position in the history of linguistics, see Jankowsky 1979. An excellent recent anthology of Mül-ler's writings is Stone 2002.

14. The quotation is from Müller 1881, 248. On British Indophobia, see Trautmann 1997, chap. 4. On British doubts about Sanskrit, see Dowling 1982, 164–65.

15. On the *Critique* in relation to the Veda, see Müller 1881, 248–49. Müller pub-lished an English translation of the *Critique* in 1881.

16. The ethnologist J. C. Prichard went so far as to judge in 1848 that the "analytic comparison of languages furnishes more important aids to the cultivation of eth-nology than any other part of what may be termed historical research." Quoted in Burrow 1967, 189. On Prichardian ethnology, see Stocking 1987, chap. 2, esp. 56–62. On ethnology and comparative philology, see also Burrow 1967, 187–90.

17. On the Hare-Thirlwall circle at Trinity, see Valone 1996, 123; and Burrow 1967, 192–93. On the nonhistoricist, nonidealist philology then current in England, see Aarsleff 1967.

18. In Bunsen's writings from the 1850s, readers learned that language was the barrier between man and brute, that the existence of this barrier showed transmutation theories of human origins to be false, and that the reasons for the falsehood of these theories were essentially Kantian. Müller would put much the same case in his own polemics against the evolutionists. On Bunsen and Müller, see esp. Burrow 1967, 193–94.

19. On the Boden controversy, see Dowling 1982, 163–65.

20. Müller 1861, 8–11, 23–24. Thanks to such a disinterested approach to language, argued Müller (11–14), progress had been made on several problems, including the nature of mythology, the common or diverse origins of the human races, and, most pressingly in the wake of the *Origin*, man's place in nature. On the links between Müller's views on language and his views on mythology, see Olender 1992, 82–92.

21. Müller 1861, 27–40, 65–66, quotations on 27, 36–37, 65. Müller explained Grimm's law on 258–59. On the folklorist Jacob Grimm and his law, see Pedersen 1962, 37–42, 258–62; and Crystal 1987, 328.

22. Müller 1861, 4–10, 15–21, 76, quotations on 22. Whewell in the mid-1840s, after the publication of *Vestiges* (discussed later in this chapter), had argued for views on language, reason, humans, and animals virtually identical to those Müller argued now. On Whewell and the origin of language, see Ferguson 2006, 23, 42–43 n. 4. Both Humboldt and Whewell were associates-cum-patrons of Müller's in the 1840s; see Radick 2004b, 1435; and Valone 1996, 123–24.

23. Müller 1861, 342–44; see also 237–61. For a virtuoso discussion of how roots were reconceived from Bopp onward, see Foucault 1970, 280–300, esp. 287–91.

24. Müller 1861, 369. Müller's aphorisms expressed a view of language with deep roots in German romanticism, in particular the writings of Hamann, Herder, and Humboldt. See Hacking 1992a.

25. Müller 1861, 341–42, 369. The Locke quotation is from Locke 1689, 159.

26. Müller 1861, 344–56, quotations on 347, 349.

27. Müller 1861, 370–71, emphasis in original.

28. Müller 1861, 313–28, 371–77, quotation on 320.

29. Müller 1861, 377, quoting Genesis 11:1.

30. On degeneration in ethnology and biblical anthropology, see Stocking 1987, 44–45. On degeneration as a theme in Müller's work, see Burrow 1967, 201–2.

31. In 1854, at Bunsen's urging, Müller published a paper arguing that a large number of Asian and European languages belonged to a hitherto unidentified "Turanian" family. The proposal aroused a great deal of controversy, notably with the French philosopher Ernest Renan. See Stocking 1987, 58–59.

32. Müller 1861, 40–66, 273–313, 316–17, quotations on 316. In the second series of lectures (1864, 197), Müller went on to argue that "changes produced by phonetic decay must admit of a simple physiological explanation—they must be referable to a relaxation of muscular energy in the organs of speech." In Müller's view, dialectical regeneration was also, at bottom, grounded in physiology, though the nature of this grounding remained obscure.

33. Müller 1864, 92. Müller criticized Herder in Müller 1861, 344–46.

34. Herder 1772, quotations on 99, 127, 91, 113.

35. An exception was the antievolutionist ethnologist C. S. Wake. See Wake 1868, chap. 4. But even Wake's endorsement of Müller was heavily qualified.

36. See Wedgwood 1859, i–v.

37. [Wedgwood] 1862, quotations on 60. On Wedgwood's review, see Alter 1999, 52–53.

38. "Origin of Language" 1866, 100–104, quotations on 104.

39. On Darwin and the *vera causa* principle, see, e.g., Hodge 1977; Hull 2003, 175–80; and Radick 2003, 148, 157–60. On Comte and causal explanation, see Hacking 1983, chap. 3, esp. 45–47.

40. "Origin of Language" 1866, 105–24, quotations on 105, 113.

41. Müller 1861, 370–71; "Origin of Language" 1866, 132. On the impact of prehistoric archaeology in shaping the post-*Origin* view of human beginnings, see Van Riper 1993; Stocking 1987, 172–73; Bowler 1986, 21–31; and Burrow 1967, 202–3.

42. Farrar 1865, 56. See also Wedgwood 1866, 138–39. On Farrar's role in the debates over Darwinism and language, see Alter 1999, 79–96.

43. Tylor 1866b, 423–24. On Tylor and the doctrine of survivals, see Stocking 1987, 156–64. On Müller among the cultural evolutionists generally, see Stocking 1987, 305–11. Schrempp 1983 is a superb study of Müller and Tylor.

44. Tylor 1865, 15.

45. Tylor 1866a, 91–92.

46. On Darwin's notebook theorizing about language and expression, see Knoll 1987, chap. 1; and Montgomery 1985. On Darwin's reading in language theory during this period, see Beer 1996. Although a number of scholars have noted suggestive similarities between Darwin's notion of a branching tree of species descent and the earlier, comparative-philological notion of a branching tree of language descent (see, e.g., Beer 1996; Alter 1999, chap. 2), the most authoritative reconstruction of Darwin's notebook theorizing, in Hodge 2003, identifies no debt to philology on Darwin's part. On the likely independence of these branching-tree notions, see Van Wyhe 2005. On Darwin's theorizing on the origin of language considered in relation to his lifelong concerns with mind, morals, and emotions, see Richards 2003, esp. 105–6. On Darwin's life, thought, and times more generally, see Desmond and Moore 1991; Browne 1995, 2002; and the first half of Hodge and Radick 2003.

47. Darwin, M Notebook, 96–97, in Barrett et al. 1987, 542. I have preserved the often incorrect grammar and spelling of the notebook comments quoted in this section.
48. Darwin, M Notebook, 153e, in Barrett et al. 1987, 558–59.
49. Darwin, N Notebook, 18, in Barrett et al. 1987, 568.
50. Darwin, M Notebook, 150–51, in Barrett et al. 1987, 558.
51. Darwin, N Notebook, 31, 64–65, in Barrett et al. 1987, 571, 581.
52. Darwin, Old and Useless Notes, 5v, in Barrett et al. 1987, 599.
53. [Chambers] 1844. On the reading of *Vestiges* in Britain, see Secord 2000.
54. [Chambers] 1844, 312–13.
55. Darwin 1859, 4.
56. Darwin 1859, 455. The other parallels were between breeds and dialects (40) and between races and languages (422–23). In the breeds-dialects case, Darwin emphasized that neither had a definite origin. In the races-languages case, Darwin was illustrating his view that genealogical classifications were the most natural by arguing that the pedigree of races would at the same time be the pedigree of languages, and so the basis for a natural system of language classification. On language-species parallels in Darwin's work from the notebooks to the *Origin*, see Alter 1999, chap. 2.
57. Darwin 1859, 488.
58. J. D. Hooker to C. Darwin, 7 November 1862, in Burkhardt et al. 1985–2006, 10:507–10, quotation on 509. Hooker was responding to Darwin's remark in a letter of 4 November: "By the way I have just read with much interest Max Muller; the last part about first origin of language seems the least satisfactory part." Burkhardt et al. 1985–2006, 10:503.
59. A. Gray to C. Darwin, 4 and 13 October 1862, in Burkhardt et al. 1985–2006, 10:444–46, quotations on 445.
60. C. Darwin to A. Gray, 6 November 1862, in Burkhardt et al. 1985–2006, 10:505–7, quotation on 505, emphasis in original. The Darwin Correspondence editors surmise (506 n. 5) that Darwin's reference to Müller "getting the better of" earlier "sneers" near the end of the book is to Müller's description of root culling in terms of natural selection. In a subsequent letter to Gray, Darwin recommended reading Frances Julia Wedgwood's *Macmillan's Magazine* review of Müller's *Lectures*. C. Darwin to A. Gray, 23 November 1862, in Burkhardt et al. 1985–2006, 10:546–48. Darwin (546) attributed the review to a collaborative effort between father and daughter Wedgwoods. This letter is discussed in Alter 1999, 165 n. 67.
61. A. Gray to C. Darwin, 4 and 13 October 1862, in Burkhardt et al. 1985–2006, 10:444–46, quotation on 445, emphasis in original.
62. A. Gray to C. Darwin, 24 November 1862, in Burkhardt et al. 1985–2006, 10:553–55, emphasis in original. On Gray, design, variation, and the language-species analogy, see Alter 1999, 53–56, 65–68, 165.
63. C. Darwin to J. D. Hooker, 13 January 1863, in Burkhardt et al. 1985–2006,

11:35–39, quotation on 35. Alter 1999, 165 n. 66, misidentifies the work under discussion as *Man's Place in Nature*.

64. Huxley 1863b, 153.

65. Huxley 1863b, 152–54, quotations on 152, 154.

66. Huxley 1863b, 155.

67. Huxley 1863b, 155. In Charles Kingsley's fantasy *The Water Babies*, also published in 1863, natural selection and use-inheritance combine to cause a race of humans, the Doasyoulikes, gradually to degenerate and to lose their power of language, until they are transformed into howling apes, the last of which is shot by Du Chaillu (229–39).

68. Huxley, 1863a, 102–3, quotation on 102; see also 112.

69. T. H. Huxley to C. Darwin, 23 November 1859, in Burkhardt et al. 1985–91, 7:390–91. Other works in which Huxley discussed language include Huxley 1874, 574; and 1881, 103–10. For Huxley's views on language in relation to his earlier and later saltationism, see Blitz 1992, 36.

70. Lyell 1863, 469–505, quotations on 468, 504, 469. On the *Antiquity of Man*, see Richards 1987, 163–64. On Lyell's comparisons of language change and species change, see Alter 1999, 43–46, 56–64; and Taub 1993.

71. Wallace, 1870, 332–60, quotations on 340.

72. Wallace, 1870, 350–51. For a sympathetic portrayal of Wallace's post-apostasy life and work, see Fichman 2004.

73. Wright 1870, 293. Darwin quoted this sentence in the *Descent*. Darwin 1871, 2:335. On Wright, see Menand 2001. A useful collection of and about Wright's writings in evolutionary philosophy is Ryan 2000.

74. Wright 1870, 294–303, quotations on 295–96.

75. Darwin 1871, 1:53–62, 105–6, quotation on 53; 2:330–37. Both the *Descent* and the *Expression of the Emotions* deserve much more extensive coverage—in themselves, in their complex relations to each other, and in their multifaceted treatments of language—than can be given here. For an attempt to clear away some of the anachronism that has marred much scholarly commentary on these works, see Radick 2002.

76. Darwin 1871, 1:54, 58, quotation on 58.

77. Darwin 1871, 1:54.

78. Darwin 1871, 1:55–56, quotation on 56. Darwin's phrase has been resuscitated by Steven Pinker (1994, chap. 1, esp. 20).

79. Darwin 1871, 1:106.

80. Darwin 1871, 1:56.

81. Darwin 1871, 1:56–59, quotation on 58.

82. My own reading of this passage is quite different from that of Stephen Alter, whose *Darwinism and the Linguistic Image* (1999) deals with this passage at length. Alter argues that Darwin intended the parallels in part to drive a wedge between evolutionary speculation on human races and evolutionary speculation on human language; see Alter 1999, esp. 104–5. It seems to me—as argued above—that the

opposite is true. For a fuller summary and critique of Alter's interpretation, see Radick 2000b. For an expanded version of the analysis offered here, see Radick 2002, 8–10. See too the exchange between Alter and me in a forthcoming issue of *Studies in History and Philosophy of Biological and Biomedical Sciences*.

83. Darwin 1871, 1:59–61, quotations on 60 and 61.
84. Wake 1868, 101, cited in Darwin 1871, 1:61. On Wake, see Stocking 1987, 179–81.
85. Darwin 1871, 1:61–62, quotation on 61. Darwin had earlier consulted with Hensleigh Wedgwood over the matter of (in Wedgwood's words) "the complex nature of the languages of the rudest people." H. Wedgwood to C. Darwin, 1870, Papers of Charles Darwin, DAR 80:164–65, Cambridge University Library. The letter is discussed in Richards 1987, 205.
86. Darwin 1871, 2:391. Richards 2002 argues that Darwin was also influenced by August Schleicher's views on language, mind, brain, and evolution, via an epitome in Ernst Haeckel's 1868 *Natürliche Schöpfungsgeschichte*.
87. Darwin 1871, 1:57–58, quotation on 57.
88. Darwin 1871, 1:105.
89. Darwin 1871, vol. 1, chap. 3, esp. 72, 86. On Darwin's account of the evolution of the moral sense, see Richards 2003; and 1987, 206–19.
90. Darwin 1871, 2:335–37, quotations on 336. For a more in-depth discussion of Darwin on sexual selection for vocal expressiveness, emphasizing the Lamarckian character of Darwin's explanation, see Radick 2002, 10–13.
91. Leopold 1990 emphasizes the *Expression of the Emotions* in Darwin's account of the origin of language. On the *Expression* more generally, see Dixon 2003, 159–77; Richards 1987, 230–34; and Browne 1985. Dixon's discussion includes some very perceptive remarks (175–79) about the *Expression*'s anthropology as more degenerationist than progressivist in spirit.
92. Darwin 1872, 366. A number of readers criticized Darwin for his reliance on Hensleigh Wedgwood's etymological conclusions in the *Expression*. See F. Baudry to C. Darwin, 4 October 1872, Darwin Papers, DAR 160; and J. V. Carus to C. Darwin, 7 October 1872, Darwin Papers, DAR 161.
93. Darwin 1877, 473. For discussion with respect to language, see Leopold 1990, 637–39. In a letter to *Nature* in 1881 (233), Darwin wrote in a similar vein: "[There is] one other small point of inquiry in relation to very young children, which may possibly prove important with respect to the origin of language, but it could be investigated only by persons possessing an accurate musical ear: children, even before they can articulate, express some of their feelings and desires by noises uttered in different notes. For instance, they make an interrogative noise, and others of assent and dissent in different tones, and it would, I think, be worth while to ascertain whether there is any uniformity in different children in the pitch of their voices under various frames of mind." Gruber (1974, 464) has noted that Darwin's 1877 "Sketch" was a response to an article by the French intellectual Hippolyte Taine on language acquisition in children, also published in *Mind*,

and itself a response to Müller's 1873 "Lectures on Mr. Darwin's Philosophy of Language." To the extent that Taine's and Darwin's *Mind* articles initiated the scientific study of child development, that science began thus as an outgrowth of the debate on the language barrier chronicled in this chapter. A recent history that elaborates this point is Ottavi 2002, esp. 65–101.

94. Toward the end of 1872, Müller lectured on this theme before the Liverpool Literary and Philosophical Society. "Max Muller on Darwin's Philosophy of Language," *Nature* 7 (1872): 145.

95. See W. D. Whitney's notice in Whitney 1873, 268–69.

96. Müller 1873, 225–27, 232, quotation on 226, emphasis in original.

97. Müller 1873, 147–59, quotations on 159.

98. Müller 1873, 160–71, 198–233, esp. 231, quotation on 167.

99. Müller 1873, 229, emphasis in original. In the *Descent* (1871, 57), Darwin had written of the first step toward language: "it does not appear altogether incredible, that some unusually wise ape-like animal should have thought of imitating the growl of a beast of prey." Darwin proceeded from here as if answering a persistent skeptic. Why are there no such wisely imitative animals among the higher apes at present? Because, wrote Darwin (59), the existing apes are not intelligent enough to use their vocal organs for language. But why should the apes be too dim for speech, given the supposed premium on intelligence in the struggle for life? Darwin's rather feeble answer came in the second, 1874 edition of the *Descent* (90): "If it be asked why apes have not had their intellects developed in the same degree as that of man, general causes only can be assigned in answer, and it is unreasonable to expect anything more definite, considering our ignorance with respect to the successive stages of development through which each creature has passed."

100. Müller 1873, 215, 217, emphasis in original.

101. C. Darwin to F. Max Müller, 3 July 1873, in Darwin and Seward 1903, 45. In the letter accompanying the "Lectures," Müller had described them as "an open statement of the difficulties which a student of language feels when called upon to explain the languages of men, such as he finds them, as the possible development of what has been called the language of animals." F. Max Müller to C. Darwin, 29 June 1873, in Müller 1902, 1:452.

102. Notes on back flyleaf of offprint of Müller 1873, Darwin Papers, Darwin Offprint Collection, no. R240.

103. F. J. Wedgwood to C. Darwin, August–September 1873, Darwin Papers, DAR 87:87–89.

104. On Whitney and language, see esp. Alter 2005; Valone 1996, 122–26; and Andresen 1990, 135–68.

105. Whitney 1867, 427–28.

106. Whitney 1873, 297, emphasis in original.

107. Whitney 1873, 148.

108. Whitney 1873, 268–69.

109. Whitney 1873, 277–78.
110. Cf. Müller 1875a, 544–45.
111. C. Darwin 1874, 89 (chap. 3, n. 63).
112. In an 1871 article on August Schleicher, Whitney had stated (47) that "all adherents of the prevailing modern school of historical philology . . . accept an explanation of structural growth which not only admits but demands the will of man as a determining force." On his own offprint copy of the article, Darwin wrote alongside this passage: "But without reflection on the result." Note on margin of p. 47 of offprint of Whitney 1871, in the Darwin Papers, Darwin Offprint Collection, no. 756.
113. C. Darwin to C. Wright, 3 June 1872, quoted in part in Thayer 1878, 240.
114. C. Wright to C. Darwin, 29 August 1872, in Thayer 1878, 240–46, quotations on 242, 244–45, emphasis in original.
115. On Wright's visit, see C. Darwin to C. Wright, 31 August 1872, quoted in part in Thayer 1878, 246; C. Wright to S. Sedgwick, 5 September 1872, in Thayer 1878, 246–48, esp. 247; and C. Darwin to C. Wright, 6 September 1872, summarized in Burkhardt and Smith 1994, 369. Wright took up the problem further in Wright 1873. On Wright's work and the tradition of American pragmatism, see Kuklick 1977, chap. 4; and Menand 2002. On debate in the late 1860s and early 1870s about the role of will in language change, see Alter 1999, 105–7.
116. Whitney 1874, 67–68, 83–88.
117. Whitney 1874, 86, 87.
118. Whitney 1874, 68.
119. Whitney 1874, 87, 88.
120. C. Darwin to J. T. Knowles, 31 July 1874, summarized in Burkhardt and Smith 1994, 412.
121. J. T. Knowles to C. Darwin, 4 August 1874, Darwin Papers, DAR 169.
122. C. Darwin to J. T. Knowles, 5 August 1874, summarized in Burkhardt and Smith 1994, 412.
123. C. Darwin to W. D. Whitney, 5 August 1874, Papers of William Dwight Whitney, group 555, series 1, box 21, folder 21/556 (1874, Aug. 1–12), Sterling Memorial Library, Yale University.
124. Note on margin of p. 30 (identical to p. 88 in published article) of offprint of Whitney 1874, Darwin Papers, Darwin Offprint Collection, no. R237.
125. Müller 1898, 176–77. Although the date of the visit is not given here, a letter from Müller to his mother on 9 October 1874 indicates the visit took place not long before then. Müller 1902, 1:468.
126. "Origin of Language" 1874, 272. On Darwin's theory of terminal additions, see Darwin 1859, 439, 444. For discussion, see Richards 1992, 153–54; and Gould 1977, 73–75.
127. C. Darwin to G. H. Darwin, 5 November 1874, Darwin Papers, DAR 210.
128. Müller 1875b, 307.
129. Müller 1875b, 313.

130. Müller 1875b, 305. In a letter to Charles Darwin that month, Müller wrote: "I hope in the course of the year to be able to place my whole argument before you." 7 January 1875, in Müller 1902, 1:476.

131. Whitney 1875a, 730–31, quotation on 730.

132. Whitney 1875b, 300.

133. Whitney 1875b, quotations on 305 and 297 respectively.

134. C. Darwin to W. D. Whitney, 8 May 1875, and E. Darwin to W. D. Whitney, 9 May 1875, Whitney Papers, group 555, series 1, box 22, folder 22/600; and C. Darwin to W. D. Whitney, 1 September 1875, box 23, folder 23/619.

135. F. Max Müller to C. Darwin, 13 October 1875, in Müller 1902, 1:495.

136. Müller 1875a, 546–49. The list covered matters ranging from "[w]hether . . . I was impelled by an overmastering fear lest man should lose his proud position in the creation" (546) to "[w]hether I invented the terms vivârasvâsâghoshâḥ and saṃvâranâdaghoshâḥ, and whether they are to be found in no Sanskrit grammarian" (547).

137. C. Darwin to F. Max Müller, 15 October 1875, in Müller 1898, 178. To judge by his annotations, Darwin's attention was drawn much more to what seemed a retreat on the question of animal concepts. "If we are right," wrote Müller (1875a, 481), "that man realises his conceptual thought by means of words, derived from roots, and that no animal possesses words derived from roots, it follows, not indeed, that animals have no conceptual thought (in saying this, I went too far), but that their conceptual thought is different in its realised shape from our own."

138. G. H. Darwin to W. D. Whitney, 21 December 1875, Whitney Papers, group 555, series 1, box 23, folder 23/631.

139. Conway 1900, 892–93.

140. Whitney 1876. Müller responded in the next issue. Müller 1876.

141. Comment on the Whitney-Müller controversy, *Academy* 9 (1876): 215.

142. A large sheaf of newspaper and magazine clippings, "Whitney's Controversy with Max Müller, 1875–76," is held in the Whitney Papers, group 555, series 2, box 53, folder 53/064.

143. Whitney 1892. Of the many controversies in the science of language, wrote one reviewer of Whitney's latest diatribe, "[p]erhaps no controversy . . . has ever found its way so extensively among the more studious portion of the people as has that between Professors Max Müller of Oxford and William D. Whitney of Yale." *Sunday School Times*, 9 April 1892, clipping, Whitney Papers, group 555, series 2, box 53, folder 53/064. For further discussion of the Whitney-Müller controversy, see Alter 2005, chap. 8; and Valone 1996.

144. Müller (1878, 491) explained that "when people work together, when peasants dig or thresh, when sailors row, when women spin, when soldiers march, they are inclined to accompany their occupation with certain more or less vibratory or rhythmical utterances. These utterances, noises, hummings, songs, are a kind of reaction against the inward disturbance caused by muscular effort." Another thinker who argued around this time for communal labor as the spur to the emer-

gence of language was Friedrich Engels. See Engels 1876, 173–74; for discussion, see Stam 1976, 298. See also note 121 in chapter 7 in this book.

145. See, e.g., Galton 1887; Argyll 1887; "Thought without Words" [Müller-Galton correspondence], *Nature* 36 (1887): 100–101; and Müller 1887a, 1888a, 1888b.

146. Müller 1891a, 429, emphasis in original.

Chapter Two

1. F. Bateman to C. Darwin, 31 March 1871, Darwin Papers, DAR 160. The *Descent* was published in late February 1871. See Freeman 1977, 129. Although the date on Bateman's letter is 31 March 1870, I venture that this date was written by mistake, and that the correct date of the letter is 31 March 1871. Two pieces of evidence in particular point toward the later date. First, Bateman quotes from the *Descent* in his letter ("I see in the 2nd chapter of your 1st volume, you call attention to the fact that 'we might have used our fingers as efficient instruments'"; see Darwin 1871, 1:58). Second, Bateman's own book carries a preface dated May 1870—*after* the date on the letter enclosed with the book. The only evidence possibly favoring 31 March 1870 rather than 31 March 1871 as the date of Bateman's letter is Darwin's reference to *On Aphasia* in his marginal comments on another letter, dated 11 March 1871, from the Shakespeare scholar John Jeremiah Jr. (Darwin Papers, DAR 87:101–2). But this annotation ("see Bateman Aphasia on Imitation p 110") is itself undated. It was probably scribbled onto the Jeremiah letter some weeks or months after the letter arrived.

2. F. Bateman to C. Darwin, 31 March 1871, Darwin Papers, DAR 160, emphasis in original.

3. Morgan 1892a, 44. For later statements of the canon in Morgan's work, see Newbury 1954, 70; and Singer 1981, 268.

4. Galef 1996, 9.

5. Manning 1998, 4, discussing Morgan 1868, 258–70.

6. Cf. "Anecdotalism and anthropomorphism were not immediately exorcised—Darwin himself . . . showed not infrequent lapses—but gradually they gave way before the advance of the more rigorous methodology of science." Hearnshaw 1964, 88.

7. Boakes 1984, 32–44.

8. Richards 1987, 380–85. Richards (385 n. 171) lists several contemporaries of Morgan's who had independently proposed canonlike rules for comparative psychology.

9. Costall 1993.

10. Gall 1835, quotations from 3:97–99, cited in Young 1990, 63, 70. On Gall and his legacies, see esp. Van Wyhe 2004; also Finger 1994, 32–36, 374–75; Eling 1994, 1–27; Young 1990, chap. 1; Clarke and Jacyna 1987, 220–44; and Clarke and Dewhurst 1972, chap. 10. Gall distinguished the faculty of verbal memory (in

organ 14) from the faculty of spoken language (in organ 15). Both faculties were common to humans and animals. See Gall 1835, 5:7–46.

11. Flourens 1842, 1846. On Flourens, see Finger 1994, 35–36; Young 1990, 54–74; and Clarke and Jacyna 1987, 244–302.

12. Bouillaud 1825, quotation on 44, cited in Young 1990, 137–38; and Head 1926, 1:13. On Bouillaud, see Finger 1994, 37, 375–77; Young 1990, chap. 4; Harrington 1987, 36–37; Clarke and Jacyna 1987, 302–7; and Schiller 1979, 171–75.

13. Clarke and Dewhurst 1972, chap. 11.

14. Broca's key papers are reprinted in translation in Eling 1994, 29–58. On Broca and the debates around his claims about brain function, see Finger 1994, 37–38, 377–79; Young 1990, chap. 4; Harrington 1987, chap. 2; and Schiller 1979, chap. 10. On aphasia in nineteenth- and early-twentieth-century British scientific and medical thought, including Bateman's, see Jacyna 2000.

15. On medical London in the 1840s, see Desmond 1994, 11–35.

16. Auziaz-Turenne 1849.

17. On Bateman, see "Sir Frederic Bateman," *British Medical Journal*, 20 August 1904, 412–15, quotation on 414. On Trousseau, see Harrington 1987, 47–48, 55–57; and Finger 1994, 392. On the Sainty case, see Bateman 1865; and 1870, 65–73.

18. P. P. Broca to unknown correspondent, 20 April 1868, WMS/ALS, Wellcome Institute for the History of Medicine Library, London. Cited in Schiller 1979, 203–4.

19. "M. Broca at Norwich," 1868.

20. Bateman 1870, 99.

21. Norwich meeting, *Lancet*, 19 September 1868, 386–87, quotation on 386.

22. Norwich meeting, *Lancet*, 386–87. See also Norwich meeting, *British Medical Journal*, 5 September 1868, 259; and Young 1990, 206.

23. Norwich meeting, *British Medical Journal*, 259. See also Schiller 1979, 203–4.

24. Ellegård 1990, 81–83; Desmond and Moore 1991, 559–60. On Vogt, see Tort 1996b, Pilet 1976, and, for disparaging commentary on Vogt's performance at the Norwich meeting, [Mivart] 1874, 45. The importance of the doctrine of localization within the new physical anthropology of Vogt, Broca, and their English supporter James Hunt has been little noted. In an 1868 article (333) on the localization of the language faculty, Hunt wrote: "Should the mental faculties be localised in different parts of the brain, and should the practical physiognomist be able to discern their relative sites, then, and not till then, shall we become free from the assumptions of the past and present ages, and have a solid foundation, on which we may confidently base a real science of man." See also Hovelacque 1877, chap. 2.

25. Norwich meeting, *Lancet*, 387.

26. Bateman 1870, 169.

27. C. Vogt to C. Darwin, 17 April 1867, Darwin Papers, DAR 180:11 (letter in

French). In a letter later that year, Darwin thanked Vogt for his book. C. Darwin to C. Vogt, 7 August [1867], summarized in Burkhardt and Smith 1994, 252. On atavisms and evolutionary theory in the nineteenth century, see Gould 1977, chaps. 4 and 5. Vogt wrote introductions to the French editions of *Variation of Animals and Plants under Domestication* and the *Descent*. On Darwin's use of Vogt's work, see Tort 1996b, 4488.

28. Darwin 1871, 1:56–57, citing Vogt 1867, 169.

29. Vogt 1867, 171–86.

30. Vogt 1867, 181–82.

31. Schleicher 1983.

32. Haeckel 1876, 2:301, a translation of Haeckel 1868, 509.

33. Di Gregorio 1990, 359, 360. On Darwin's reading of Haeckel and Schleicher, see Richards 2002. On the sources of Haeckel's views on the origin of language, and their congruence with his evolutionary-hierarchical understanding of human races, see Di Gregorio 2002. Other authors whom Darwin read on language and the brain include Huxley and Wallace. See Huxley 1863b, 154–55; 1863a, 102–3; and Wallace 1870, esp. 332–37.

34. Note in margin of p. 50 of offprint of Whitney 1871, Darwin Papers, Darwin Offprint Collection, no. 756.

35. F. Bateman to C. Darwin, 31 March 1871, Darwin Papers, DAR 160.

36. Bateman 1870, chaps. 1–3, pp. 130–60.

37. Bateman 1870, chap. 6.

38. Bateman 1870, chaps. 4 and 5.

39. Bateman 1870, 174–77.

40. Bateman 1870, 173–74, 178.

41. Darwin 1871, 1:58. In a famous marginal note in Abercrombie's *Inquiries*, Darwin wrote: "By Materialism, I mean, merely the intimate connection of kind of thought, with form of brain.—like, kind of attraction with nature of element." Di Gregorio 1990, 1. On this note, see Gruber 1974, 104.

42. See Abercrombie 1838, 150.

43. From Darwin 1871, 1:58.

44. F. Bateman to C. Darwin, 31 March 1871, Darwin Papers, DAR 160, citing Bateman 1870, 110–11.

45. For Darwin's marginalia in *On Aphasia*, see Di Gregorio 1990, 35/36.

46. Darwin 1872, 356–57.

47. C. Darwin 1874, 88.

48. Bateman 1874.

49. "Scientia Scientiarum" 1867, 5.

50. On the Victoria Institute, see Moore 1979, 53.

51. Bateman 1874, 74–76.

52. Bateman 1874, 78–81.

53. Bateman 1874, 89.

54. Bateman 1874, 78–79, 87–91, quotations on 88, 91.

55. Bateman 1874, 89–90.
56. Bateman directed readers to the "Man versus Ape" controversy in the *Eastern Daily Press* between 27 March and 13 July 1872. Bateman 1877, 148–51. On the discussion aroused by Bateman's paper, see Romanes 1888, 134.
57. Bateman 1874, 92–93.
58. Bateman 1874, 93. The English-speaking triumphs of the Fuegians were recorded by the *Beagle* captain Robert FitzRoy in FitzRoy 1839, 2:2, 121, 189. At one point the Norwich geologist F. W. Harmer entered the fray to defend Darwin generally against the charge—not made by Bateman himself—that no intermediate form or "missing link" had been found in the lamentably incomplete geological record. F. W. Harmer to C. Darwin, 28 August 1872, Darwin Papers, DAR 166.
59. Müller 1873, 193–96. In his lecture at the Victoria Institute, Bateman (1874, 81) had cited Müller on behalf of the view that language was a difference in kind between man and brute. Müller in turn referred readers of his lectures to "Dr. Bateman's book on Aphasia" for fuller treatment of the topic. Like Bateman, Müller (1873, 196) emphasized that there was nothing about aphasic phenomena that showed mind to be a function of the brain: "I shall not be suspected, I hope, of admitting that the brain, or any part of the brain, secretes rational language, as the liver secretes bile."
60. Note on back flyleaf of offprint of Müller 1873, Darwin Papers, Darwin Offprint Collection, no. R240.
61. "Origin of Language" 1874, 273.
62. Bateman 1877, chaps. 1–3.
63. Bateman 1877, 200.
64. Bateman 1877, chaps. 4–7.
65. Bateman 1877, 135.
66. It seems that Bateman became consulting physician to the asylum in the early 1870s. His position with the asylum is listed after his name on the title page of the 1877 *Darwinism* book, but not on that of the 1870 *Aphasia* book.
67. Porter 1997, 506–7; also Gladstone 1996, esp. 137–42. Founded around 1850 as Essex Hall, the institution in Colchester became the Eastern Counties Asylum in 1859. It is remembered today chiefly because Lionel Penrose conducted an important inquiry into the causes of mental deficiency among the Colchester population in the 1930s. See Kevles 1985, chap. 10.
68. Bateman 1882, 1–12. A revised and expanded second edition was published in 1897.
69. Bateman 1882, 34.
70. Bateman 1882, 35, 42.
71. Müller 1873, 172.
72. Müller 1873, 173, emphasis in original.
73. Müller 1873, 211.
74. See Möbius 1873. In a footnote, Möbius wrote: "Councillor for Economics Amtsberg in Stralsund ordered that the interesting experiments with the pike should

be carried out, and it is to him that I owe a written description of these experiments as well as the kind permission to publish them." Although Darwin's copy of Möbius's lecture (Darwin Papers, Darwin Offprint Collection, no. 850) is inscribed "Herr Ch Darwin von K. Möbius," Darwin does not appear otherwise to have corresponded with Möbius, nor to have owned or cited any of Möbius's many other writings. I suspect that Darwin first read about the pike experiment in Müller's lectures, and then wrote to Möbius to ask for the source lecture, which Darwin cited thereafter whenever he discussed the pike experiment. In Darwin's copy of Möbius's lecture, the pike passage (11) is marked out. The explanatory footnote is on 19. On Möbius and Darwin's use of his work, see Tort 1996a. For helpful discussion of Möbius, I am grateful to Lynn Nyhart.

75. Müller 1873, 212–13.
76. Note on back flyleaf of offprint of Müller 1873, Darwin Papers, Darwin Offprint Collection, no. R240. In the text itself, Darwin added: "so it wd be w. a negro." Note on margin of p. 33 (corresponding to p. 193 of Müller 1873).
77. Note on back flyleaf of offprint of Müller 1873, Darwin Papers, Darwin Offprint Collection, no. R240, emphasis in original.
78. C. Darwin 1874, 75, 76.
79. C. Darwin 1874, 77.
80. Darwin 1877, 471. Darwin also referred to the experimentally revealed pike mind in his final major scientific book, on the earthworm (1881). For the passage, see the entry on pike in Pauly 2004, which also assembles the other pike passages discussed here. For Edward Thorndike's rather Darwinian view of the relationship between association and reason, see chapter 6 of this book.
81. C. Darwin 1874, 83.
82. C. Darwin 1874, 83 n. 44.
83. Stephen 1873, 81–82, quotations on 82. Darwin (1874, 89) quoted these lines in the second edition of the *Descent*, but reversed their order, and misquoted Stephen as comparing a dog with "a philosopher." Stephen's essay first appeared (minus the commentary on Müller) in *Fraser's Magazine*.
84. Stephen 1873, 82, 81.
85. Whitney 1874, 68–72, quotations on 69, 70, 72.
86. F. J. Wedgwood to C. Darwin, August–September 1873, Darwin Papers, DAR 87:87–89.
87. "Origin of Language" 1874, 261–62; the quotations are from Müller's translation of Möbius. Remembrance of Amtsberg's pike experiment carried forward into the early twentieth century, when (as discussed in chapter 6) the science of animal behavior professionalized around the experimental study of associative learning in animals. See Triplett 1901—effectively an attempt to replicate the experiment, published under the title "The Educability of the Perch"—and, for discussion, Washburn 1908, 248; and Thorpe 1956, 275.
88. "Origin of Language" 1874, 269–70.
89. [Mivart] 1871, 372. On Mivart, see Richards 1987, 353–63.

90. Mivart 1873, 192.
91. Mivart 1875, 772–73.
92. Mivart 1875, 773. Cf. Darwin 1871, 1:79, writing of the evolution of the moral sense, and the pleasure of acting instinctively: "What a strong feeling of inward satisfaction must impel a bird, so full of activity, to brood day after day over her eggs."
93. Huxley 1881, 107. It was in 1878, however, that Romanes first addressed the matter of animal intelligence squarely and publicly, in an evening lecture at the BAAS meeting in Dublin in August. The lecture was later published in the *Nineteenth Century*; see Romanes 1878; and for analysis, Schwartz 2002, 141–42; also White 2005, 72. On Romanes's life and work, see Richards 1987, 346–53; Knoll 1987, chap. 3; 1986, 12–17; and Boakes 1984, 24–32. Thanks to Joel Schwartz for helpful discussion of Romanes.
94. Romanes 1882, 10.
95. Romanes 1883, 337, quoted in Richards 1987, 347. On *Animal Intelligence*, see Richards 1987, 347–48; and Boakes 1984, 25–27.
96. Romanes 1888, 273–74.
97. Romanes 1888, chaps. 3 and 4.
98. Romanes 1888, 274–81.
99. Romanes 1888, chaps. 4–6.
100. Romanes 1888, 439.
101. Romanes 1888, 126–27.
102. Romanes 1889, 160.
103. Müller 1891b, 585–86. On Müller versus Romanes, see Knoll 1986.
104. Romanes 1892, 67–68.
105. Carus 1892, 94. On differences of kind arising from differences of degree, see 85–86.
106. Evans 1891, 300.
107. Evans 1891, 309.
108. Morgan 1932, 248. For biographical details on Morgan, see also Radick 2004a; Costall, Clark, and Wozniak 1997; and Clarke 1974.
109. See Romanes 1879, 1882, 1883. In what follows I present Morgan as arguing throughout the 1880s for a comparative psychology with radically deflated ambitions. Other historians, however, have detected a major divide in Morgan's thought during this period. In their view, Morgan first argued that comparative psychology was impossible, then, perhaps on becoming a monist, reversed himself and argued that it was possible after all. See Richards 1987, 380; and Costall 1993, 119. I see at least three problems with this view. First, the sole witness to this supposed shift toward Romanes was Romanes—hardly a neutral Morgan watcher. Second, Morgan never denied that comparative psychology of any sort was possible. What he denied was that Romanes's kind of comparative psychology was objective. Third, it was not the strong thesis of monism, or neurosis-psychosis identity, that anchored Morgan's belief that animals had minds, but the much

weaker theses of neurosis-psychosis parallelism and evolutionary kinship between humans and animals—both of which Morgan publicly backed in the same 1884 paper in which he pronounced comparative psychology "impossible."

110. Morgan 1882, 523. Romanes had made the claim about a "good for eating" concept in dogs in Romanes 1879. He used the example again in *Mental Evolution in Man*, 1888, 27, attracting a sarcastic comment from Mivart in *Origin of Human Reason*, 1889, 49–50. See Knoll 1986 n. 63.

111. An almost identical analysis of language, abstraction, general ideas, and animal minds can be found in the ethnologist C. S. Wake's *Chapters on Man, with the Outlines of a Science of Comparative Psychology* (1868, 87–96). Wake was among the first writers to call the new science "comparative psychology." Wake too quoted Müller quoting Locke (34).

112. Morgan 1882, 524. Morgan argued that while some animals had a kind of language, the confines of the animal mind limited animal communication to feelings and actions in the here and now. True language liberated its users from the tyranny of the present.

113. Morgan 1884, 371.

114. Morgan 1886, 177–78. The Müller quotation is from Müller 1873, 211. Müller quoted his own antianthropomorphic credo in Müller 1875b, reprinted in Müller's *Chips from a German Workshop*—the source that Morgan cites.

115. Morgan 1884, 372. See also Morgan 1886, 185.

116. Morgan 1886, 179–80, 182. The Romanes-Morgan debate on animal intelligence between 1882 and 1886 became the subject of an unusually prolonged and wide-ranging correspondence in *Nature*. See Mergoupi-Savaidou 2004, 34–35.

117. Morgan 1923, vii. Morgan reported the conversation ending thus (viii): "In conclusion, as he [Huxley] answered a knock at the door, he dismissed a mere neophyte with the encouraging: 'You might well make all this a special field of enquiry.'" On Huxley and Mivart, see Richards 1987, 225–30, 354–57.

118. F. M. Müller to C. L. Morgan, 27 September 1887, Papers of Conwy Lloyd Morgan, 128/32, Bristol University Library. For Morgan on language and abstract ideas, see Morgan 1885, 14–15.

119. Morgan 1890–91, 358. On Lubbock's contribution to the Victorian debate on mental evolution, emphasizing his studies of animal communication, see Clark 1998.

120. Morgan 1890–91, 348–50, 376. Morgan reaffirmed (349) that he was "prepared to say, with John Locke, that [such] abstraction 'is an excellency which the faculties of brutes do by no means attain to.'"

121. Müller 1887b, title page, 200.

122. Morgan 1890–91, 373. In *Animal Life and Intelligence* Morgan also discussed *The Science of Thought* with respect to Müller's views on monism (467–72).

123. Romanes 1888, 135.

124. Romanes (1888, 135) argued that the Darwinian needed only that there be some kind of cerebral seat for language, however large or diffuse. It seems that Vogt

held a roughly similar view. Bateman (1870, 169) reported that in Vogt's letter on apes, microcephali, and Broca's region, Vogt "express[ed] a doubt about whether we shall ever be able satisfactorily to assign 'the divers functions' which compose language, to special parts of the brain, until we have a physiological analysis of articulate language, similar to that which Helmholtz has given of sight and hearing."

125. Jackson 1958, 1:39, 50–51, cited in Young 1990, 206–7.

126. Jackson 1868, 12.

127. Jackson's writings on evolution and speech are collected in Jackson 1958, 2:3–205. On Jackson and the brain, see esp. Smith 1992; also Critchley and Critchley 1998; Young 1990, chap. 6; Harrington 1987, chap. 7; and Greenblatt 1977, 1970, 1965.

128. Young 1990, chaps. 6–9. Cf. Star 1989, chaps. 5 and 6.

129. For contemporary summaries of research on Broca's region in ape and microcephalous brains, see Hale 1886, 309–10; and MacNamara 1908, chap. 10. See also Schiller 1979, 270–71.

130. The American neurologist William A. Hammond wrote a critical review of Bateman's *Darwinism* book for the *Journal of Nervous and Mental Disease*. Hammond 1877. Hammond later (1878) gave a talk defending Darwinism against Bateman at the New York College of Veterinary Surgeons. For discussions of Hammond and Bateman, see Blustein 1991, 143–44. As compendia of the literature on aphasia, Bateman's books have proved useful to investigators from his day to the present. See, e.g., Freud 1953, 41; and Rosenfield 1988, 28–29.

131. Evans 1891, 311.

132. On aphasia and the brain, including a schematic drawing of the left hemisphere of the brain with Broca's and Wernicke's regions identified, see James 1890a, 1:49–50. On James and evolution, see Richards 1987, chap. 9.

133. James 1890a, 2:973–83, esp. 980–83. Morgan's 1886 article is cited on 979. James earlier sketched his views on the differences between human and animal minds in James 1878.

134. James 1890a, 2:980–83. Wright's 1873 article is cited on 982.

135. James 1890a, 2:977, 973.

136. Morgan reviewed both James's *Principles of Psychology* and its subsequent abridgement, *Psychology: A Briefer Course* (1892). See Morgan 1891 and 1892d. It was only in the 1892 book that James used the phrase "stream of consciousness" (chap. 11), having written previously of the "stream of thought" (1890a, vol. 1, chap. 9).

137. Morgan 1892a.

138. What Costall 1993 describes as Morgan's embrace of Romanesian anthropomorphism would much more accurately be described as an embrace of Jamesian introspectionism. It was as if James the brilliant writer had climbed into his own head and come back with descriptions of the mental gulf Morgan knew to be there from the start. As for Morgan's adoption from James of the "stream of conscious-

ness" and the "feeling versus perception of relations" themes, I should add that neither was original to James. Morgan had encountered both themes earlier: the former in William Clifford's work, the latter in Herbert Spencer's. On Clifford, Morgan, James, and the "stream" theme, see Richards 1987, 378. On Morgan's reading of Spencer in the 1880s, see Morgan 1932, 247. For Spencer on relations, see Spencer 1870–72, vol. 1, part 2; vol. 2, part 6. Nevertheless, it was Morgan's reading of James around 1891 that pushed these themes to the center of Morgan's thinking. In his preface to his *Introduction to Comparative Psychology*, Morgan wrote of his "indebtedness to the work of Herbert Spencer, whose description of relations as 'the momentary feelings accompanying transitions' in consciousness contains the germinal idea from which my own thinking . . . has developed." But James quoted exactly the same passage (favorably) in his *Principles*. See Morgan 1894, x; James 1890a, 1:242; and Spencer 1870–72, 1:64.

139. Morgan 1892a, 44–47.
140. Morgan 1893b, 755.
141. Morgan 1892b, 417.
142. Morgan 1893a, 239, emphasis added.

Chapter Three

1. "Simian Talk," *Punch* 100 (1891): 289, emphasis in original.
2. Jones 1912, 185. I am grateful to John van Wyhe for this reference. Butler also referred to Garner's work in an 1894 lecture concerning Müller's views on thought, language, and evolution; see Butler 1904, 212.
3. Sanford 1892.
4. "In the Monkey-House," *Punch* 103 (1892): 153, emphases in original.
5. On the triumph of evolutionism in the later nineteenth century, see, e.g., Ruse 1999a, chaps. 8 and 9.
6. On amateurs in the scientific community of the United States in Garner's day, see Goldstein 1994, esp. 574–75, for a helpful survey of the secondary literature.
7. For biographical details on Garner, see "Garner, Richard Lynch" 1906, 1968. A journalist described the adult Garner as having "dark eyes, dark complexion and a very strong face. He is broad-shouldered and of great physical strength, weighing about two hundred pounds." "Can Make Monkeys Talk," *New York World*, 20 December 1891, 22.
8. On antebellum Abingdon and the surrounding region, see Garland 1987.
9. In Garner's day, for example, the present Barter Theatre, then owned by the Sons of Temperance, "was used for lectures, grange meetings, theatrical productions and . . . [from] 1855 a male school met in the basement." *Abingdon* n.d.
10. Garner made two attempts at autobiography, both centering on his Abingdon boyhood. I have drawn the above mainly from an unpublished manuscript, "Just as It Happened: A True and Unvarnished Statement of Events in the Life and Experience of R. L. Garner," Papers of Richard Lynch Garner, box 4, folder "Writ-

ings: 'Jack-o'-Lantern Farm'–'Just as It Happened,'" National Anthropological Archives, National Museum of Natural History, Smithsonian Institution. On Garner's father, see 2; on Garner's precocious agnosticism, see 3–6, quotation on 3. Between January and May 1904, Garner wrote a series of autobiographical letters, published by his son Harry as *Autobiography of a Boy: From the Letters of Richard Lynch Garner* (1930). On the water mill owned by Garner's father, see 7–8.

11. On the wartime activities of Garner and his family, see, in addition to the biographical articles cited above, an article on Garner from around December 1916, in the *Philadelphia Evening Bulletin*, held (in typescript) in the Urban Archives, Samuel Paley Library, Temple University, Philadelphia; and a letter from Garner's son, Harry, to Garner's biographer J. P. Harrington, 22 July 1941, Garner Papers, box 7, folder "Biographer's Papers." On Forrest's raids, see Starr 1979. On the Virginia-Tennessee border region in the war, see Phillips 1997.

12. After Garner visited the English newspaperman W. T. Stead in the summer of 1892, Stead (1892) wrote thus about "one of the few Americans who has recently become familiar throughout the world": "Mr. Garner is a tough customer, having served in the Confederate Army, and thus gained practical personal experience of the hardships endured by Confederate prisoners in about a dozen different northern prisons. He then spent four years in the plains campaigning against the Indians. He may therefore possibly return from Africa alive." I am grateful to Gowan Dawson, then of the Leeds-Sheffield SciPer project, for this reference and all subsequent references to Stead's journal, the *Review of Reviews*.

13. See Bell 1867, 20–21, quotation on 21. On "the nineteenth-century infatuation with phonetics," see Rée 1999, 255–65, quotation on 265. On Garner's "hard work for years as a village school-master, where he paid most attention to the study of phonetics," see Phillips 1891.

14. R. L. Garner, "The Speech of Animals," unpublished MS, 2, Garner Papers, box 6, folder "Writings: 'Sanctuary to Women'–'Spider Webs.'"

15. On the vogue of Spencer, see Hofstadter 1992, chap. 2.

16. On Garner's reading, see the 1916 *Philadelphia Evening Bulletin* article cited above, and R. L. Garner, "Critics and Croakers," unpublished MS, 1, Garner Papers, box 3, folder "Writings: 'Civilised Savagery'–'Critics and Croakers.'" In *The Speech of Monkeys* (1892f, 205), Garner wrote of the relations between language and reason: "I beg to be allowed to stand aside and let Prof. Max Müller and Prof. Whitney, the great giants of comparative philology, settle this question between themselves; and I shall abide by the verdict which may be finally reached."

17. See Garner 1891b, 314–15.

18. In "The Simian Tongue [I]" (1891b, 315), Garner wrote that, after the Cincinnati epiphany, he pursued his translation studies without much success "in the gardens of New York, Philadelphia, Cincinnati, and Chicago, and with such specimens as I could find with the travelling menagerie museum, or hand organ, or aboard some ship, or kept as a pet in some family."

19. R. L. Garner to Prof. Brown Goode, 30 January 1888 (one letter) and 31 July 1888 (two letters), RU 189, box 43, folder 7, Smithsonian Institution Archives, Smithsonian Institution (SIA). On Garner's study of Maya glyphs, see Phillips 1891 and the contributor's note on Garner in the May 1892 *Cosmopolitan*, 128.

20. Note to Prof. Brown Goode, 7 August 1888, RU 189, box 43, folder 7, SIA. Garner tried again the following year. R. L. Garner to Prof. Brown Goode, 2 July 1889, RU 189, box 43, folder 7, SIA.

21. Garner 1892f, 221–22. Garner claimed (222) that, at the time of his Maya work, he was not familiar with Bell's *Visible Speech*. Garner later accused the Bureau of American Ethnology's Garrick Mallery of pirating and exploiting this method of decipherment. R. L. Garner to W. H. Holmes, 6 December 1900, Garner Papers, box 1, folder "Outgoing Letters." On Mallery's earlier research into Amerindian "gesture speech," and relations between this speech and the undeciphered glyphs, see Rée 1999, 285–92, esp. 285.

22. "The phonograph has made more progress in Washington, in office and home-use, than in any other city thus far." "The Columbia Phonograph Co., Washington, D.C.," *Phonogram* 1 (1891): 88–92, quotation on 89.

23. Phillips 1891.

24. On Edison's invention of the phonograph, and the importance of the phonograph in turning Edison into "the wizard of Menlo Park," see Israel 1998, 142–56. On forerunners of the phonograph, see Read and Welch 1976, chap. 1.

25. On the rapid rise and fall of the tin-foil phonograph, see Gelatt 1977, chap. 1; and Read and Welch 1976, 21–26. For a lucid explanation of how early "acoustic" phonographs work, see Brady et al. 1984, 4–6.

26. Israel 1998, chaps. 10–12; Read and Welch 1976, 26.

27. On Edison's perfecting of the phonograph, see Israel 1998, 277–91; Millard 1990, chap. 4; and Gelatt 1977, chap. 2.

28. "A Retrospect," *Phonogram* 1 (1891): 187.

29. On Edison's visit to the Paris Exposition of 1889, where the perfected phonograph was first exhibited on the Continent, becoming "the most notable and popular device at the exhibition—every day, some 30,000 people heard twenty-five phonographs talking in dozens of languages," see Israel 1998, 369–71, quotation on 371.

30. Gelatt 1977, 43–44.

31. Israel 1998, 287–90; Gelatt 1977, 45.

32. Israel 1998, 291; Gelatt 1977, 43–44.

33. "Columbia Phonograph Co.," 89–90.

34. On the nickel-in-the-slot phonograph, see Musser and Nelson 1991, 38–40; and Gelatt 1977, 44–57.

35. On the Bureau of American Ethnology, see Hinsley 1981.

36. Fewkes 1890, 267, 268, 269.

37. F. Baker to R. L. Garner, 8 December 1890, RU 74, box 7, folder 2, SIA. On the National Zoological Park and Baker's role there, see "National Zoological Park"

1996, esp. 187. As discussed in the introduction to this book, Garner's first experiments had also involved Bell's graphophone rather than Edison's phonograph. See "Can Monkeys Talk? A Queer Scientific Experiment at the Smithsonian Institute," *St. Louis Daily Globe-Democrat*, 21 September 1890. Two months later, Garner spoke again to a reporter from the same newspaper about new experiments in Chicago and Cincinnati. "Monkey Talk of the Future," *St. Louis Daily Globe-Democrat*, 22 November 1890. I am grateful to Pamela Henson of the Smithsonian Institution Archives for these references.

38. F. Baker to R. L. Garner, 8 December 1890, RU 74, box 7, folder 2, SIA.

39. "Novel Experiments with the Phonograph," *Phonogram* 1 (1891): 85.

40. Garner 1891b, 314. The article was published in the United States in *Littell's Living Age*.

41. James 1890b, 2:980; 3:1470.

42. "Apes and Men," *Spectator* 66 (1891): 787–88, quotation on 788. Garner's article also appeared in condensed form in the *Review of Reviews* and the *Phonogram*.

43. For an overview of science in British periodicals during the nineteenth century, see Brock 1994 and, much more extensively, the SciPer trilogy: Cantor et al. 2004, Cantor and Shuttleworth 2004, and Henson et al. 2004. For an overview of science in American periodicals during the same era, see Mott 1957, 306–10.

44. On the *New Review*, see Houghton and Houghton 1979, 303–8.

45. On McClure, see Gale 1999. On McClure's connections in English publishing, see esp. Wilson 1970, 48–55. On McClure's placing of Garner's works with Grove's *New Review*, see Garner's remarks on his 1893 *McClure's Magazine* article, in R. L. Garner, "Index to Manuscripts with Notes on Publications, Transmission, etc.," Garner Papers, box 2, folder "Writings: Indexes" (hereafter cited as "Index to Manuscripts with Notes").

46. The earliest evidence I have found of the relationship between Garner and McClure is a letter from late 1891. R. L. Garner to S. S. McClure, 24 November 1891, collection no. 9040, Alderman Library, University of Virginia. In his autobiography (1914, 244), McClure remarked on how rare were the interesting scientific experts, and how valuable to publishers for the general public. Garner was paid $65 for the first installment of "The Simian Tongue," and $35 and $65 for the subsequent two. Garner, "Index to Manuscripts with Notes."

47. See various letters and announcements in the Letter Copybooks from 1891–92 and 1892–96, Papers of Samuel Sidney McClure, Lilly Library, Indiana University, Bloomington. I am grateful to Dina Kellams for locating these documents.

48. "Can Make Monkeys Talk," 22.

49. Phillips 1891.

50. "Speech in the Lower Animals," *Scientific American*, 27 February 1892, 129.

51. Garner 1892d, 1892g.

52. On the origin of and profits from *The Speech of Monkeys*, see Garner, "Index to Manuscripts with Notes." Previously Garner had privately published two books

of verse. In *Nancy Bet: The Story of Sloomy Perkins and His Transaction in Real Estate* (1889), the narrator tells of having married ugly Nancy forty years before, in hopes that her ailing father would pass away and leave them his farm. Alas, the old man improved, and Nancy turned out to be an insufferable shrew. After much mayhem ("She'd banged my eyes till they went black, / And poured hot ashes down my back"), they divorced. The poems collected in *The Psychoscope* (1891a) are far more serious and contemplative, with somber titles such as "Perfidy," "Despair," and "The Church Militant." I discuss the religious sensibility of these poems in chapter 4.

53. Garner 1892f, 43, 103, 3.
54. "Mr. Garner Talks with Monkeys," *New York Herald*, 10 December 1891, 14.
55. Garner 1891b.
56. Garner 1892e, 328–29.
57. Garner 1891b, 321.
58. "I think their speech, compared to their physical, mental, and social state, is in about the same relative condition as that of man by the same standard." Garner 1891b, 320. On the craniofacial angle and gnathic index, see Garner 1891c, 325.
59. Garner 1892e, 331.
60. Garner 1891b, 320, emphasis in original.
61. Garner's variants on Müller's maxim appear in Garner 1891c, 327. Müller's own variant appears in Müller 1861, 369.
62. On savage and civilized tongues, see Garner 1891b, 320; on simian monophones, see Garner 1891c, 326, emphasis in original.
63. Garner 1891b, 320; 1891c, 327.
64. The following analysis owes much to Ian Hacking's writings on experiment. See esp. Hacking 1992b, 1990, and 1983, chap. 13.
65. Garner 1892f, 221.
66. Garner 1892f, 211.
67. Garner described the procedure in Garner 1892f, 210–12.
68. Garner 1892f, 73.
69. Garner 1892f, 111.
70. Cf. Garner 1892f, 201: monkeys showed they understood the meanings of monkey utterances even when those utterances were "imitated by a human being, by a whistle, a phonograph, or other mechanical devices, and this indicates that they are guided by the sounds alone, and not by any signs, gestures, or psychic influence."
71. Garner 1892f, 61–62.
72. Garner 1892f, 210.
73. Garner 1892f, 209–10.
74. For other instances of phenomena creation, see Garner 1892f, 181–82, 209, 212, 215–16, 219–21.
75. Garner 1892f, 211–12. Garner continued to explain what he took to be unorthodox views on the nature of sound waves on 212–14.

76. Garner 1892f, 213. On the phoneidoscope, see Taylor 1878, which identifies the manufacturer of the phoneidoscope as S. C. Tinsley and Co., Philosophical Instrument Makers, London. A phoneidoscope is held in the Whipple Museum, Cambridge.
77. Garner 1892f, 24–26.
78. Garner 1892f, 26–30.
79. Garner 1892f, 32–33.
80. Garner 1892f, 158.
81. Garner 1892f, 160–61. See also Garner 1896, 128, 180–83.
82. Garner 1892f, 158, 164–65, 159, 89.
83. Garner 1892f, 153–55.
84. Garner 1892f, chap. 10.
85. Garner 1892f, 46.
86. Garner 1892f, 81–82. Garner further argued (174–75) that the presence in certain creatures of functioning vocal organs indicated either that speech was useful for these creatures now, or had been useful in the recent past, for, as "a law of evolution and progress, all organs are imparted to animals for use and not for ornament."
87. Garner 1892f, chaps. 17, 18, 20, and 25.
88. Garner 1892f, e.g., 171–76, 194.
89. Garner 1892f, chaps. 3 and 18.
90. Garner 1892f, 134–35. Garner gave the fullest exposition of the law of cranial projection (now so named) in Garner 1900, 8–11.
91. Garner 1892f, chap. 15. "It is an interesting fact," wrote one of the journalists attending Garner's experiments in Central Park in December 1891, "that he [Garner] is not a Darwinian. He believes in evolution, but not in Darwin's theory as to the origin of species. He has an origin of species theory of his own." "Can Make Monkeys Talk." On the proliferation of non-Darwinian theories of evolution between 1870 and 1930, see Bowler 1983.
92. Garner 1892f, 206–7.
93. Garner 1892f, 109–10. I have omitted a paragraph break.
94. Garner 1891c, 325. In his memoir of the expedition, *Gorillas and Chimpanzees* (1896, 14–15), Garner wrote of selecting equatorial West Africa because both gorillas and chimpanzees could be found there, "in a state of freedom." Among them, he planned to test what, on the basis of his work with captive monkeys, "it was logical to infer[:] that the anthropoid apes, being next to man in the scale of nature, must have the faculty of speech developed in a corresponding degree."
95. "Can Make Monkeys Talk."
96. "Can Make Monkeys Talk."
97. Phillips 1891.
98. On Garner's departure from New York, see the letter from G. N. Morison to S. F. Moriarty, 18 July 1892, Papers of Thomas Alva Edison, microfilm/online ed., reel 133, frame 332 (hereafter cited as, e.g., "TAEM 133:332"). On Garner's depar-

ture from Liverpool, see the letter from R. L. Garner to H. Garner, 28 September 1892, Garner Papers, box 1, folder "Outgoing Letters."

99. Garner 1892b, 1892h. On the phonograms, see "Going to Monkeyland to Talk with Monkeys," *New York Herald*, 6 July 1892. The *North American Review* article was summarized ("the oddest article in the magazines for the month") in the *Review of Reviews* 6 (July 1892): 49. "The scheme in itself is notable enough," commented the reviewer, "but it is raised to the veriest limit of fantasy by the developments which it has undergone in the ingenious brain of the Professor. I should like to hear what Mr. Stanley or any other African traveller would say to the museum of scientific knicknacks by which the Professor proposes to rise superior to the difficulties of African travel."

100. Phillips 1891.

101. Garner 1892b, 289, 290.

102. Garner 1892f, 223.

103. Garner 1892h, 715.

104. Garner 1892b, 289–90, 292.

105. Garner 1892b, 291–92. Rebecca Herzig (2005) has recently explored the theme of "suffering for science" in the American science of Garner's era. It is striking both how strenuously the pre-expedition Garner linked (as in the quoted passage) his claims to truth and his willingness to suffer for it, and how badly—as we will see in the next chapter—the postexpedition Garner's credibility was affected when West Africa proved rather less arduous than expected.

106. R. L. Garner to S. S. McClure, 24 November 1891, collection no. 2040, Alderman Library, University of Virginia.

107. Open letter from R. L. Garner, 31 March 1892, TAEM 133:324–25, with signatures and comments of all but Alexander Melville Bell on frame 325. For Bell's participation in the scheme, see the letter from A. M. Bell to R. L. Garner, 13 April 1892, Garner Papers, box 1, folder "Incoming Letters." Garner dedicated *The Speech of Monkeys* (1892f, v) to his wife and to twelve "earnest friends . . . who have opened their purse as they opened their hearts, and afforded me that aid which made it possible for me to continue my researches."

108. R. L. Garner from F. S. Church, 22 April 1892; from F. W. Putnam, 7 May 1892; and from Mrs. Wilcox, n.d., emphases in original; Garner Papers, box 1, folder "Incoming Letters."

109. "Professor Garner has the special patronage of ex-President Grover Cleveland and Mr T. A. Edison in his unique undertaking." *Yorkshire Post*, mid-September 1892, quoted in E. H. T. 1893.

110. R. L. Garner to T. A. Edison, 27 July 1891 and 25 August 1891, TAEM 131:851–53 and 857–58, respectively; quotation on frame 853.

111. T. A. Edison's notes for G. Lathrop, n.d., TAEM 128:749. On *Progress*, see Israel 1998, 365–68. I am grateful to Paul Israel for this reference.

112. A. O. Tate to R. L. Garner, 31 August 1891, TAEM 142:758–59, quotation on frame 759.

113. R. L. Garner to T. A. Edison, 21 December 1891 (with Edison annotation dated 24 December 1891), TAEM 131:868–70. In the letter, Garner mentioned that it was "Prof. Bell" himself who had advised Garner on switching from batteries to springs.

114. "Mr. Edison's Forecast," *Phonogram* 2 (1892): 1–2, quotation on 2.

115. "The Phonograph to Aid in Establishing the Darwinian Theory," *Phonogram* 2 (1892): 16.

116. See Israel 1998, 286.

117. See, e.g., Israel 1998, 132–41, 234–37.

118. "Mr. Edison's Forecast," 2.

119. "Speech in the Lower Animals."

120. Garner 1892b, 285–86.

121. G. N. Morison to T. A. Edison (with Edison annotation), 1 April 1892, TAEM 133:329.

122. Garner 1892f, 208–9.

123. "Going to Monkeyland"; "Important and Novel Quest," *Phonogram* 2 (1892): 156–57.

124. G. N. Morison to S. F. Moriarty, 18 July 1892, TAEM 133:332.

125. "Babies, Born with Speech Instinct, Have Own Language—Prof. Garner," *New York Evening Journal*, 4 December 1919, Garner Papers, box 1, folder "Newspaper Clippings."

126. Phillips 1891.

127. The letter was quoted at the end of a four-page promotional brochure, dating from around 1912, for Garner's lectures on his life and work. "Professor Richard L. Garner: The Renowned Traveler, Anthropologist, and Student of Animal Language," brochure, Garner Papers, box 7, folder "Biographer's Papers." I have not found the original letter from Whitney, but I have found references to it in two other letters. The first is in a letter from Page to Whitney, 6 July 1892, asking Whitney's opinion as to the scientific value of Garner's work. Whitney Papers, group 555, series 1, box 42, folder 42:1302. (The Whitney letter reproduced in Garner's brochure is addressed to Page, and written as a response to Page's letter of 6 July.) The second reference is in a letter from Garner to Page, 14 September 1892, thanking Page for forwarding Whitney's letter about Garner. "It is very gratifying to me," wrote Garner, "to know that such men are in sympathy with my efforts." Papers of W. H. Page, bMS Am 1090 [406], in the Houghton Library, Harvard University.

128. "Speech in the Lower Animals."

129. R. L. Garner, "Critics and Croakers," unpublished MS, 1, Garner Papers, box 3, folder "Writings: 'Civilized Savagery'–'Critics and Croakers.'"

130. On Brinton, see Darnell 1988, 1976.

131. Brinton 1888, quotation on 225. For Brinton's views on race, language, and the brain, see Brinton 1890, 60–67.

132. Brinton dedicated his book *Races and Peoples* (1890) to Hale.

133. On Hale, see Gruber 1967, and, in this book, chapter 5 and the conclusion.

134. Hale 1888, 38–39.

135. For a superb introduction to the "new Darwinism" of Galton and his followers, see Depew and Weber 1995, chap. 8. For further discussion, see the conclusion in this book.

136. Hale 1886, 322–23. On Hale's views in the context of nineteenth-century anthropological debate about primitive languages and related topics, see Henson 1974, chap. 1. For further discussion, see chapter 5 in this book.

137. Cf. Hale 1891, 111–12: "Unless it can be clearly shown that man is separated from other animals by a line as distinct as that which separates a tree from a stone or a stone from a star, there can be no proper science of anthropology. . . . Anthropology begins where mere brute life gives way to something widely different and indefinitely higher. It begins with that endowment which characterizes man, and distinguishes him from all other creatures. The real basis of the science is found in articulate speech, with all that this indicates and embodies."

138. Hale 1892, 382.

139. Conn 1892; [Cornish] 1892; Hubrecht 1892; Jastrow 1892; Morgan 1892c; and the reviews in the *Athenaeum* (100:821–22), the *Critic* (21:161–62), the *Nation* (55:267), and *Popular Science Monthly* (42:270); also in the London *Times*, 5 August 1892, 3; *New York Times*, 4 September 1892, 18; *New-York Daily Tribune*, 17 November 1892, 6; and *Brooklyn Daily Eagle*, 11 September 1892, 17. Hale 1892 has already been noted.

140. Garner 1899, 1905a.

141. London *Times*, 5 August 1892, 3.

142. J. T. R. of Baltimore, letter to the editor, *New York Times*, 17 September 1892, 17.

143. *Spectator* 69 (1892): 228. In all likelihood, the *Spectator* reviewer was C. J. Cornish, as the review appears verbatim in the chapter "The Speech of Monkeys" in Cornish's 1895 book on Regent's Park Zoo, *Life at the Zoo*. I am grateful to Sofia Åkerberg for this reference.

144. Afterword of W. Marshall, Garner 1905a, 152. In addition to the afterword, Marshall's translation of *The Speech of Monkeys* (which went through two editions) includes a new foreword by Garner and extensive footnotes by Marshall.

145. Review of Garner, *The Speech of Monkeys*, *Critic* 21 (1892): 162.

146. Morgan 1892c, quotations on 509, 510. On this review in the context of Morgan's wider efforts to discredit nonprofessional students of the mind, see Costall 1998, 22–26. I am grateful to John Clark for the Costall reference.

147. Jastrow 1892, 215–16. Like the more positive *Critic* reviewer, however, Jastrow (215) described the experiments testing "general intelligence" as "the most valuable in the book. The method of testing the counting powers of monkeys is ingenious, and some of the tales illustrating their successes in adapting means to ends form welcome contributions to our stock of observations." The Dutch reviewer Hubrecht (1892), at least as translated in a brief excerpt in the *Review of Reviews*,

took an unrelievedly dim view of Garner and his project: "When Garner, in his semi-naive communications, not only sells the skin before catching the bear, but tries to coin money out of the various apparatus with which the capture is to be effected, we see in this a new proof of his want of that seriousness and critical sense which is absolutely indispensable to the success of any scientific investigation."

148. Evans 1893, 438.

149. Evans 1893, 438.

150. Evans 1897, 315.

151. *The Speech of Man and Holy Writ* (1894). Discussing the spate of scientific writings tending toward extreme skepticism about the supernatural, the author wrote: "Of the latter class there has recently been issued in this country a very remarkable book, entitled *The Speech of Monkeys*, written with great spirit and seeming earnestness, which has attracted considerable attention, but up to the present time does not appear to have received any corrective notice; it has therefore gone forth uncontradicted among the general mass of readers" (3–4). The tenth and eleventh chapters of *The Speech of Man and Holy Writ* are entitled, respectively, "The Ape and Evolution" and "The Monkey and the Phonograph."

152. *The Speech of Man and Holy Writ*, 142–43, 147. Garner's reviewer in the *Leisure Hour* (1893, 205), a British evangelical weekly published by the Religious Tract Society, wrote in a similar spirit that Garner's claims were "absurd to all, except extreme evolutionists who believe in the development of man from monkey" (205). Thanks to Jim Secord for this reference.

Chapter Four

1. E. H. T. 1893, 254–55, emphasis in original. Prefacing the verse was a notice reprinted from the *Yorkshire Post* about Garner's departure for Africa.

2. Verne 1901. For a reading of Garner's works alongside H. G. Wells's *Island of Dr. Moreau* (1896) and other late-Victorian fictional and/or antivivisectional writings dealing with talking animals, see Ferguson 2006, chap. 4. On the Wells-Garner connection in particular, see McLean 2002.

3. For a fine appreciation of Verne's work in the context of his life and times, see Weightman 1997.

4. I quote in what follows from I. O. Evans's 1964 translation, entitled *The Village in the Treetops*. See Verne 1964, 85.

5. Verne 1964, 85: "The efforts made by the American Professor Garner may still be remembered—his scheme for studying the language of the monkeys and of giving his theories experimental verification. The professor's name, the articles and books which he had published in America and Europe, could not have been forgotten by the people of Africa."

6. Verne 1964, 87. There seems to have been at least one real-life imitator of Garner's cage exploits—but with fish. See the curious report in *Country Life Illustrated*, 3 February 1900, 132, telling of the attempt of a Naples professor to learn the lan-

guage of fish by lowering himself into the Mediterranean in a cage equipped with sound-recording equipment. Thanks to Alan Longbottom for this reference.

7. On the military metaphor, see esp. Moore 1979, chaps. 1–4.

8. The classic sociological analysis is Turner 1978. Cf. Desmond 1997, 250: "There is no doubt that Frank Turner is right: the 'war' between science and religion was a professional territorial dispute."

9. R. L. Garner, "Mr. Garner and the Language of Monkeys," letter to the editor, London *Times*, 25 August 1892, 5. In the *Journal of Sectional Proceedings* no. 5 (Monday, 8 August 1892), Garner's name and lecture title appear in the third slot for the 11 a.m. meeting of the anthropological section. Papers of the British Association for the Advancement of Science, Dep. BAAS 179, Annual meeting–Edinburgh 1892, Bodleian Library, Oxford University.

10. Review of Garner, *The Speech of Monkeys, Athenaeum* 100 (1892): 821, 822.

11. Ursula's letter to Madge, *Truth* 32 (1892): 367. "Madge" was the Victorian etiquette expert Mrs. Humphrey. I am grateful to Barbara Onslow for this identification.

12. Comment on Garner at the BAAS, *Natural Science* 1 (1892): 495.

13. Garner, "Mr. Garner and the Language of Monkeys," 5.

14. *New York Times*, 28 August 1892, 5. Garner's appearance before a meeting of the Balloon Society of Great Britain in early September went much more smoothly. See *Nature* 46 (1892): 451.

15. R. L. Garner, "Notes and Comments . . . ," typescript, 1, in *The Record of Idle Moments*, a notebook in the Garner Papers, box 2, folder "Writings: 'The Record of Idle Moments.'" I discuss this notebook at length below.

16. On the history of the Edison phonograph business in Britain in the late 1880s and early 1890s, see Martland 1998, 4–6; and Andrews 1986, chaps. 1 and 2.

17. *New York Times*, 28 August 1892, 5.

18. Garner 1892a. The article also appeared, with minor variations and a different title, in the September 1892 *New Review*. Most of the article dealt with Garner's psychological experiments in Regent's Park with Jim the orangutan. As before, boxes with holes featured prominently (Garner claimed the particular experiments he conducted with Jim derived from work carried out with children and idiots by an American surgeon and psychologist, Walter Kempster). One experiment involved two boxes, a green one with a round hole, and a yellow one with a square hole. These two boxes each came with a matching plug or "block" of the right shape and size to fit into the hole. After several trials over several sessions, in which Garner presented the boxes and blocks in different orders, rewarding Jim with a bit of apple whenever he matched them up correctly, Garner arrived at a rather curious conclusion. Jim's overriding concern seemed to be keeping the blocks together. "No matter which block I gave him [first]," Garner later wrote, "he would generally fit it into the hole of the same shape, but seemed determined always to place the second block in the same hole. . . . In a few cases where I compelled him to put the second block in the right hole he endeavored

to recover the first block which he had deposited in order to place it in the same hole."

19. It was reported that "[t]he famous cage in which he proposes to dwell was erected on the quayside at Liverpool, for the admiration of beholders and to give the *Graphic* artist a chance." From an item in *Lightning: The Popular and Business Review of Electricity*, 22 September 1892, 181. Thanks to Graeme Gooday for this reference.

20. R. L. Garner to T. A. Edison, 4 November 1892, TAEM 133:336–38, quotation on frames 336–37.

21. S. S. McClure to T. A. Edison (with Edison annotation), 11 January 1893, TAEM 134:501.

22. On his return to the United States in March 1894, the *New York Times* quoted Garner thus: "Now, about those phonographs. I did not have such an instrument with me. I had fragments of one, but they were absolutely useless for reproducing sound. I ordered one from Rochester in May, 1893, and have not received it even yet." "Thinks Well of Gorillas," *New York Times*, 26 March 1894, 1. Cf. Desmond 1979, 254: "Garner . . . built himself a stout cage, in which he lived for 112 days . . . , recording on the latest wax-cylinder phonograph the bewildered reactions of a wealth of animals (including gorillas)."

23. R. L. Garner to W. H. Page, 14 September 1892, Page Papers, bMS Am 1090 (406).

24. M. Garner to W. H. Page, 15 January 1893, Page Papers, bMS Am 1090 (406).

25. R. L. Garner to F. W. True, 15 December 1892, RU 189, box 43, folder 7, SIA. What Mark Barrow (2000) has called "entrepreneurial natural history" was in boom time in Garner's America.

26. F. W. True, Memorandum, n.d., RU 189, box 43, folder 7, SIA.

27. R. L. Garner to T. H. Huxley, 8 January 1894, and T. H. Huxley to R. L. Garner, 9 January 1894, Papers of Thomas Henry Huxley, 17.14 and 17.14 verso, respectively, Imperial College, London.

28. The travel writer Caroline Alexander (1990, 225) has suggested that Gabon is probably *more* densely forested today than in the 1890s, mainly due to the draining away of people from the small villages that once dotted the interior, as the main towns along the coast have grown ever more important economically.

29. I take my taxonomy of Europeans in 1890s West Africa from Alexander 1990, 8, who, I presume, took it in turn from Kingsley 1897, 13. On the French Congo more generally, see, in addition to these works, Gardinier 1994 and West 1972.

30. Garner 1893, 364. Garner later recalled that he had sent the article to McClure for publication in Archibald Grove's *New Review*, but that McClure "dissuaded Grove from publishing it[,] and published it himself in . . . *McClure's Magazine*." Garner, "Index to Manuscripts with Notes."

31. Kingsley arrived in West Africa just as Garner was preparing to leave, in August 1893, although most of her *Travels* narrate a longer subsequent journey of 1894–95. On her first journey, see Kingsley 1897, 14–17.

32. Kingsley 1897, 35–36.
33. In the 1860s, the African explorer W. Winwood Reade wrote of the region (1863, 77–78): "The trade of this part of Africa is conducted on very primitive principles. A factory in the Gaboon is not a mammoth building, which vomits endless smoke and resounds with perpetual machinery. It is a ground-floor house built of a kind of palm, vulgarly called bamboo. There is a spacious piazza floored with deal, where the factor and his subordinates take their meals. On the left hand a door leads to the kitchen; on the right hand, to the parlour and a bedroom or two; in the centre is the store. Here you may see bales of Manchester cloth, American tobacco in leaf, barrels of coarse powder, casks of Coast o' Guinea rum, and Birmingham trade-guns, long as the ancient matchlocks, their stocks painted in bright red."
34. On Horn, see, e.g., Alexander 1990, 77–80, 104–8.
35. Garner 1893, 365. On Kingsley and trade, see West 1972, 56. Cf. Kingsley 1897, 135: a successful West African trader has "to be a 'Devil man.' They [the natives] always kindly said they recognised me as one, which is a great compliment."
36. Kingsley 1897, 162.
37. On the controversy surrounding the proposal, see Alexander 1990, 132–36. Garner (1896, 258) reported discussing gorillas with the author of the proposal, the missionary Adolphus Clemens Goode.
38. Garner 1893, 365.
39. Garner 1893, 367.
40. On the Spiritans and their missionary work in Africa, see Koren 1983, chap. 12. On the history of the mission of St. Anne, see Gaulme 1974.
41. Kingsley 1897, 41ff.; Garner, "Notes and Comments," 1. For Garner's comment to Edison, see R. L. Garner to T. A. Edison, 4 November 1892, TAEM 133:336–38, quotation on frame 338. In mentioning the weather to Edison, Garner was not just making small talk (frame 338): "Will you kindly give me your opinion as to the safety of my cage from lightning, when charged and when not charged with electricity[?]"
42. Garner 1893, 364–65. Noting an old and decrepit steamer anchored at the mouth of the creek leading from the Ogowé to Fernan Vaz, Kingsley explained (1897, 43) that Hatton & Cookson used the steamer as a depot because, "during the height of the dry season, the *Mové* cannot get through the creek to supply the firm's Fernan Vaz factories." Of the journey from the Ogowé to Fernan Vaz, she wrote (43): "some say it is six hours' run, others that it is eight hours for a canoe; all agree that there are plenty of mosquitoes."
43. Kingsley 1897, 41.
44. Garner 1893, 364.
45. Garner 1893, 364–66, quotation on 366, emphasis in original.
46. Garner 1893, quotations on 366, 372, 370. Such is its immediacy that at points Garner's field dispatch reads like a real-time radio transmission; e.g., "At night I frequently have a leopard or bush-cat visit me; it is then too dark to shoot them,

but my interest is centr—s-s-st!—s-s-st! Oh, the precious moment! I have just had a new and grand experience. I am a trifle nervous, but I must tell you. While writing the last few lines above . . ." (368). In brief, Garner reported seeing a female gorilla sneak up on a dog from the nearby mission, and then leave at the sound of Garner cocking his gun.

47. Garner 1893, 371. Garner did write at one point of an invisible gorilla "uttering a low murmuring sound which seems to express pleasure, but I am not yet able to translate it into English. Time and patience, however, will accomplish that, and much more" (367). He continued: "It is a fact worthy of notice that some of the sounds uttered by the gorilla and chimpanzee are identical with certain sounds in the native language. . . . One word in N'kami, meaning *yes* or *assent*, is exactly the same as one sound that is much used by the chimpanzee. . . . The same is true of the word for *five* in one dialect of Kroo speech" (367–68, emphases in original).

48. "Studying the Monkey Language," *New York Times*, 16 November 1893, 9; "One Less to Talk With," *New York Times*, 18 December 1893, 4. See also Garner 1896, chaps. 8 and 11.

49. "Studying the Monkey Language," 9.

50. "The Language of Monkeys," *New York Times*, 17 November 1893, 4.

51. "Professor Garner on the Gorilla," *Westminster Gazette*, 20 February 1894, 8. The paleontologist S. S. Buckman later wrote of his good fortune in coming across this report, as he found in it confirmation of his own, independently derived view, based on his studies of the speech of children, that "man's language is the greatest perfection of simian speech." See Buckman 1897, 798. On Buckman's recapitulationism, see Gould 1977, 137–39.

52. *Illustrated London News*, 3 March 1894, 1.

53. "To a Gorilla Girl," *Punch* 106 (1894): 97. Garner's cage also featured in a spoof, "Personal Experiences in Monkey Language," published that summer in the *Pall Mall Magazine*; see Nye 1894. Thanks to Sally Shuttleworth for this reference.

54. In what follows I shall treat Labouchere the author as identical with Labouchere the person. Certainly Garner and observers of the Garner-*Truth* controversy treated them this way. But I have no direct evidence for this identity. On the contrary: from the 1880s, Labouchere apparently wrote little of the copy attributed to him, according to insiders at the time and present-day scholars of journalism. Rather, the illusion of Laboucherean authorship was cultivated by the staff of *Truth*, who worked to ensure the voice throughout sounded like Labouchere's. See Hirshfield 1984, 427.

55. [Labouchere] 1894a, 480, 481.

56. [Labouchere] 1894c.

57. On Labouchere's life and times, see Pearson 1936. For Labouchere as Guy Fawkes, see *Punch*, 23 June 1894, 300. For Labouchere the "feisty old Etonian maverick" as precursor of more recent (and successful) reformers of the House of Lords, see D. McKie, "House of Cards," *Guardian Weekend*, 27 March 1999, 10–17, esp. 13–14, quotation on 13.

58. H. Labouchere quoted in Weber 1993, 39. On Arnold's assessment of the "new journalism," see Wiener 1988.

59. Bourne 1887, 310. On the reader-friendly format of *Truth*, see Weber 1993, 37.

60. For a superb discussion of *Truth*'s combination of society journalism and what we now call investigative journalism, see Weber 1993, quotation from Joseph Hatton on 41. The literary historian Barbara Onslow has suggested to me that *Truth* in its day was much like the British periodical *Private Eye* today: hard-hitting, gossipy, amusing, and the beneficiary of publicity generated through libel trials.

61. On *Truth*'s techniques of community building, see Weber 1993, 38, 40.

62. [Labouchere] 1894a, 480. He continued (480–81): "No sooner did Professor Garner land in Africa, than a reader of *Truth* marked him for his own, and proceeded to send particulars of his proceedings."

63. [Labouchere] 1894a, 481.

64. "Thinks Well of Gorillas," 1; "Studying the Gorilla Language," *New-York Daily Tribune*, 26 March 1894, 7. See also "Monkey Talk," *Brooklyn Daily Eagle*, 26 March 1894, 4; and "Three Months with Gorillas," *New York Herald*, 26 March 1894, 3. The *New York Times* reporter wrote that Garner first "relieved his mind" about "some performances with which he has been credited" before telling his story as such. According to the reporter, Garner was especially exercised about "a long letter said to have been written by the Professor to his brother in Australia," and printed in full, first in a New South Wales paper, and then in a New York paper. "Prof. Garner says that he has no brother in Australia, never wrote such a letter, and that it is a deliberately planned fabrication."

65. Garner 1894.

66. [Labouchere] 1894d, 887, 888.

67. "Le cas du Docteur Garner," *Figaro*, 5 May 1894, 1; reported in the *Nation* 58 (1894): 450. Dybowski confirmed that Garner never received a phonograph while in Fernan Vaz.

68. Remarks made to American newspapers, quoted in [Labouchere] 1894e.

69. [Labouchere] 1894e.

70. [Labouchere] 1894b, 1513. Labouchere reported that a photograph of the IOU had been included with H. T.'s letter.

71. [Labouchere] 1894e. In the *Figaro* report of Dybowski's Paris lecture, it was not Buléon but another missionary, Bichet, who was identified as Dybowski's informant. "Le cas du Docteur Garner," 1. Dybowski published his account of his explorations in Dybowski 1893.

72. [Labouchere] 1894b, 1513.

73. [Labouchere] 1894b, 1514.

74. "Written on return from my second voyage from Africa while remaining in London. Published by Osgood, McIlvaine & Co. . . . It was not published in America, by reason of some failure on the part of their associate publisher in New York." As with Garner's first scientific book and publisher, so with his second: Garner got his advance (about $250), then the publisher went bust. Garner, "Index to

Manuscripts with Notes." The book included a large number of photographs and illustrations, including a frontispiece of Garner standing with helmet and gun. As for the second voyage, it was reported that Garner departed for West Africa from Europe in August 1895, intending to stay only a few months—just long enough to procure several young apes and bring them back to the United States. *New York Times*, 25 August 1895, 9; *New-York Daily Tribune*, 25 August 1895, 2.

75. Garner 1896, 12.

76. On the intensive effort to map the evolutionary, "consanguineous" links between species in the later nineteenth century, see Bowler 1996. On that effort in relation to humankind specifically, see Bowler 1986.

77. Garner 1896, 48–49.

78. Garner 1896, chap. 6. The nine words were for "food," "calling," "affection," "good/thanks," "alarm," "cold/discomfort," "drink," "illness," and "dead/death." Garner noted that the alarm call increased in loudness and frequency as danger grew nearer, thus showing the chimpanzees understood the meaning and value of the sound. He added that "the native tribes often use the same word in the same manner for the same purpose" (72). Cf. "'I have learned,' said he [Garner], 'probably a dozen at most of the words of the chimpanzee tongue. . . . I think there must be forty or fifty words available for their utterances, that is by combining the sounds . . . to mean different things.'" "Studying the Gorilla Language," 7. In *Gorillas and Chimpanzees*, Garner also wrote (73–74) of some gestural chimpanzee signs, including a headshake for no and an arm extension for yes.

79. Garner 1896, 93.

80. Garner 1896, 95.

81. Cf. "'I, of course, made a close study of all the sounds they [the gorillas] uttered, but it is utterly futile to attempt to reproduce them with letters. They cannot be formulated into words. I made memoranda at the time, and was afterward puzzled when attempting to reproduce the sounds by the letters I had made.'" "Thinks Well of Gorillas," 1.

82. Garner 1896, 67–71.

83. On Garner's arrival in New York in 1894, the *Brooklyn Daily Eagle* (26 March 1894, 4) reported that he "has learned twelve words of the chimpanzee language and six words of the gorilla language. . . . He discovered that the gorillas speak a different language from that used by the chimpanzees and are an inferior race." I do not know when the chimpanzee overtook the gorilla within scientific consensus as humankind's nearest ally among the apes. In *The Speech of Monkeys*, 1892f, 133–34, Garner described his view of the chimpanzee's nearer position as unorthodox (e.g., "I am aware that this view is not in strict accord with that of Professor Huxley, who assigns the gorilla the highest place next to man in the order of Nature, and the chimpanzee next below him").

84. Garner 1896, 237. Put another way (see the note above): Garner expected the gorilla's speech to be lower than the chimpanzee's, but not quite as low as it seemed to be.

85. Of the kulu kambas, often considered to be gorilla-chimp hybrids, the primatologist A. F. Dixson (1981, 6) writes that these "very odd creatures . . . continued to stalk the pages of zoological texts until quite recently." Reports still surface; see, e.g., Young 2004. For further comments on the kulu kamba, see Leach 1996, 32, 158–59.

86. Garner 1896, 185. Still, there were occasions for optimism about future translation success. In one village in the Eshira country, Garner encountered a sort of ape-savant (144–53). At the bidding of the chief's son (in the chimp and native tongues—some of the words were nearly identical), this ape filled a gourd with water, gathered firewood from the forest, and brought in various villagers. "I do not know to what extent they played upon my credulity," wrote Garner, "but, so far as I could discern, their statements concerning the animal were verified" (152). (An ominous tag: "I proposed to buy the ape, but the price asked was nearly twice that of a slave, and I could have bought any child in the town at a smaller cost" [153].)

87. On Aaron and Elishiba, see Garner 1896, chaps. 10 and 11. On Othello, see Garner 1896, chap. 16.

88. Garner 1896, 95.

89. Garner 1896, 96–98. Garner also reported conducting psychological experiments with Moses, of the sort he had already used with his simian subjects in the United States and the Regent's Park orangutan. Cf. Boakes 1984, 177: "There was no immediate sequel to Romanes' work and the nineteenth century ended before any further attempts were made to test the intelligence of apes."

90. Garner 1896, 99–102. "Moses will live in history," wrote Garner. "He deserves to do so; because he was the first of his race that ever spoke a word of human speech; because he was the first that ever conversed in his own language with a human being; and because he was the first that ever signed his name to any document; and Fame will not deny him a niche in her temple among the heroes who led the races of the world" (101).

91. Garner 1896, chap. 2, quotation on 15–16.

92. Garner 1896, 15, 28.

93. Garner 1896, 20. Writing again of watching the gorilla try to attack the mission dog, Garner emphasized how important the cage was for objective observation: "From the time this ape came into view until she departed was about four minutes, and during that time I was afforded an opportunity of studying her in a way that no one else has been able to do. I watched every movement of her body, face and eyes. I could sit with perfect composure and study her without fear of attack. . . . I believe that no other white man has ever seen an equal number of these animals [twenty-two] in a wild state, and it is certain that no other has ever seen them under as favorable conditions for study" (245–46). On the need to study the habits of animals in a free state, see Garner 1896, 213.

94. Garner 1896, 20, 20, 82, 102, respectively.

95. Garner 1896, 81 and 97, 117, 25 and 245, 128, respectively.

96. [Labouchere] 1896, 1232, complaining of reviews in the *Globe*, the *Daily News*, the *Liverpool Post*, and the *Spectator*. I have not located the first three of these. For the *Spectator* review, see "Animal Language," 24 October 1896, 550–51, discussed in Ferguson 2006, 107. The only other review I have found is in the *Westminster Review* (146 [1896]: 691–92). "Although no very striking discoveries were made," wrote the reviewer, "several obscure points in the history of these apes were cleared up. For instance, the author found that both chimpanzees and gorillas sleep on the ground and not in trees as had been previously supposed." Cf. Garner 1896, 50, 228–29.

97. Consider, for example: (1) Garner's triumphant return to the Edison laboratory, now with designs on Edison's new protomovie camera, the kinetograph. See Dickson and Dickson 1895, 30–31. (2) Reports of Garner working for a rather shadowy and well-financed business-cum-scientific-society, the Africa Fund of Chicago. See the *New-York Daily Tribune*, 27 April 1896, 6; and [Labouchere] 1895. (3) Reports of Garner organizing a big-game hunt in Africa. See the *New York Times*, 10 April 1896, 9.

98. Evans 1897, 331. Evans (330–32) is my predecessor in the attempt to sort out what happened in Fernan Vaz in 1892–93. His disappointment at Garner's "utter failure to accomplish what he set out to do" (331) is palpable. Another disappointed contemporary was the Regent's Park chronicler C. J. Cornish. See Cornish 1895, 240, 247.

99. "Myths about Monkey Talk Exploded," *New York Times*, 22 November 1896, 17.

100. R. L. Garner, *The Record of Idle Moments*, Garner Papers, box 2, folder "Writings: 'The Record of Idle Moments." For the remainder of this chapter I shall refer to this notebook as the *Record*. At several points I have altered the spelling and punctuation in quotations from the *Record* for ease of reading. For the dating of the notebook, cf. Garner's remark that "it is now 16 years since these things happened." "Notes and Comments," 7, in the *Record*.

101. R. L. Garner, "What the Monkeys Say," typescript, 1, 8, 10, in the *Record*. Of the unnamed newspaper, Spud said: "It poses as the friend and champion of the working classes. So do the rum-sellers and both for like reasons. How many public houses would be open as loafing places for the working men if they did not buy swill to degrade themselves and blight their homes? And how long would this self-appointed guardian of the poor man carry on his comic opera if the rabble did not applaud him and buy his weakly [*sic*] trash at ta-penny a throw?" (4).

102. R. L. Garner, "Preface to Booklet," typescript, quotation on 1, in the *Record*.

103. The first mention I have discovered is in one of Labouchere's 1894 *Truth* articles (1894e), in which Labouchere quotes from a report in the *Morning Journal* of New York, as follows: "'I want to be emphatic right at the start,' said Garner, 'by saying that this fellow McLaughlin is a liar.'"

104. Garner, "Notes and Comments," 1–2, quotation on 2.

105. Garner, "Notes and Comments," 2.

106. Garner, "Notes and Comments," 2.
107. Garner, "Notes and Comments," 2–3.
108. Garner, "Notes and Comments," 3. As we have seen, Garner mentioned Buléon as a gorilla-lore informant in the 1893 *McClure's* dispatch, and in 1894, Labouchere (1894e) quoted Garner's remarks to an American newspaper reporter about Buléon's desertion of Garner in the jungle. On Buléon's life and career, see the obituary in the *Bulletin Général de la Congrégation du Saint-Esprit* 20 (1899–1901): 527–31. Wolfgang is mentioned (as is Garner, but not McLaughlin) in the mission's diary of activities over the relevant period; see the *Journal de Communité Ste Anne de Fernan-Vaz*, 1893, 95–101, held at the Archives Générales Spiritaines, Paris, no. 4J2.3.
109. Garner, "Notes and Comments," 1, 4, quotation on 4.
110. Garner, "Notes and Comments," 4–5.
111. Garner, "Notes and Comments," 4–5.
112. Garner, "Notes and Comments," 5–6.
113. Garner, "Notes and Comments," 6.
114. Garner, "Notes and Comments," 6–8, quotation on 7–8.
115. Garner, "Notes and Comments," 8–9, emphasis added; ms 33, 35.
116. Garner 1891a, 34.
117. On the "pervasive, conservative Protestantism" of the South, see Mathews 1979, quotation on 1046. On anti-Catholicism in Garner's America, see Higham 1955, 77–87. On the anti-Catholicism in Draper's version of science-religion warfare, see Moore 1979, 24–26.
118. A letter from Garner to his son, Harry, in May 1909 tells of Garner being "homeless, penniless, and friendless." He admitted to a general darkening of his mood, and to feeling greatly disillusioned with his fellow men. R. L. Garner to H. Garner, 18 May 1909, Garner Papers, box 1, folder "Outgoing Letters." At this point Garner had been in Fernan Vaz continuously for about five years.
119. Although these manuscripts are undated, the circumstances they describe closely resemble those recorded in Garner's correspondence from around 1909, while the sentiments they express closely resemble those recorded in this correspondence and in the Fernan Vaz testament, independently dated to around 1909. Cf. "My anathema is hurled at those cormorants that perch upon the boughs of other people's trees and fatten on the stolen fruit before it is mature. That some candid minds accord to me the credit due my efforts and success I am convinced." R. L. Garner to H. Garner, 18 May 1909, Garner Papers, box 1, folder "Outgoing Letters."
120. R. L. Garner, "What Next?" unpublished MS, 7, Garner Papers, box 6, folder "'What Next?'–'Women of Other Lands.'"
121. R. L. Garner, "A State of Mind and a State of Facts," unpublished MS, 1, Garner Papers, box 6, folder "'A State of Mind and a State of Facts'–'Superstitions of the West African Tribes.'"

122. "Le langage des singes," *Figaro*, 4 March 1895, 1. Reports on Buléon's lecture on Garner at the museum on 26 February also appeared in the *Bulletin du Muséum d'Histoire Naturelle*, no. 2 (1895): 27–28; and *La Nature*, 6 April 1895, 294. I am grateful to Marion Thomas for locating these and other documents relating to Buléon.

123. Garner, "Notes and Comments," 8–9.

124. Le Roy 1895, 173–74. Le Roy (175) identifies the photographer of the Garner-cage-boy photo (see fig. 4.4) as Buléon. Le Roy wrote that Garner in person seemed anything but a revolutionary. "From the point of view of religion—since he brought it up—M. Garner said that he was born a Protestant, but that reflection inclined him now towards Catholicism, though later on he declared himself a heathen. In any event, during his stay at Fernan Vaz, he never missed any of the mission's religious services" (174). Le Roy also reported Garner's being dismissive of the Darwinian theory of descent (174).

125. One last account of Garner's time among the missionaries appeared, in a book about the mission of St. Anne published in 1896, *Under the African Sky (Sous le ciel d'Afrique)*. It consists mainly of excerpts from Buléon's diary and letters and includes a section on "the joys and sadnesses of the missionary," the last item of which is "the language of the monkeys: M. Garner." The author of these six pages is not Buléon, but another missionary, "S. F.," whose account is taken more or less directly from the *Figaro* news report and Le Roy's *Cosmos* articles. See [Buléon] 1896, 149–54 (note that the pages were printed out of order).

126. Cf. Dixson 1981, 6.

127. Garner 1896, 248.

128. Cf. "About twenty [gorillas], I should judge, came around the cage. They would stalk close up to the bars, gaze steadfastly at me for a minute or two, and gravely walk off. I don't really believe that I made a very favorable impression upon any of them, at least, they did not act as if I had. They did not mention any of their objections, however." "Thinks Well of Gorillas," 1.

129. A 1910 American newspaper article on Garner's then-recent work devoted a paragraph to how, in the 1890s, "Labouchere, editor of the London *Truth*, sought to expose Garner as a 'faker.'" "Prof. Garner's Talking Monkeys," Garner Papers, box 1, folder "Newspaper Clippings."

130. For the path quotation, see Garner 1893, 368.

131. Cf. "The steel cage which Professor Garner took with him . . . was not a complete protection. . . . One safeguard was the hollow arrows filled with prussic acid and impelled through a blow-gun. Another precaution against savage gorillas was a reservoir of concentrated ammonia whose fumes would kill any animal. There was no need, however, of these implements of defence, or for the other arms Professor Garner carried." "Studying the Gorilla Language," 7.

132. Garner, "Notes and Comments," 8.

133. Garner, "Notes and Comments," 10.

Chapter Five

1. Garner 1902, 369.
2. *Liverpool Courier*, 9 June 1902, Garner Papers, box 1, folder "Newspaper Clippings."
3. "Forthcoming Lectures at Cambridge on Primitive Custom and Belief," *Journal of the African Society* 1 (3) (1902): 368.
4. *Cambridge University Reporter*, 14 January 1903, 352.
5. Missionaries are absent from the account of anthropology's curricular progress at Cambridge in this period in Stocking 1995, 116, though Gunson (1994, 303) notes that after the expedition Haddon, from a Congregational/Baptist family, "took an active role in promoting anthropological training for missionaries," as did Haddon's fellow expeditionist William H. R. Rivers. On the importance of missionaries to the success of the expedition itself, see Herle and Rouse 1998, 13–14.
6. The historiography on anthropology in this period is large and rich. For a concise introduction to the anthropological long run, see Stocking 1992e. A classic collection of essays covering the same broad area is Stocking 1968d. A good if opinionated primary-sources anthology is Montagu 1974. For an overview of national anthropological traditions in Britain, Germany, France, and the United States, see Barth et al. 2005. For nation-specific studies of anthropology in the second half of the nineteenth century and its legacies, see, for Britain, Stocking 1987, 1995; also Kuklick 1991; for Germany, Zimmerman 2001; for France, Hecht 2003; and for the United States, Hinsley 1981 and Darnell 1998. For excellent articles on all aspects of the history of physical anthropology, see Spencer 1997.
7. On American anthropology in the years of Hrdlička and Boas—the main terrain of this chapter—seen in its wider scientific, cultural, and political contexts, two books are outstanding: Cravens 1978 and Degler 1991.
8. Haddon 1898b, xv–xx, 491–94, quotations on xv and xvii. See also Brinton 1895 and, for discussion, Darnell 1988, 37–39.
9. Haddon 1890, 638. On Haddon's life and career up to ca. 1890, see esp. Stocking 1995, 98–104. Useful biographical articles include Blackman 2004 and Fleure 2004.
10. On the Dublin anthropometric laboratory, see Cunningham and Haddon 1892. On the Cambridge lecturing, see Stocking 1995, 107.
11. On Geddes and Haddon, who first met at the Naples marine biological laboratory in 1879, at the start of their careers, see Fleure 2004 and Stocking 1995, 107. On Geddes's scientific development in this period, see Radick and Gooday 2004 and Renwick 2005.
12. On Haddon's role in the Ethnographic Survey, see Urry 1984 and Schaffer 1994, 5–6.
13. On the Cambridge Torres Straits expedition, see Kuklick 1994 and the papers collected in Herle and Rouse 1998. Also useful are the discussions in Schaffer 1994, esp. 33–40; and Stocking 1995, 107–15. Later Seligmann dropped the final *n* and

the Torres Straits the final *s* from their names; see Herle and Rouse 1998, 1 nn. 1, 2.

14. Haddon 1898a. On Seligmann's anthropometric activity, see Urry 1998, 207.

15. For a brief account of what was done when and where on the expedition, see Haddon 1899.

16. Haddon 1901–35. For a chart summarizing the contents of these volumes, see Herle and Rouse 1998, 4.

17. Haddon 1900; for discussion, see Urry 1998, 211.

18. Quotation from Haddon 1898b, xxiv. On the extensive psychological tests with natives as, in line with Haddon's claim, unprecedented, see Stocking 1995, 109. On the expedition's role in the history of British anthropological fieldwork from Tylor onward, see Stocking 1992c. Stocking notes (27) that it was Haddon who introduced the word "field-work" into anthropology from natural history. On the history of the field sciences generally, see Kuklick and Kohler 1996.

19. The child, wrote Haddon in his 1898 *Study of Man* (xxviii), "repeats in its growth the savage stage from which civilized man has so recently emerged. . . . There is not only parallelism to some extent in physical features between children and certain savages, but there is in children a persistence of savage psychological habit, and in the singing games of children a persistence of savage and barbaric practice." On Haddon's recapitulationism as a legacy of his Cambridge embryology teacher F. M. Balfour, see Kuklick 1998, 168–70; and for background MacLeod 1994. On Rivers's experiments as "informed by the same principles that guided Haddon's research," see Kuklick 1998, 170–72. For more on Rivers's recapitulationism, see note 23 below.

20. Haddon 1898b, xxv–xxvi.

21. Rivers 1901, 96. For discussion, see Kuklick 1994, 351–55; and Schaffer 1994, 33–40.

22. Although there was more to Haddon's evolutionary analysis of art, including ideas about variation, selection, and geographical distribution (see Alvarez Roldán 1992 and Herle 1988, 84–86), the notion of forms changing according to a constantly changing balance of regressive homogenizing and progressive differentiating forces was distinctively Spencerian; see Ruse 2004. Kuklick and Richards both emphasize how Spencerian were the evolutionary views of Haddon and Rivers; see Kuklick 1998, 167, 171–72; and Richards 1998, 137–38.

23. Rivers 1901, 44–45. On Rivers's time with Jackson and its influence, see Kuklick 1998, 167, 171–72; 1994, 359–60. On the Riversian (Jacksonian, Spencerian) view, the developing nervous system recapitulated its evolutionary history, with more advanced and finely differentiated functional capacities developing progressively out of primitive and coarsely differentiated ones. When trauma, physical (nerve damage) or psychic (shell shock), disrupted functioning at a higher level, a previously submerged lower level would be revealed. The primitivity of savages was due to their conditions of life not permitting higher capacities ever to develop.

24. On the "weakening of the holistic vision of anthropology after 1900," see Urry 1984, 97–101, quotation on 99. For Haddon's post-1900 publications, including *The Races of Man and Their Distribution* (1909, 1924), see Herle and Rouse 1998, 238. On the leaders of the functionalist social anthropology of the next generation as students of Haddon, Rivers, and Seligmann, see Kuklick 1994, 355–59.

25. For many psychologists, the Torres Straits expedition demonstrated how impossible it was to get good experimental-psychological data in the field. Virtually no one followed it up. See Richards 1998, esp. 145.

26. See figure 5.1. Another photo of a phonograph in use on the expedition can be seen in Herle and Rouse 1998, 228.

27. On the phonograph in American anthropology of the late nineteenth and early twentieth centuries, see Brady 1999, esp. chap. 3. See 122 for the figure cited for the number of cylinder recordings made.

28. On the life of cylinder phonographs after the arrival of the disk, see Brady 1999, 24–26. On their use in what became ethnomusicology, see, e.g., Shelemay 1991. I have learned much on the latter topic from unpublished essays kindly provided by Gerald Fabris and Alexander Rehding.

29. Koch 1955. On his first recording of a songbird in 1889, at age eight, using a wax-cylinder phonograph brought back from the fair in Leipzig by his father, see 15–16. This recording can be heard in the permanent exhibition at the British Library.

30. Boutan's research is discussed in chapter 6. On Ditmars, see Sparkes 1920. Ditmars was described as "a strong advocate of the theories of Professor Garner."

31. Huxley and Koch 1938, 5–6, quotation on 5. See also Koch 1955, 52, 67. Huxley was quite taken with these activities: "Students of animal behaviour will be able to make experiments on the differences between related species, on the share of what is innate and what is learnt in determining reaction (or absence of reaction) to specific sounds, and so on. In any case, it is clear that an interesting field, with both practical and theoretical sides, is here opened up by the advance of technique" (6). On Huxley, see chapter 7.

32. The occupation stated on Garner's passport was "scientific research." Garner Papers, box 7, folder "Legal Records."

33. Garner's efforts on behalf of the New York Zoological Park culminated in the arrival in 1914 of Dinah the gorilla—one of the first live gorillas to be successfully exhibited in the United States. See Bridges 1974, 346–50. For Garner's relationship with the park and its irascible director, William T. Hornaday, see the voluminous correspondence preserved in the Wildlife Conservation Society Archives, Bronx Zoo.

34. "Everyone knows the scandal of the Congo Free State as ruled by the Belgians, but few know that the French Congo is but a repetition of all that is hideous and cruel there. Prof. Garner does not lay all the blame on the officials, the trouble lies

with the 'system.' But an audience must be aroused when he lays the onus of the crime on the United States." Brochure for "The R. L. Garner Lectures," ca. 1912, Garner Papers, box 7, folder "Biographer's Papers."

35. The African Society was founded in 1901 in honor of the recently deceased Mary Kingsley.

36. Garner 1902, 369–71.

37. Garner 1902, 371–75, quotations on 372.

38. Garner 1902, 373.

39. Garner 1902, 373.

40. Although not discussed in Brady's history of the enthnographic phonograph, Garner appears in a cartoon reproduced there (1999, 17).

41. Garner wrote that he posted the article from Fernan Vaz to his agent in the United States on 18 March 1907. Garner, "Index to Manuscripts with Notes."

42. R. L. Garner, "The Phonograph among the Savages," unpublished MS, 2, Garner Papers, box 5, folder "Writings: 'The Phonograph among the Savages'–'The Psychoscope.'"

43. Garner, "Phonograph among the Savages," 4, 6, 10. Erika Brady has suggested that such stories of ludicrous first encounters with the phonograph were common, indeed that the stories have a mythic structure; see Brady 1999, 28–31.

44. For biographical details on Harrington, see Walsh 1999 and H. Lawton's foreword to Laird 1975, xv–xxii.

45. J. P. Harrington, "He Spoke: Biography and Scientific Work of Richard Lynch Garner," 1941, unpublished MS, Garner Papers, box 7, folder "Biographer's Papers." Hereafter I shall refer to this manuscript as "He Spoke."

46. Harrington 1912, 196–97. Harrington also recommended the use of kymographs, mouth-measurers, and artificial palates. On Harrington's "Notes" and the "problems of realism and artifice" then being confronted in a range of sciences moving from the laboratory to the field in this era, see Schaffer 1994, 44–45. On the 1912 revision of *Notes and Queries* more generally, see Stocking 1992c, 36–40.

47. Harrington, "He Spoke," 3.

48. Harrington, "He Spoke," 29–30.

49. Lecture Notes on Linguistics [1910–14], frames 0518–0920, vol. 8, reel 22, Papers of John Peabody Harrington (microfilm version), National Anthropological Archives, National Museum of Natural History, Smithsonian Institution; see esp. frames 0522–0549.

50. Lecture Notes on Linguistics, frame 0522. Harrington cited as his source Moore's *Universal Kinship*, 157.

51. Laird 1975, 1–2, emphases in original.

52. Laird 1975, 4.

53. Laird 1975, 6.

54. Laird 1975, 11. According to Laird, "while he [Harrington] thought that anything primitive was to be admired and emulated, he did not approve of intermarriage

with Indians or with any non-Nordic race. If the races intermarried, he said, everybody would look alike, everybody would have a sort of muddy-brown complection [*sic*] and a mediocre intellect" (22).

55. Carobeth Laird's 1975 memoir of her time with Harrington carries one of his poems as an epigraph: "Give not, give not the yawning graves their plunder, / Save, save the lore, for future ages' joy; / The stories full of beauty and of wonder / The songs more pristine than the songs of Troy, / The ancient speech forever to be vanished— / Lore that tomorrow to the grave goes down! / All other thought from our horizon banished, / Let any sacrifice our labor crown."

56. J. P. Harrington to H. E. Garner, 20 March 1937, Harrington Papers, vol. 9, reel 013, frames 0043–0045.

57. Harrington, "He Spoke," 29.

58. The Piltdown-man fraud of the 1910s well illustrates the shift in intellectual climate documented here—a shift away from living apes and toward the fossil record as the most important and promising storehouse of evidence for human origins. For the Piltdown discovery, see Reader 1990, chap. 4; and, more extensively, Spencer 1990. On the history of paleoanthropology generally, see Reader 1990, Bowler 1986, and, for background, Van Riper 1993. The accumulated fossil evidence is exhibited handsomely in Johansen and Edgar 1996.

59. Hale 1886, 304–6, 309. On the nineteenth-century Neanderthal controversies, see, e.g., Reader 1990, chap. 1; and Bowler 1986, 33–34.

60. Hale 1886, 306–7. Cf. De Mortillet 1883, 244–45.

61. Hale 1886, 306–8. Cf. De Mortillet 1883, 249–51. For a modern description of genial tubercles, see Aiello and Dean 1990, 35.

62. Hale 1886, 308, referring to Baume 1883.

63. Hale 1886, 309.

64. Hale 1886, 309–10.

65. On the identification and description of the Cro-Magnon race, see Bowler 1986, 32.

66. Hale 1886, 311–12, quotation on 311.

67. Hale 1886, 312–14.

68. Hale 1886, 317–18, quotations on 318.

69. Dubois 1888, 163, quoted in English in Theunissen 1989, 33. In the same year, the German language theorist Heymann Steinthal concluded, also on the basis of the Naulette and Schipka jawbones, that their possessors could indeed have spoken. See Steinthal 1888, 264–70, cited in Brinton 1888, 212. Steinthal's volume offers a still-useful survey of the views of German-language writers on the language-origin question, from the Romantic trio of Herder, Hamann, and (Wilhelm von) Humboldt through to Müller's sources Schelling, Heyse, and Noire and on to Wundt, with discussions as well of Darwin, Hale, Brinton, and others.

70. On Dubois, and the discovery of the Java man generally, see Theunissen 1989 and Shipman 2001. A useful brief account can be found in Reader 1990, chap. 3.

71. The article is summarized in Theunissen 1989, 32–35.
72. Theunissen 1989, 39, 53–61.
73. On how *Pithecanthropus erectus* got its name, see Theunissen 1989, 59–60. Cf. "We may therefore distinguish a special (24th) stage in the series of our human ancestors, namely, Speechless Man (*Alalus*) or Ape-man (*Pithecanthropus*), whose body was indeed formed exactly like that of man in all essential characteristics, but who did not as yet possess articulate speech." Haeckel 1876, 2:406.
74. Theunissen (1989, 114–15) argues that the Java-man discovery triggered a rediscovery of the known Neanderthal specimens as objects of evolutionary, as opposed to merely racial, significance. On the debate over the Java man between 1894 and 1900, see Theunissen 1989, chap. 4.
75. Theunissen 1989, 14.
76. Theunissen 1989, 13.
77. Cf. Theunissen 1989, 31. Two famous evolutionist illustrations make the point. In 1863, Huxley showed a row of increasingly humanlike ape skeletons behind a human skeleton. In 1898, Haeckel showed *Pithecanthropus* and Neanderthal skulls nested between an innermost chimpanzee skull and an outermost Cro-Magnon skull. Both illustrations are in Bowler 1986, 65 and 69, respectively.
78. On apes in post-Java-man theorizing about human evolution, see Bowler 1986, 68–74.
79. Hrdlička 1914, 494.
80. On the Heidelberg jawbone and speech, see Reader 1990, 56; on the La Quina skull and speech, see Anthony 1913.
81. Hrdlička 1914, 494.
82. On Hrdlička's life and career, see Spencer 1999, Montagu 1944, and Stewart 1940 (which includes an extensive bibliography). See also A. Hrdlička, "Main Biographical Data," Papers of Aleš Hrdlička, box 145, folder "Aleš Hrdlička–Main Biographical Data" [no date], National Anthropological Archives, National Museum of Natural History, Smithsonian Institution. On Hrdlička's role in the development of academic physical anthropology in the United States, see Spencer 1981. Readers unfamiliar with Czech pronunciation may wish to know that his name is pronounced AH-lesh HIRD-lich-kah, with the *r* in *hird* rolled.
83. Hrdlička, "Main Biographical Data," 3. On Lombroso's anthropology, see Gould 1997, 151–73. Anthropometry was a common interest among the medical men of the period; in *Heart of Darkness* (Conrad 1902, 10), the doctor who examines Marlow before his departure for Africa asks permission to measure his head. "Rather surprised, I said Yes, when he produced a thing like calipers and got the dimensions back and front and every way, taking notes carefully. . . . 'I always ask leave, in the interests of science, to measure the crania of those going out there,' he said."
84. On Manouvrier's role in the Java-man debates, see Theunissen 1989, 81–82, 99–101.

85. On Hrdlička and Manouvrier, see esp. Hecht 2003, 241–42.

86. On Hrdlička and the head in question, W. H. Holmes, see Hinsley 1981, 281–82; and Stewart 1940, 11–12.

87. Hrdlička 1908, esp. 37–38, 42–43.

88. Hrdlička 1918, 21.

89. Hrdlička 1908, 39–41.

90. For a list of anthropologists who looked to Hrdlička for training between 1914 and 1920, see Spencer 1981, 359. Perhaps the most important was Harvard's Earnest Hooton, whose devotion to anthropometry eventually provoked his student Washburn's behaviorally oriented "new physical anthropology"; see chapter 8 in this book. It should be noted that Hooton, for all the bad press from Washburn, was a major supporter of primate behavioral studies within physical anthropology; see Giles 1997.

91. Hrdlička 1947. The first edition was published in 1920. Hrdlička dedicated the book to Manouvrier.

92. On the disciplinary significance of the journal and the research network that formed around it, see Spencer 1981, 358–59.

93. On the importance and popularity of Hrdlička's paleoanthropological surveys, see Johnson and Mann 1997, 1075–76. On a tour of European collections of human and protohuman fossils in 1912, Hrdlička tried and failed to persuade the reclusive Dubois to allow examination of the Java-man fossils. The story is entertainingly told in Shipman 2001, 342–46.

94. On Hrdlička's theories, see Spencer 1999, 377; and Bowler 1986, 106–7. For Hrdlička as "primarily responsible for documenting that human evolution had *not* occurred in the New World, thus concentrating the attention of Americans on the discoveries abroad," see Johnson and Mann 1997, 1075, emphasis in original.

95. Hrdlička 1908, 41.

96. Hrdlička 1940.

97. R. L. Garner to A. Hrdlička, 7 July 1918, typescript, 1, RU 208, box 29, folder 9, SIA.

98. *Chicago Herald*, 26 November 1916, clipping, RU 45, box 12, folder 12, SIA.

99. Garner to Hrdlička, 7 July 1918, typescript, 1.

100. Garner to Hrdlička, 7 July 1918, typescript, 3. Garner continued: "Certainly the champion of the Mosaic story of the creation of man has got a lot of stumps in his row; and the psychologist who attempts to define the line between the realms of *animal psychics*, as they call it, and human psychics has to be a skillful hair-splitter to make up his case" (emphasis in original).

101. A. Hrdlička to R. L. Garner, 2 October 1918, Hrdlička Papers, box 26, folder "Garner, R. L. Dr., 1918–20." Garner's letter appeared in part under the title "Some Observations on Diseases and Mental Characteristics of Apes," *American Journal of Physical Anthropology* 2 (January–March 1919): 75–77. I am grateful to Mandy Rees for this reference.

102. J. W. Jefferis, "A Hermit in the Jungle," ca. June 1919, incomplete typescript, Garner Papers, box 7, folder "Biographer's Papers."

103. Jefferis, "A Hermit in the Jungle." Further evidence that Garner in the 1910s saw extinct hominids as filling a large ape-human gap comes from magic-lantern slides preserved in the Smithsonian, and undoubtedly from that decade. One of the slides illustrates "comparative intelligence" in the primates, and shows a "hiatus" between humans (occupying degrees 68 to 100) and apes (32 to 44). Right in the middle is "Pithecanthropus," with "Neanderthals" at 68, just inside the human sector. Lantern slides, photo lot 81–58A, SIA. By contrast, in *The Speech of Monkeys* (1892f, 151–52), Garner had commented briefly on what he considered the inconclusive "fossil proofs" bearing on the connection between apes and humans.

104. Clipping ca. 6 June 1919, Garner Papers, box 1, folder "Newspaper Clippings."

105. "Denies Garner Heard Apes," *New York Times*, 8 June 1919, 10.

106. See, e.g., "The Professor's Honour Vindicated," *Bellman* 26 (1919): 708.

107. A. Hrdlička to R. L. Garner, 11 June 1919, Hrdlička Papers, box 26, folder "Garner, R. L. Dr., 1918–20."

108. R. L. Garner to A. Hrdlička, 3 January 1920, Hrdlička Papers, box 26, folder "Garner, R. L. Dr., 1918–20." Garner's "Adventures in Central Africa" (1920) appeared posthumously in the *Century*.

109. A. Hrdlička to R. L. Garner, 5 January 1920, Hrdlička Papers, box 26, folder "Garner, R. L. Dr., 1918–20."

110. R. L. Garner to A. Hrdlička, 13 January 1920, Hrdlička Papers, box 26, folder "Garner, R. L. Dr., 1918–1920."

111. "Famed 'Monkey Man' Dies," *Richmond Times-Dispatch*, 24 January 1920, Garner Papers, box 1, folder "Newspaper Clippings." Other obituaries appeared in (among other places) the *New York Sun*, *New York Herald*, *Baltimore Sun*, *Washington Post*, *New York Times*, London *Times*, *Chicago Daily Tribune*, *New York Tribune*, *Washington Herald*, *Baltimore Star*, *New York American*, and *Philadelphia Evening Bulletin*.

112. Garner obituary, *American Journal of Physical Anthropology* 13 (1920): 314. I am grateful to Mandy Rees for this reference.

113. A. Hrdlička, introduction to Garner 1930.

114. For a representative and widely read Hrdličkian account of the origin of language (including a brief mention of Garner), see Hooton 1947, 160–69, with Garner's work discussed on 160.

115. On this notion as characterizing anthropological thinking in the later nineteenth century, see Stocking 1968b.

116. On the basic character of Boasian anthropology, see Stocking 1974a and Darnell 1998, chap. 14. In her influential study of symbolism, *Philosophy in a New Key* (1942, 1957), the philosopher Suzanne Langer registered the bafflement of a post-Boasian inquirer into the origin of language: "If we find no prototype of speech in

the highest animals, and man will not say even the first word by instinct, then how did all his tribes acquire their various languages? Who began the art which now we all have to learn? And why is it not restricted to the cultured races, but possessed by every primitive family, from darkest Africa to the loneliness of the polar ice? Even the simplest of practical arts, such as clothing, cooking, or pottery, is found wanting in one human group or another, or at least found to be very rudimentary. Language is neither absent nor archaic in any of them." She went on, explicitly in relation to the Sapir encyclopedia article discussed later in this chapter: "The problem is so baffling that it is no longer considered respectable." Langer 1957, 108–9.

117. For discussion of Whitneyesque features of the Boasian view of language, see Andresen 1990, 211–20.

118. On the Hale-Boas relationship in general, and the similarities in anthropological outlook in particular, see Gruber 1967, 18–34, esp. 31–34.

119. Cole 1999, 99.

120. Boas 1888, quotation from Boas 1897, 40. The unsigned 1888 *Science* article, "The Origin and Development of Language," published in the 27 August issue, is, so far as I know, new to Boasian scholarship. It is not listed in an otherwise comprehensive bibliography of Boas's works (Andrews et al. 1943). My attribution is based on Hale's letters to Boas of 15 September and 8 October 1888, in which Hale expressed thanks, as he put it in the second letter, "for bringing my theory so forcibly before the readers of Science." Papers of Franz Boas, American Philosophical Society, Philadelphia, microfilm edition, reel 1, letters 121 and 132. Typically, Boas's endorsement was heavily qualified; he was especially dubious (1888, 145) about whether similarities across language stocks could be explained by supposing that they were formed by children at the same developmental stage. Even Hale, it seems, was too much the evolutionist for Boas.

121. Boas 1889, esp. 76 on alternating sounds as "in no way a sign of primitiveness of the speech in which they are said to occur," but merely "alternating apperceptions of one and the same sound." Cf. Hale 1884 and, e.g., Brinton 1888, 217–18. For a deft analysis of similarities and differences between Hale and Boas on alternating sounds, see Stocking 1992a, 66 n. 2. On the importance of Boas's paper, see Stocking 1968c, 159.

122. On the common saltationism, see this book's conclusion. I think Gruber (1967, 32 n. 91) overstated when he declared Hale a greater influence on Boas than his German antecedents. It is remarkable how "Boasian" Hale could sound, however—as in an 1889 letter to Boas complaining of the "mere speculation" that "primitive language was monosyllabic, and that, as civilization progressed, it becomes first agglutinative and finally inflected. We know now that this was mere *a priori* theorizing, and that the languages of primitive races are of all descriptions—monosyllabic, polysyllabic, agglutinative, and inflected, in every variety. In like manner, it will be found, I believe, that primitive tribes have almost every variety of social institutions." 13 July 1889, quoted in Gruber 1967, 33.

123. On Boas's German upbringing and education and their legacies, see Stocking

1968c; Cole 1999, chaps. 1–5; and the essays in Stocking 1996. On the Kantian element in Boas's culturalism, see Freeman 1999, chap. 1, in particular his discussion (24–25) of Boas's acknowledged debt to Theodor Waitz. On the alternating-sounds paper in relation to German psychophysics and the "laboratory culture of self experimentation" in which Boas trained, see Schaffer 1994, 41–42, quotation on 42; and Stocking 1968c, 157–59. On Virchow's criticisms of claims that apparently atavism-prone races were lower in organization than others, and that language and culture progressed as anatomy did, see Boas 1902, esp. 39–40. On the anti-Darwinism driving Virchow and the German anthropological scene he dominated, see Zimmerman 2001, chap. 3.

124. For a concise overview of Boas's career, see Voget 1970.

125. Boas 1899. On Haddon as, in "general anthropological outlook," the closest thing to a "British Boas," see Stocking 1995, 125.

126. On Boas's efforts to establish his science in the academy and their wider context, see Cravens 1978, 92–105. On his leaving the museum, see Schaffer 1994, 42–43.

127. Boas 1911b, chap. 1. Cf. Haddon 1898b, xxvi: "While the white man may, for example, be nearer the ape in the character of his hair than the Mongol or the Negro, the usual short body and long legs of the latter also remove him farther from the ape, to whom, in this respect, the other groups are more closely allied"– though the Boasianism of the point would have been strengthened if, as I suspect he meant to, Haddon had written "former," not "latter."

128. Boas 1911b, 97–99, quotations on 98, 99.

129. Boas 1911b, 95–97, quotations on 96.

130. On Boas's "critique of evolutionary linguistics" as manifest in the *Handbook of American Indian Languages*, also published in 1911, see Stocking 1992a, 74–80.

131. Boas 1911b, 146–47. Cf. Powell 1880, 55, 69–74. On the Spencerian, developmentalist strain in Powell's thought, see Hinsley 1981, chap. 5.

132. Boas 1911b, 142–48, quotation on 142. In the language of the Eskimo, wrote Boas, "we find one word expressing 'snow on the ground'; another one, 'falling snow'; a third one, 'drifting snow'; a fourth one, 'snow drift.' . . . [W]here it is necessary to distinguish a certain phenomenon in many aspects, which in the life of the people play each an entirely independent rôle, many independent words may develop, while in other cases modifications of a single term may suffice" (146–47). Largely due to the linguist Benjamin Lee Whorf's elaboration of Boas's example and claim, the semantic largesse of Eskimos talking about snow came to be a staple feature of arguments for extreme linguistic relativism in the twentieth century. For a history and critique of "the great Eskimo vocabulary hoax," see Pullum 1991. Pullum misses the antievolutionist, antiracist point of Boas's introducing the example.

133. Boas 1911b, 146–47.

134. Boas 1911b, 149–50.

135. Boas 1911b, 150–51. For Alfred Russel Wallace, of course, the existence of such latent linguistic and mental capacities in the primitive mind was evidence of humans' having a supernatural origin.

136. Boas 1911b, 194.

137. Quotation from Degler 1991, 188. On the role of the Boasians in promoting the "nurture" side in the nature/nurture controversies in the United States of the first half of the twentieth century, see, in addition to chaps. 3 and 4 of Degler's study, Cravens 1978, chap. 3; and Freeman 1983, chap. 3; as well as 1999, chaps. 14 and 15. Following Freeman, it is tempting to define the Boasian era as the one that made a popular bestseller and then a required university text of Boas student Margaret Mead's cultural-determinist *Coming of Age in Samoa* (1928). For a wonderfully readable journalistic profile of Boas as a scientific force for antieugenical, antiracist good, see Pierpont 2004.

138. In the late 1970s, the Bloomfieldian linguist Charles Hockett (1980, 99) recalled a saying from thirty years before: "if you know Latin, you're educated; if you know Greek, you're cultured; if you know Sanskrit, you're learned; if you know an American Indian language, you're scientific (unless you're an American Indian)." On Bloomfield and Boas see, e.g., Fries 1963, 217–19. On Boas's contribution to the descriptive study of language, see Stocking 1974b, 157–59. On the egalitarianism of Boasian-Bloomfieldian linguistics, see Newmeyer 1986, 39–47, who notes that Bloomfield founded the Linguistic Society of America in 1924 in part to "counteract resistance to the idea that the languages of highly civilized people were on a par with those of 'savages'" (41).

139. On the Boasian takeover, see Darnell 1998, chaps. 9 and 13; also Cravens 1978, 102–5. On the shift "from ethnology to cultural anthropology" in the interwar years, as Boas's students rose to prominence, see Stocking 1992d, 134–47.

140. On the *American Journal of Physical Anthropology*'s founding as a reaction to Boasian dominance of the *American Anthropologist*, see Spencer 1981, 357–59. Racism was not the issue. A major source of resentment was that in his anthropometric studies, Boas used a statistical idiom that Hrdlic̆ka did not have the training to follow and about which he was deeply skeptical. Cf. Hrdlic̆ka 1918, 15; and Montagu 1944.

141. Kroeber 1917, 169–76, quotations on 169, 173. See Desmond 1979, 71–79, for an interesting discussion of Kroeber's article as unwitting stimulus to the ape language experiments later in the century.

142. Kroeber 1948, 41. On the "lack of speech among apes" specifically, see Kroeber 1928, 329–30.

143. Sapir 1921, 8. On the role of Sapir's *Language* in "synthesizing the Boasian paradigm," see Darnell 1990, 96–106.

144. Sapir 1933, 158–59.

Chapter Six

1. Reeve 1909, 98.

2. For other contemporary reckonings of Darwinism's impact on psychology, see Angell 1909, Baldwin 1909, Hall 1909, and Thorndike 1909. On the 1909 celebration, see Richmond 2006.

3. Reeve 1909, 95.

4. Reeve 1909, 99–100, quotation on 99. For Boas and Thorndike, see Thorndike 1936, 268. There is another, more fundamental if roundabout debt to Boas. Thorndike decided on animal experiments as his thesis project at Harvard only after authorities refused permission for him to continue conducting mind-reading experiments with children, and that refusal most likely expressed a concern to avoid the scandal that erupted when Boas, while at Clark in the early 1890s, undressed some Boston schoolchildren in the course of anthropometric research. See O'Donnell 1985, 164, 166. For Thorndike's use of the (conventional) language of actions "stamped in and out," see Thorndike 1898, e.g., 36; and, for discussion, Jonçich 1968, 142–43.

5. Reeve 1909, 99 (Yerkes), 101–2 (Watson).

6. Reeve 1909, 100–101 (Kinnaman), 102 (Hobhouse and Haggerty). The researches of Yerkes, Watson, Hobhouse, and Haggerty will be discussed later in the chapter. On Kinnaman, see Boakes 1984, 149.

7. Reeve 1909, 95–98, 102–3.

8. Reeve 1909, 100. That both men had declared the higher apes to be at the same mental level as a year-old human child was, it seems, a sign for Reeve of just how robust the conclusions of the new experimental study of mental continuity could be. Reeve 1909, 96, 98 (quoting Garner) and 100 (quoting Thorndike). The source in Thorndike's work—the statement as to monkey-child parallelism is rather more qualified than Reeve indicated—is Thorndike 1901b, 239 in the 1911 reprint.

9. For a concise review of comparative psychology around 1900, emphasizing its diversity, see Dewsbury 2000a. An outstanding survey of the science of animal mind and behavior from Darwin to B. F. Skinner is Boakes 1984, though ideally it should be read alongside Burkhardt 2005, since the two books—the former a history of animal psychology, the latter a history of ethology—have almost completely nonoverlapping casts of behavior-minded characters. On animal-behavior studies in France—which, in 1900, was arguably poised to be the world leader in the field—see Thomas 2003. On animal-behavior studies in Britain after 1900, and the reasons they did not flourish as they did in the United States, see, respectively, Wilson 2001, 2002b.

10. In 1938, the leading comparative psychologist Edward Tolman remarked that "everything important in psychology . . . can be investigated in essence through the continued experimental and theoretical analysis of the determiners of rat behavior at a choice-point in the maze." Cited in Boakes 1984, 236. The classic analysis of the rise of rat learning as reflected in psychological journals is in Beach 1950, 116–19. For important qualifications, however, see Dewsbury 1989a, 103–4.

11. Beach 1950, quotations on 119, 120. In Beach's view (119–20), rat learning had become the experimental subject of choice for two related sets of reasons: first, most psychologists in America cared only about humans, and regarded learning

as the most important aspect of human behavior; second, rats were good lab animals whose associative learning resembled associative learning in humans. Beach
went on, however, to praise those psychologists and, especially, zoologists who
had managed to buck the trend; see 122–24. For more on the interdisciplinary
nature of animal behavior studies in the United States, see Dewsbury 1989a. For
a rich picture of the non-rat-learning sides of American animal psychology in the
first half of the twentieth century, see also Dewsbury 1992a, where he points out
(209) that rats were first run through a maze at the turn of century "because such
a task mimicked what rats learn in nature."

12. On the long-run history of psychological studies of apes, see Mitchell 1999.
13. Morgan 1893a, 239.
14. Morgan refers to the canon thus in Morgan 1894, 287, 377.
15. Morgan 1894, 54–59. On the greater parsimony of explanations of animal behavior in terms of reasonable inference, and on some of the cases in which "the simplest explanation is not the one accepted by science," see Morgan 1894, 54–55,
quotation on 55. Curiously, Richards and Boakes attribute to Morgan a rather
different evolutionary argument: that natural selection will never favor more complex mental processes if simpler mental processes will suffice to adapt creatures to
their conditions of life. See Boakes 1984, 40; and Richards 1987, 395. This is an
argument for the canon based on the *limits* of the power of evolution—an interesting argument, but not, it seems to me, Morgan's own.
16. Morgan 1894, 377. For a modern critique of the 1894 justification, see Sober
1998, 228–35.
17. Cf. the much simpler views in Morgan 1890–91, 374–76.
18. Morgan 1894, 241–304, quotations on 246, 283.
19. Morgan 1900, 198, 204–5, quotation on 204. See also Morgan's unpublished
paper "The Beginnings of Speech," 1–3, 9, 14, 17, Morgan Papers, DM 612.
20. Morgan 1900, 205. See also Morgan 1903b, 261, where, drawing on the recent
studies of Kinnaman and Hobhouse, he further threw doubt on "the so-called
speech of monkeys."
21. For Thorndike's life and work, the main sources are Thorndike 1936 and Jonçich 1968. For a concise overview, see Von Mayrhauser 2000. For discussion of
Thorndike's work on animal intelligence, see Boakes 1984, 68–73; Dewsbury
1998; Galef 1998; and Stam and Kalmanovitch 1998.
22. On Thorndike's turn to psychological experiments with animals, see Thorndike
1936, 264–65; and Jonçich 1968, 86–89. Although Thorndike recalled taking up
his chick experiments only after being forced to abandon mind-reading experiments with children (see note 4 above), Jonçich shows that this cannot be quite
how it happened. Most likely the shutting down of the child experiments led him
to upgrade the chick experiments he was already doing or had recently done into
his thesis project. On Wundtian experimental psychology and its formative influence on the new academic psychology, in the United States and elsewhere, see
Smith 1997, chap. 14. For Wundt on association in animals, see Wundt 1894,

340–66. On James's teaching of it, see Jonçich 1968, 86. For James on association in animals, see James 1890a, 2:348–60.

23. On Thorndike's debt to Morgan, and the latter's trip to Harvard, see Jonçich 1968, 131–33. On Morgan's staying with James on that trip, see Richards 1987, 398. On the puzzle boxes as experimentalizing Morgan's account of his gate-opening dog, see Boakes 1984, 69. For that account—introduced explicitly in support of the canon—see Morgan 1894, 287–90. Thorndike himself wrote that he got the puzzle-box idea from John Lubbock's placing obstacles in the way of ants; see Burnham 1972, 166, citing Thorndike 1899c, 412.

24. On this first period of animal research, and the migrations of Thorndike and his animals, see Thorndike 1936, 264–65; and Jonçich 1968, 88–148; also 262 in Jonçich for Yerkes's report that, when he arrived at Harvard not long after Thorndike's departure, "the scent of his experimental chicks still hung about the James cellar." On the boxes themselves, see Burnham 1972. Thorndike used the puzzle boxes with dogs and cats only. For the chicks, he used labyrinthlike pens; see Thorndike 1898, 61–64. Throughout the 1898 monograph, Thorndike used the phrase "time-curve," not "learning curve"; see, e.g., 38.

25. Thorndike 1898, 74. Cf. 122: "Probably the idea of the look of the loop or lever or thumb latch never entered the mind of any one of my cats during the months that they were with me, except when the front end of the association containing it was excited by putting the cat into the box." Thorndike explicitly excluded birds from his generalization about nonimitation in animals; see 76–77.

26. Thorndike quoted Morgan so extensively "because he has taken the most advanced stand along the line of the present investigation [and] because my differences from him are in the lines of his differences from other writers." Thorndike 1898, 126. Morgan's work played a similarly honored and criticized role in Thorndike's report on his earlier Harvard study of instinct in chicks; see Thorndike 1899a.

27. For the quotations praising Morgan, see Thorndike 1898, 80, 100, 126. For criticism of Romanes as representing "an attitude of investigation which this research will, I hope, render impossible for any scientist in the future," see 68.

28. It should be noted that, without proposing an interpretative rule as such, Wundt took a "canonical" line on animal actions, recommending a search for associative explanations in the first instance, in a book published in English in the same year as Morgan's *Introduction*; see Wundt 1894, 358–62. Nowhere in his publications on animals did Thorndike mention Wundt, however.

29. "The best explanation of even the most extraordinary performances of animals has been that they were the result of accident and association or imitation." Thorndike 1898, 67. On parsimony and the matter of representations in the animal mind, see 108–16, quotation on 109.

30. Thorndike 1898, 80, 126–27.

31. Thorndike 1898, 122, 127, quotation on 127.

32. Morgan 1898c. Morgan's paper (1898a) at the September 1898 Bristol meeting of

the BAAS was on Thorndike's research, as was an article he published in October in *Natural Science* (Morgan 1898b). Summaries also appeared in Morgan's 1900 book *Animal Behaviour* (147–54, 183–93) and the 1903 revised edition of his *Introduction to Comparative Psychology* (298–300).

33. Morgan 1898c, 250.

34. See, e.g., Morgan 1898b, 265; 1900, 147. In 1899, *Nature* (1 June, 112) reported the critique of the comparative psychologist Wesley Mills, who "points out that in investigating the psychology of animals, care must be taken to observe them under conditions as nearly approaching their normal surroundings as possible. He maintains that to place a cat in a box, as has been done, and then to expect it to act naturally, is about as reasonable as to enclose a living man in a coffin, lower him, against his will, into the earth, and attempt to deduce normal psychology from his conduct." See Mills 1899 and, in reply, Thorndike 1899c. For discussion, see Dewsbury 2000a, 1122.

35. Morgan 1898c, 250.

36. Thorndike 1936, 265; Jonçich 1968, 149–92, 261–62. At Woods Hole, where Thorndike spent time in the summers of 1899 and 1900, he adapted the puzzle-box method to study intelligence in fish housed in an aquarium there; see his brief report, Thorndike 1899b.

37. Thorndike 1898, 95–96.

38. Thorndike 1898, 96; 1901b, 176.

39. On Thorndike's monkey monograph, see Boakes 1984, 73.

40. Thorndike 1901b, 211–22. Morgan disseminated this finding in the second edition of his *Introduction to Comparative Psychology*. Morgan 1903b, 300–301.

41. Thorndike 1901b, 182–94, quotations on 189, curves on 185–86.

42. Thorndike 1901b, 190–92. The change from breaking out to breaking in represented a concession to his critics; see 191. See also the discussion on 206–8 of generous versus parsimonious interpretations of a different set of experiments, testing the monkeys' abilities to associate food with one signal and the absence of food with a different-but-not-very-different signal. Cf. Thorndike 1898, 73, where he wrote that "a sudden vertical descent in the time-curve" for an animal learning something even a little complex would warrant an inference of animal inference.

43. Thorndike 1901b, 236–40. He spelled out this theory of mental evolution at slightly greater length that same year in the *Popular Science Monthly*; see Thorndike 1901a.

44. "I would have gladly continued the work with the higher apes, but could not afford to buy or maintain them," Thorndike later recalled (1936, 265). The 1898 dissertation had picked out studies with apes and children as the most important for the future; see Thorndike 1898, 151, 154–55.

45. For a good overview of Thorndike's research in education and the links to his earlier animal-learning research, see Beatty 1998.

46. Sigmund Koch, quoted in Smith 1997, 660. For a similar judgment, see, e.g., Burnham 1972, 166.

47. In the 1911 book collecting his 1898–1901 animal-experiment papers, Thorndike included two new, protobehaviorist essays, one recommending the study of behavior over the study of conscious states, the other discussing the laws that connect situations with responses and thereby make behavior predictable. Thorndike stated two such laws: the law of effect (roughly, that rewarded actions become more probable, unrewarded ones less so) and the law of exercise (roughly, that connections get stronger as they recur). See Thorndike 1911, chaps. 1 and 6. He believed far too strongly in hereditary influences on character and intelligence, however, to go along with Watson's radical environmentalism. On Thorndike's hereditarianism and support for eugenics, see, e.g., Beatty 1998, 1147, 1151.

48. On Washburn and her textbook—the fourth edition was published in 1936—see Boakes 1984, 148–49; and Scarborough 2000.

49. Washburn 1908, 9–12, 24–26. Washburn wrote both about "the Labyrinth Method" and "the Puzzle-box Method" for studying learning, with the former guiding old impulses and the latter forming new ones. She nevertheless regarded each method as a variation on a theme—the dropping-off of useless movements—and credited the invention of both to Thorndike. Washburn 1908, 219–44, esp. 219, 232.

50. Washburn 1908, 3–4.

51. Washburn 1908, 25–26.

52. On adaptive, graduated "intelligence" between the animal psychology laboratory and the mass classroom at the turn of the century, see Stam and Kalmanovitch (1998, 1139), who note that Thorndike in 1901 introduced "learning" as a psychological category. ("Intelligence" too was relatively new—the older, humans-only term was "intellect"—as was "behavior" used in connection with animal action.)

53. Colloquially, the phrase "steep learning curve" has come, of course, to refer to arduous and gradual learning—the opposite of Thorndike's usage. On the educational role of animal psychology laboratories in the early twentieth century, see Stam and Kalmanovitch 1998, 1140; also Dewsbury 1992b, 5–6; and 2000a, 750.

54. In 1908, Watson published a review of a book about the Clever Hans investigations, recommending it "as an antidote henceforth and forever" to tendencies to describe animal behavior "in glowing anthropomorphic terms." Watson 1908, 331. The book came out in English translation in 1911; see Pfungst 1911.

55. The comparative psychologist Harvey Carr (1927, 88) recalled how the term "anthropomorphic" became "a choice epithet" in debate because it conveyed "the somewhat delicate suggestion that the individual in question was obtuse in intellect, a logical pervert, and an insipid sentimentalist."

56. Washburn 1908, e.g., 16 (on evolutionist anthropomorphism) and 247–50 (on instinct inhibition). The emphasis on mechanical tropisms derives from the immensely influential work of Jacques Loeb; see Boakes 1984, 137–39.

57. On behaviorist psychology as contributing, along with Boasian anthropology, to the never-quite-complete triumph of cultural "nurture" explanations of behavior

over biological "nature" ones in the American academy from the 1920s to the 1960s, see Degler 1991, esp. 152–66, 216. Again, Thorndike himself stood somewhat to the side of changes that his work was partly responsible for; consider, for instance, the first volume of his three-volume *Educational Psychology* (1913), entitled *The Original Nature of Man* and dealing at length with instinct, heredity, and evolution.

58. Hobhouse 1901, 1902. On Hobhouse's comparative psychology, see Boakes 1984, 179–84; and Hearnshaw 1964, 101–4.

59. Hobhouse 1901, 142–55, quotation on 142.

60. Hobhouse 1901, 150–51; for a perception-with-reward experiment of Hobhouse's, see 179, where he described how, after taking a stopper out of a jar and letting his dog get at the meat inside, the dog subsequently jawed out the stopper for himself at the next trial.

61. See Hobhouse 1901, 170, 174, 244, for illustrations of some of the puzzle boxes used; 236–37 for the chimp/stick experiment; and 248–51 for the monkey/stool experiments.

62. Hobhouse 1901, 266–69.

63. Hobhouse 1902, 37, emphasis in original; see also Hobhouse 1901, 145–47.

64. L. Hobhouse to Rosalind Frances Howard, 18 October 1886, Papers of Gilbert Murray, MSS. Gilbert Murray 529, fols. 148–53, Bodleian Library, Oxford University. Words in brackets are unclear in the original.

65. For Hobhouse's claim that the "animal intelligence" parts of the book served to test his thesis about the general trend of evolution, see Hobhouse 1901, v. On that trend as "the growth of mind," see 5. On "the higher development of the human mind in society" as the next subject to be dealt with, see v. Hobhouse—who held the chair in sociology at the London School of Economics from 1907 on—stressed both the systematic nature of his thought and its Spencerian beginnings in the 1880s; see Hobhouse 1913, xv–xvi, xxiv–xxv. How far Hobhouse's mature thought and activities deserve the adjective "Spencerian" merits fuller discussion than can be provided here. But on this question, see Owen 1974 and Collini 1979.

66. In a letter of 7 November 1900 to his editor at the *Manchester Guardian*, C. P. Scott—a letter that concludes "My experiments are going on well"—Hobhouse reported on his own participation in a movement "protesting against the present methods of war, and taking the line that as the alternative, negotiations ought to be opened with the Boer leaders," and his sister Emily's work on behalf of "a relief fund for the Dutch women and children whose farms have been burnt, and for those who are imprisoned." Papers of C. P. Scott, 132/102, John Rylands Library, University of Manchester.

67. Indeed, Hobhouse was remarkably free of the professionals' anxiety about making comparative psychology look as "scientific" as possible. He was doubtful, for instance, about the whole business of charting learning curves, indeed of timing the learning process generally. "The time of each trial was not regularly taken," he wrote of his own experiments. "Times are apt to be quite as misleading as in-

structive. . . . What an animal does is far more important for our purposes than the time which it takes to do it." Hobhouse 1901, 154–55.

68. Morgan 1903a; see also Morgan 1903b, 303–4.

69. Washburn 1908, esp. 245–46, but also 239–40, 243.

70. On Thorndikian experimental situations as enabling "an entirely new way of 'seeing' the animal," a "form of knowledge production [which] became a privileged and definitive source of knowledge about animals in psychology," see Stam and Kalmanovitch 1998, 1140; see also 1142–43.

71. Köhler 1925. Köhler conducted his experiments in 1914 at an ape research station on Tenerife. For his criticism of Thorndike's work, see 26–28. For his acknowledgment of Hobhouse's precedent, see 34. For discussion of Köhler's work, including the context of general opposition of the Gestalt psychologists to the associationist analysis of mental life into discrete elements, see Boakes 1984, 184–96. For a reading of the bananas-and-crates experiment as nevertheless profoundly reductive, designed to reveal thoughts about nothing other than how to get bananas, see the novelist J. M. Coetzee's *Lives of Animals* (1999), 27–30. As concerns language in nonhuman primates, neither Hobhouse nor Köhler believed it rose above emotional expression; see Hobhouse 1901, 286–92, esp. 291; and Köhler 1925, 258–59. On studies of insight learning after Köhler, see Thorpe 1956, 99–107.

72. Hobhouse 1901, 234. "Thus, that a bolt must be pushed back is a crude idea; that it must be pushed back so as to clear a staple, a relatively articulate one, implying a distinction between the parts of the object perceived (the bolt and its staples), and an appreciation of the relation between them."

73. R. M. Yerkes to L. Hobhouse, 5 January 1916, Papers of Robert Mearns Yerkes, group 569, box 23, folder 407, Sterling Memorial Library, Yale University. The two men had corresponded the previous year as Hobhouse worked on his revised edition; see the letters from Hobhouse to Yerkes of 8 March and 8 May 1915, Yerkes Papers, group 569, box 23, folder 407.

74. Vygotsky 1934, chap. 4. On Vygotsky and his context, see Smith 1997, 791–93. Vygotsky's book came belatedly to widespread psychological attention in the West in the 1960s thanks to its affinities with the work of behaviorism-rebelling cognitive scientists. The first English translation, published in 1962, came with a laudatory preface from Jerome Bruner, cofounder of the Center for Cognitive Studies at Harvard. On the "cognitive revolution" and its role in reinvigorating interest in animal language, see chapters 7–9 in this book. A small irony: Pavlov's work first came to widespread Western attention thanks to a 1909 notice coauthored by Yerkes (Yerkes and Morgulis, 1909). On the distorting influence of Yerkes's interests in sensory discrimination on the account he gave of Pavlov, see Boakes 1984, 152–53. Yerkes was also responsible for spreading the word about Nadezhda Kohts's investigation into chimpanzee intelligence and expression; see Kohts 1935, Yerkes and Yerkes 1929, Yerkes 1936, and, for discussion, Boakes 1984, 201. Donald Dewsbury (2000b, 296) reports that "Russian scientists placed a bust of Yerkes in the Darwinian museum of Moscow," where Kohts did her work.

75. On Yerkes's life and career, see an autobiographical essay (Yerkes 1932) and an unpublished book-length autobiography from 1950, known under the title "Testament" or "The Scientific Way," Yerkes Papers, group 569, box 146, folders 2311–17. Useful biographical articles include Dewsbury 2000b, Reed 1999, and Burnham 1976. For a virtuoso interpretation of Yerkes's career as zeitgeist-channeling, see Haraway 1989, chap. 4. A comprehensive bibliography can be found in Hilgard 1965. For Yerkes's early career in comparative psychology and its context, see Boakes 1984, 150–58. For Yerkes's own account of his scientific work at this time, see his "Testament," 112–16. For the graduate students who worked with him (he named sixteen), see Yerkes 1943b, 76. On the *Journal of Animal Behavior* and Yerkes's role, see Burkhardt 1987, esp. 227 for the provenance of research published there. For a historical census of courses and research laboratories in comparative psychology at American universities in the early twentieth century, see Warden and Warner 1927, esp. 200–201.

76. Yerkes 1943b, 74.

77. On Royce, see Yerkes 1932, 388; "Testament," 86, 89. Yerkes later drew on Royce's *Outlines of Psychology* (1903); see Yerkes 1905, 143–44.

78. E. L. Thorndike to R. M. Yerkes, 25 February 1899, Yerkes Papers, group 569, box 47, folder 919.

79. E. L. Thorndike to R. M. Yerkes, n.d., but certainly spring 1899, Yerkes Papers, group 569, box 47, folder 919.

80. Yerkes 1901, esp. 551, with the gradually sloping learning curve on 548. Thorndike reported Yerkes's confirmatory work in, among other places, the 1901 monkey monograph; see Thorndike 1901b, 240. Yerkes spent two summers at Woods Hole with Thorndike, the first (1899) as a student in his course, the second (1900) as an assistant. In September 1899, the solicitous Thorndike asked Yerkes if he would like his name printed in the Woods Hole prospectus for next year "as an assistant in Comparative Psychology." E. L. Thorndike to R. M. Yerkes, 4 September 1899, Yerkes Papers, group 569, box 47, folder 919. In March 1900, he asked Yerkes if he would be interested in taking up a paid assistantship at Columbia; E. L. Thorndike to R. M. Yerkes, 8 March 1900, Yerkes Papers, group 569, box 47, folder 919.

81. The thesis research was supervised by the completely unsuitable Wundtian introspectionist Hugo Münsterberg, and Yerkes came to agree with Jacques Loeb's harsh judgment of it. Yerkes credited his training in laboratory studies in animal behavior to Harvard biologist Charles Davenport, under whom he examined reactions to light (phototropism) in the water flea. See Yerkes "Testament," 94–98.

82. In a 1905 article, "Animal Psychology and Criteria of the Psychic" (esp. 145, 147), Yerkes allowed for the possibility that some animals enjoy conscious experience sufficiently complex to support inventiveness, and not just associative learning.

83. See Yerkes 1910, 123. The argument was, of course, more than a little self-interested. In later life, Yerkes was frank about his youthful reckoning that as a

biologist his professional chances were better in psychology, where the competition was lower, than in biology. Yerkes, "Testament," 92.

84. Yerkes's notebook can be found in Yerkes Papers, group 569, box 155, folder
2437. The lecture sequence ran "Instinct," "Habit and Association," "Reason in
Animals," "Imitation in Animals," and "Inner Life of Animals."

85. For his first statement to this effect, see Yerkes 1906, esp. 385, 388.

86. E. Titchener to R. M. Yerkes, 11 January 1907, Yerkes Papers, cited in O'Donnell
1985, 182. Chapter 10 of O'Donnell's book is an outstanding analysis of the marginal status of the science of Yerkes and Watson up to the First World War.

87. On the fate of Yerkes's students, see O'Donnell 1985, 195–97, 281 n. 68. On the
attitudes of administrators, see 185.

88. On Yerkes's difficulties at Harvard, see Yerkes 1932, 390–91; and O'Donnell
1985, 193–95.

89. On Yerkes's work at the hospital, see Yerkes 1932, 393; "Testament," 140–44.
I see Yerkes as making a smaller concession to his Harvard paymasters than does
O'Donnell, who notices, in addition to the new human-oriented position at the
hospital, a more generous attitude on Yerkes's part at around the same time toward the scientific study of consciousness. See O'Donnell 1985, 195, 198. The
trouble with this interpretation is that, in line with his Thorndikian training, Yerkes had been equally generous in the 1905 and 1906 articles discussed above
(and not considered by O'Donnell). Similarly, Boakes (1984, 153) and Dewsbury
(2000a, 752) make much of Yerkes, in his 1911 psychology textbook, allowing
for consciousness in plants. But he had already stated that view in 1906 (387).

90. Beer, Bethe, and Uexküll 1899, summarized in Yerkes 1906, 380–81; and discussed in Burkhardt 2005, 155; also Boakes 1984, 136.

91. Yerkes 1906, 385. For the word "objectivism," see 389.

92. Yerkes saw objectivism and behaviorism as continuous; see Yerkes 1932, 396.
For Watson's prediction that psychology refocused on stimuli and responses could
expunge such terms as "mind," "consciousness," and so on "without running
into the absurd terminology of Beer, Bethe, Von Uexküll, Nuel, and that of the
so-called objective schools generally," see Watson 1913, 31–32.

93. Watson 1913. For the word "behaviorism" (emphasized in original), see 31. On
the intellectual and institutional changes that created a sympathetic professional
audience for Watson's general idea of a behaviorally focused, practically oriented
psychology, if not for Watson's specific doctrines, see O'Donnell 1985 and Smith
1997, 650–59.

94. For an overview of Watson's research up to 1913, see Boakes 1984, 144–49.

95. On the emergence of Watson's critical attitude toward introspectionism, see
O'Donnell 1985, 189–90. For brief overviews of Watson's life and work generally, see Watson 1936 and Buckley 2000. For book-length biographies, see Cohen
1979 and Buckley 1989.

96. Watson 1913, 33.

97. Watson 1913, 26–27. Although Watson does not name Yerkes in the quoted pas-

sage, reference to the attempt among some students of behavior to frame "criteria of the psychic," along just the structural and functional lines set out in Yerkes's 1906 article with that phrase in its title, leaves little doubt about his role as target.

98. For the flavor of the Watson-Yerkes correspondence, see O'Donnell 1985, chap. 10, which quotes from it extensively, especially on professional discontents.

99. Yerkes and Watson 1911; for discussion, see Boring 1950, 628.

100. See Burkhardt 1987, esp. 225–26

101. J. B. Watson to R. M. Yerkes, 2 October 1907, Yerkes Papers, group 569, box 50, folder 975. The work under discussion was Watson 1907, reviewed in Yerkes 1907b.

102. J. B. Watson to R. M. Yerkes, 12 December 1907, Yerkes Papers, group 569, box 50, folder 975.

103. "The orangs are the most humanly interesting animals I have seen," Haggerty wrote to Yerkes from New York in August 1908. "They are phlegmatic in temperament and work slowly but with *evident purpose to accomplish an end and adapt their means to their purpose.* I have no doubt that they have Thorndike's '*Free ideas,*' and Hobhouse's 'Articulate ideas.'" 23 August 1908, Yerkes Papers, group 569, box 23, folder 412, emphases in original. On Yerkes and Haggerty, see Boakes 1984, 196–97. On Haggerty's career, see O'Donnell 1985, 196; see also 230–31 on how Haggerty came to think of his efforts in pure comparative psychology and applied educational psychology as expressions of an underlying commitment to the "laws of learning."

104. Hamilton 1907, esp. 341, for the relationship between his experimental work and "the assumption of degenerative reversion in certain psychoses." For discussion of the paper (misdated as 1908), see Boakes 1984, 200–201, who credits Hamilton with thereby inventing the matching-to-sample task, since the dog had to find the pedal with the color or odor matching the one on a board on the side of the cage. On the psychiatric motives behind Hamilton's comparative psychology, and the Yerkes-Hamilton relationship generally, see Thomas 2006.

105. Hamilton 1911. For discussion, see Boakes 1984, 153–55. An important departure from the 1907 experiments is that now it was impossible for the animal to predict which door was the correct one. On Hamilton's psychiatric interest in assaying the reactions to "bafflement," see Thomas 2006, esp. 283.

106. W. Köhler to R. M. Yerkes, 17 April 1914, Yerkes Papers, group 569, box 57, folder 1090. Yerkes told of his contacts with the Tenerife station in Yerkes 1916a, 1–2. On the history of the station, see Boakes 1984, 185.

107. On Hamilton's generosity toward Yerkes and his family, see Yerkes 1916a, 2–7; "Testament," 156; and Thomas 2006.

108. For ground plans, photographs, and description of the Montecito laboratory, see Yerkes 1916a, 5–7. On the laboratory's multiple-choice apparatus and its use, including a ground plan and photographs, see 11–20. Yerkes first described his multiple-choice method in a 1914 *Science* article where he introduced it alongside

"the quadruple choice method of Hamilton" as enabling comparison for the first time between human and nonhuman reactive tendencies. As first instantiated, Yerkes's method took the form of a desktop keyboard. See Yerkes 1914, 631; 1921, photos on 375–76. Its first subjects were human ones at the Boston Psychopathic Hospital, though crows, pigs, and other animals were also tested with other versions; see Yerkes 1916a, 9; 1916b; Yerkes and Coburn 1915; and Coburn and Yerkes 1915. For discussion of the pig multiple-choice experiment, including a photograph (the setup was very similar to the Montecito one), see Boakes 1984, 155–57. Yerkes contrasted his approach as "genetic psychologist" and Hamilton's approach as "psychopathologist" to animal problem-solving experiments—the former asking, how does the animal solve it? the latter asking, how does the animal react when the problem is made insoluble? See Yerkes, "Testament," 157–58.

109. Yerkes 1916a, 88–98, quotation on 97. Yerkes stressed that these and similar experimental tests were not original to him, but had been used by several investigators, included Haggerty, Köhler, and their pioneer, Hobhouse; see Yerkes 1916a, 128–29, 131–32. Yerkes compared his three-year-old son David's performance on the same test; see Yerkes, "Testament," 159.

110. Yerkes 1916a, 63–68, quotation on 68, cited in Boakes 1984, 198. For the learning curve from this experiment, see figure 6.4 in the text. An interesting feature of Yerkes's curves here as compared with Thorndike's is that the y-axis for the former represents, not time elapsed, but number of errors made before the problem was solved. It is not clear to me whether Yerkes, Hobhouse, or some other figure first used the term "insight" to mean the ability to perceive and act on relations. For some acute remarks on the term, see Vygotsky 1934, 39.

111. Yerkes 1916a, 132. Although Yerkes immediately qualified that judgment, adding that, compared with human mental life, orang mental life was "poverty stricken," he did not qualify enough to satisfy Vygotsky; see Vygotsky 1934, 36, 38.

112. J. B. Watson to R. M. Yerkes, 12 May 1916, Yerkes Papers, group 569, box 51, folder 987, cited in Boakes 1984, 199.

113. R. M. Yerkes to J. B. Watson, 16 May 1916, Yerkes Papers, group 569, box 51, folder 987. In fact, Yerkes had gone into print with his dissatisfaction over Watson's manifesto; see Yerkes 1913, 581–582—published in the same year, we should note, that Yerkes invented his multiple-choice method, designed to draw out the relation-perceiving mental powers that the Thorndikian puzzle box kept hidden. On the multiple-choice method as an advance on the puzzle-box method, see Yerkes 1916a, 127.

114. On Watson during and after the war, see Boakes 1984, 218–28. On the surprisingly independent academic careers of rat-learning research and behaviorism—the former much more successful than the latter until the 1930s, when they converged—see Boakes 1984, 228–41. On Yerkes's activities during the war, see Yerkes 1932, 397–402; and Boakes 1984, 199–200. Just before the war, Yerkes resigned from Harvard and accepted a post at the University of Minnesota, but

he never took it up, and after the war, once it was clear that he would remain in Washington, he resigned from Minnesota too. On Yerkes's wartime and postwar work (later notorious) on human intelligence and its hereditary basis, and the relationship with his older and enduring preoccupations with animal intelligence and the status of psychology, see Reed 1987. On the purchase and study of Chim and Panzee, see Yerkes, "Testament," 245–46; and Yerkes and Learned 1925, 15–16. In late September 1923, Yerkes returned to Washington, taking the chimps with him.

115. Boutan 1913. On Boutan's work with Pépée, including innovative—and distinctively French—experiments comparing Pépée's performance solving a puzzle box with the performances of human children, see Thomas 2005, esp. 455 for Yerkes's admiring correspondence with Boutan. Yerkes wrote approvingly of Boutan's work in Yerkes 1925, 169–71, though he added (170) that the distinction Boutan drew between inherited pseudolanguage and learned true language was likely oversimplified. Another, later admirer was Peter Marler; see Marler 1959, 202–3.

116. Witmer 1909; Furness 1916. Surveys of ape language experiments routinely cite one or both of these as the first; see, e.g., Wallman 1992, 10; and Ristau 1996, 645. For discussion, see Desmond 1979, 61–71; and Candland 1993, 196–207 (including several of the photographs of Peter from Witmer's paper). For Yerkes's discussion, see Yerkes 1916a, 129; 1925, 176ff. Candland dates Witmer's first encounter with Peter to September 1908 rather than 1909 (196); internal evidence across the two papers, however, suggests that the earlier date cannot be right. On Witmer's life and career, see McReynolds 2000 and, more extensively, 1997.

117. Witmer 1909, 179. On Furness as witness, see 184.

118. Furness 1916, 281–82, quotation on 281.

119. Quotation in Witmer 1909, 179–80. I cannot resist a literary speculation here. "A Report to an Academy" is a famous, enigmatic story by Franz Kafka, published in 1917. It tells of a chimpanzee called Red Peter, standing on stage in a theater, dressed impeccably, telling the scientific assembly about his transformation from monkey into something like a man. In *The Lives of Animals* (1999, 2), J. M. Coetzee has the novelist Elizabeth Costello suggest that Köhler's book on his Tenerife experiments with apes, published in German in 1917, was a source of inspiration for Kafka's story. But, timing aside, Witmer's paper seems a closer fit, with its tale of a near-humanized ape named Peter, in three-piece suit, on the stage and under the scientific gaze, attempting speech—though how the paper would have come to Kafka's attention I cannot say.

120. Witmer 1909, 184, 199.

121. Furness 1916, 283–85.

122. For Learned's assessment of the chimpanzee nonlanguage, see Yerkes and Learned 1925, 60. For "gahk," see 63. For the "fruit motive," "uttered . . . very frequently in connection with the eating of oranges, apples, and bananas," see 78. For the thirty-two-word inventory, see 154–56.

123. Yerkes and Learned 1925, 53–56, quotation on 53; Yerkes 1925, 179–80. Talking with journalists about his work, Yerkes drew out the implications for the origin of language. "That man went on beyond [the apes] . . . and developed, probably many hundreds of thousands of years ago, the ability to speak and to understand a real language is probably due, Professor Yerkes thinks, to the fact that men have a natural mental tendency to imitate sounds." "Science Trying to Find a Way to Make Animals Talk," *Pittsburgh Gazette Post*, ca. 1925, Yerkes Papers, group 569, box 60, folder 1134. In light of her own chimp studies, Kohts—who also catalogued (albeit only verbally) the natural vocalizations of the animal—fully endorsed Yerkes's conclusion about the lack of a tendency to imitate human speech. Kohts 1935, 198–202, esp. 201. Vygotsky, however, thought Yerkes mistaken about how fundamental the lack of vocal imitation was; more important, in Vygotsky's view, was the lack of an intellect capable of using signs. See Vygotsky 1934, 36–40.

124. C. L. Morgan, offprint of review of Yerkes and Learned 1925, source and date unknown, Morgan Papers, DM 612.

125. Yet another Bronx Zoo connection: Learned supplemented her vocalization studies of Chim and Panzee with studies of three adult chimps in the zoo; see Yerkes and Learned 1925, 144.

126. On the use of the calipers, see Yerkes and Learned 1925, 21. In characterizing and measuring the chimps' physical traits, Yerkes relied on Hrdlička's manual *Anthropometry*.

127. He tried four methods, all of them using spoken sounds in Chim's presence as a kind of magic sesame: first, "ba, ba," as a cue for the arrival of pieces of banana through a chute in the wall of the observation room; second, "co, co," making banana pieces appear on a little table in a box in Chim's cage; third, "na, na," springing open the door of a shut box with a banana visible inside; fourth, letting Chim get the banana in the shut box even when he did not say "na, na." Yerkes and Learned 1925, 54–56.

128. "Science Trying . . . to Make Animals Talk."

129. "Two men, whose investigations into mental phenomena have made them famous, have accepted invitations to speak at the annual dinner of the Massachusetts Fish and Game Protective Association. . . . These men are Professor Richard L. Garner, who has made three expeditions into the African wilderness, in an attempt to learn the monkey language and establish the Darwinian theory of the descent of man, and Professor Hugo Münsterberg of Harvard, the well-known psychologist, who, besides his studies of mental phenomena of the human, has done important research work regarding the mental life of animals. As it is expected that Professor Garner will speak of the results of his investigation, and that Professor Münsterberg will tell of his experiments, one of the most interesting and instructive meetings in the history of the association is confidently predicted." "Professor Garner to Speak," *Philadelphia Evening Bulletin*, 9 March 1903, Urban Archives, Samuel Paley Library, Temple University, Philadelphia.

130. J. B. Watson to R. M. Yerkes, 29 March 1907, Yerkes Papers, group 569, box 50, 975.

131. Watson 1913, 27.

132. Watson 1914, 323. On Watson's argument in this chapter, "Man and Beast," and his distinctive view that thought, of which humans only were capable, was nothing but the subvocal speech of which humans only were capable (thanks to a vocal apparatus reacting in uniquely varied ways to uniquely varied stimuli), see Boakes 1984, 171–72.

133. Boutan 1913, 20–25. See also 7–8. I am grateful to Marion Thomas for drawing these passages to my attention.

134. "Garner Made Monkey Talk," *Philadelphia Evening Bulletin*, 22 August 1910, Urban Archives. An article from the same day or the next ("Chimpanzee Says One Human Word") is more definite: "Susie is to be lent to the psychological department of the University of Pennsylvania for a series of experiments." An article on 23 August ("Doubts Susie Can Speak") stated: "Susie's accomplishments are of interest in this city because she is expected to come here. The Professor is said to intend showing her to the animal-psychology students at the University of Pennsylvania. The latter examined Peter, another celebrated monkey, who is an actor, and will welcome a chance to put Susie to all the tests they have for educated or unusually intelligent beasts."

135. See, e.g., Garner's 1905 article in the *North American Review*, "Psychological Studies of the Chimpanzee," where he wrote on experiments testing a chimp's sense of form and color, using apparatus—shaped pegs and holes, painted cubes, and so on—"similar in principle to those used in training children of feeble minds, but . . . much less complex, as they were designed to reach a lower mental horizon than that of children." The series of sounds he was attempting to teach her, he reported, "have been selected for me by experts in teaching deaf-mutes and masters of phonetics." Garner 1905b, 275, 280.

136. Garner 1910 (*The Independent*, 8 September), esp. 518–19; "Professor Garner Gives Susie Her Debut," *Philadelphia Evening Bulletin*, 16 September 1910, Urban Archives. In his article, Garner indicated that his use of the phonograph was ongoing and that, over the years, he had come to see that many of his original translations were correct as far as they went, but too narrow, for many simian words embrace a larger number of meanings than he had at first understood. What he had identified as the *Cebus* word for "food," "hunger," or "to eat," for instance, he now regarded as better translated "I want" or "want" (518–19).

137. "Prof. Garner Urges School for Apes," *Philadelphia Evening Bulletin*, 16 May 1912, Urban Archives. Middleton 1912, 571–72, recounts a demonstration by Garner in Philadelphia of his ability to converse with monkeys in their own language.

138. Yerkes 1914, 632–33. The paper began (625) as a contribution to a symposium held in June 1913.

139. Yerkes 1916a, 130, including both Garner's *Speech of Monkeys* and *Gorillas and*

Chimpanzees, and a 1900 book largely combining the two, *Apes and Monkeys: Their Life and Language*.

140. Yerkes 1925, 168–69.

141. R. M. Yerkes to H. E. Garner, 5 December 1925, replying to Garner's letter of 4 December 1925, Yerkes Papers, group 569, box 20, folder 364. Yerkes asked if he could call on Garner when next in Washington, as Learned had expressed an interest in his father's phonograph records of gorillas and chimpanzees. So far as I know, none of these records have survived.

142. Yerkes and Yerkes 1929, 164. The Yerkeses complained especially (164, 178) about Garner's low estimate of orangutan mental ability. They also took issue with, among other things, Garner's early claims about chimpanzees sleeping on the ground, rather than in nests up in trees (223). "As regards Garner, I concur with Yerkes: he knew his primates but misunderstood them," wrote Alfred Kroeber around the same time. On apes and language, Kroeber judged that the negative findings of Boutan, Furness, and Yerkes and Learned were "conclusive." Kroeber 1928, 342.

143. On the Orange Park facility and the background to its establishment, see Yerkes 1943a, esp. 214–20, 289–301; "Testament," 266–72, 281–95; and Dewsbury 2006.

144. See the checklist of ambitions, from September 1929, in Yerkes 1932, 402–3.

145. For a survey of work done at Orange Park relating to language and symbolism, including a well-known study by Yerkes and Henry Nissen, see Yerkes 1943a, chap. 10. The Yerkes-Nissen paper, "Pre-linguistic Sign Behavior in Chimpanzee" (1939), described experiments attempting to reveal whether chimps can mentally represent or symbolize an object—here, an apple-containing colored box, presented first to the left of a differently colored box, then to the right, then above, and so on. Consistently correct choices, it was supposed, would be evidence for symbolism. In the investigators' view, the chimps' successes indicated some symbolic capacity, though it was "rudimentary and ineffective" (150).

146. Kellogg and Kellogg 1933, esp. preface (xi for "humanizing") and chaps. 1 and 12. For summary and discussion, see Candland 1993, 269–86. On the Orange Park experiment in the context of Kellogg's life and career, see Benjamin and Bruce 1982. Up to the late 1960s, the Yerkes lab—by then the Yerkes Regional Primate Research Center, based at Emory University in Atlanta, Georgia—had been host to four chimp home-raising experiments, three of them dealing with speech; see Kellogg 1968, 423. On the Hayeses' efforts in the 1950s, see chapter 8 of this book. On the two-and-a-half-year experiment of G. Finch, the results of which (never published) were the same regarding speech as in the Kelloggs' experiment, see Yerkes 1943a, 192; and Kellogg 1968, 424. On the discussion in the 1920s and early 1930s of the Indian "wolf children," and Kellogg's participation in it, see, respectively, Candland 1993, 53–68; and Benjamin and Bruce 1982, 466. On the Kelloggs' experiment as a reversal of the impossible feral-children cases, see Kellogg and Kellogg 1933, 11, quoting from Yerkes 1925, 180. For

Yerkes's recommendation about teaching sign language, see Wallman 1992, 11. After the signing successes of Washoe in the late 1960s (see chapter 8 of this book), Yerkes was belatedly applauded for his prescience—though Wallman also quotes similar suggestions in the writings of Kipling and Pepys.

147. The Kelloggs, it should be noted, described in some detail Gua's natural vocalizations, including their development and Donald's imitation of some of them; see Kellogg and Kellogg 1933, 282–86.

148. C. R. Carpenter to R. M. Yerkes, 12 February 1931, Yerkes Papers, group 569, box 10, folder 160. On Carpenter's life and career generally, and his field primatology of the 1930s in particular, see Haraway 1983 (a more detailed version of the better-known material in Haraway 1989, chap. 5) and Montgomery 2005.

149. On these earlier expeditions—Harold Bingham's to the gorillas of the Congo and Henry Nissen's to the chimpanzees of French Guinea—see Sussman 1997, 843; and Montgomery 2005, 500–503.

150. Timing is everything: less than a month before Carpenter wrote to him, Yerkes received a letter from the ornithologist Frank Chapman, painting a vivid picture of the opportunities the howlers of Barro Colorado Island afforded for field study; see Yerkes's foreword to Carpenter 1934. The island had been a favorite of American biologists for some time; see Montgomery 2005, 511.

151. Carpenter 1934, table on 108–9.

152. Carpenter 1934, 110.

153. On Craig 1908, and Craig's life and work generally, see Burkhardt 2005, 33–59, esp. 39–42. On Carpenter's primatology as drawing in several ways on his previous ornithological experience, see Haraway 1989, 88–90. On Carpenter's debt to Craig's 1908 paper, see Carpenter 1969, 45; and Montgomery 2005, 514–15.

154. In McDougall's popular textbook, *An Outline of Psychology* (1923; 4th ed., 1928), which had extensive coverage of instinct, he recommended Craig's pigeon papers in the strongest terms (97). On Carpenter's work with McDougall, see, e.g., Haraway 1989, 86. On McDougall, see Boakes 1984, 206–11, 226. On the psychologist McDougall's theory of instinct as rather similar to that of the ethologist Lorenz—as Craig pointed out in correspondence with Lorenz—see Burkhardt 2005, 153.

155. De Laguna 1927, quotation on 19. For her extensive, laudatory comments on Craig's 1908 paper, see 24–35. On her correspondence with Craig, see 30. On her behaviorism—different from Watson's, she emphasized (obviously so in her phylogenetic interests)—see xi, 123–39. On Carpenter's debt to De Laguna, see Carpenter 1969, 46; and Montgomery 2005, 515–16.

156. Here I take issue with the interpretation in Montgomery 2005, esp. 516–20, which represents Carpenter's playbacks as central to his research enterprise.

157. Carpenter's assistant on the Asiatic Primate Expedition was a young physical anthropologist from Harvard, Sherwood Washburn. He would go on to make behavioral field studies of apes and monkeys an integral part of physical anthropology; see chapter 8 in this book. On the expedition and Carpenter's role in it, see Montgomery 2005, 516–17.

158. Carpenter 1940, 169–82, tables on 171, 177.
159. Montgomery (2005, 517) claims that Carpenter in Siam used playback with gibbon vocalizations "to verify their functions," but cites no evidence in support. Carpenter himself stressed how difficult and uncertain were his functional ascriptions; see Carpenter 1940, 178.
160. Carpenter 1940, 180. Coolidge was leader of the Asiatic Primate Expedition and a member of the Harvard Museum of Comparative Zoology.
161. Carpenter 1939, 325. Carpenter was much more effusive twenty-five years later, in his contribution to a conference and then volume on animal communication. Field playback was now a commonplace, and he credited himself with "originat[ing] in miniature" in Siam a procedure that could be used "for validating calls and for checking their fidelity in terms of responses as well as providing a means of controlled study of the functions of different sound signals" (Carpenter 1969, 55).
162. De Laguna 1927, 3–43.
163. De Laguna (1927, 28–29) especially called for studies of the nonhuman primates that would improve on what Garner had done.
164. In time-honored fashion, Carpenter looked to his recordings to capture subtle aspects of the action that he might not have detected in the moment. He wrote, for instance, that his recordings showed how "two animals of the same group may call simultaneously though not synchronously." Carpenter 1940, 175.
165. On the laboratory work he hoped would complement his fieldwork, see Carpenter 1940, 170.
166. In Haraway's influential writings on Carpenter, she attributes his emphasis on animal communication as social control to his contact with the behaviorist analysis of language due to the Chicago unity-of-science movement, especially the "semiotics" of Charles Morris. See Haraway 1989, 97–99; and the slightly more expansive treatment in Haraway 1983, 167–70. Although Haraway rightly notes (1989, 98) that Morris's 1938 book appears in the bibliography of Carpenter's 1940 monograph, Carpenter there in fact made no use whatsoever of any of Morris's distinctive categories—"designators" and so on. It is surely not Morris, but De Laguna (not mentioned by Haraway), to whom Carpenter was principally indebted. Indeed, it was not until well after the war, and the work of Peter Marler in particular, that Morrisian semiotics had an impact on the analysis of animal communication; see chapter 8 in this book.
167. For a lively collection of essays on technological determinism, see Smith and Marx 1994.
168. In his gibbon monograph (1940, 203), Carpenter cited Garner's 1900 book *Apes and Monkeys: Their Life and Language.*

Chapter Seven

1. For biographical information, see Marler 1989. For an overview of Marler's scientific contributions, see Hauser 1996, 53–60.

2. "Come friendly bombs and fall on Slough! / It isn't fit for humans now, / There isn't grass to graze a cow. / Swarm over, Death!" (J. Betjeman, "Slough," 1937). A further stanza begins: "It's not their fault they do not know / The birdsong from the radio." On British natural history after the First World War, and the 1940s as "a veritable golden age," see Allen 1978, 244–71, quotation on 266. See also Macdonald 2002, on the self-consciously scientific and nationalist character of British amateur birdwatching in this period.

3. On the scientific pursuits of a similarly inclined—though more privileged—London boy at around the same time, see Sacks 1996, xiv; also 2001.

4. Webster and Marler 1952.

5. On the Edward Grey Institute, and the distinctive ecological-evolutionary cast of Oxford zoology at that time, see Morrell 1997, 268–304, esp. 298ff. On Lack, see Lack 1973 and Thorpe 1974, which includes a complete bibliography. Two short obituaries are quite useful: Mayr 1973 and Hardy 1973.

6. Here I follow Ernst Mayr's strictures on nomenclature, in Mayr 1982, 454–55.

7. Both Lack and Kettlewell produced classroom-friendly digests of their research in *Scientific American*; for Lack's, see Lack 1953, quotation on 67. Several commentators have noted how well these empirical studies fitted the new, "modern synthesis" theoretical climate in biology, bringing together Mendelian population genetics and Darwinian natural selection. See, e.g., Larson 2001, 167–72 (on Lack); Hooper 2002, 146–49, 165–68, 241–42; and Hagen 1999, 47–51 (on Kettlewell); and Depew and Weber 1995, 323 (on both).

8. Lack 1947, esp. 113–14. Although predisposed from his student days toward the theory of natural selection, Lack did not at first interpret the Galápagos finch-beak variations as adaptive. Lack attributed his change of mind to an encounter with the Russian ecologist G. F. Gause's ideas about what is now known as "competitive exclusion." See Lack 1947, 62; 1973, 423, 425, 429–30. For discussion, see esp. Kingsland 1995, 146–75; Mayr 1973, 433; Wiener 1994, 54–56; and Larson 2001, 169–72, 287 n. 83. Stephen Jay Gould (1985, 87–88) cited Lack's theoretical switch as an instance of a more general "hardening" of midcentury evolutionary biology into a doctrinaire adaptationism.

9. On the songs of the Galápagos finches, see Lack 1947, 45–46. In his 1943 book on the robin, Lack devoted a whole chapter to song. Although emphasizing the role of singing in fending off rival robins, he also touched on specific distinctiveness, noting that songs tend to be distinct from those of other species in a region (since male-female recognition gives a survival advantage), but occasionally to be similar to those of species not found in a region (since no survival penalty thereby accrues). Recalling birds in the Galápagos and East Africa whose songs reminded him of other species in England, he wrote: "Bird song has considerable subjective associations, and it was curious in the heat and dust of the tropics to be suddenly transported back to a Devon spring." Lack 1946, 21–22, quotation on 22 (the passage is unchanged from the 1943 edition [33–34]).

10. Lack and Southern 1949, 615.

11. Lack and Southern 1949, 607.

12. Marler and Boatman 1952.

13. In a late-career autobiographical memoir (1989, 318), Marler underplayed the theoretical bent of his early studies of birdsong, describing them as "still aesthetic rather than scientific, more on a par with my excursions into the Vedas and The Cloud of Unknowing."

14. His Oxford zoological colleague Alister Hardy (1973, 435) wrote that Lack was "something very much more than an ornithologist," for he brought "the study of birds into the forefront of a much wider biological context—that of ecology, behaviour, and particularly evolution. He showed how bird research could throw a powerful light on the fundamental problems of life in general." Lack's impact on the British scene, not least in priming the warm reception that Continental ethology received after the war, has not been sufficiently noticed. For Marler's part, by 1948, he was not just an accomplished amateur ornithologist, but a trained plant ecologist, and this ecological perspective—much more thoroughly absorbed among the botanists than the zoologists in Britain at that time (see Allen 1978, 260–63)—would have made Lack's distinctive approach especially congenial.

15. Marler and Boatman 1951, 93. Cf. Lack's similar arguments, summarized in note 9 above. The form as well as the content of the paper departs little from Lack and Southern's model, notably in the sequence of topics: avifauna, vegetation, habitat differences, song-and-call differences, annotated list of species. On Lack's role in stimulating the Pico expedition, see Marler and Boatman 1951, 98; and Marler 1989, 318.

16. In *Darwin's Finches*, Lack famously included an evolutionary tree for the Galápagos finches. See Lack 1947, 100–106.

17. Marler and Boatman 1951, 96.

18. Marler reiterated the point in his 1952 paper (467) on Chaffinch variation: "The song of the Azores Chaffinch is simpler than in other areas. . . . Whether this is to be considered as a characteristic persisting from the original stock, or as a reversion to an earlier evolutionary form, is not certain, but there is some evidence from the behaviour of other species that the latter may be closer to the truth." The blue tit in the Canaries and the goldcrest in the Azores "show song characteristics which are clearly related to the peculiar conditions of competition to which they are subject. It is conceivable that in the absence of some of the finch species with which it is usually associated in Europe, the survival value of distinctiveness in the song of the Chaffinch may have been reduced."

19. Marler and Boatman 1951, 94. Although the quasi-musical notation appears elsewhere in the paper, there are no comparably quantitative treatments.

20. Marler 1952, 460–62, 471. On this period of "chasing chaffinches," see Marler 2004, 3–4, quotation on 3. On the chaffinch breeding/singing season, see Marler 1956e, 70.

21. Marler 1952, 458–69, quotation on 469. In a later paper, Marler wrote: "It is

as though all the towns of Yorkshire had been scattered across Britain." Marler 1956e, 73.

22. Marler 1952, 469–70, quotation on 470. Peter Marler emphasized to me in correspondence that what is said here about grasshopper warblers and Baltimore orioles does represent conventional opinion at the time but, so far as he knew, has never been thoroughly checked.

23. Marler 1952, 467.

24. Marler 1952, 470. In recent autobiographical reflections (2004, 8–10), Marler has written of this sentence that the conclusion it expressed was reached before he went to Cambridge. That may be, but the paper was not submitted until September 1951 (see 458), by which time Marler had been at Cambridge for several months. Also, W. H. Thorpe and his colleague at Cambridge, Robert Hinde, are among those thanked in the paper's acknowledgments for discussion and criticism (see 470).

25. On Haldane's contributions to, among other things, mathematical population genetics, the experimental study of biosynthetic pathways, the calculation of mutation rates, and linkage analysis of pedigree data, see Sarkar 1992. For general biographical background, see Clark 1968.

26. Haldane divided nature-nurture interactions into four types. Debuting in a 1936 issue of *Erkenntnis*, this analysis appeared often in Haldane's public and professional writings. See Haldane 1936; 1938, 34–42; 1941, 25; and 1946. For Marler's recent, positive assessment of Haldane's 1946 paper, in the *Annals of Eugenics*, see Marler 2004, 27. Of Haldane, Marler writes (25): "He is reminding us that an animal's genetic constitution influences how it responds to the environment, echoing a central tenet of ethological theory." Cf. note 29 below.

27. On Haldane's role as a leading scientific critic of mainstream eugenics, see Kevles 1985, 118–27.

28. Haldane 1941, 37–38, quotation on 38.

29. Quotation from Haldane 1941, 11. On Haldane's never-repudiated eugenical ideas, and the wider context of "left" eugenics in Britain, see Paul 1983, 30–33. In 1938 (31), he gave a concise statement of what, precisely, is inherited: "We cannot always speak of the inheritance of a character; in many cases we can speak of the inheritance of a constitution which in a particular environment will give such and such a range of characters."

30. On Haldane's response to Lysenkoism, see Paul 1983, esp. 20, 24–30.

31. Based in Moscow when the paper was published (1930), Promptov was one of the original members of Sergei Chetverikov's population genetics group at the Koltsov Institute. For information on Promptov, I am grateful to Mark Adams and especially Nikolai Krementsov, who is presently at work on a major study of Promptov.

32. Thorpe 1940, 344–45.

33. Theodosius Dobzhansky also mentioned it, as Margaret Nice noted. "Song is one of the biological isolating mechanisms that serve as species recognition marks. In his book on 'Genetics and the Origin of Species,' Dobzhansky (1941: 264) empha-

sizes this: 'The importance of song in the courtship in birds is well known; songs may differ not only in different species but in races of the same species (Promptoff 1930).'" See Nice 1943, 148.

34. Mayr 1942, 53–55, quotations on 53, 55.

35. Huxley 1942, 308–9, quotations on 309. Organic selection now tends to be called "the Baldwin effect," after one of its first theorists, J. M. Baldwin. Huxley gave a characteristically lucid account: "modifications repeated for a number of generations may serve as the first step in evolutionary change, not by becoming impressed upon the germ plasm, but by holding the strain in an environment where mutations tending in the same direction will be selected and incorporated into the constitution. The process simulates Lamarckism but actually consists in the replacement of modifications by mutations" (304). On organic selection and birdsongs and calls, see esp. 305–7.

36. Marler 1952, 458, 471.

37. Poulsen 1951, esp. 218–19; Thorpe 1951, 264–67, esp. 265n; and Marler 1952, 470. Poulsen too had noted that the acquired basis of the terminal phrase fitted his own observations of geographical variation in chaffinch song; see 218.

38. On the emergence of the fellowship, see Marler 1989, 319. On Nicholson, see Allen 1978, 255ff., esp. 267 on his role at the Nature Conservancy. Marler arrived in Cambridge around May 1951. See Marler 1956f, 231; 1956a, 1. In 1952, he spent eight weeks in the Azores studying the chaffinch (Marler 1956a, 1).

39. Tinbergen 1951.

40. On the origins and history of ethology, the most comprehensive study is Burkhardt 2005, to which the account here is indebted. On the word "ethology," and especially Heinroth's use of it to describe his program, see Durant 1981, 160–77, 190–91. On the concept of the umwelt, see Uexküll 1934; for its roots in the romantic holism of Wilhelmine Germany, see Harrington 1996, 34–71, esp. 38–48. For brief autobiographical statements from Lorenz and Tinbergen, see Lorenz 1989 and Tinbergen 1989. Recent biographies include, for Lorenz, Taschwer and Föger 2003, and for Tinbergen, Kruuk 2003. Useful collections of papers are Lorenz 1970–71 and Tinbergen 1972–73; for a broad sampling, see Houck and Drickamer 1996. Lorenz introduced "imprinting" in Lorenz 1935, 245–46. While most commentators consider imprinting a kind of learning, Lorenz here insisted it was not, because, in his view, the process was irreversible and time-dependent, hence totally unlike conventional learning (246).

41. The founding of two societies in 1936 did much to foster ethological studies: in Berlin, the Deutsche Gesellschaft für Tierpsychologie, and in London, the Institute for the Study of Animal Behaviour. See Durant 1986, 1601.

42. Lorenz 1941 was an especially important paper. On the prewar work of Lorenz and Tinbergen, separately and together, see Burkhardt 2005, chaps. 3 and 4. Tinbergen recounted his experiments with Lorenz on escape responses, and their debt to Portielje, in Tinbergen 1953, 214–17. On the egg-rolling experiments, see Lorenz [and Tinbergen] 1938.

43. But on quasi-ethological aspects of much work in comparative psychology in the first half of the twentieth century, see Dewsbury 1992a. In an interesting analysis of popular articles in the postwar period, Dewsbury has suggested that the truly sharp contrasts can be found not in the research itself but in its public presentation. Where the ethologists celebrated nature's wonder in an upbeat and ahistorical way, comparative psychologists situated themselves within a grand but sadly declining tradition of objective science. See Dewsbury 1997. It should be noted, however, that Lorenz and Tinbergen in their early days were at least as concerned to differentiate themselves from subjectivistic "animal psychologists," such as Bierens de Haan, as from the objectivistic comparative psychologists in America. See Burkhardt 2005, chaps. 3 and 4.

44. Tinbergen 1942, 41–42, quotations on 42. Tinbergen (42) also adduced a nicely self-mocking joke from the Berkeley neobehaviorist Edward Tolman, who dedicated his 1932 book *Purposive Behavior of Animals and Man* to M(us) N(orvegicus) A(lbinus)—the white rat, habitué of Tolman's mazes. Tinbergen expanded on these points in his preface to an influential English-language collection of writings by, among others, Uexküll, Lorenz, and himself, published in 1957. It begins: "Whenever I meet American behaviorists I am struck by the very great difference in approach between them and us." Tinbergen 1957, xv.

45. Lorenz 1950, 230.

46. Lorenz 1950, 239–40.

47. See, e.g., Lorenz 1935, 115–16, 120–21. For an excellent discussion of Lorenz's indifference toward variability, bordering on distaste, see Burkhardt 1983, 438–39.

48. On Lorenz's activities in the Nazi era, see Burkhardt 2005, chap. 5; see also Radick, in press a.

49. Consider the attitudes of T. C. Schneirla and his famous student Daniel Lehrman, both at the American Museum of Natural History in New York. See Burkhardt 2005, 363–69, 384–90.

50. A letter circulated in early 1936, seeking support for what became the Institute for the Study of Animal Behaviour, made plain the discontent with behaviorism: "Methods which permit animals to unfold the whole variety of their activities and do not restrict them to responses to the narrow and pre-determined conditions of a laboratory experiment promise to be of inestimable value. . . . The student of Animal Behaviour should seek first to study his objects in their own environment, not transport them straightway into surroundings specially adapted for human requirements and standards." Quoted in Durant 1986, 1606. Early members of the institute included a number of highly placed figures, including Huxley, Thorpe, and the Oxford physiologist Solly Zuckerman. On the wider British reluctance to embrace American comparative psychology in its behaviorist mode, see Wilson 2002b.

51. On the 1947 reorganization of Oxford field zoology, see Morrell 1997, 304; and Burkhardt 2005, 332–33. It should be borne in mind, of course, that part of the

attraction of ethology in Britain was precisely its novelty. "It was once said of certain investigations of population dynamics that the animals concerned did not *behave* at all: they merely numbered off from the right," recalled the Glasgow zoologist S. A. Barnett in 1957 (118). "Certainly, undergraduates studying zoology in the 1930s, and even the 1940s, often completed their courses without having to give any systematic attention to the behaviour of the animals they studied" (emphasis in original).

52. Thorpe was already involved editorially with an English-language ethological journal, the bulletin of the Institute for the Study of Animal Behaviour. It became the *British Journal of Animal Behaviour* in 1953 and finally *Animal Behaviour* in 1958. See Durant 1986, 1611–15.

53. See Burkhardt 2005, 342.

54. On Thorpe, see Thorpe 1979, 118–23; Hinde 1987; Burkhardt 2005, esp. 337–45; and Radick, in press b.

55. Thorpe 1940, 354–55, quotation on 355.

56. See Burkhardt 2005, 340, quoting from W. H. Thorpe, "The Evolutionary Significance of Habitat Selection," *Journal of Animal Ecology* 14 (1945): 67–70, quotation on 69. Thorpe discussed locality imprinting earlier in Thorpe 1943–44, 67–69.

57. Thorpe 1951, 14, 26. See also Thorpe 1979, viii–x.

58. Thorpe (1951, 266–67) highlighted the importance of the study of imprinting, and of song learning as an instance of imprinting.

59. See Thorpe 1979, 122–23.

60. Biographical information on Hinde is taken from Hinde 1989, and an interview with Robert Hinde, 25 July 2002.

61. Lack 1946, 153–54; also in Lack 1943, 144–45. Lack wrote that his own inspiration was the experiments of two American ornithologists, A. A. Allen and F. M. Chapman, published in 1935 (see 1946, 148, 209; 1943, 139, 193).

62. Lack 1946, 158; also in Lack 1943, 149. See Lack 1940 for a brief, lightly critical note on Lorenz's concept of releasers in bird behavior. In the *Robin* book, Lack wrote that he preferred Tinbergen's term "signal" to Lorenz's term "releaser" for the response-eliciting situations in a bird's environment (Lack 1946, 161; 1943, 152).

63. Quoted in Thorpe 1974, 277; most likely the source is D. Lack, "The Position of the Edward Grey Institute, Oxford," *Bird Notes* 23 (1949): 231–36.

64. Marler 1989, 318–19. Marler recalled getting his "first intoxicating taste of ethology" (319) from Margaret Morse Nice's monograph on the life history of the song sparrow, published in 1943—the same year as David Lack's life history of the robin. Nice was an early supporter and translator of Lorenz, and her monograph contains much on inheritance and learning, including of song; see Nice 1943, 133–50. Marler described as "equally momentous" (319) his encounter with the published papers from the 1949 Cambridge conference.

65. On song as Thorpe's dominion, see Marler 1989, 323.

66. See Burkhardt 2005, 341, crediting P. Chavot's 1994 doctoral dissertation on the history of ethology. In his Royal Society memoir of Lack, Thorpe (1974, 277) lauded the "sound decision" to dedicate the institute to population studies, leaving studies of instinct to others.

67. Watson 1970, 25. Watson further commented: "In France, where fair play obviously did not exist, these problems would not have arisen. The States also would not have permitted such a situation to develop. One would not expect someone at Berkeley to ignore a first-rate problem merely because someone at Cal Tech had started first. In England, however, it simply would not look right." Watson arrived in Cambridge in the autumn of 1951 (19), and almost became a member of Marler's college, Jesus College (118). Such was the organization of the biological sciences at that time that, despite his considerable interest in heredity, Marler does not recollect being aware of Watson and Crick's discovery during his Cambridge years.

68. Thorpe seems to have arrived at an understanding with Tinbergen as well—Oxford ethology had dominion over studies of survival value and evolution, Cambridge ethology over studies of behavioral causation and development. See Durant 1986, 1612.

69. Marler 1989, 320, 325.

70. The Madingley station and its equipment are described in Thorpe 1954, 465–66, and in the Hinde interview. There were around sixty metal, six-foot-cubed aviaries, built in the summer of 1950 by Hinde and an assistant, Gordon Dunnett. There were also some larger aviaries, purchased from the Duke of Bedford.

71. See Marler 1989, 320, 323.

72. On Lack's radar studies, see Hardy 1973, 435; and Thorpe 1974, 275–76. For a superb discussion of how scientific knowledge after the Second World War became a product as well as a producer of war weapons such as radar, see Hacking 1999, 163–85.

73. The laboratories at this time were based in New York City. On their history, see Bernstein 1987, esp. 1–12; and Riordan and Hoddeson 1997. For a contemporary appreciation, see Kelly 1950. Recalling Bell Labs' wartime contribution, one of Bernstein's interviewees, who had tested equipment in a navy laboratory, said: "When somebody from Bell Labs came along with a piece of equipment there was this tremendous feeling of competence. Nothing crummy ever came out of Bell Labs. When they did something they really 'Cadillaced' it. They produced a clean piece of equipment that was GI-proof and also proof against *us*, which was something" (128).

74. The date is given in Potter, Kopp, and Green 1947, 4.

75. See, e.g., McKendrick 1897, esp. 26–32; and Scripture 1902, esp. 38–39, 49–50, 55–61. For the former reference I am grateful to Peter Martland. On sound-visualization technologies in the early twentieth century, see Beyer 1999, 212–16; and Thompson 2002, 85–87. Not even the encyclopedic Beyer mentions the phoneidoscope.

76. On the oscilloscope as one of several electrical devices in the acoustician's armamentarium from the 1920s onward, see Thompson 2002, 96.

77. Potter 1946, 8. I am grateful to Ed Eckert at the Lucent Archives for this reference. Potter was director of transmission research at Bell from 1934 to 1955; see his entry in *American Men of Science*, 10th ed., 1961, Physical and Biological Sciences series, 3227.

78. For useful primers on reading sonograms, as they are now called, see Catchpole and Slater 1995, 14–15; Marler 2004, 6; and Rothenberg 2005, chap. 5 (an especially accessible account that quotes from an interview with Marler). For the same words displayed on an oscilloscope and a sound spectrograph, see Potter, Kopp, and Green 1947, 4. For a similar comparison, again to the spectrograph's advantage, involving one of the earlier, gramophone-tracing devices, see 315–17. The first sound spectrographs even distinguished sounds that were the same pitch and intensity but different in resonance—the *a* in "cat" versus the *a* in "father."

79. Potter 1930. See also Potter 1946, 7, 10; and, for a helpful summary of the underlying principles, Beyer 1999, 278–79. In simple terms, there were two main features of the sound spectrograph: a tape loop, with the sound of interest recorded on it; and a filter that blocked out all but a select range of frequencies. As the tape loop went round and round, the filter setting gradually shifted, so that, at the end of the process, the entire frequency range was mapped visually. In nonsimple terms, Thorpe (1961, 12) wrote that the sound spectrograph "produces essentially a Fourier analysis of frequency against time." According to another ornithological enthusiast, C. E. G. Bailey, however, the spectrograph made use of major revisions to Fourier's theory dating from the 1930s. See Bailey 1950, 115–17.

80. Potter, Kopp, and Green 1947. On the sound spectrograph as maintaining a longstanding commitment at Bell to the deaf, see the foreword to Potter, Kopp, and Green 1947, xv–xvi, by the president of the laboratories, Oliver Buckley.

81. "[M]ilitary interests resulted in a considerable improvement of the first experimental models by the Laboratories." Potter 1946, 7–8. Although Potter in several places mentioned the war-project rating, and the temporary shifting of priorities toward military needs for sound visualization (e.g., Potter, Kopp, and Green 1947, 4–5), he nowhere, to my knowledge, specifies what those military needs were. Within the bioacoustics community, there is a widespread belief that the sound spectrograph was put to use as described here. See, e.g., Marler 2004, 1. The only direct supporting evidence I have found is in a 1946 summary technical report on the "recognition of underwater sounds," from the Sub-Surface Warfare Division of the National Defense Research Committee. In a chapter on "characteristics of target sounds and noise background," there are a number of oscillograms and what are called "time-frequency-intensity analyses" (National Defense Research Committee 1946, 38, 40, 44–45). They show the acoustic signatures of a steam freighter, a diesel freighter, a large freighter, an old destroyer, an old surfaced submarine, a new surfaced submarine, a torpedo, a five-knot cargo vessel, a ten-knot cargo vessel, and the propeller of a submerged submarine. The "time-frequency-

intensity displays" look like spectrograms, and were produced "by sampling the various frequencies very rapidly with a 45-cycle band-pass filter" (44)—an exact description of the sound spectrograph of the era. For a survey of underwater acoustics up to 1950, with particular attention to American developments in the world wars, see Lasky 1977 (the sound spectrograph is not discussed, however). Stephen Crump (pers. comm., November 2005), whose father started Kay, told me that he recalls the sound-spectrograph work at Bell Labs being done not for antisub purposes but for codebreaking—though he was not at all sure, and added that such projects were often "compartmentized" for security purposes, so it was not always clear even to those who worked on them what they were for.

82. The training began in the autumn of 1943. See Potter, Kopp, and Green 1947, 5.

83. Potter 1945; "Visible Speech," *Time*, 19 November 1945, 50; "Pictures of Speech: New Device Makes It Possible for the Deaf to Hear with Their Eyes," *Life*, 26 November 1945, 91–94.

84. On the older "visible speech" and the new, see Potter, Kopp, and Green 1947, 3. Recall that Garner and Melville Bell were mutual admirers.

85. See the company history and profile at www.kayelemetrics.com (accessed 30 April 2004). One of the earliest models was called the Vibralyzer.

86. Potter, Kopp, and Green 1947, 411. Here Potter also drew attention to the first surprising ornithological discovery made with the spectrograph: some birds produce two separate sounds simultaneously. For discussion, see Simms and Wade 1953, 201–2; and Marler 2004, 3. A communication engineer who "went native" as a student of insect communication was George Pierce of Harvard; see Pierce 1948, esp. 7ff. On the further history of the spectrograph's uses in the study of speech, see Baken and Daniloff 1991.

87. On the postwar boom in sound recording with the new wire and tape recorders, see Morton 2000, 136–59. On tape recorders and the sound spectrograph together as fomenting modern bioacoustics, see Falls 1992, 13; and Owings and Morton 1998, 35–36. The latter even attribute the species-level emphasis of mid-century ethological studies of animal communication to "the invention of field tape recorders and spectrographs at the same time as the biological species concept was formulated" (21). The International Committee of Bio-Acoustics was inaugurated at an American conference in 1956. The published proceedings, together with those from a second meeting in 1958, raised the profile of the new field. See Busnel 1963; Lanyon and Tavolga 1960; and, for discussion, Falls 1992, 14. Marler has papers in both volumes. Professional scientists were, it should be noted, but one of the groups using the new technologies to record birds and other animals at this time. Morton's book includes (142) a charmingly goofy photograph of an American couple out birdwatching, the woman holding a pair of binoculars, the man holding a microphone connected to a tape recorder worn on a strap around his shoulder.

88. On the greater usefulness of the sound spectrograph to the ornithologist than to the entomologist (since birds, like humans, are far more sensitive to frequency

modulation than to amplitude modulations, whereas arthropods have opposite sensitivities), see Thorpe 1961, 12.

89. Simms and Wade 1953. A photograph of a contemporary investigator (Donald Borror) in the field with a parabolic reflector and portable tape recorder can be found in Busnel 1963, 32.

90. Borror and Reese 1953; Collias and Joos 1953. For discussion, see Marler 2004, 3. On Joos's use of the spectrograph from the mid-1940s on, see the notes in Joos 1948, x, 42. I am grateful to Peter Marler for this reference.

91. This account of Thorpe and the spectrograph draws on the one in Burkhardt 2005, 343–44, itself based on Thorpe's annual reports to the university on the Madingley station's progress. Less detailed but corroborative versions can be found in Thorpe 1954, 465; and Sparks 1982, 170 (based on an interview with Thorpe). On the timing of Thorpe's United States trip, see Thorpe 1956, 418. Marler, however, recalls a spectrograph being at the Madingley station when he arrived. See Marler 1989, 319; also 2004, 7. Another minor discrepancy in the historical record concerns the Teddington spectrograph. Burkhardt and Sparks follow Thorpe in describing the secret admiralty spectrograph as the only one in the country at that time. But in a 1950 *Ibis* paper promoting the use of the spectrograph in the study of birdsong, the electrical engineer C. E. G. Bailey wrote that the only machine in the United Kingdom was at the research station of the General Post Office (then in charge of telegraphs and telephones in Britain). Moreover, according to Bailey, this machine was available for ornithological use.

92. Thorpe 1954, 465, 467, 469, emphasis in original. It was also found that chaffinches in their mature songs do not imitate the songs of other, cohabiting species (466).

93. Thorpe 1954, 468.

94. Thorpe 1954, 465. On the impact of the paper, see Marler 2004, 8 (in a volume including a CD with some of Thorpe's original chaffinch recordings). Owings and Morton (1998, 35) have indeed judged that, from midcentury, "auditory signals began to replace visual ones as the most commonly studied form of communication."

95. Marler 1989, 319–20.

96. Marler 1989, 320; 2004, 8. The BBC transferred a complete set of its birdsong recordings in 1953, with periodic supplements arriving thereafter. See Thorpe 1961, vii.

97. Marler 1955a. On Haldane's enthusiastic response to the research, see the next section. The paper was published the following year in German translation in the *Journal für Ornithologie*, at the invitation of its editor, the Berlin ornithologist Erwin Stresemann (another of Lorenz's mentors). See Marler 1956d and, for discussion, Marler 1986. Marler summarized the paper in a more popular forum in Marler 1956e, esp. 78–84. In 1959, Tinbergen summarized the paper in his contribution to the *Origin* centenary celebration at Chicago; see Tinbergen 1960, 607. John Maynard Smith—who studied under and then collaborated with Hal-

dane at UCL from the late 1940s onward, overlapping with Marler—drew heavily on Marler's paper to illustrate how kin selection can promote the spread of altruistic traits such as alarm calling; see Maynard Smith 1965, and, for his Haldanian education, Maynard Smith 1989, 348–52. By 1986, Marler's paper had been cited 120 times in the scientific literature; see Marler 1986.

98. Collias and Joos, in their 1953 paper on the spectrographic analysis of chicken vocalizations, had distinguished between sounds that attract small chicks and sounds that warn of danger, noting their contrasting acoustic properties. But they had not related this difference to the different means and ends of location. Marler showed how well their data fitted his new generalization.

99. Marler 1955a, esp. 6 for the Azores work in relation to the present analysis. As noted above, Marler had made a second, longer trip to the Azores for ornithological purposes. See Marler 1956a, 1.

100. Marler 1955a, 8.

101. See Marler 1989, 321–22, where he also discussed subsequent challenges and tests to the 1955 generalizations. For Broadbent's ambition to bring studies of perception and behavior together, via investigations of responses to simultaneous stimuli, see Broadbent 1958, 7–10. For his contribution to Thorpe's seminar, see Broadbent 1961.

102. Marler 1954.

103. Shannon and Weaver 1949 consists of a long technical article of Shannon's, published in a Bell journal in 1948, together with an explanatory essay by Weaver, originally published in abridged form in *Scientific American* in 1949. Another book of Wiener's (1950), aimed at a general audience, also helped spread the word. (*Cybernetics* was, indeed, Wiener's word, coined in the summer of 1947 from the Greek for "steersman"; see Wiener 1948, 11–12. Unbeknownst to Wiener, the word had been invented before; see Kay 2000, 84. But the recently proliferating *cyber-* terms, in connection with computers, derive from Wiener.)

104. On Wiener, Shannon, and Weaver, see Kay 2000, 78–102; for Weaver's preference, see esp. 83, 88, 98. On Shannon's contributions to information theory crowding out Wiener's more generally, see, e.g., Pierce 1972, esp. 32–33, 40–41; and Kay 2000, 95.

105. One of the clearest expositions of the theory is still Weaver's contribution to Shannon and Weaver 1949. For a useful and entertaining recent effort, keyed to the equations for information content and channel capacity, see Aleksander 2002. On the origins and etymology of "bits," see Shannon and Weaver 1949, 32.

106. For surveys of the conceptual background to midcentury information theory, as well as the contributions of theorists besides Wiener and Shannon, see Aspray 1985 and the still-useful Cherry 1952. For Wiener's and Shannon's acknowledgments of their debts to prior work at Bell Labs, see Wiener 1948, 4; Shannon and Weaver 1949, 31–32; and, for discussion, Aspray 1985, 120–23; and Mindell 2002, esp. chap. 4. Another important resource for information theory was the theory of entropy, especially in its statistical, Boltzmannian form. See, e.g., Shan-

non and Weaver 1949, 3, 12–13, 28; and Aspray 1985, 124. On Shannon's and Wiener's divergent interpretations of information as positive and negative entropy, respectively, see Kay 2000, 94.

107. On the war work of Wiener and Shannon, and its relations to information theory, see Kay 2000, 78–82, 85–86, 92–93; and Aspray 1985, 119. On Wiener's war in particular, see also Wiener 1948, 3–12; and Galison 1994.

108. Cherry 1955, 48. Shannon himself was ambivalent about the vast extension of his ideas. He told the communication theorist John Pierce about once, at a conference, "watching a man who spoke with his eyes turned toward the heavens, as if he were receiving divine revelation. Shannon was disappointed that he could not make any sense out of what the man was saying." Pierce 1972, 32.

109. An instructive figure here is the Harvard psychologist George Miller, later a colleague of Marler's at the Rockefeller University (see chapter 9, note 19). Miller is widely acknowledged as a leader in academic psychology's reengagement with the mind and mental processes from midcentury—in the "cognitive revolution" that brought behaviorism's dominance to an end. In two illuminating interviews, he explained how work during the war on the jamming of transmitted speech (his own project) and related problems put psychologists into contact with mathematicians and engineers, from whom psychologists imported new ideas about feedback mechanisms, signal detection, and information processing and storage. It was these ideas that relegitimized the use of such mentalistic terms as "purpose" (from cybernetics) and "expectation" (from information theory). See Miller 1983, 12–28, esp. 23–27; and Baars 1986, 198–223, esp. 201–7. On wartime scientific collaboration and its consequences, see Heims 1991, 7; and Haraway 1981–82, 249ff. On how behaviorist taboos collapsed under the exigencies of war, see also Galison 1994, 263.

110. See, e.g., Evans 1955, 4; and Quastler 1958, 4–5.

111. On the Macy conferences, see Heim 1991.

112. Ayer et al. 1955. On the early history of the center, see Evans's introduction. Symposia on information theory were held in 1950 (at the Royal Society), 1952 (at the Institute for Electrical Engineers), and 1955 (at the Royal Institution). Two hundred and fifty people, from fifteen countries, attended the 1955 meeting. See Cherry 1956, v, xi.

113. N. Wiener to J. B. S. Haldane, 22 June 1942, quoted in Kay 2000, 80. Wiener described Haldane as his "old friend" in Wiener 1948, 23.

114. Wiener 1948, 23. On Haldane's commitment to a scientifically safeguarded future as the coordinating theme in his bogglingly varied life and work, see Adams 2000, esp. 476, 480. Wiener's visits with the Communist Haldane attracted the attention of the FBI and MI-5; see Conway and Siegelman 2005, 261–62.

115. Wiener 1948, 94. Cf. Haldane 1941, 60, discussed in Kalmus 1950, 19.

116. Wiener 1948, chaps. 5 and 8.

117. J. B. S. Haldane to N. Wiener, 12 November 1948, quoted in Kay 2000, 87. On Haldane's further correspondence with Wiener, see Kay 2000, 87, 350 n. 54.

118. Haldane 1955b. I am not wholly certain of 1953 as the date of these lectures. In his introduction to the volume, Ifor Evans wrote (1955, 7): "It was felt desirable in the first year in which the Centre was operating to hold a series of lectures to invited audiences. It is these papers which are collected in this volume." On 1, he reports 1953 as the year the center was proposed, though when exactly the center and the inaugurating lectures began is unstated. Neither Haldane's, Cherry's, nor Young's chapters in this volume cite sources from later than 1953.

119. The use of the language of "ritual" to describe certain instinctive behaviors in animals goes back at least to Julian Huxley's famous 1914 paper on courtship in the great crested grebe; see Huxley 1914 and, for discussion, Lorenz 1952b, 302–3. According to Lorenz, the choice of the term "ritualization" was "undoubtedly influenced by the analogy of these phenomena to certain types of human behavior. This analogy consists, first, in the fact that a sequence of movements has forfeited its original, everyday meaning and acquired a new one as a means of expression, and secondly, that an originally variable sequence of actions has become a rigid, unchangeable 'ceremony.'" Aside from Lorenz and Tinbergen, Haldane drew on the Cambridge-based ornithologist Edward Armstrong's 1947 book *Bird Display and Behavior* for his evidence about animal rituals.

120. Haldane 1955b, 35.

121. Haldane 1953, 71; see also Haldane 1952. For an interesting discussion of Haldane's "hortatory" theory of the origin of language as reflecting trends in British theorizing about instinct and experience, see Griffiths 2004, esp. 619–22. Haldane himself placed his theory in the tradition of the nineteenth-century theorists Ludwig Noiré and Friedrich Engels, who saw language as arising through collective work. "Its critics," wrote Haldane (1955a, 399), "called it the 'Yo-heave-ho' theory." F. Max Müller was an adherent of the theory near the end of his life (and was the first to dub origin-of-language theories with silly names). See chapter 1.

122. On Von Frisch's marginal role in the building of ethology as a scientific discipline, see Burkhardt 2005, 6. Von Frisch published a book in English on his dance-language research in 1950, based on lectures given in the United States in 1949. On the origins of and debates over that research, see Munz 2005.

123. Haldane 1955b, 36, referring to the decipherers of three ancient human languages, respectively, Egyptian hieroglyphics, Sumerian cuneiform, and Mycenaean "Linear B."

124. A useful summary of their findings up to the mid-1950s is Frisch and Lindauer 1956. Commenting at UCL on Lindauer's work, which he thought would become mandatory reading in political science, Haldane quipped (1955b, 39): "Let no one say that communist insects use force rather than persuasion on their fellows."

125. Haldane and Spurway 1954. Von Frisch and Lindauer took the waggling suggestion seriously enough to begin making slow-motion film studies of the bee dance, in order to count the number of waggles at different distances; see Frisch and Lindauer 1956, 46. On the influence of the paper on an up-and-coming student

of insect communication, Edward O. Wilson, see Haraway 1981–82, 259–60. A little-noted feature is Haldane and Spurway's clear articulation of what came to be known as kin selection (277): "It is possible that the constructional and communicatory capacities of the social insects [bees and termites], and of man, have arisen because these three groups have evolved social organisations which permit non-reproducing members of a community to contribute to the Darwinian fitness of animals having similar genotypes to themselves."

126. Haldane and Spurway 1954, 254–56, quotation on 254. Strictly speaking, theirs was not the first attempt to discuss animal communication within a cybernetic-informational framework. Aside from Wiener's brief remarks in *Cybernetics*, there was the American comparative psychologist Herbert Birch's paper "Communication between Animals," presented at the 1951 Macy conference, and concentrating mainly on the bee dance; see Birch 1952. On the other hand, Haldane and Spurway almost certainly deserve the accolade that Donna Haraway bestowed on the American primatologist Stuart Altmann, of being "the first . . . to apply the 1948 Shannon-Weaver equation [*sic*] for analyzing sequences of messages to animal behavior"; see Haraway 1981–82, 263; also 1989, 104. Altmann was an admirer of what he called the "pioneering paper" of Haldane and Spurway; see Altmann 1967b, 341. For a recent appreciation of Haldane and Spurway's calculation, see Dawkins 2003, 111. On the history of information theory and the study of ant behavior, see Sleigh 2007.

127. It should be noted that the biological sciences figured in the making as well as the reception of information theory. Wiener, for instance, had learned much about homeostasis from the Harvard physiologist Walter B. Cannon (pictured holding Chim and Panzee in figure 6.5), and Shannon, during his MIT years, had looked at the mathematics of population genetics. See Kay 2000, 79, 92. I am grateful to Graeme Gooday for helpful discussion of these matters.

128. Kalmus 1950, discussed in Kay 2000, 87.

129. Young 1951, esp. 14–23, 28–33, quotation on 14.

130. Young 1953.

131. Young 1955.

132. Young 1954, discussed in Marler 1956f, 245, 257–58.

133. Cherry 1953; Broadbent 1958, 22ff.; and, for discussion, Gardner 1987, 91–93. Cherry's and Broadbent's experiments on selective attention in hearing, some of which involved sending different samples of recorded speech into the left and right sides of a subject's headphone, have become immortalized in psychology textbooks as illuminating "the cocktail party problem": how do people at a party tune out ambient speech to register just a select set of voices? The phrase does appear in Cherry's paper, in quotes (976). But his research was sponsored by the military (975). Broadbent too was working to a military agenda, specifically the need to understand problems arising in naval communication centers; see Broadbent 1980, 54. For Marler, the "dichotic listening" experiments, as they came to be known, were important for what they revealed about how sound localization

depends on differences—in timing, loudness, and phase—between what reaches the left ear and the right ear; see Marler 1956e, 80.

134. Broadbent 1956, 354. Cherry had also contrasted his own information-processing approach with behaviorism; see Cherry 1953, 976. By training, Cherry was an electrical engineer, and he was based in the electrical engineering department at Imperial. But, wrote Broadbent (1980, 59), "his students were often doing work that might just as well have been in a psychological laboratory." Cherry had spent 1952 on sabbatical leave at MIT, where he had worked with Norbert Wiener. See Asa Briggs's foreword to Cherry 1985, x.

135. Broadbent 1958, esp. 5, 297–307; the chart is on 299. For its pioneering status, see Gardner 1987, 92. According to Broadbent, the conviction that psychologists had much to learn from communication engineering was widespread in the post-war Cambridge psychology department, thanks largely to the influence of Kenneth Craik and the work he did during the war. Craik's 1943 book *The Nature of Explanation* was, in Broadbent's view, "one of the first statements of the cybernetic point of view of human beings"; see Broadbent 1980, 47. The cybernetic perspective had representatives in the Cambridge life sciences as well, notably John Pringle, Marler's other consultant on sound localization. See Pringle 1951.

136. Marler 1954. The Institute for the Study of Animal Behaviour, begun in 1936, was renamed the Association for the Study of Animal Behaviour in 1949. See Durant 1986, esp. 1613. See also notes 50 and 52 above.

137. On the Paris conference on instinct, see the published proceedings, in Autuori et al. 1956, and for discussion, Burkhardt 2005, 390–94. On Haldane and Spurway's subsequent criticisms of Lorenz, see Griffiths 2004, 615–19; and Burkhardt 2005, 565 n. 47. In January 1954, Haldane gave a lecture at the Sorbonne on "la signalisation animale," published that same year, and largely summarizing his other papers on animal communication. See Haldane 1954, 89. An anonymous referee has suggested to me that Lorenz had provoked Haldane's ire not by having the affair with Spurway but by ending it!

138. See Marler 1989, 322.

139. Haldane 1955a, esp. 392–94.

140. Marler 1989, 322. In his UCL days, Marler had been an "anonymous presence" (322) in Haldane's genetics lectures, so their new friendship was a reunion from only one side.

141. Marler 1956f. In the same year, he published a rather different article under a nearly identical title in the popular Penguin *New Biology* series. See Marler 1956e. The *New Biology* article largely summarized Marler's research on song variation, song learning, and call localization, with just a brief discussion (78) of the information conveyed in chaffinch vocalizations.

142. Nice 1943, 274. On Marler and Nice, see note 64 above.

143. Promptov and Lukina 1945, discussed in Haldane 1953, 68; 1955b, 35; and 1955a, 394. Marler knew the paper; see, e.g., Marler 1956a, 169–70.

144. Tinbergen 1953, 7–12.

145. Marler 1956f, 244. When major variations in the fourteen basic calls were taken into account, the chaffinch had twenty-one calls (259).

146. This is not to say, however, that cybernetic information theory failed utterly to register. At the 1954 Paris instinct conference, for instance, Lorenz recommended "cybernetics" as the source of an "important hint" toward understanding the central nervous system as a mechanism composed of discrete structures with particular functions, some of them computational (like the honeybee dance). See Lorenz 1956, 63. And in Thorpe's 1956 book *Learning and Instinct in Animals* (143–55, esp. 153–55), a summary of Pringle's 1951 paper on parallels between learning and evolution (see note 135 above) prompted some brief reflections on "the implications of cybernetic theories for ethology and psychology" (155). But, to my knowledge, in none of Thorpe's publications through the 1950s did he demonstrate anything like the acquaintance with information theory evident in his 1961 book on birdsong, where his discussion drew on Marler's work. See Thorpe 1961, 7–11.

147. Tinbergen 1953, 8. Lorenz (1935, 243–44) invoked a lock-and-key analogy in explaining the relationship of releasers and innate releasing mechanisms.

148. Lorenz 1952a, 76–91, esp. 76–79, quotation on 76.

149. On the information-theoretic approach in Marler's 1956 paper as "a significant departure from the more mechanical ethological releaser concept," see Owings and Morton 1998, 36–37, quotation on 37. It should nevertheless be noted that some features of the lock-and-key view of call function lent themselves well to an information-theoretic treatment—for instance, Lorenz's view that the structure of calls, qua releasers, balanced maximal simplicity with maximal improbability. See Lorenz 1935, 106–7, 245.

150. Marler 1956f, 245–46, 257, quotation on 246, emphasis in original.

151. Marler 1956f, 246–52, quotations on 249, 250.

152. Marler 1956f, 252.

153. Marler 1956f, 257–59, quotation on 258.

154. Marler 1956f, 257–59, quotation on 258–59. The Haldane papers cited were Haldane 1953 and 1954.

Chapter Eight

1. Schwidetzky 1932, vii. On the beginnings of his research, see 6–7.

2. Schwidetzky 1932, 44–47.

3. Schwidetzky 1932, 20. On the need to go to the animals to trace the origins of language, see 6. On Boutan's debt to Garner, see 47–48. On the desirability of further experiments of this kind, see 113–14. On Schwidetsky's society, publisher of his first book and, by 1938, a number of other books on related topics, see Burkhardt 2005, 260–61, 541 n. 87. I am grateful to Richard Burkhardt for bringing Schwidetzky to my attention. On the disquieting history of plans to crossbreed apes and humans in order to hybridize a "missing link" into being, much discussed in Europe between 1900 and 1920, see Rooy 1995.

4. I quote Burkhardt's translation of Lorenz's letter to Heinroth, 29 January 1940. Burkhardt notes that this passage was omitted from a version of the letter in the published Heinroth-Lorenz correspondence. For the letter, and the Schwidetzky affair generally, see Burkhardt 2005, 261–62, 541 nn. 88–92. The passage from the letter is on 261–62, as is Heinroth's labeling of people such as Schwidetzky as "dangerous," in his letter to Lorenz of 27 January 1940.

5. Schwidetzky died in 1952, never having obtained a university post or affiliation. His tale does intersect once more with that of the primate playback experiment. His daughter Ilse Schwidetzky became an anthropologist who, unlike her father, flourished during the Nazi period, and after. (It is said that she produced viciously anti-Semitic scientific articles on race during the war.) In 1973, she published an edited volume on the evolution of speech, mostly collecting, and presenting in German translation, recent English-language work on the topic, including papers by Marler, Hockett, and Premack (all discussed later in this chapter). See Schwidetsky 1973. On the life and work of Georg Schwidetzky, see Anderson 2001. I am most grateful to Stephen Anderson for sharing this paper and other Schwidetzkyana with me, including very helpful biographical information provided by Clement Knobloch.

6. Marler 1956a.

7. Marler 1955c, 1955d, 1956b, and 1957b.

8. Marler 1956c.

9. Marler 1957a. Lack is among those thanked for comments on the paper (13).

10. Marler 1961a. Marler noted the overlap with Broadbent's interests (165).

11. Marler 1955b.

12. See Marler 2003, 1; also 1989, 325.

13. From the mid-1950s on, the Universities Federation for Animal Welfare (UFAW), an organization with links to the Institute for the Study of Animal Behaviour and its successors, and with a flourishing branch at Cambridge, began pressing for more humane experimentation on animals. See Wilson 2002a, 243–44, and, for Thorpe's involvement with animal welfare issues, 249.

14. In one set of aggression experiments, Marler dyed the breast feathers of female chaffinches red, in imitation of male breast feathers. He found, among other results, that the sight of the dyed females triggered the same avoidance responses in other females as the sight of males triggered. "The red breast of the male Chaffinch," he concluded, thus "conforms with the definitions of Lorenz (1935) and Tinbergen (1948) of a 'social releaser.'" Marler 1955d, 144.

15. On these differences, see chapter 7. A major finding of Marler's aggression studies, that chaffinches harbor no innate aggressive tendencies, later put him at odds with another of Lorenz's views. See Marler 1957b; also 1989, 323.

16. For Lorenz's dictum that, with releasers, "similarity *always* means homology," see Lorenz 1935, 249, emphasis in original; see also Lorenz 1941, 19. He sometimes defended this assumption by way of an analogy with comparative philology; see Lorenz 1941, 19; also 1950, 242. Lack picked up on the analogy; see Lack 1943,

45. On nineteenth-century antecedents to such language-species comparisons, see chapter 1 in this book.

17. See Marler 1957a, 13, 15. Lorenz later conceded the point; see Lorenz 1970–71, 2:338 n. 4. Dissent from Lorenzian views within ethology was by no means unique to Marler, of course. Paul Griffiths (pers. comm., April 2004) has noted that, like Marler, and at the same time, Tinbergen was moving away from Lorenz in functionalist directions—a convergence due, it seems, to Marler's and Tinbergen's independent contacts with the Oxford tradition of adaptationist natural history.

18. On the history of the playback experiment, see Falls 1992. Entomologists also figured among the midcentury playback enthusiasts. The American student of ants T. C. Schneirla, mentor to Daniel Lehrman, called attention in 1950 to the experimental possibilities of field playbacks under selected conditions. Schneirla 1950, 1022–44. I am grateful to Mandy Rees for this reference. Other proponents were less high-minded, such as the exterminators who used playbacks to flush rats out into the open, or the hunters who did the same to deer. See Morton 2000, 142–43.

19. Collias and Joos 1953, 177. Going further, Collias and Joos analyzed the attractive sounds into separate acoustic components, recorded organ tones mimicking each of these elements, then played the recordings to the chicks to see which elements had the most pulling power; see 184–87. For discussion, see Falls 1992, 23; and chapter 7 in this book.

20. Haldane 1955a, 393; Marler 1956a, 88; and 1956f, 250. Marler also reported playbacks with chaffinches inducing aggressive displays (in Marler 1956c, 497) and evoking subsong (in Marler 1956f, 237–38). Haldane (392) took a lively interest in the technical problems surrounding the artificial reproduction of animal sounds as cues for animal responses. In interview (9 September 2003), Marler emphasized how constrained playbacks at the time were technically. He recalled that his primary aim in using playbacks of songs with unmated females was to see whether they would respond preferentially to their home dialects. Although birds tolerate greater distortion in their songs than in their alarm calls, he nevertheless found responses so variable that the experimental data "had no statistical potency whatsoever."

21. In collaboration with Judith Stenger Weeden, Falls used field playbacks to territorial male ovenbirds to show that they recognize individual differences in song, reacting more strongly to a stranger's song than to a neighbor's song. See Weeden and Falls 1959. For discussion of this and related experiments, see Falls 1992, 14, 23; Catchpole and Slater 1995, 134; and Terborgh 1996, 42–43.

22. Hinde 1958; Thorpe 1958. On the Song Tutor, see Thorpe 1958, 543. It was introduced, wrote Thorpe, because the "prospect of spending several hours a day piping to Chaffinches on a bird flageolet or a mouth organ did not appeal to us" (558). On the pioneering status of Hinde's and Thorpe's papers, see Falls 1992, 23.

23. The interests of quite a few scientists specializing in birds, including Hinde and

Collias, grew to include primates. On the multiple points of overlap between ornithology and primatology in the twentieth century, see Haraway 1989, 391 n. 11.

24. The citation to Zuckerman 1932 is in Marler 1957b, 34; the one to Carpenter 1942 is in Marler 1955d, 144. On Zuckerman's primatology, see Burt 2006.

25. Marler 1959; see esp. 170–72 (discussing Kohts 1935), 202–3 (discussing Boutan 1913, Carpenter 1934, and Yerkes and Yerkes 1929). On Kohts's research, see Boakes 1984, 201–2; see also chapter 6 of this book, note 74.

26. Marler 1959, 205, 204.

27. Marler 1959, 203.

28. Marler 1959, 206.

29. Marler 1961b, 295–96, citing Cherry 1957. Cherry was among those thanked for comments on an earlier draft of the 1961 paper; see 316.

30. Hockett gave an account of his early life and career in Hockett 1980. Also useful for biographical background is the preface to Agard et al. 1983 and an obituary notice, *Cornell Chronicle*, 16 November 2000, http://www.news.cornell .edu/Chronicle/00/11.16.00/obit.html (accessed April 2004). For Hockett's bibliography up to 1976, see Hockett 1977, 323–29. As a linguist, Hockett was identified especially closely with Bloomfield, and indeed went on to edit an anthology of Bloomfield's writings; see Hockett 1970. One recent historian of twentieth-century linguistics has described Hockett at midcentury as "the Bloomfieldian boy-wonder"; see Harris 1993, 43. On Bloomfield and the Boasian tradition, see chapter 5 in this book. On the emphasis in that tradition on training in American Indian languages, see Hockett's own testimony in note 138.

31. Hockett 1948, 9 for "sociobiology." The page reference is to the 1977 reprint, where Hockett added a note stating that, as far as he knew, E. O. Wilson's more famous coinage of "sociobiology" had been independent. On the history of the term, see Haraway 1989, 394 n. 39.

32. On linguistics and sociobiology, see Hockett 1948, 10–16, esp. 12.

33. Hockett directed readers to Bloomfield's *Linguistic Aspects of Science* (1939)—his contribution to the famous *International Encyclopedia of Unified Science*. See Hockett 1948, 17, citing Bloomfield 1939. On the unity-of-science movement, see Reisch 2005, esp. 8–21. On Bloomfield's behaviorism, see Fries 1963, 204–9. Bloomfield in certain moods could sound almost Spencerian, e.g., "language is to the social organism what the nervous system is to the individual." Bloomfield 1942, 396.

34. For Kroeber's earlier views on animal communication, see chapter 5. In 1952, he published a paper on "Sign and Symbol in Bee Communications," arguing that, on the evidence then available, the impressive communication system of the bees seemed at most "pseudo-symbolic" (757), lacking as it did the plasticity of true symbol systems. But he looked forward to further research that might clarify the matter. See also Kroeber 1955, which Hockett later praised (1958, 586) as suggesting "a new and important type of investigation of animal behavior in comparison with human." While at the Center for Advanced Studies in the Behavioral

Sciences, as Hockett recalled, he produced a first paper on the origins of language, intended for an issue of *Language* in honor of Kroeber, but later judged the paper unripe and withdrew it from publication; see Hockett 1977, 124. He did, however, dedicate a successor to Kroeber; see Hockett 1959, 32.

35. On Hockett's time at the center, see his remarks in Hockett 1960b, 48; also 1977, 124.

36. Hockett 1956, reviewing, among other books, Simpson 1950. Hockett also (461) referred readers to Julian Huxley's *Evolution: The Modern Synthesis* (1942). On the modern synthesis and the unity-of-science ideal, see Smocovitis 1996, esp. 100–153.

37. By 1964, *A Course in Modern Linguistics* had gone through seven printings. The editors of the 1983 Festschrift for Hockett described Hockett's *Course* as "a book which initiated many of us into the profession." Agard et al. 1983, x.

38. Some fifteen years later (1973), Hockett published a textbook with the same name.

39. Hockett 1958, 573.

40. When Hockett gave a briefer version of the same definition in Hockett 1959, 35, he cited a 1942 essay of Bloomfield's, "Philosophical Aspects of Language." Recalling his first reading of Bloomfield's *Language*, Hockett reckoned that it had led him to see "analogically conditioned trial and error as the mechanism not only of speaking but of all human (or even organic) action"—something he still believed more than forty years later. See Hockett 1977, 1.

41. Hockett 1958, 574–80.

42. Hockett 1958, 574.

43. Hockett 1958, 581.

44. Hockett 1958, 581–84, quotation on 584.

45. Hockett 1958, 583.

46. Hockett 1958, 582.

47. Hockett 1958, 583. In his 1956 review (462) of recent biological literature bearing on the origin of language, Hockett cited the chimp-rearing Hayeses for the related view that chimpanzees "developed neither culture nor language nor larger brains because they needed none of these for survival in their particular ecological niche." On the Hayeses, see the discussion later in this chapter.

48. Hockett 1959, see 32. The phrase "design features" was one Hockett had used before; see, e.g., Hockett 1955a, 106; 1958, 137–38 (where, however, it names not the seven properties of language, but the five systems of which languages are said to be composed: grammatical, phonological, morphophonemic, semantic, and phonetic).

49. Appropriately, the symposium's sponsor was the Section of Animal Behavior and Sociobiology of the Ecological Society of America. See Lanyon and Tavolga 1960, preface. The source of this use of "sociobiology" was not Hockett, however, but the comparative psychologist John Paul Scott, who had used it a couple of years before Hockett. See Haraway 1989, 394 n. 39.

50. Hockett 1960a, 392–93.
51. Hockett 1960a, definition of "semanticity" quoted from 409.
52. Critics did wonder, however, whether the "vocal-auditory channel" criterion had defined language out of the reach of those who communicate by signs—including, from the late 1960s, signing chimpanzees. See Lyons 1977, 143–44.
53. Hockett 1960a, 409.
54. Marler 1960.
55. For the revised grid, see Hockett 1960a, 428. The compared systems had expanded as well as the design features, to include cricket songs, western meadowlark songs, instrumental music, and "paralinguistic phenomena"—vocal sounds accompanying but not strictly part of language.
56. See esp. Hockett 1960a, 399.
57. Hockett 1960a, 398, 414.
58. Hockett 1960a, 392.
59. Marler thanked Hockett for comments on a draft of Marler 1961b; see 316.
60. Hockett 1960b. In 1964, in collaboration with the anthropologist Robert Ascher, Hockett published a long paper, with extensive commentary from other scholars, elaborating his account of the origins of language. See Hockett and Ascher 1964. Meanwhile, the number of design features in Hockett's scheme continued to expand, reaching, in its settled mid-1960s form, a total of sixteen. For the expanded list, see his collaborative paper with the primatologist Stuart Altmann, who had been working independently along similar lines. Hockett and Altmann 1968, esp. 63–64. At the end of the seventies, the British linguist John Lyons (1977, 143–45), a critic, complained about how often Hockett's list was used to grade communication systems.
61. Marler 1961b, 296–97. Marler did not let Hockett completely off the hook, however, arguing that Hockett's "interchangeability" criterion was met much more satisfactorily by nonhuman communication systems than by human ones (303–4). Marler sometimes struck a more conciliatory note toward comparative psychology; see, e.g., Marler and Hamilton 1966, 22.
62. Marler 1961b, 298–309. On semiotic and its levels, see Cherry 1957, 219–26 (in a chapter entitled "On the Logic of Communication [Syntactics, Semantics, and Pragmatics]"). As Cherry pointed out (219, second note), John Locke coined "semiotic" to name "the doctrine of signs" in his 1689 *Essay Concerning Human Understanding*, book IV, chap. 21. The Genevan philologist Ferdinand de Saussure, widely regarded as the guiding spirit of the culturally engaged science of signs that flowered in Paris and elsewhere in the 1960s, proposed what he called "semiology," defined as a "science that studies the life of signs within society," in his lectures published posthumously in 1916 as his *Course in General Linguistics*. See Saussure 1974, 16. For Morris on semiotics (he used the plural form), see Morris 1938 and 1946. The 1938 volume was part of the same *International Encyclopedia of Unified Science* to which Bloomfield contributed (see note 33 above), and Morris indeed presented semiotics as a kind of master science, as all

the special sciences dealt with signs in one way or another. On Morris's unificatory project and its influence, see Haraway 1989, 97–99—though, as I noted earlier, I think Haraway has exaggerated the impact of Morrisian semiotics on Carpenter's primatology.

63. Shannon and Weaver 1949, 27.
64. Cherry 1957, 226; also 168.
65. Marler 1961b, 301.
66. According to Cherry (1957, 242–43), "'Information' in most, if not all, of its connotations seems to rest upon the notion of *selection power*. The Shannon theory regards the information source, in emitting the signals (signs), as exerting a selective power upon an ensemble of messages. . . . Again, signs have the power to select responses in people, such responses depending upon the totality of conditions" (emphasis in original).
67. Marler 1961b, 299. In the discussion of animal pragmatics on the same page, Marler wrote: "We . . . require a means of inferring information content from the nature of the response given. We may note in passing that the information theory developed by Wiener & Shannon (Shannon & Weaver 1949) is of no help to us here since it operates only 'at the syntactic level' (Cherry 1957)." The latter part of the paper, on animal syntactics, contained Marler's first sustained reflections on continuous versus discrete signals, the selectional pressures favoring one or the other, and whether animals combine signals in ways that vary their information content. See Marler 1961b, 309–15, esp. 309–13; and, for brief earlier remarks, Marler 1959, 164–65.
68. Marler 1961b, 306–8, esp. 308.
69. Thorpe 1956, 369.
70. Marler and Tamura 1962, esp. 375; also Marler 1963a, 795.
71. The assembly and matériel of the lab are described in Marler, Kreith, and Tamura 1962, 12–13.
72. The best-known paper from the era is Marler and Tamura 1964. On Marler's group at Berkeley, and the work done by his students (whose interests were more varied than I have indicated here), see Marler 1989, 327–29.
73. For Marler on Nottebohm and Konishi's contributions, see Marler 1967a, 12–13; also 1989, 329, 334. Nottebohm's discovery of neurogenesis in bird brains in the 1980s is now seen as having sparked an explosion of work on that subject. Cures for Alzheimer's, Parkinson's, and other neurodegenerative nightmares may well be among the legacies of the Marlerian tradition in song-learning science. On the Nottebohmian present and future, see Specter 2001—which, however, presents Nottebohm as utterly self-formed.
74. Marler and Hamilton 1966, 686–89, 696–701. For an instance of later textbook treatment, see Campbell 1993, 1168.
75. Washburn 1951, 155. A table summarizing Washburn's views on the differences between the two anthropologies can be found in Washburn 1952, 716. On the 1950 symposium, see Bowler 1986, 240–44. For "racial quality" as discussed in

the most popular work of Washburn's Harvard teacher, Earnest Hooton, see Hooton 1947, 658–61. On Washburn's life and work, see Washburn 1983; Haraway 1989, chap. 8; and esp. Strum, Lindburg, and Hamburg 1999, which includes reprintings of several key papers, testimonials from students and colleagues, and a bibliography of his writings.

76. Washburn 1952, 719.

77. Washburn 1952, 723–24. In later life, Washburn (1983, 3, 6–7) credited his conversion from anthropometrical anthropology to two sets of influences: his mentors in medical (hence functional) anatomy, W. T. Dempster at Michigan and S. R. Detwiler at Columbia; and his reading of three books, Percy Bridgman's *Logic of Modern Physics* (1927), which "gave clear guides to the relations of technique to analysis," and thus revealed biometry as designed to describe not man, but bones; C. K. Odgen and I. A. Richards's *Meaning of Meaning* (1923), which "helped one to see the nature of words and avoid at least some of the common errors"; and Bronislaw Malinowski's *Argonauts of the Western Pacific* (1922), which "read like a great novel," and so taught by example that "[h]uman behavior cannot be reduced to trait lists."

78. Washburn 1983, 2–6. Washburn noted (4) that Carpenter had set up camp a few miles from where Schultz was based. The separation was more than just geographical. According to DeVore, Washburn tried and failed repeatedly to interest the senior men in each others' work, with Schultz complaining to him at one point: "Sherry, that man calls a tree limb a 'behavioral platform.' I have no patience with him!" Interview with I. DeVore, 12 September 2002. On the expedition more generally, see chapter 6 in this book. En route to Asia, Washburn worked as a lab assistant to the era's other well-known student of primate behavior, Solly Zuckerman; see Ribnick 1982, 59.

79. See Washburn 1983, 16–17.

80. Interview with DeVore. See also Haraway 1989, 219–20.

81. Washburn and DeVore 1961, 102.

82. On Marler's interactions and collaborations with Washburn, see Marler 1989, 328, 330. On the primate colony at the station, see Washburn 1983, 21; and Ribnick 1982, 61. For many ethologists, including Marler, Frank Beach represented the acceptable face of comparative psychology. On Beach's classic critique of the anthropocentrism of his own discipline, see chapter 6. Beach, Hinde, Tinbergen, and Washburn had been among the participants in important meetings in the mid-1950s on behavior and evolution; see Roe and Simpson 1958.

83. Marler 1963b.

84. In a 1966 letter of recommendation for DeVore, Marler wrote that DeVore "is more responsible than anyone else for my present enthusiasm for primate behaviour." Peter Marler to Douglas Oliver, 27 January 1966, Peter Marler Papers, University of California at Davis.

85. On the invitation and its consequences, see Marler 1989, 330–31.

86. DeVore 1965, ix.

87. Observers then and later have noted the stimulus that increasing use of primates in postwar medical research gave to field studies of these animals. "The combined effort of anthropologists, psychologists, and zoologists, which stems in part from demands made by the health-oriented sciences for better information, has instigated a wide range of field studies of primates," wrote Marler in his 1963 review of Schaller's gorilla study (Marler 1963b, 1081). Funding for the Primate Project indeed came from the National Institutes of Health (DeVore 1965, viii). On the founding in these years of seven regional primate centers in the United States by NIH and other bodies, see Dukelow 1995 and the useful review in Weidman 1996. Medicine played a less direct role as well, through the development of antibiotics and other drugs, which made fieldwork in tropical climates less dangerous; see Ribnick 1982, 55. For historical surveys of primatological fieldwork in this and other periods, see Ribnick 1982; Haraway 1989, chap. 6; and Rees 2006.

88. On the early days of Madingley primatology, see Hinde 2000, 106–7. On Goodall and Leakey, see, e.g., Haraway 1989, 151; and Goodall 2001, 1.

89. On Altmann's early work, see Haraway 1989, 101, 104–8; and Wilson, 1994, 308–11. Wilson here credits the first stirrings of sociobiology to his 1956 visit with Altmann to the rhesus colony on Cayo Santiago, near Puerto Rico. (The colony was NIH's—see note 87.) Altmann's melding of animal communication and information theory was independent of Marler's; see Haraway 1989, 401 n. 11, on the "two streams of communications analysis," one in the United States, the other in the United Kingdom and Europe. Indeed there were still more—a French stream, for instance (Moles 1963), and a Russian one too (Zhinkin 1963).

90. Marler 1965, 558. Schaller, Goodall, the Reynolds, DeVore and Hall, and Jay all have papers in the same volume, DeVore 1965. For the relevant papers by the others, see the references in Marler's table. On postwar Japanese anthropology, see, e.g., Ribnick 1982, 55–57. For vivid—and typically quantitative—testimony to the explosion of primatological field studies since the mid-1950s, see the graph in Altmann 1967a, xi. On a study of the vocalizations of semicaptive baboons in the Soviet Union of the era, see Zhinkin 1963, 150ff.

91. Marler 1965, 583. DeVore stressed in interview that his impression watching baboons—and Washburn's too—was that vocalizations are mere "exclamation points" to what is expressed by face and gesture.

92. Marler 1965, 558–65, with discussion of Rowell 1962 on 559–61.

93. Marler 1965, 583.

94. Marler 1965, 565–66.

95. Marler 1965, 584. See also 567, and Marler 1963b, 1082 (of mountain gorillas: "there is no evidence that certain sounds are associated with particular objects in the environment").

96. Marler 1965, 584. For rather different, Hockett-inspired reflections on the same set of Primate Project signaling data, see Bastian 1965 (for the debt to Hockett, see 604).

97. Lancaster 1968, 441.

98. Marler 1965, 545–46.

99. Falls 1963. There was, of course, a precedent in Collias and Joos's 1953 paper; see note 19 above. The most famous experimental studies in this genre from this era are those of the French scientist J.-C. Bremond, credited with discovering the "syntactical rules" governing the construction of the song of the European robin. See Catchpole and Slater 1995, 123–27; and Terborgh 1996, 43.

100. Marler 1965, 569.

101. The colobines studied were the red colobus (Marler 1970c) and the black-and-white colobus (Marler 1969b; 1972). The guenons—the genus including the vervets—were the red-tailed monkey and the blue monkey (Marler 1973a). On the difficult field conditions, see Marler 1972, 177; 1973a, 225. Marler emphasized to me in correspondence that another major obstacle was the noisy, broadband tonal quality of many primate calls, making sonograms harder to read and categorize by eye. On the year in Uganda, see Marler 1989, 331–32. The fieldwork with monkeys—and perhaps especially that reported in Marler 1969b—also generated a kind of Primate Project postscript, surveying the roles of primate signals in increasing, maintaining, or decreasing space between groups and within them; see Marler 1968. It is a strikingly classical paper, deliberately modeled on Tinbergen's herring-gull studies (423), and dealing not in "information" but in "functions."

102. On Marler's time with Goodall, see Marler 1969c, 94; Marler and Hobbett 1975, 97–98; Marler 1989, 335; and Goodall 2001, 43ff., with the letter to Leakey on 54, emphasis in original. To another correspondent she reported (54): "Flint [a chimp] has a thing about Peter Marler. He runs up & hits him at the slightest provocation! Doesn't like his beard!" Goodall described the baiting system in Van Lawick-Goodall 1968, 317, with a photo on 357. I learned of the prohibition on playbacks from interviews with Marc Hauser, Phyllis Lee, and Peter Marler.

103. On the Institute for Research in Animal Behavior in its early days, see Marler 1989, 332–33; and Griffin 1989, 139–40.

104. In his 1969 *Science* paper on the black-and-white colobus, Marler compared spectrograms of the roar of adult male black-and-white colobus monkeys from Uganda with Nottebohm-recorded roars of adult male red howler monkeys from Trinidad. So similar were the roars in physical structure, wrote Marler (1969b, 95), that "it is tempting to think of it as an example of convergent evolution in behavior."

105. For much of the biographical detail here, I am indebted to Thomas Struhsaker (pers. comm., autumn 2004). On Struhsaker's route to Amboseli and his research there, see Marler 1989, 333; and Struhsaker 1967a, 281–82. On the Amboseli vervets' ecology, habits, and rapid habituation to humans, see Cheney and Seyfarth 1990, 19–24.

106. Struhsaker 1967a, table on 314–17, with discussion on 304–12. For Carpenter's tables, from his 1934 howler monkey paper and his 1940 gibbon paper, see Carpenter 1964, 76–77 and 248, respectively.

107. Struhsaker himself identified the paper's major contribution as its demonstration

that—contrary to the views of Marler and Altmann—the functional contexts of calls can and should be taken into account from the beginning in devising call taxonomies; see Struhsaker 1967a, 318–19, 323. Altmann did make a connection between the vervet alarm calls and questions of semanticity. By now in collaboration with Hockett, with whom he shared interests in information-theoretic approaches to animal communication, Altmann reviewed the Hockettian design features for language and their applicability to primate communication systems. Under "semanticity," he wrote that alarm calls and food calls were "instances of semantic messages" and that Struhsaker had "described a variety of alarm calls in vervet monkeys." See Altmann 1967b, 338, 339.

108. The primatologist Alison Jolly, who was not present but had a paper read, got the impression that Struhsaker had identified five alarm calls; see Jolly 2000, 77.

109. The picture is even more complex than indicated here. For full details, see Struhsaker 1967a, 304–12. For his conclusions about gestural communication, social structure, and ecology among the vervets and other cercopithecines, see respectively Struhsaker 1967b, 1967d, and 1967c, esp. 899–901 on predators.

110. Marler 1967b, 771–72. In fact, Struhsaker reported that whether they were flying or perched, eagles evoked the "rraup" call; see Struhsaker 1967a, 308. On his published account, it was not eagle behavior, but the duration of an encounter with an eagle, that led to a change in the call elicited, from "rraup" to "chirp." But things are even more complicated than this. It was, according to Struhsaker, only adult females and juveniles who rrauped on initial perception of eagles and chirped subsequently. Subadult and adult males gave "threat-alarm-barks" in response to eagles from beginning to end. See 311 for Struhsaker's summary.

111. Marler 1967b, 771–72. Again, Struhsaker's own report (1967a, 308) made little fuss about how the vervets' contexts shaped their responses to "rraup." He wrote merely that "immediately subsequent to the *Rraup* call, the vervets ran or dropped into the dense thickets from the tree branches and the open, short-grass areas."

112. Smith 1963, 1965. For biographical background, see the contributor's note in Sebeok 1968, xv.

113. Hockett was also present. On the symposium date and participants, see Sebeok 1968, v. For a report on the symposium, see Ramsay 1966. For Smith's paper, see Smith 1968. The symposium organizer was Thomas Sebeok, a linguist based at Indiana University, in the process of launching himself with great energy into what he called "zoosemiotics." Sebeok was to become a central figure in organizing skeptical reactions to the ape language projects: see chapter 9 in this book. Apart from Marler's paper, the symposium papers and others appeared in two volumes: Sebeok 1968 and Sebeok and Ramsay 1969.

114. See, e.g., R. M. Seyfarth and D. C. Seyfarth, Proposal to the National Science Foundation, December 1979, 3–5, Marler Papers; Seyfarth, Cheney, and Marler 1980b, 1090–92.

115. See Smith 1965, 406, 408.

116. Following Cherry (as, previously, had Marler), Smith (1965, 406) argued "that

meaning be identified as the response selected by the recipient from all of the responses open to it." But what is "the response"? On Smith's interpretation, a "rraup" call that elicits crouching has a different meaning from a "rraup" call that elicits running. It follows for the Smithean that *rraup* has no determinate meaning independent of context. For the postplayback Cheney, Seyfarth, and Marler, by contrast, what mattered was not the ultimate escape behavior but the immediate response. On their interpretation, *rraup* means "eagle" because the call consistently evokes looking up, irrespective of the listener's situation. Context enters only in determining a vervet's course of action once it has been put on the alert for an eagle, thanks to a call representing "eagle." See Seyfarth, Cheney, and Marler 1980b, 1092.

117. Smith 1965, 405.

118. Smith 1977, 181.

119. On Smith's contextualism and its impact, see Hauser 1996, 60–62.

120. Lancaster 1968, 446. On her dissertation, and the date of the second Wenner-Gren meeting (September 1965), see her online CV, at the Web site of the Department of Anthropology, University of New Mexico: http://www.unm.edu/~anthro/faculty/cvs/lancastercv.pdf (accessed 12 September 2007).

121. Lancaster 1968, esp. 446–54, 439; Geschwind 1965, esp. 272–77; but see also the more schematic, language-focused Geschwind 1964. Geschwind (1965, 641) called the angular gyrus the "association area of association areas." For a lucid discussion of the Geschwindian picture and its sources, see Rosenfield 1985, 49–51. For Lancaster's debt to discussions with Geschwind, see Lancaster 1968, 439.

122. Lancaster 1968, 453–54.

123. On the vervets, see Lancaster 1968, 444–46; on "the difficulty with which a monkey or ape learns any rudiment of what might conceivably be called language," see 453 n. 5, 451.

124. On object naming, tool use, and human evolution, see Lancaster 1968, 453 n. 5, 456–57. For Washburn on the same themes, see Washburn 1959, esp. 27–29, where he drew on Wilder Penfield's experimental research on the proportions of motor cortex devoted to different parts of the body—the best-known localization study of the previous generation—to argue that the "reason that a chimpanzee cannot learn to talk is simply that the large amounts of brain necessary for speech are not there" (28). For Penfield's research, conducted on patients undergoing brain surgery under a local anesthetic, see Penfield and Rasmussen 1950; for discussion, see Fancher 1996, 101–5. Washburn and Lancaster later published a paper together on the evolution and origin of language; see Lancaster and Washburn 1971.

125. On Paul Flechsig's turn-of-the-century studies of association cortex, see Finger 1994, 308–9. On the identification of the limbic system, see Finger 1994, 286–90; and esp. Durant 1985. On the disenchantment with localization studies in the pre-

Geschwind era, see Rosenfield 1985, 49; Robinson 1967a, 138; and Geschwind 1974, ix.

126. On the archival dimension of the 1965 paper, see the quotation from Geschwind in the obituary by Galaburda (1997, 296). See also Rosenfield 1985, 49–51. For Geschwind's papers on the history of studies of aphasia, see Geschwind 1974. Another historically minded aphasiologist of the period with an interest in animal communication and the origin of language was MacDonald Critchley. See Critchley 1958, 1960, and 1970. His archly discursive papers had little impact on the wider discussion, however.

127. Robinson 1967a.

128. Ploog 1967. Ploog and his colleagues went on to do a great deal of research along these lines.

129. For many testimonials to the impact of Geschwind and his work on the rising generation, see Schachter and Devinsky 1997. Note too the titles of two journals begun in 1966: *Experimental Brain Research* and *Physiology and Behavior*.

130. Magoun in Darley 1967, 18. Geschwind was present at the meeting.

131. Robinson 1967b, 353. He wrote that in his experimental study of rhesus monkeys, "several hundred neocortical sites were stimulated without the production of any type of sound" (353).

132. The best brief introduction to the language-learning experiments discussed in this section and chapter 9 is Wallman 1992, chap. 2. For a detailed comparative table of the various "ape language cognition projects" in the post-Viki period, see Ristau 1996, 670–76. Two good journalistic treatments from the 1970s are Linden 1974 and Desmond 1979. An opinionated but nevertheless useful history is Candland 1993, 286–351. For an excellent reflective essay on the history of primate communication studies more generally, see Lestel 1995.

133. Hayes 1952, quotation on 146. After a frustrating day spent trying to film lab chimps vocalizing, Keith Hayes (146) "came home muttering dolefully, 'With men who know chimpanzees best, it's white rats two to one.'" On Catherine Hayes's background, see Premack 1976, 29. On Viki's death, see Desmond 1979, 28.

134. Hayes 1952, 240–41.

135. Hayes 1950; Hayes and Hayes 1951.

136. Hayes 1952, 170. On the same tour, Viki met Robert Yerkes (164) in New Haven and the Kelloggs in Bloomington (171).

137. Hayes and Hayes 1954, esp. 297–98, 301.

138. For biographical information, see Premack 1973.

139. Premack 1976, 79.

140. Premack and Schwartz 1966; for the phrase "mother surrogate," see 319. See also Premack 1976, 78.

141. On the difficulties Premack and his collaborator Arthur Schwartz had in using the joystick, see Premack and Schwartz 1966, 346.

142. See Premack 1971; on symbolization and Sarah, see 820. For more accessible ac-

counts of the experiments, see Premack and Premack 1972; and Premack 1976, chaps. 6 and 7. On the use of rewards in Sarah's training, see Wallman 1992, 16.

143. On the Gardners as teachers and people, see Fouts 2003, 28, 59–61. See also Linden 1974, 16.

144. On films of Viki as a spur to the Gardners' experiment, see Linden 1974, 15–16, 244; and Fouts 2003, 27.

145. The most famous of these graduate students, Roger Fouts, has published a moving memoir of his "conversations with chimpanzees": Fouts 2003. On life with Washoe in Reno, see chaps. 1–5. As Fouts explains, Washoe was acquired through the same military program that supplied "chimponauts" for the first American space flights.

146. Gardner and Gardner 1969, 664–65. Fouts, who made guidance in Washoe's education his PhD dissertation topic, considers guidance a kind of rewardless teaching by imitation, wholly distinct from the instrumental shaping of behavior with rewards. But the Gardners described guidance as a variety of instrumental shaping. See Fouts 2003, 76–84; and Gardner and Gardner 1969, 672.

147. Gardner and Gardner 1969. On their attitude to methods, see esp. 668 ("a minimal objective of this project was to teach Washoe as many signs as possible by whatever procedures we could enlist," etc.).

148. Consider an illustrated sequence in a 1973 children's book on communication (Bear 1973, 38–39). The groovy scientist (most likely Fouts) sits cross-legged on the floor opposite a chimp, Lucy. He signs: "Who are you?" She signs: "Lucy." He signs: "What do you want?" She signs: "Tickle."

149. Hill 1980, 351.

Chapter Nine

1. For overviews of the projects, contemporary and retrospective, see chapter 8, note 132. The ape language projects as scientific and cultural phenomena still await their historian. Documentaries include *The First Signs of Washoe* and *Koko: A Talking Gorilla*; see Gardner 1980, 3. Magazine pieces include a long *New Yorker* article on "Washoese" (Hahn 1971) and a *National Geographic* cover story on Koko, "Conversations with a Gorilla" (Patterson 1978). The *Reader's Guide to Periodical Literature* reveals the period 1975 to 1980 as the projects' print-media boom years. Popular science books included three cited in the last chapter: Linden 1974, Premack 1976, and Desmond 1979. Novels featuring signing apes included Peter Dickinson's *Poison Oracle* (1974) and John Goulet's *Oh's Profit* (1975); see Sebeok 1980, 426. For a critic's recollection of the love affair between popular culture and the ape language projects, see Pinker 1994, 335–36. There had, of course, been critics of the projects from early days, but their complaints had had little public impact. It should be noted as well that the period was also the heyday of popular interest in attempts to communicate with dolphins and whales; see, e.g., Crail 1981.

2. "Are Those Apes Really Talking?" *Time*, 10 March 1980, 50, 57. For Terrace's case, see Terrace 1979, chap. 13. The anthology—still a valuable resource—is Sebeok and Umiker-Sebeok 1980; see 433 for a remark of Chomsky's like the quoted one. Not long after the *Time* article, Martin Gardner, famous scourge of pseudoscience, published a joint review of these books in the *New York Review of Books* (1980), savaging the ape language projects.

3. Entitled "The Clever Hans Phenomenon: Communication with Horses, Whales, Apes, and People," the conference was held on 6–7 May 1980; see Sebeok and Rosenthal 1981. On much of the proceedings as "summed up by a participant who said that 'the tide had turned' against the ape-language studies," see the *New Yorker*'s "Talk of the Town" piece "Conference" (26 May 1980, 29). For Sebeok's charge—repeated many times since (e.g., Seyfarth and Cheney 1992, 122)—that the signing apes were like circus-trained animals, see the report on the conference in *Science* (Wade 1980). For the reminiscences of one of the few ape language researchers to participate besides Terrace, see Savage-Rumbaugh and Lewin 1994, 49–57. For a scholarly analysis of the images of science operative in Penny Patterson's book on her pupil Koko and Sebeok's negative review of it, see Prelli 1989.

4. For Fouts on Terrace's accusations and the damage they caused to the reputation of the projects, see Fouts 2003, 273–78, quotation on 275, emphasis in original. On Terrace's coinvestigator, see Patterson 1981.

5. Schmeck 1980, reporting on the experiments described in Seyfarth, Cheney, and Marler 1980a.

6. Marler 1959, 206, referring to Darwin 1871, 1:62.

7. Lyons 1970a. The book sold well enough to go into a second printing the following year.

8. Marshall 1970, 234–36, with a further, approving summary on 240 of Geschwind's picture of the language-capable human brain, where the angular gyrus region supports the cross-modal connections needed for object naming, and the language-incapable brains of the "sub-human" primates, who lack the angular gyrus, and enjoy only limbically mediated sensations.

9. The best introduction to Chomsky's life and ideas up to the late 1960s comes from that time: Lyons 1970b. For a more recent, richly detailed account of the same, attending especially to the controversies among Chomskyans in the 1960s and 1970s, see Harris 1993. Chomsky's version can be sampled in the interview in Baars 1986, 338–51; and in Barsky 1997—an uncritical biography full of quotations from Chomsky's letters to its author.

10. Chomsky 1957b, 15–17, naming and shaming the statistical model in the opening pages of Hockett's *Manual of Phonology* (1955). See also Chomsky's long, hostile review of Hockett's book in the *International Journal of American Linguistics* (1957a, esp. 223–26). Although Hockett was by no means the only proponent of statistical models, he was crucial in bringing them to the attention of linguists, publishing an enthusiastic notice of Shannon and Weaver's 1949 book in *Language* in 1953.

11. Chomsky 1959, reviewing Skinner 1957. Chomsky set out his views on an inborn "language-acquisition device" more fully in Chomsky 1965, 47–69, quotation on 55. On the Skinner review's misrepresentation of its subject, see Joseph 2002, esp. 175–76. On the historical importance accorded the review within cognitive science, see Gardner 1995. For the position of the Chomskyan revolution within a wider cognitive revolution, see Gardner 1987, esp. chap. 7.

12. Chomsky 1968, 66–71, quotation on 70. See also Chomsky 1967, 73–74, 85. Though he dismissed the animal origins of language as a topic of serious inquiry, Chomsky was not indifferent to questions about language and the brain; see, e.g., Chomsky 1959, 43 n. 31; 1967, 81; and his contribution to the Harvard neuroscientist Eric Lenneberg's book (1967) on the biology of language. I develop the analysis sketched here of Chomsky's program as anti-Hockett in a paper in preparation.

13. See Lyons 1970b—the author was also editor of *New Horizons in Linguistics*—and Allen 1975.

14. "The development of human speech represents a quantum jump in evolution comparable to the assembly of the eukaryotic cell." Wilson 1975, 556, with discussion of Chomsky on 558–59.

15. See Lieberman 1968, and other articles collected in Lieberman 1972. For Wilson's discussion, see Wilson 1975, 559. Lieberman's recollections of his work from this time—his low estimate of Neanderthal speech proved especially controversial—can be found in Lieberman 1998, 45–67. For a summary appraisal of his work on the evolution of speech and language up to the mid-1990s, see Hauser 1996, 43–47.

16. See Wilson 1975, 556, summarizing the more extensive treatment in Wilson 1972, 60. Other appraisals in a similar spirit include Bronowski and Bellugi 1970 and, from a Chomskyan perspective, Fodor, Bever, and Garrett 1974, 440–51.

17. For Singer's comments, see Singer 1975, 13–14. For Chomsky on Descartes, see Chomsky 1966.

18. Washburn thought the experiments of the Gardners and Premack lent impressive support to the main finding of the new neuroscience: the humans-only "sound code" of language is anatomically and evolutionarily separate from symbolic thinking, which the nonhuman primates, indeed all mammals, share. See Washburn 1973, 46–47. For his critical remarks on Struhsaker's large list of meaningful vervet sounds, see 46. See also note 159 below.

19. In a pair of famous papers from 1970, one summarizing the white-crowned sparrow research, the other drawing out parallels between birdsong and language, Marler discussed Chomsky's work on inborn predispositions for language learning in humans; see Marler 1970b, 365; 1970a, 672. From the late 1960s, Marler had been postulating inherited "auditory templates," of a broadly Chomskyan kind, to account for song learning in certain birds; see, e.g., Moser 1974, 164. It should also be noted that around this time, Marler was in frequent contact with Chomsky's early supporter George Miller, who had joined the Rockefeller faculty

and who ran a seminar with Marler on the biology of speech; see Marler 1989, 340.

20. Interview with P. Marler, 11–12 September 2003. For Goodall's observations, see Van Lawick-Goodall 1968. A Dutch observer of chimpanzees in the field, Adriaan Kortlandt, was also reporting extensive gestural communication at around the same time. See Fouts 2003, 85.

21. Marler 1995, 13.

22. Marler 1969a, 50–59, quotations on 59. In 1972, Marler spoke in Reno alongside the Gardners and Norman Geschwind at a symposium, "Animal Communication and Human Language: A Discontinuity in Approach or in Evolution?" For a vivid account of the occasion (when Marler unsurprisingly argued that the discontinuity was in approach), see Linden 1974, 240–70. In Marler's solicitous attitude toward the experiments of the ape language psychologists, he was following in the footsteps of his ethological mentor W. H. Thorpe, who in 1952 had spent time with the Hayeses and Viki. For Thorpe's positive assessment of their work, see Thorpe 1972, 35–36. For his views on whether Washoe and Sarah were crossing the Chomskyan syntax border, see 36–47.

23. Marler 1969c, 99, citing Itani 1963. Itani was a student of the totemic figure in Japanese primatology, Kinji Imanishi; see De Waal 2001, 113.

24. Interview with P. Marler, 27 October 2003. The conference circuit brought the two men together fairly regularly in the mid-1970s; see their papers in, e.g., Kavanagh and Cutting 1975 and Harnad, Steklis, and Lancaster 1976. The latter collects the proceedings from a mammoth 1975 conference, "Origins and Evolution of Language and Speech." As a snapshot of the state of the debate, it is invaluable.

25. Marler 1995, 13.

26. Interview with P. Marler, 27 October 2003.

27. Premack and Schwartz 1966, 301. Specifically, the psychological evidence he cited favoring discontinuity was the strict environmental control of primate calls versus the independence from such control of human speech, while the neurological evidence was the structural and functional separation between the parts of the brain mediating calls and speech. Premack here sided against Hockett, and with Geschwind's neurological predecessor Magoun.

28. Premack 1974, quotation on 348; reprinted with an additional concluding section as Premack 1975. Marler would later reproduce the quotation several times, as part of a gallery of statements showing how entrenched was the view that animal communication in nature merely expresses emotions; see, e.g., Marler 1985b, 212; 1992, 225. As Premack's paper made clear, his idea of the "field" here was the one-acre outdoor enclosure in which Emil Menzel in the early 1970s conducted experiments on communication in chimpanzee groups; see Premack 1974, 347–48. According to Menzel, "the entire enclosure system can be thought of as an enormous choice apparatus for releasing and testing groups of large animals." See Menzel 1974, 89, with a photograph on 90. His basic technique was

to show one animal a hidden object, then watch as that animal's actions gradually informed the rest of the group about the object—thus, he argued, instancing non-symbolic communication about the environment. On the experiments, see Menzel 1971, 1974, 1975. For discussion, see Desmond 1979, 231–34; and Wallman 1992, 129.

29. In interview (10 September 2002), Dorothy Cheney and Robert Seyfarth recalled Premack once telling them—he was a colleague at the University of Pennsylvania—that there was absolutely nothing of interest he could imagine learning from watching a chimpanzee in its natural habitat. A rather different attitude, however, can be found in a popular 1976 book on the ape language projects by Premack's wife, Ann, who wrote (1976, 8): "it is entirely possible we might be missing something in accepting the traditional analysis of calls [as emotional], considering how specific are some of the calls of vervets. Struhsaker worked very closely among the vervet monkeys, far more closely than a human can get to a wild chimp. Perhaps in the future a disguised human may impose his presence upon a troop of wild chimps and discover a semantic key to the calls of chimpanzees."

30. Marler 1974, 47. The paper was published in French translation the previous year; see Marler 1973b.

31. Marler 1974, 43; see also 34–35, 42–49.

32. Marler 1977b. See also, e.g., Marler 1977a, 54–55, 59–60, 65–66; Marler and Tenaza 1977, 1024–25; and Green and Marler 1979, 127–34.

33. Lana, the name of the first chimp enrolled in the Atlanta program, was an acronym for "language analogue." The language she learned was called Yerkish. On the project, see Rumbaugh 1977. For Marler's recent reflections on a successor to Lana, the bonobo Kanzi, see Marler 1999.

34. Marler 1977b, quotation on 224. Premack was among those thanked for prior discussion of the paper. For restatements, with variations, of the analysis sketched here, see Marler 1977c, 28–29; 1978, 115–16. Although, as noted earlier, Marler seems to have popularized the idea that there were just three vervet alarm calls, for snake, eagle, and leopard, in these papers he offered a rather fuller taxonomy, more akin to Struhsaker's initial presentation.

35. Frederick Seitz [Rockefeller president] to R. Seyfarth, 16 September 1976; and P. Marler to F. Seitz, 21 September 1976; Marler Papers.

36. P. Marler to J. Lederberg, 19 June 1979, Marler Papers. See also Marler 1989, 338.

37. Much of the rest of this section is based on interviews with Seyfarth and Cheney (esp. 9–11 September 2002) and Hinde (25 July 2002). Subsequent sections draw much more fully on letters, grant applications, and other private documents from the time, preserved in Peter Marler's files. Where possible, I have tried to use information from interviews with Seyfarth, Cheney, Marler, and Phyllis Lee to amplify, qualify, and fill gaps in this primary archival record.

38. See esp. Trivers's papers on reciprocal altruism (1971) and parental investment

(1972). It was principally the latter, and its implications for feminist dreams of equality of the sexes, that Seyfarth recalls discussing with Trivers.

39. On Trivers and his role in the sociobiological revolution, see Segerstråle 2000, 79–84.

40. A note on names: from this time until the mid-1970s, she was "Dorothy Cheney Seyfarth." From the mid-1970s, she has been "Dorothy L. Cheney." Throughout the period I am discussing, Robert and Dorothy were more commonly referred to as "the Seyfarths" than as "Seyfarth and Cheney." I have largely followed contemporary practice in the text. Also, and in accord with what the Seyfarths have told me, I have generally attributed joint authorship to their field letters, grant applications, and so forth, indicating stated authorship in the associated citations.

41. On the Gombe Stream project, and the delayed research visa, see R. Seyfarth, Research Fellowship Application to the National Institute of Mental Health, December 1975, Marler Papers (hereafter called NIMH application 1975). This application also includes a record of Seyfarth's Harvard coursework, including the independent study with Trivers in 1969–70.

42. See Altmann 1974, esp. 242–47 for focal-animal sampling. On the background to and significance of Altmann's paper, see Rees 2006. The Seyfarths used other methods of sampling as well; see, e.g., Seyfarth 1976, 917–18.

43. Data collection in accord with Altmann's recommendations began in January 1973. The Seyfarths sometimes described their research in South Africa as lasting eighteen months or fifteen months, depending on whether they counted the previous, less systematic period of observations. For a fifteen-month reckoning, from a January 1973 start, see Curriculum Vitae of R. Seyfarth, ca. summer 1975, Marler Papers. For the baboon research as beginning in September 1972, see Seyfarth 1976, 917.

44. For Hinde on interactions, relationships, and social structure, the key statement is Hinde 1976; see 16 for an acknowledgment of the Seyfarths' help. See also the autobiographical discussion in Hinde 2000, 110–12, which touches on how the interest in the individual's social behavior was shared with Rowell, Goodall, Fossey, and Trivers. Trivers (1974) appealed to the Madingley rhesus data in another of his classic sociobiological papers of the early 1970s, on parent-offspring conflict.

45. The division is reflected in their thesis titles: "The Social Relationships among Adults in a Troop of Free-Ranging Baboons" (Robert) and "The Social Development of Immature Male and Female Baboons" (Dorothy). See Seyfarth 1975 and Cheney Seyfarth 1976.

46. See Hinde's testimonial in his letter to Marler, 16 May 1979, Marler Papers. Hinde's textbooks were Hinde 1970, 1974.

47. "I feel," wrote Seyfarth in his PhD thesis (1975, 1.1–1.2), "that the social organization and behaviour of free-ranging primates are the result of natural selection acting on individuals." For the grooming model, see Seyfarth 1977.

48. R. Hinde to P. Marler, 11 July 1975, Marler Papers.
49. Seyfarth 1977.
50. Seyfarth 1978a, 1978b.
51. Seyfarth 1976; Cheney 1977.
52. Cheney and Seyfarth 1977; Cheney 1978a, 1978b.
53. See, for instance, Struhsaker's 1975 monograph on the red colobus monkey, and the papers cited therein.
54. T. Struhsaker to P. Marler, 4 September 1975, Marler Papers.
55. P. Marler to R. Seyfarth, 18 September 1975, Marler Papers.
56. P. Marler to R. Seyfarth, 7 October 1975, Marler Papers.
57. R. Seyfarth to P. Marler, 15 October 1975, Marler Papers. The Seyfarths recall the concerns expressed about safety in Uganda as more than anything a means to the end of getting Marler to agree to a Botswana project.
58. P. Marler to R. Seyfarth, 12 November 1975, Marler Papers. In a previous letter (7 October), Marler had sketched the comparative prospects thus: "On the one hand the black-and-white colobus has a small group, tending to be one-male, so that a young animal has limited opportunity for sibling play. By contrast, the group size in red colobus is, as you know, much larger, multi-male, and siblings have many opportunities to play with one another, as well as ample contact with adults of both sexes. There are all kinds of possibilities for exploring correlations between infant environment and adult personality there."
59. R. Seyfarth to P. Marler, 18 November 1975, Marler Papers. It should be said that Marler, in his 12 November letter, had solicited counterarguments, and urged the Seyfarths to write to Struhsaker about any concerns.
60. On the founding and character of the Field Research Center for Ecology and Ethology at Millbrook—the move was completed in 1972—see Marler 1989, 336, 339; and Griffin 1989, 140.
61. In her methodological survey (Altmann 1974, 229–31), Jeanne Altmann presented systematic sampling as what field observers of behavior could do in lieu of experimental manipulations, since sampling increased the "internal validity" of an interpretation of observational data (since rival interpretations can be more confidently excluded) without diminishing its "external validity" (since more trustworthy generalizations could be made about the behavior of unmanipulated animals). She did, however, acknowledge a few attempts, like Menzel's, to blend laboratory and field methods.
62. Green 1975b. In the same year, Green published a paper (1975a) on the evidence for regional dialects among Japanese macaques—a new primatological variation on an old ornithological theme in the Marler lab. For a useful biographical sketch of Green, see Poquette 2000. On Green's work at the Rockefeller, see Marler 1989, 335–36.
63. For Green's tributes to the Japanese research he built upon, see Green 1975b, 3, 83–84. On the pioneering establishment of long-term study sites in Japan, and the

intellectual character of the work done there, see Takasaki 2000, 152–53; and Asquith 2000, 167–69.

64. See Green 1975b, esp. 15, 37–41. On Darwin, see 91–92.

65. "In the Japanese monkey, subtle aspects of motivational state may be communicated by small differences in sounds." Green 1975b, 94. There was considerable debate at the time about how to tell graded calls from discrete but variable ones—an issue of some importance, for example, when trying to compare repertoire sizes in different species. See, e.g., Green 1975b, 84–85; and Struhsaker 1967b, 1197.

66. Green 1975b, 85–86, quotation on 86.

67. Waser 1975, quotation on 57. He noted in a fuller report on his experiments (1976, 39), which involved other forest monkeys besides mangabeys, that "playback methods have been infrequently attempted with primates, and then with little success. Perhaps the most important reason for this is that primates are adept at distinguishing, and at learning to distinguish, real from imitation stimuli." Once due precautions were taken, Waser emphasized (1975, 57–58), the playback technique permitted the student of natural vocal communication unprecedented control over the effects of variables—who was vocalizing, where the vocalizations were heard, and so on. In his studies, he had found, against expectation (though in line with observation), that a test group responded no differently to foreign whoopgobbles heard in the middle of its home range and at the edge. The response in both cases was retreat.

68. Underway as well was another primatological playback-based investigation, making use of field recordings, though not set in the field. Beginning in 1974, Marler, Green, and others at Millbrook worked in collaboration with members of a psychoacoustic laboratory at the University of Michigan on experiments in the perception of natural speech sounds by captive Japanese macaques. Out of this research came the first persuasive evidence in a nonhuman species for cerebral lateralization and perceptual specialization for the processing of same-species vocalizations. See Peterson et al. 1978; and Zoloth et al. 1979. For discussion, see Marler 1989, 336.

69. On the dates of the Indiana lectures, see P. Marler to R. Seyfarth, 26 November 1975, Marler Papers; and the organizer Thomas Sebeok's remarks in Sebeok 1978, vii. For Marler's paper, entitled "Affective and Symbolic Meaning: Some Zoosemiotic Speculations," see Marler 1978.

70. R. Seyfarth to P. Marler, 18 December 1975, Marler Papers.

71. Seyfarth, NIMH application 1975. On the feedback from Marler and Green, see P. Marler to R. Seyfarth, 23 December 1975, and R. Seyfarth to P. Marler, 26 December 1975, Marler Papers.

72. This is not to deny that the alarm calls could be approached with social questions in mind. As we shall see, from at least December 1976, the Seyfarths were interested in finding out whether alarm calls from certain individuals or classes of individual were less effective in eliciting responses than alarm calls from others.

Later in their vervet studies, they considered the issue of whether vervets ever deliberately give false alarms—using alarm calls not to inform, but to manipulate.

73. Seyfarth, NIMH application 1975.

74. See Green 1975b, 86–95, esp. 88–89. Green (91) welcomed the recent neuroanatomical research on the limbic nature of primate vocalizations as entirely in keeping with his findings on the macaque coos.

75. See, e.g., D. Cheney, Application to the National Geographic Society, March 1976, Marler Papers.

76. The sequence can be traced in the following letters, all in the Marler Papers: R. Seyfarth to P. Marler, 15 June 1976; R. Seyfarth to P. Marler, 30 June 1976; [P. Marler] to R. Seyfarth, 6 July 1976; R. Seyfarth and D. Cheney to P. Marler, 7 July 1976; and P. Arnott to R. Seyfarth, 9 July 1976.

77. R. Seyfarth and D. Cheney to P. Marler, ca. 31 August 1976, Marler Papers. At the conference, the Seyfarths presented a paper coauthored with Hinde on social interactions and social structure in primates; see Seyfarth, Cheney, and Hinde 1978.

78. See the account by Dorothy's sister Margaret Cheney (2001).

79. Quotations from R. Seyfarth, Research Fellowship Application to the NIMH, December 1976, Marler Papers (hereafter called NIMH application 1976). See also D. Cheney, Postdoctoral Fellowship Application to the NSF, December 1976, Marler Papers. The two applications outlined overlapping but not identical projects, and Dorothy's (rather shorter) application had nothing on alarm calls per se, though it did gesture toward the prospect of determining if calls "are solely a consequence of affective state, or if there is also a sense in which they may be said to function symbolically."

80. Seyfarth, NIMH application 1976. Waser (1976, 70) had stressed the potential for playbacks to unpack the kinds of information conveyed in primate vocalizations. The mangabey whoopgobble, he reported, transmits information on individual identity, group identity, group location, and potentially even the size of a group (since the rate of calling can vary with group size). Green (1977, 219), in a paper summarizing an autumn 1976 conference discussion on comparative aspects of vocal signaling, noted that, as regards information content, there had not been "playback experiments, even with the vervets, to determine how detailed are the vocal designations of external referents." Green's paper is cited in Robert's application.

81. R. Seyfarth and D. Cheney [?], n.d., script for talk with diagrams (including a table of vervet alarm calls), Marler Papers.

82. Seyfarth, NIMH application 1976.

83. R. Seyfarth [and D. Cheney] to P. Marler, 20 March 1977, Marler Papers. In general, on the Seyfarths' letters, the first name in the signature indicates the writer (as, often, does the handwriting).

84. R. [and D.] Seyfarth to P. Marler, 17 June 1977, Marler Papers. Marler came with his wife, Judith; son, Christopher (who helped a great deal with the field experi-

ments); and daughters, Cathy and Marianne. The letter also reported: "A student of Stuart Altmann's is doing some nice work on lapwings, which he thinks have different alarm calls for different predators." Much later, the Seyfarths would investigate the responses of Amboseli vervets to the predator alarm calls of another bird, the superb starling. See Cheney and Seyfarth 1990, 158–64. The Seyfarths remained well disposed to good living in the field; see Hauser 2004, 157–58.

85. See Seyfarth, Cheney, and Marler 1980b, 1071, 1076. The skill had survival as well as scientific value. In interview (17 September 2002), Phyllis Lee, the Seyfarths' first "carry-over" assistant on the vervet site, recalled once jumping out of the way on hearing a vervet snake call, just as a deadly mamba slithered past where her feet had been.

86. Seyfarth, Cheney, and Marler 1980b, 1072–73.

87. For the full description of the experimental methods and design, see Seyfarth, Cheney, and Marler 1980b, 1079–81.

88. The Seyfarths did the dubbing the way any teenager of the era would have. They lay their two tape recorders side by side on their bed; wound the original tape to just before the call; started recording with the new tape; played the call on the original tape; then, the moment the call had finished, turned the gain down on the receiving machine.

89. As noted in chapter 8, although Struhsaker had identified what could be called a snake alarm (a high-pitched chutter) and an eagle alarm (a low-pitched staccato grunt or rraup), he did not pick out a leopard alarm—and indeed, Marler and the Seyfarths' "leopard alarm" referred to one or the other of two of Struhsaker's calls: the multiunit "threat alarm bark," given by adult males to "major mammalian predators," and the single-unit "chirp," given by adult females and juvenile males to those predators; see Seyfarth, Cheney, and Marler 1980b, 1073. But see also 1075–76 for data suggesting that, on balance, the multiunit leopard call—a series of short tonal calls, sometimes compared with a dog's barking—is the more representative, since adult females (when the leopard is close) and juvenile males (when older) do give it. For leopards and martial eagles as the sole actual vervet predators in their respective predator classes, and so appropriately regarded as the candidate referents of the alarms given to animals in those classes, see Seyfarth, Cheney, and Marler 1980b, 1071–73. Alarms to baboons and humans were not included in the playback trials because not enough were heard for recording (1073). Typical examples of the three main predator alarm calls of the vervet monkeys can be heard at the Web site of Marc Hauser's lab at Harvard: http://www.wjh.harvard.edu/~mnkylab/media/vervetcalls.html (accessed 20 January 2005).

90. See Seyfarth, Cheney, and Marler 1980b, 1081, which reports 94 playbacks versus more than 100 aborted ones.

91. Seyfarth, Cheney, and Marler 1980b, 1079–81.

92. D. [and R.] Seyfarth to P. Marler, 13 September [1977], Marler Papers. I am not completely sure about "elastically."

93. P. Marler to R. Seyfarth and D. Cheney, 20 September 1977, Marler Papers.

94. The person eventually found was, as noted, Phyllis Lee—like the Seyfarths before her, an American graduate student at Madingley working under Robert Hinde, and in search of a field site for studies of baboon behavior. Lee had had plenty of field experience before she arrived in Amboseli in the spring of 1978, however, having spent time at Gombe Stream as an undergraduate and having even helped to establish a field site elsewhere in Tanzania.

95. See note 89 above for some complications regarding the identity of the leopard call.

96. Both Cheney and Seyfarth had discussed the relevant papers by Premack in their most recent grant proposals. See Seyfarth, NIMH application 1976; and D. Cheney, Postdoctoral Fellowship Application to the NSF, December 1976, Marler Papers.

97. For their statement in the published papers, see Seyfarth, Cheney, and Marler 1980b, 1091.

98. To make sense of the structuralist view being discarded here, think of water. As water is heated, it passes through three phases, from solid to liquid to gas, for reasons to do not with some functional advantage that accrues from being solid at minus ten or whatever, but from the physical makeup of water. Along similar lines, one might imagine the vervet vocal system as having the interesting physical property that, under low levels of the sort of nervous stimulation brought on by fear, the system makes the sound associated with the presence of snakes; that when a certain fear threshold has been passed, the system abruptly switches to the sound associated with eagles; and that when a further threshold has been passed, the system switches to the sound associated with leopards—all the while, analogous to temperature in water, with the amplitude and length of the calls increasing.

99. R. Seyfarth to P. Marler, 11 October 1977, Marler Papers. For the exploitation of natural variation in the alarm calls to construct the longer versions of the calls for playback, see Seyfarth, Cheney, and Marler 1980b, 1080.

100. R. Seyfarth to P. Marler, 11 October 1977, Marler Papers.

101. D. Cheney to Ms. Osmundssen, 16 November 1977, Marler Papers.

102. Humphrey 1976. The argument's author, Nick Humphrey, was a friend of the Seyfarths from their Madingley days. The "social intelligence hypothesis," as it is now known, is now credited both to Humphrey and to Alison Jolly, who had published similar ideas a decade earlier. On the hypothesis, its origins, and its impact on primatology, see Steklis 1999, 50–51.

103. D. Cheney to Ms. Osmundssen, 16 November 1977, Marler Papers. The report described two intragroup vocalizations that, when given by a subordinate male to a dominant male, seemed to affect the subsequent interaction. One vocalization appeared to signal submissiveness and the desire to avoid attack; the other, to signal defiance and the willingness to risk attack.

104. R. Seyfarth [and D. Cheney], 20 November 1977, Marler Papers, emphasis in original.

105. On Lee, see notes 85 and 94 above. Much of the correspondence from here deals with technical or practical matters. But occasional glimpses of the research do surface. A 24 November letter, for instance, reports the successful acquisition of a stuffed leopard, martial eagle, and python. An 11 December letter vividly describes watching "a martial eagle carry off one of our juveniles. The monkey was taken on open ground, apparently killed instantly by talons piercing its neck." A 20 April letter recounted the first playbacks to elicit alarm calling—the only one in the whole series, though many more were expected (see Seyfarth, Cheney, and Marler 1980a, 803 n. 9). A 20 May letter tells of the strong likelihood of over ninety-five playbacks being done before the end, and of some long-wished-for recordings of males alarming at snakes. All letters in the Marler Papers.

106. Seyfarth and Cheney 1980.

107. Cheney and Seyfarth 1980; see 362 for the starting date.

108. Cheney 1981.

109. Cheney and Seyfarth 1982.

110. Seyfarth 1980.

111. Cheney and Seyfarth 1981.

112. Cheney and Seyfarth 1981, 34–36. I am grateful to Dorothy Cheney for some extra telephone tuition on these experiments.

113. Seyfarth, Cheney, and Marler 1980b, 1980a. For vervet alarm calls as "rudimentary semantic signals," variation in the acoustic structure of which "was the only feature both necessary and sufficient to explain response differences," see the *Science* paper, Seyfarth, Cheney, and Marler 1980a, 803. The *Science* paper was basically a digest of the long *Animal Behaviour* paper on the alarm-call playbacks (Seyfarth, Cheney, and Marler 1980b) and the *Zeitschrift* paper on ontogeny (Seyfarth and Cheney 1980).

114. I must stress that, even more so than elsewhere in this chapter, the interpretation that follows is my own.

115. Other, observational evidence adduced in support of call semanticity included the arbitrary (in the sense of noniconic) acoustic structure of the alarm calls—a traditional symptom of symbolhood; and the many occasions when vervets giving alarms did not act on them and vervets hearing alarms did not give them in return—signs, on Marler's view, that alarm-call behavior is not an indivisible, inflexible, physiological whole. See Seyfarth, Cheney, and Marler 1980b, 1090–92.

116. Seyfarth, Cheney, and Marler 1980b, 1070.

117. The question of whether an affect-only theory would *have* to predict, as Seyfarth, Cheney, and Marler put it (1980b, 1090), a blurring of the qualitative distinctions between escape responses when all the calls are presented at the same length and amplitude, is an interesting one that no one involved in the debate over the vervet results seems to have raised.

118. See also the data and discussion in Seyfarth, Cheney, and Marler 1980b, 1088–91. Struhsaker (1967a, 310) had reported his impression that the "threat-alarm-bark" to "major mammalian and avian predators" (what Marler and the Seyfarths called the "leopard alarm") got louder, with more units more closely packed, the closer the predator was observed. In his commentary on Struhsaker's paper, Stuart Altmann had observed, in Hockettian fashion, that the vervet calls thus appeared to have both arbitrary and iconic features; see Altmann 1967b, 339.

119. For a literalist Smithean reading of the vervet alarms, see Marler 1967b, 772, as discussed in chapter 8. It should be noted that, in their 1976 grant applications, Robert and Dorothy were both very positive about Smith's "message + context = meaning" perspective as a framework for their studies. See Seyfarth, NIMH application 1976; and D. Cheney, Postdoctoral Fellowship Application to the NSF, December 1976, Marler Papers.

120. Seyfarth, Cheney, and Marler 1980b, 1092. The same passage appears more or less verbatim in the *Science* article, Seyfarth, Cheney, and Marler 1980a, 803 n. 11.

121. D. Seyfarth to Ms. Osmundssen, 16 November 1977, Marler Papers.

122. On the occasional failure of playbacks to elicit responses other than looking, see Seyfarth, Cheney, and Marler 1980b, 1085. For a taxonomy of "wrong" responses, see Seyfarth and Cheney 1980, 48. For a table exhibiting the relevant data, anomalous responses and all, see Seyfarth, Cheney, and Marler 1980b, 1082.

123. Seyfarth, Cheney, and Marler 1980b, 1083.

124. Incidentally, it is also hardly what one would expect if vervet alarm calls conformed to one of Marler's first famous generalizations—that alarm calls tend, for Darwinian reasons, to have acoustic structures rendering them unlocalizable. See Cheney and Seyfarth 1981, 56. In correspondence, Marler pointed out, however, that his generalization was never meant to apply to *all* alarm calls; so-called mobbing calls, for instance, are adapted to be easy to locate.

125. Seyfarth, Cheney, and Marler 1980a, 802.

126. Seyfarth, NIMH application 1976. It is interesting to note that although the fourth, longest question, about whether the age/sex class of the caller affects response, was likely the most intriguing to the Seyfarths before entering the field, it turned out to have a not very interesting answer: no. See Seyfarth, Cheney, and Marler 1980b, 1086–88.

127. There are, however, two moments in the *Animal Behaviour* paper that look suggestively like vestiges of this earlier, different conception of context, responses, and semanticity. On 1080, in the table displaying the number of times each type of alarm was played, there is a finer breakdown of context—closed canopy/open canopy in the case of tree-bound vervets hearing eagle alarms, tall grass/short grass in the case of ground-bound vervets hearing snake alarms—than is made use of in the rest of the paper. And on 1084, it is noted that, in general, when

"animals were in trees, results were less conclusive." See Seyfarth, Cheney, and Marler 1980b.

128. I say "arguably," for perhaps humans sometimes fail to understand words used outside accustomed circumstances. If so, then nothing follows about signal semanticity from anomalous responses in atypical contexts.

129. Seyfarth, Cheney, and Marler 1980b, 1085. On the different vulnerabilities to predation, see Cheney and Seyfarth 1981, 49–50.

130. Seyfarth, Cheney, and Marler 1980b, 1085. It was also noted that, as with adult males, there were generally too few infants around to make secure generalization. On "wrong" responses and their distribution according to age, see also Seyfarth and Cheney 1980, 48–50. Another source of ambiguity in the data on junior animals was the tendency to flee to the mother as refuge whatever the danger—an adaptive response that was not alarm-specific enough to be interesting.

131. Seyfarth, Cheney, and Marler 1980b, 1085.

132. Seyfarth and Cheney 1980, 49–50.

133. A letter of 20 May 1978—near the end of the Seyfarths' Amboseli research— perhaps suggests that even at this date, the operative conception of semanticity and context was the earlier, typicality-versus-atypicality one. In the letter, the Seyfarths wrote that "for the snake-on-the-ground trials we've added an extra environmental context, distinguishing between long grass/thick bush and short grass/open plains." R. Seyfarth [and D. Cheney] to P. Marler, 20 May 1978, Marler Papers. In the *Animal Behaviour* paper, as noted earlier (see note 127, above), this distinction survives only in a lonesome table of total playbacks performed (Seyfarth, Cheney, and Marler 1980b, 1080).

134. Although, in interview, Cheney and Seyfarth never mentioned a change of mind about alarm-call semanticity, they read through a draft of the account above and offered the following reconstruction. In their words, "while we were in the field we feared that *any* context-dependence in our data could be used to argue against a semantic interpretation. Once we returned to Millbrook and considered our data more carefully, however, we realized that *the kind of context-dependent responses we'd obtained* argued in favor of a semantic interpretation, not against it. It makes sense, for example, to run out of a tree if you interpret the eagle alarm as designating something like 'bad bird.' If you're in a bush, however, you're safe." They mention the eagle example especially as helping them along to the later, semanticity-friendly interpretation of context dependency. Letter to author, 7 February 2005, emphases in original. For the reasons already given, however, I find this reconstruction unpersuasive. But there is another dimension worth noting, mentioned by Dorothy in a subsequent e-mail to me (24 March 2005): "I think we were trying to dampen Peter's enthusiasm, lest he expect too much. I, at least, held Peter in considerable fear (I still do), and we didn't want his expectations to be too high."

135. As a control, another person not involved in the experiments scored the films

independently. On the film analysis, see Seyfarth, Cheney, and Marler 1980b, 1081–82.

136. Seyfarth, Cheney, and Marler 1980b, 1089–90.

137. The idea of natural categories was emerging as an explicit coordinating theme in Marler's work at around this time. His first paper discussing the Seyfarths' research on the ontogeny of vervet alarm-call behavior was entitled "Avian and Primate Communication: The Problem of Natural Categories" (1982). In light of the now-customary distinction drawn between classical and cognitive ethology—I shall make use of it shortly—it is interesting to note the effortless way Marler assimilated the new to the old: "The young [vervets] are in fact behaving as though they already possess a set of rules for the perceptual analysis and classification of predatory animals. The rules are sufficient to divide them into three lumped classes and yet inappropriate for the separate classification of different predator species. While the developmental work remains to be done, the general findings are nevertheless consonant with those that ethologists have discovered in other perceptual domains. Rather unspecific, generalized innate release mechanisms set the path of perceptual development of the young organism on certain species-specific trajectories." Marler 1982, 88 (I have removed two citations). The paper also discusses Marler's recent work on song learning in swamp and song sparrows as well as the Rockefeller-Michigan research on own-species sound perception in macaques.

138. Herrnstein and Loveland 1964 (cited in Seyfarth and Cheney 1980, 55) and more recently Herrnstein, Loveland, and Cable 1976 (cited in Seyfarth, Cheney, and Marler 1980a, 803). For an accessible discussion of Herrnstein's work on pigeon categories, see Gould and Gould 1994, 172–74.

139. E.g., Johnson-Laird and Wason, *Thinking: Readings in Cognitive Science* (1977), cited in Seyfarth and Cheney 1980, 54–55.

140. For citations to the ape language writings of the Gardners, Fouts, Premack, Rumbaugh and Savage-Rumbaugh, Patterson, Terrace, and Ristau, see Seyfarth and Cheney 1980, 54–56, with discussion esp. on 38, 51–52.

141. Quotation from Seyfarth and Cheney's letter to the author, 7 February 2005. For the diagrams, see Seyfarth, Cheney, and Marler 1980a, 802, and more extensively Seyfarth and Cheney 1980, 45–47. Cf. Anglin 1977, 32, 44. I am grateful to Seyfarth and Cheney for drawing the Anglin connection to my attention.

142. For the rethinking of older work as about thinking, see, for instance, the brief note Seyfarth published in 1981 in the key cognitive science journal *Behavioral and Brain Sciences* on whether monkeys rank each other. Describing his earlier research on baboon grooming choices, he wrote that females "revealed that they had gone beyond the simple discrimination of 'dominant to me vs. subordinate to me' and created a true rank hierarchy of individuals" (447). No such classificatory language appears in the original papers.

143. Cheney and Seyfarth 1980, esp. 363, 365–66.

144. Seyfarth, Cheney, and Marler 1980a, 801. The subtitle in fact settled rather late. On 4 February 1980, Marler wrote to the Seyfarths (now in Amboseli again) that

after "worrying at length about the question of 'natural categories'" and finding he "still felt unclear in my own mind as to whether we can be sure that this
is the appropriate concept to use," he "chickened out and substituted 'predator
classification,'" on the grounds that, with competition for publication in *Science*
so fierce, every step needed to be taken to guard against negative comments. The
new phrase, it seemed to him, "is more neutral, and still captures the most interesting point," though he awaited their judgment. P. Marler to R. Seyfarth and D.
Cheney, 4 February 1980, Marler Papers.

145. On Griffin's life and work, see Griffin 1989.

146. On the bee dance controversies and Griffin's role, see Munz 2005.

147. It is said that the philosopher Thomas Nagel's choice of example in his famous
paper "What Is It Like to Be a Bat?" (1974), on the mind-body problem, was
inspired by hearing Griffin lecture on bat perception during a year spent at the
Rockefeller. See Hauser 2000, 255. Griffin in turn credited the stimulus of Nagel's
visit. Griffin 1976, vii.

148. Griffin 1976, 105. Besides Von Frisch and the honeybees, another main example
of admirably ethological work on symbolic, two-way communication with animals was the ape language projects; see, e.g., 16–19, 89, 101.

149. Of the "open door" reviewers, see, e.g., Mason 1976, Krebs 1977, and Humphrey 1977.

150. See, e.g., the passage arguing that the time has come to re-examine whether animal communication is wholly affective, or whether at least some signals also include specific environmental information about objects and events. Griffin 1976,
98, citing Marler.

151. Seyfarth, Cheney, and Marler 1980b, 1091. Marler later wrote (1989, 338): "the
subtleties of the [vervets'] responses, such as looking into the sky for the eagle
that the tape recorder told them must be there, and into the long grass for the
nonexistent snake, convinced us that Don Griffin had been right to chide us about
neglecting notions of mental imagery in animals."

152. P. Marler to R. Fox and L. Tiger, 3 October 1979, Marler Papers. The most influential linking of the couple's research with cognitive ethology was the philosopher Daniel Dennett's widely discussed 1983 article "Intentional Systems in Cognitive
Ethology: The 'Panglossian Paradigm' Defended," where he exhibited the vervet
alarm research as a real-world case where the attribution of rational beliefs to animals was turning out to be more productive than stimulus-response conservatism
would be. In its original form as a target article in *Behavioral and Brain Sciences*,
the article drew extensive commentary, from figures including B. F. Skinner, Herbert Terrace, Richard Dawkins, Niles Eldredge, John Maynard Smith, Richard
Lewontin, Emil Menzel, Alison Jolly, and Nicholas Humphrey. Dennett met Seyfarth at a 1981 conference organized by Don Griffin on human and animal minds
(for a photo of Dennett, Seyfarth, and Marler in their session group, see Griffin
1982, 390). Dennett's subsequent visit with Cheney and Seyfarth in the field is
described in Dennett 1988.

153. Nottebohm 1971, 1972. For discussion, see Marler 1989, 335.

154. Gould 1975. For discussion, see Griffin 1976, 27–29; and Munz 2005.

155. For citations and discussion, see note 68 above.

156. P. Marler to R. Seyfarth and D. Cheney, 13 November 1980, Marler Papers.

157. Cheney and Seyfarth 1982.

158. There were, however, observations showing that "the link between alarm call and escape behaviour, while real, can be severed, as might be expected if animals were capable of operating in a 'symbolic' mode." Seyfarth, Cheney, and Marler 1980b, 1091–92. As we saw in chapter 8, the limbic system was associated with behaviors that were automatic and inflexible as well as affective.

159. In a well-known article, Jacob Bronowski and Ursula Bellugi had concluded that Washoe's successes counted against the Washburnian picture of naming as "biologically confined to humans" (1970, 104–5, quotation on 105). But Washburn did not see things that way. In a 1973 paper, he reviewed the neuroanatomical evidence that had accumulated since the mid-1960s showing that "the human sound code is unique." Nonhuman primates should be expected, he continued, to have small repertoires of affective signals. He mentioned Struhsaker's paper on vervet vocalizations as an example of a misguided tendency among field primatologists to attend closely to sounds that meant little to the monkeys themselves. As for Washoe and Sarah, they confirmed what was already known, that the capacity for symbolization is a thing apart from the human sound code. See Washburn 1973, 46–47, quotation on 46, and for a strong restatement in the year the alarm-call papers were published, Washburn and Moore 1980, 172–78. In February 1980, now in search of jobs, the Seyfarths wrote to Marler from Amboseli that, according to "the Anthropologists' grapevine, . . . Washburn was against our even being asked to come for an interview" at Berkeley. R. [and D.] Seyfarth to P. Marler, 7[?] February 1980, Marler Papers. Hinde referred (very positively) to the couple's talks at Berkeley in May 1979 in a letter that month to Marler. 16 May 1979, Marler Papers. For further discussion of Washburn's stance, its roots in his disciplinary synthesis, and the dim prospects for Darwinism as a unifying force in the behavioral sciences, see Radick 2006, esp. the conclusion.

160. Sebeok in interview was quoted as saying of the vervet papers recently published: "I was surprised those studies made the news they did. This is not new." Sebeok insisted that the calls meant not "leopard," "eagle," and "python," but "danger sideways," "danger up," and "danger down." Crail 1981, 190–92, quotations on 191.

161. Clippings, Papers of Robert Seyfarth and Dorothy Cheney, University of Pennsylvania, Philadelphia.

162. "Descent of Man," *Wall Street Journal*, 1 December 1980, Seyfarth Papers.

163. Silcock 1980, clipping, Seyfarth Papers.

164. Segment on NBC's *Today* program, 30 April 1981, transcript by Burrelle's T.V. Clips, Seyfarth Papers.

165. Seyfarth 1982, quotations on 13, 15, 18.
166. The project psychologists did not necessarily take this lying down. Herb Terrace, for instance, in his commentary on Dennett's 1983 *Behavioral and Brain
Sciences* article, wrote that, when it came to the vervet alarm calls, he (Terrace)
found himself "bogged down at the killjoy zero-order intentionality that Dennett
seeks to transcend: 'Tom (like other vervet monkeys) is prone to three flavors of
anxiety or arousal: leopard anxiety, eagle anxiety, and snake anxiety. . . . The effects on others of these vocalizations have a happy trend, but it is just a tropism,
in both utterer and audience.'" Terrace 1983, 378. Terrace had participated with
Seyfarth and Dennett at the 1981 Animal Mind–Human Mind conference, as had
Sue Savage-Rumbaugh.

Conclusion

1. Jardine 1991; Hacking 1999, esp. 163–66.
2. It should also be noted that ethology from its earliest days was a technophile
science, enthusiastically partnered to movie cameras and other recording and detection devices, and to that extent as well a natural disciplinary setting for the
primate playback experiment. On ethological technophilia, see Mitman 1999,
chap. 3.
3. In what follows I explore some themes developed in complementary ways in
Radick 2003, 2005a, 2005b. In conversation, Bob Batterman suggested that the
question I am groping toward here might be usefully reformulated in the language of complexity theory, as whether or not the primate playback experiment
is an attractor in the space of possibilities for the scientific study of animal communication. For an extended and much more general analysis along these lines,
examining "self-similarity" or pattern repetition as a stable feature of disciplinary
change, see Abbott 2001.
4. For a wide-ranging meditation on "the disappearance of useful sciences"—the
phrase is, appropriately for this book, adapted from the title of an anthropological
essay by the Torres Straits expeditionist W. H. R. Rivers, on the cultural persistence or otherwise of useful techniques—see Schaffer 2000–2001.
5. See, e.g., Marler 1985a, 505.
6. See Barkan (1992) on what he has called "the retreat of scientific racism"; on the
evolutionary biologists specifically, see, e.g., the latter chapters of Ruse 1996.
7. The history of the primate playback experiment thus stands as a corrective to
Robert Kohler's picture of biological field experiments as invariably falling short
of the standards of good laboratory science and good field science. See Kohler
2002, chap. 5.
8. On the saltationism rife during the turn-of-the-century "eclipse of Darwinism,"
see Bowler 1983, chap. 8; Gould 2002, 342–51, 396–466; and Gillham 2001,
chap. 20. The phrase "saltatory evolution" seems to have been introduced by the

Johns Hopkins morphologist W. K. Brooks in 1883 (see Gillham 2001, 287–88), mindful no doubt of reversing, in Darwin's words, "the canon, 'Natura non facit saltum' [Nature makes no leaps]." Darwin 1859, 460.

9. Hale 1886, 317–18, emphasis added. It was T. H. Huxley, in his review of the *Origin*, who introduced the Ancon sheep as illustrating nature's capacity for leap-making; see Bowler 1983, 187.

10. Hale 1886, 318–19. "Heterogenesis" was often spoken of, from the 1850s to the 1870s, especially in connection with the spontaneous generation of microbes from decaying plant and animal matter; see Strick 2000.

11. The dichotomy is Ernst Mayr's. See, e.g., Mayr 1982, 45–47.

12. An important exception here is Allen 1989, which offers an analysis very much complementary to the one offered here; see esp. 81–82. For the range of commentators who have lauded Boas for breaking with nineteenth-century typological thinking on race, see Allen 1989, 79. An especially egregious recent example is Lewis 2001, 382.

13. Boas 1911a. Boas's analysis of the data in this study has been the subject of well-publicized recent debate; see Pierpont 2004, 63.

14. Boas 1911b, 94: "What we have said . . . in regard to the overlapping of variations among different races and types, and the great range of variability in each type, may also be expressed by saying that the differences between different types of man are, on the whole, small as compared to the range of variation in each type."

15. On the later glossing, see Lewontin 1972, quoted in Edwards 2003, 798, and making the same point in relation to genetic variation but concluding that racial classification therefore has "no justification." I am grateful to Anthony Edwards for drawing my attention to his paper (which argues that Lewontin's influential conclusion is fallacious).

16. "I confess I do not consider such a result [the convergence of racial types among New York immigrants] likely, because the proof of the plasticity of types does not imply that the plasticity is unlimited." Boas 1911a, 332, with a near-identical statement in Boas 1911b, 64, and, in each case, skeptical remarks about selection theory in the paragraph preceding.

17. See Lewis 2001, 387 n. 15.

18. Franz Boas, "The Relation of Darwin to Anthropology," 1909 (?),Boas Papers, collection 5, box 3.

19. Boas 1911b, 248, summarizing arguments distributed throughout chapter 5. The argument makes sense only if one follows Boas in supposing, first of all, that the mental powers at issue came into being well before the languages currently spoken, making these languages the products of thought, not the reverse; and second, that the various racial types existed in relative isolation for such a long time that the associated language families can be assumed to reflect the mental powers of the associated racial types.

20. On the resistance to Darwinism among German anthropologists accustomed to

treating human racial types as timeless and absolutely bounded expressions of the human form, see Zimmerman 2001, 62–70. On the "racial liberalism" of German anthropology under Virchow's leadership, see Massin 1996, 86–94, quotation on 86.

21. On Galton's discussion of the "stability of types" (his phrase) in the final chapter of *Hereditary Genius*, see Gould 2002, 344. A catalogue of Boas's library (Boas Papers, collection 2) shows that he owned Galton's principal works on biometrical methods and saltationist evolutionary theory, *Hereditary Genius* (1869, 1882; Boas owned the later edition) and *Natural Inheritance* (1889). Unfortunately these books have since vanished from the shelves of their home institution, Northwestern University Library, so there are no marginal clues to take up. In Boas's rather considerable anthropometric output of the 1890s and 1910s, we find both Galtonian topics and explicit references to Galton as having provided important new tools. The best survey of Boas's debt to Galton's work remains George Stocking's superb 1968 essay "The Critique of Racial Formalism," from which I have quoted the line about "inconsistencies and obscurities" in Boas's application of his own critique; see Stocking 1968a, 187–88, quotation on 188.

22. Yerkes 1905, 147.

23. Watson 1914, 320–23.

24. It is especially intriguing, in this connection, to consider the activities of the London-born physical anthropologist Ashley Montagu. One of Boas's last PhD students, Montagu went on to collaborate with a number of biologists during and after the war, including Theodosius Dobzhansky—perhaps the most important figure in the modern synthesis—in establishing consensus on the absence of a scientific basis for race bias. On Montagu, see, e.g., Marks 2000 and Barkan 1992, 342 n. 4.

25. On Chomsky's exiguous but consistent remarks along these lines, see Newmeyer 1998, 306.

26. Bickerton 1990, esp. chap. 7. On the maverick Bickerton's life and work, see Byrne 1991, esp. 1–3; Bickerton 1992; and Berreby 1992.

27. The key papers on punctuated equilibrium theory were Eldredge and Gould 1972 and Gould and Eldredge 1977. On the theory and its scientific reception, see Ruse 1999b, 136–42, 146–52, including citation data showing a 1980s peak. On Bickerton's encounter with and subsequent drawing upon the theory, see Bickerton 1992, 104–5, and 1990, 260. For his use of "catastrophic," see, e.g., Bickerton 1990, 165, 174. Gould later returned the compliment with a wonderful essay linking his own work on snails on Curaçao with Bickerton's work on creoles; see Gould 1996, 344–55.

28. Bickerton 1990, 12–16.

29. Bickerton (1990, 100) entitled one of his subsections "The Long yet Straight Road to Language and Consciousness." On "the crucial mutation," see 190–97.

30. Bickerton 1990, 197.

31. Bickerton 1990, chap. 6, esp. 155–56.

32. Bickerton 1990, 10.

33. Bickerton 1990, chap. 5. Bickerton first came to wider professional and popular attention in the early 1980s for his proposal about the origin of creole languages, in his book *Roots of Language* (1981). On the book's impact, see Byrne 1991, 3. On Bickerton's proposal as confirmed by later work, including a well-known study on the emergence of a syntactic sign language out of a pidgin predecessor in a single generation among deaf children in Nicaragua, see Pinker 1994, 32–39.

34. Cheney and Seyfarth 1990, 151–58; for an especially vivid and accessible account of these experiments, see also Seyfarth and Cheney 1992, 125–27. On the importance of the hab/dishab experiments, including discussion of the methodological concerns they raise and attempts to resolve them, see Hauser 1996, 527–28, 623–32.

35. Pinker and Bloom 1990, quotations on 725 and 726. The sciences to be synthesized are listed on 727. On the paper's impact , see Aitchison 1998. For Pinker's criticisms of Bickerton's catastrophism, see Pinker 1992 and, in a more irenic spirit, Pinker 1994, 366 ("The languages of children, pidgin speakers, immigrants, tourists, aphasics, telegrams, and headlines show that there is a vast continuum of viable language systems varying in efficiency and expressive power, exactly what the theory of natural selection requires").

36. The shift away from catastrophism in Bickerton's theorizing has itself been gradual. See, e.g., Bickerton 1998, 2000; and Calvin and Bickerton 2000, esp. chap. 15. In a recent book (2004, 313–18), another Chomskyan linguist, Stephen Anderson, endorses Bickerton's protolanguage-to-language picture while noting that a "number of logically possible paths" from the one to the other "are outlined in the literature, and new ones seem to appear all the time" (318).

37. Interview with M. Hauser, 12 September 2002. On Hauser's time with Cheney and Seyfarth, see Hauser 2004. For Hauser on the current state of the debate on the vervet alarm calls, see Hauser 2000, 188–90.

38. Hauser, Chomsky, and Fitch 2002, 1576.

39. For Cheney and Seyfarth's summary report on their playback-intensive baboon research, see Cheney and Seyfarth 2007. A richly stimulating survey of the results of recent ape language–instruction experiments is Savage-Rumbaugh, Shanker, and Taylor 1998. For an overview of the current state of scientific discussion, the best guides are a series of edited collections: Hurford, Studdert-Kennedy, and Knight 1998; Knight, Studdert-Kennedy, and Hurford 2000; Wray 2002; Christiansen and Kirby 2003; and Tallerman 2005.

40. B. Blanshard to R. M. Seyfarth, 13 March 1982, Yerkes Papers, group 569, box 181, folder 12.

41. Borges 1970.

42. Darwin 1859, 488.

43. Huxley 1863a, 112.

Bibliography

Archives

Alderman Library, University of Virginia.

Archives Générales, Congrégation du Saint-Esprit, Paris.

British Association for the Advancement of Science. Papers. Bodleian Library, Oxford University.

Boas, Franz. Papers. Primary collection and microfilm edition. American Philosophical Society, Philadelphia.

Carpenter, Clarence Ray. Papers. Penn State University Archives.

Darwin, Charles. Papers. Cambridge University Library.

Edison, Thomas Alva. Papers. Primary collection and microfilm edition. Edison National Historic Site, West Orange, NJ, and online at http://edison.rutgers.edu.

Garner, Richard Lynch. Papers. National Anthropological Archives, National Museum of Natural History, Smithsonian Institution.

Harrington, John Peabody. Papers. National Anthropological Archives, National Museum of Natural History, Smithsonian Institution.

Hrdlička, Aleš. Papers. National Anthropological Archives, National Museum of Natural History, Smithsonian Institution.

Huxley, Thomas Henry. Papers. Imperial College, London.

Marler, Peter. Papers. Private collection, University of California at Davis.

McClure, Samuel Sidney. Papers. Lilly Library, Indiana University.

Morgan, Conwy Lloyd. Papers. Bristol University Library.

Müller, Friedrich Max. Papers. Bodleian Library, Oxford University.

Murray, Gilbert. Papers. Bodleian Library, Oxford University.

Page, Walter Hines. Papers. Houghton Library, Harvard University.

Scott, C. P. Papers. John Rylands Library, University of Manchester.

Seyfarth, Robert, and Cheney, Dorothy. Papers. Private collection, University of Pennsylvania.

Smithsonian Institution Archives, Smithsonian Institution.

Urban Archives, Samuel Paley Library, Temple University.

Washburn, Sherwood. Papers. Bancroft Library, University of California at Berkeley.

Wellcome Institute for the History of Medicine Library, London.

Whitney, William Dwight. Papers. Sterling Memorial Library, Yale University.
Wildlife Conservation Society Archives, Bronx Zoo.
Yerkes, Robert Mearns. Papers. Sterling Memorial Library, Yale University.

Interviews

Cheney, Dorothy. Philadelphia. 5 April 2002; 10 September 2002; 8 March 2004.
DeVore, Irven. Cambridge, MA. 12 September 2002.
Hauser, Marc. Cambridge, MA. 12 September 2002.
Hinde, Robert A. Cambridge. 25 July 2002.
Lee, Phyllis C. Cambridge. 17 September 2002.
Marler, Peter. Davis. 9–12 September 2003; 27 October 2003 (telephone); 3–5 April 2005.
Seyfarth, Robert. Philadelphia. 5 April 2002; 9–11 September 2002; 8 March 2004.
Struhsaker, Thomas. By e-mail. Autumn 2004.

Books and Articles

Aarsleff, Hans. 1967. *The Study of Language in England, 1780–1860*. Princeton: Princeton University Press.
———. 1982. "An Outline of Language-Origins Theory since the Renaissance." In *From Locke to Saussure: Essays in the Study of Language and Intellectual History*, 278–92. Minneapolis: University of Minnesota Press.
Abbott, Andrew. 2001. *Chaos of Disciplines*. Chicago: University of Chicago Press.
Abercrombie, John. 1838. *Inquiries Concerning the Intellectual Powers and the Investigation of Truth*. 8th ed. London: John Murray.
Abingdon: A Self-Guided Walking Tour of Abingdon's Downtown Historic District. n.d. Abingdon, VA: Abingdon Convention and Visitors Bureau.
Adams, Mark B. 2000. "Last Judgment: The Visionary Biology of J. B. S. Haldane." *Journal of the History of Biology* 33:457–91.
Agard, Frederick B., Gerald Kelley, Adam Makkai, and Valerie Becker Makkai, eds. 1983. *Essays in Honor of Charles F. Hockett*. Leiden: E. J. Brill.
Aiello, L., and C. Dean. 1990. *An Introduction to Human Evolutionary Anatomy*. London: Academic Press.
Aitchison, Jean. 1998. "On Discontinuing the Continuity-Discontinuity Debate." In Hurford, Studdert-Kennedy, and Knight 1998, 17–29.
Aleksander, Igor. 2002. "Understanding Information, Bit by Bit: Shannon's Equations." In *It Must Be Beautiful: Great Equations of Modern Science*, ed. Graham Farmelo, 213–30. London: Granta Books.
Alexander, Caroline. 1990. *One Dry Season: In the Footsteps of Mary Kingsley*. New York: Knopf.
Allen, David Elliston. 1978. *The Naturalist in Britain: A Social History*. Harmondsworth: Penguin. First published in 1976 by Allen Lane.

Allen, John S. 1989. "Franz Boas's Physical Anthropology: The Critique of Racial Formalism Revisited." *Current Anthropology* 30 (1): 79–84.

Allen, Woody. 1975. "The Whore of Mensa." Reprinted in *Without Feathers*, 35–42. New York: Random House, 1986.

Alter, Stephen G. 1999. *Darwinism and the Linguistic Image: Language, Race, and Natural Theology in the Nineteenth Century.* Baltimore: Johns Hopkins University Press.

———. 2005. *William Dwight Whitney and the Science of Language.* Baltimore: Johns Hopkins University Press.

Altmann, Jeanne. 1974. "Observational Study of Behavior: Sampling Methods." *Behaviour* 49:227–67.

Altmann, Stuart A. 1967a. *Social Communication among Primates.* Chicago: University of Chicago Press.

———. 1967b. "The Structure of Primate Social Communication." In Altmann 1967a, 325–62.

Alvarez Roldán, Arturo. 1992. "Looking at Anthropology from a Biological Point of View: A. C. Haddon's Metaphors on Anthropology." *History of the Human Sciences* 5:21–32.

Amigoni, David. 1995. "Proliferation and Its Discontents: Max Müller, Leslie Stephen, George Eliot, and *The Origin of Species* as Representation." In *Charles Darwin's The Origin of Species: New Interdisciplinary Essays*, ed. David Amigoni and Jeff Wallace, 122–51. Manchester: Manchester University Press.

Amsterdamska, Olga. 1987. *Schools of Thought: The Development of Linguistics from Bopp to Saussure.* Dordrecht: D. Reidel.

Anderson, Stephen R. 2001. "Do You Speak Chimpanzee? A Neglected Moment in the Study of Language Origins." Paper delivered at a conference on language origins, Institute for Advanced Study, Princeton, May.

———. 2004. *Doctor Dolittle's Delusion: Animals and the Uniqueness of Human Language.* New Haven: Yale University Press.

Andresen, Julia T. 1990. *Linguistics in America, 1769–1924: A Critical History.* London: Routledge.

Andrews, Frank. 1986. *The Edison Phonograph: The British Connection.* London: City of London Phonograph and Graphophone Society.

Andrews, H. A. 1943. "Bibliography of Franz Boas." Ed. Bertha C. Edel. *American Anthropologist*, n.s., 45 (July–September): 67–109.

Angell, J. R. 1909. "The Influence of Darwin on Psychology." *Psychological Review* 16:152–69.

Anglin, Jeremy M. 1977. *Word, Object, and Conceptual Development.* New York: W. W. Norton.

Anthony, R. 1913. "L'encéphale de l'homme fossile de La Quina." *Bulletin et Mémoires de la Société d'Anthropologie de Paris* 6:117–95.

Argyll, Duke of. 1887. "Thought without Words." *Nature* 36:52.

Armstrong, Edward A. 1947. *Bird Display and Behaviour*. London: Lindsay Drummond.

Aspray, William. 1985. "The Scientific Conceptualization of Information: A Survey." *Annals of the History of Computing* 7:117–40.

Asquith, Pamela. 2000. "Negotiating Science: Internationalization and Japanese Primatology." In Strum and Fedigan 2000, 165–83.

Autuori, M., M.-P. Bénassy, J. Benoit, R. Courrier, Ed.-Ph. Deleurance, M. Fontaine, K. von Frisch, et al. 1956. *L'instinct dans le comportement des animaux et de l'homme*. Paris: Masson.

Auziaz-Turenne, J. A. 1849. *Cholera and Its Treatment: A Short Essay*. Translated by F. Bateman. London: Simpkin, Marshall.

Ayer, A. J., J. B. S. Haldane, Colin Cherry, Geoffrey Vickers, J. Z. Young, R. Wittkower, T. B. L. Webster, Randolph Quirk, and D. B. Fry. 1955. *Studies in Communication*. Introduction by B. Ifor Evans. London: Martin Secker and Warburg.

Baars, Bernard J. 1986. *The Cognitive Revolution in Psychology*. New York: Guilford Press.

Bailey, C. E. G. 1950. "Towards an Orthography of Bird Song." *Ibis* 92:115–22.

Baken, Ronald J., and Raymond G. Daniloff, eds. 1991. *Readings in Clinical Spectrography of Speech*. San Diego: Singular Publication (with Kay Elemetrics).

Baldwin, James Mark. 1909. "The Influence of Darwin on Theory of Mind and Philosophy." *Psychological Review* 16:207–18.

Barkan, Elazar. 1992. *The Retreat of Scientific Racism: Changing Concepts of Race in Britain and the United States between the Wars*. Cambridge: Cambridge University Press.

Barnett, S. A. 1957. "The New Ethology." In *New Biology*, no. 24:118–24. Harmondsworth: Penguin,

Barrett, P. H., P. J. Gautrey, S. Herbert, D. Kohn, and S. Smith, eds. 1987. *Charles Darwin's Notebooks, 1836–1844: Geology, Transmutation of Species, Metaphysical Enquiries*. London: British Museum (Natural History) and Cambridge University Press.

Barrow, Mark V. Jr. 2000. "The Specimen Dealer: Entrepreneurial Natural History in America's Gilded Age." *Journal of the History of Biology* 33:493–534.

Barsky, Robert F. 1997. *Noam Chomsky: A Life of Dissent*. Cambridge, MA: MIT Press.

Barth, Frederik, Andre Gringrich, Robert Parkin, and Sydel Silverman. 2005. *One Discipline, Four Ways: British, German, French, and American Anthropology*. Chicago: University of Chicago Press.

Bastian, Jarvis R. 1965. "Primate Signaling Systems and Human Languages." In DeVore 1965, 585–606.

Bateman, Frederic. 1865. "On Aphasia, or Loss of the Power of Speech." *Lancet*, 20 May, 532–33.

———. 1870. *On Aphasia, or Loss of Speech, and the Localisation of the Faculty of Articulate Language*. London: John Churchill and Sons; Norwich: Jarrold.

———. 1874. "Darwinism Tested by Recent Researches in Language." *Journal of the Transactions of the Victoria Institute* 7:73–95.

———. 1877. *Darwinism Tested by Language*. Preface by Edward Meyrick Gouldburn. Norwich: Rivington's.

———. 1882. *The Idiot: His Place in Creation and His Claims on Society*. London: Jarrold and Sons.

———. 1890. *On Aphasia, or Loss of Speech, and the Localisation of the Faculty of Articulate Language*. 2nd edition. London: Churchill / Jarrold.

Baume, R. 1883. *Die Kieferfragmente von La Naulette und aus der Schipkahöle als Merkmale für die Existenz inferiorer Menschenrassen in der Diluvialzeit*. Leipzig: Felix.

Beach, Frank A. 1950. "The Snark Was a Boojum." *American Psychologist* 5:115–24.

Bear, John. 1973. *Communication*. London: Macdonald / Educational.

Beatty, Barbara. 1998. "From Laws of Learning to a Science of Values: Efficiency and Morality in Thorndike's Experimental Psychology." *American Psychologist* 53 (10): 1145–52.

Beer, Gillian. 1996. "Darwin and the Growth of Language Theory." In *Open Fields: Science in Cultural Encounter*, 95–114. Oxford: Clarendon Press.

Beer, Thomas, Albrecht Bethe, and Jakob von Uexküll. 1899. "Vorschläge zu einer objectivierenden Nomenclatur in der Physiologie des Nervensystems." *Biologisches Zentralblatt* 19:517–21.

Bell, Alexander Melville. 1867. *Visible Speech: The Science of Universal Alphabetics, or Self-Interpreting Physiological Letters, for the Writing of All Languages in One Alphabet*. London: Simpkins, Marshall.

Benjamin, Ludy T. Jr., and Darryl Bruce. 1982. "From Bottle-Fed Chimp to Bottlenose Dolphin: A Contemporary Appraisal of Winthrop Kellogg." *Psychological Record* 32:461–82.

Bernstein, Jeremy. 1987. *Three Degrees above Zero: Bell Labs in the Information Age*. Cambridge: Cambridge University Press.

Berreby, David. 1992. "Kids, Creoles, and the Coconuts." *Discover*, April, 44–53.

Beyer, Robert T. 1999. *Sounds of Our Times: Two Hundred Years of Acoustics*. New York: Springer-Verlag / AIP Press.

Bickerton, Derek. 1981. *Roots of Language*. Ann Arbor: Karoma.

———. 1990. *Language and Species*. Chicago: University of Chicago Press.

———. 1992. "The Creole Key to the Black Box of Language." In *Thirty Years of Linguistic Evolution*, ed. Martin Pütz, 97–108. Amsterdam: John Benjamins.

———. 1998. "Catastrophic Evolution: The Case for a Single Step from Protolanguage to Full Human Language." In Hurford, Studdert-Kennedy, and Knight 1998, 341–58.

———. 2000. "How Protolanguage Became Language." In Knight, Studdert-Kennedy, and Hurford 2000, 264–84.

Birch, Herbert G. 1952. "Communication between Animals." In *Cybernetics: Circular Causal and Feedback Mechanisms in Biological and Social Systems*, ed. Heinz von Foerster, 134–72. New York: Josiah Macy Jr. Foundation.

Blackman, Helen J. 2004. "Haddon, Alfred Cort." In Lightman 2004, 2:872–73.

Blitz, David. 1992. *Emergent Evolution: Qualitative Novelty and the Levels of Reality.* Dordrecht: Kluwer.

Bloomfield, Leonard. 1939. *Linguistic Aspects of Science.* In *International Encyclopedia of Unified Science*, ed. Otto Neurath, vols. 1 and 2, *Foundations of the Unity of Science*, no. 4. Chicago: University of Chicago Press.

———. 1942. "Philosophical Aspects of Language." In Hockett 1970, 396–99. First published in *Studies in the History of Culture: The Disciplines of the Humanities*, 173–77. Menasha, WI: George Banta.

Blustein, Bonnie E. 1991. *Preserve Your Love for Science: Life of William A. Hammond, American Neurologist.* Cambridge: Cambridge University Press.

Boakes, Robert. 1984. *From Darwin to Behaviourism: Psychology and the Minds of Animals.* Cambridge: Cambridge University Press.

Boas, Franz. 1888. "The Origin and Development of Language." *Science*, 28 September, 145–46.

———. 1889. "On Alternating Sounds." Reprinted in Stocking 1974b, 72–77. First published in *American Anthropologist* 2:47–53.

———. 1897. "Horatio Hale." *Critic*, 16 January, 40–41.

———. 1899. "Fieldwork for the British Association, 1888–1897." Reprinted in Stocking 1974b, 88–107. First published as "Summary of the Work of the Committee in British Columbia," *Report of the British Association for the Advancement of Science for 1898* (London), 667–82.

———. 1902. "Rudolf Virchow's Anthropological Work." Reprinted in Stocking 1974b, 36–41. First published in *Science* 16:441–45.

———. 1911a. "Changes in Bodily Form of Descendants of Immigrants." Reprinted in Montagu 1974, 321–32. Extract from Boas's *Abstract of the Report on Changes in Bodily Form of Descendants of Immigrants.* Washington, DC: Immigration Commission / Government Printing Office.

———. 1911b. *The Mind of Primitive Man.* New York: Macmillan.

Borges, Jorge Luis. 1970. "Pierre Menard, Author of the *Quixote*." In *Labyrinths*, 62–71. London: Penguin.

Boring, Edwin G. 1950. *A History of Experimental Psychology.* 2nd ed. New York: Appleton-Century-Crofts.

Borror, Donald J., and C. R. Reese. 1953. "The Analysis of Bird Songs by Means of a Vibralyzer." *Wilson Bulletin* 65:271–303.

Bouillaud, J.-B. 1825. "Recherches clinique propres à démontrer que la perte de la parole correspond à la lésion des lobules antérieurs du cerveau, etc." *Archives Générales de Médecine* 8:25–45.

Bourne, H. R. F. 1887. *English Newspapers: Chapters in the History of Journalism.* Vol. 2. London: Chatto and Windus.

Boutan, Louis. 1913. "Le pseudo-langage: Observations effectuées sur un anthropoide: Le Gibbon (*Hylobates leucogenys*—Ogilby)." *Actes de la Société Linnéenne de Bordeaux* 67:5–80.

Bowler, Peter J. 1983. *The Eclipse of Darwinism: Anti-Darwinian Evolution Theories in the Decades around 1900.* Baltimore: Johns Hopkins University Press.

———. 1986. *Theories of Human Evolution: A Century of Debate, 1844–1944.* Baltimore: Johns Hopkins University Press.

———. 1996. *Life's Splendid Drama: Evolutionary Biology and the Reconstruction of Life's Ancestry, 1860–1940.* Chicago: University of Chicago Press.

———. 2003. *Evolution: The History of an Idea.* 3rd ed. London: University of California Press.

Brady, Erika. 1999. *A Spiral Way: How the Phonograph Changed Ethnography.* Jackson: University Press of Mississippi.

Brady, Erika, Maria La Vigna, Dorothy Sara Lee, and Thomas Vennum. 1984. *The Federal Cylinder Project: A Guide to Field Cylinder Collections in Federal Agencies.* Vol. 1. Washington, DC: American Folklore Center / Library of Congress.

Bridges, W. 1974. *Gathering of Animals: An Unconventional History of the New York Zoological Society.* New York: Harper and Row.

Brinton, Daniel G. 1888. "The Language of Palaeolithic Man." *Proceedings of the American Philosophical Society* 25:212–25. Reprinted in *Essays of an Americanist*, 390–409. Philadelphia: Porter and Coates, 1890.

———. 1890. *Races and Peoples: Lectures on the Science of Ethnography.* New York: N. D. C. Hodges.

———. 1895. "The Aims of Anthropology." *Proceedings of the American Association for the Advancement of Science* 44:1–17.

Broadbent, Donald E. 1956. "The Concept of Capacity and the Theory of Behaviour." In Cherry 1956, 354–60.

———. 1958. *Perception and Communication.* London: Pergamon Press.

———. 1961. "Human Perception and Animal Learning." In *Current Problems in Animal Behaviour*, ed. W. H. Thorpe and O. L. Zangwill, 248–72. Cambridge: Cambridge University Press.

———. 1980. Autobiography. In *A History of Psychology in Autobiography*, ed. Gardner Lindzey, 7:39–73. San Francisco: W. H. Freeman and Co.

Brock, W. H. 1994. "Science." In *Victorian Periodicals and Victorian Society*, ed. J. D. Vann and R. T. VanArsdel, 81–96. Aldershot, UK: Scolar Press.

Bronowksi, Jacob, and Ursula Bellugi. 1970. "Language, Name, and Concept." Reprinted in Sebeok and Umiker-Sebeok 1980, 103–13. First published in *Science* 168:669–73.

Browne, Janet. 1985. "Darwin and the Expression of the Emotions." In *The Darwinian Heritage*, ed. David Kohn, 307–26. Princeton: Princeton University Press.

———. 1995. *Charles Darwin: Voyaging.* London: Jonathan Cape.

———. 2002. *Charles Darwin: The Power of Place*. London: Jonathan Cape.

Buckley, Kerry W. 1989. *Mechanical Man: John Broadus Watson and the Beginnings of Behaviorism*. New York: Guilford Press.

———. 2000. "Watson, John B." In Kazdin 2000, 8:232–35.

Buckman, S. S. 1897. "The Speech of Children." *Nineteenth Century* 41:793–807.

[Buléon, J.] 1896. *Sous le ciel d'Afrique: De Sainte Anne d'Auray à Sainte Anne du Fernan-Vaz: Recit d'un missionaire*. Abbeville, France: C. Paillart.

Burkhardt, Frederick H., et al., eds. 1985–2006. *The Correspondence of Charles Darwin*. 15 vols. Cambridge: Cambridge University Press.

Burkhardt, Frederick H., and S. Smith, eds. 1994. *A Calendar of the Correspondence of Charles Darwin, 1821–1882*. Cambridge: Cambridge University Press.

Burkhardt, Richard W. Jr. 1983. "The Development of an Evolutionary Ethology." In *Evolution from Molecules to Men*, ed. D. S. Bendall, 429–44. Cambridge: Cambridge University Press.

———. 1987. "The *Journal of Animal Behavior* and the Early History of Animal Behavior Studies in America." *Journal of Comparative Psychology* 101 (3): 223–30.

———. 2005. *Patterns of Behavior: Konrad Lorenz, Niko Tinbergen, and the Founding of Ethology*. Chicago: University of Chicago Press.

Burnham, John C. 1972. "Thorndike's Puzzle Boxes." *Journal of the History of the Behavioral Sciences* 8:159–67.

———. 1976. "Yerkes, Robert Mearns." In Gillispie 1970–80, 14:549–51.

Burrow, John W. 1967. "The Uses of Philology in Victorian England." In *Ideas and Institutions of Victorian Britain: Essays in Honour of George Kitson Clark*, ed. R. Robson, 180–204. London: G. Bell and Sons.

Burt, Jonathan. 2006. "Solly Zuckerman: The Making of a Primatological Career in Britain, 1925–1945." *Studies in History and Philosophy of Biological and Biomedical Sciences* 37:295–310.

Busnel, R.-G., ed. 1963. *Acoustic Behaviour of Animals*. Amsterdam: Elsevier.

Butler, Samuel. 1904. "Thought and Language." In *Essays on Life Art and Science*, 176–233. London: Grant Richards.

Byrne, Francis. 1991. "Introduction: Innovation and Excellence within a Scholarly Tradition." In *Development and Structures of Creole Languages*, festschrift for Derek Bickerton, ed. Francis Byrne and Thom Huebner, 1–14. Amsterdam: John Benjamins.

Calvin, William H., and Derek Bickerton. 2000. *Lingua ex Machina: Reconciling Darwin and Chomsky with the Human Brain*. Cambridge, MA: Bradford Books / MIT Press.

Campbell, Neil. 1993. *Biology*. 3rd ed. Redwood, CA: Benjamin / Cummings.

Candland, Douglas Keith. 1993. *Feral Children and Clever Animals: Reflections on Human Nature*. Oxford: Oxford University Press.

Cantor, Geoffrey, Gowan Dawson, Graeme Gooday, Richard Noakes, Sally Shuttleworth, and Jonathan R. Topham. 2004. *Science in the Nineteenth-Century Period-*

ical: Reading the Magazine of Nature. Cambridge Studies in Nineteenth-Century Literature and Culture. Cambridge: Cambridge University Press.

Cantor, Geoffrey, and Sally Shuttleworth, eds. 2004. *Science Serialized: Representations of the Sciences in Nineteenth-Century Periodicals.* Cambridge, MA: MIT Press.

Carpenter, Clarence Ray. 1934. "A Field Study of the Behaviour and Social Relations of Howler Monkeys." *Comparative Psychology Monographs* 10:1–168. Reprinted in Carpenter 1964, 3–92.

———. 1939. "Behavior and Social Relations of Free-Ranging Primates." *Scientific Monthly* 48:319–25.

———. 1940. "A Field Study in Siam of the Behavior and Social Relations of the Gibbon (*Hylobates lar*)." *Comparative Psychology Monographs* 16 (5): 1–212.

———. 1942. "Societies of Monkeys and Apes." In *Levels in Integration in Biological and Social Systems*, ed. R. Redfield, 177–204. Lancaster, PA: Jacques Cattell Press.

———. 1964. *Naturalistic Behavior of Nonhuman Primates.* University Park: Pennsylvania State University Press.

———. 1969. "Approaches to Studies of the Naturalistic Communicative Behavior in Nonhuman Primates." In Sebeok and Ramsey 1969, 40–70.

Carr, Harvey. 1927. "The Interpretation of the Animal Mind." *Psychological Review* 34 (2): 87–106.

Carus, Paul. 1892. "The Continuity of Evolution: The Science of Language versus the Science of Life, as Represented by Prof. F. Max Müller and Prof. George John Romanes." *Monist* 2:70–94.

Catchpole, C. K., and P. J. B. Slater. 1995. *Bird Song: Biological Themes and Variations.* Cambridge: Cambridge University Press.

[Chambers, Robert.] 1844. *Vestiges of the Natural History of Creation.* London: Churchill. Reprinted with an introduction by J. A. Secord. Chicago: University of Chicago Press, 1994.

Chaudhuri, Nirad C. 1974. *Scholar Extraordinary: The Life of Professor the Rt. Hon. Friedrich Max Müller, PC.* New York: Oxford University Press.

Cheney, Dorothy L. 1977. "The Acquisition of Rank and the Development of Reciprocal Alliances among Free-Ranging Immature Baboons." *Behavioral Ecology and Sociobiology* 2:303–18.

———. 1978a. "Interactions of Immature Male and Female Baboons with Adult Females." *Animal Behaviour* 26:389–408.

———. 1978b. "The Play Partners of Immature Baboons." *Animal Behaviour* 26:1038–50.

———. 1981. "Intergroup Encounters among Free-Ranging Vervet Monkeys." *Folia Primatologica* 35:124–46.

Cheney, Dorothy L., and Robert M. Seyfarth. 1977. "Behaviour of Adult and Immature Male Baboons during Inter-group Encounters." *Nature* 269:404–6.

———. 1980. "Vocal Recognition in Free-Ranging Vervet Monkeys." *Animal Behaviour* 28:362–67.

———. 1981. "Selective Forces Affecting the Predator Alarm Calls of Vervet Monkeys." *Behaviour* 76:27–61.

———. 1982. "How Vervet Monkeys Perceive Their Grunts: Field Playback Experiments." *Animal Behaviour* 30:739–51.

———. 1990. *How Monkeys See the World: Inside the Mind of Another Species*. Chicago: University of Chicago Press.

———. 2007. *Baboon Metaphysics*. Chicago: University of Chicago Press.

Cheney, Margaret. 2001. "Coming Home." *Washingtonian*, December. http://www.washingtonian.com/people/cominghome.html (accessed December 2004).

Cheney Seyfarth, Dorothy. 1976. "The Social Development of Immature Male and Female Baboons." PhD thesis, Cambridge University.

Cherfas, Jeremy. 1980. "Voices in the Wilderness." *New Scientist*, 19 June, 303–6.

Cherry, Colin. 1952. "The Communication of Information: An Historical Review." *American Scientist* 40:640–64, 724–25.

———. 1953. "Some Experiments on the Recognition of Speech, with One and with Two Ears." *Journal of the Acoustical Society of America* 25:975–79.

———. 1955. "'Communication Theory'—and Human Behaviour." In Ayer et al. 1955, 45–67.

———, ed. 1956. *Information Theory*. London: Butterworths Scientific Publications.

———. 1957. *On Human Communication: A Review, a Survey, and a Criticism*. New York: MIT Press / John Wiley / Chapman and Hall.

———. 1985. *The Age of Access: Information Technology and Social Revolution: Posthumous Papers of Colin Cherry*. Ed. William Edmondson. Foreword by Asa Briggs. London: Croom Helm.

Chomsky, Noam. 1957a. Review of Hockett, *A Manual of Phonology*. *International Journal of American Linguistics* 23:223–34.

———. 1957b. *Syntactic Structures*. The Hague: Mouton.

———. 1959. Review of Skinner, *Verbal Behavior*. *Language* 35:26–58.

———. 1965. *Aspects of the Theory of Syntax*. Cambridge, MA: MIT Press.

———. 1966. *Cartesian Linguistics: A Chapter in the History of Rationalist Thought*. New York: Harper and Row.

———. 1967. "General Properties of Language." In Darley 1967, 73–88.

———. 1968. *Language and Mind*. New York: Harcourt Brace Jovanovich.

Christiansen, Morten H., and Simon Kirby, eds. 2003. *Language Evolution*. Oxford: Oxford University Press.

Clark, John F. M. 1998. "John Lubbock and Mental Evolution." *Endeavour* 22 (2): 44–47.

Clark, Ronald. 1968. *J. B. S.: The Life and Work of J. B. S. Haldane*. London: Hodder and Stoughton.

Clarke, E. 1974. "Morgan, C. Lloyd." In Gillispie 1970–80, 9:512–13.

Clarke, E., and K. Dewhurst. 1972. *An Illustrated History of Brain Function*. Oxford: Sandford.

Clarke, E., and L. S. Jacyna. 1987. *Nineteenth Century Origins of Neuroscientific Concepts*. Berkeley: University of California Press.

Coburn, Charles A., and Robert M. Yerkes. 1915. "A Study of the Behavior of the Crow *Corvus americanus* Aud. by the Multiple-Choice Method." *Journal of Animal Behavior* 5 (2): 75–114.

Coetzee, J. M. 1999. *The Lives of Animals*. Princeton: Princeton University Press.

Cohen, David. 1979. *J. B. Watson: The Founder of Behaviourism: A Biography*. London: Routledge and Kegan Paul.

Cole, Douglas. 1999. *Franz Boas: The Early Years, 1858–1906*. Vancouver: Douglas and McIntyre; Seattle: University of Washington Press.

Collias, Nicholas, and Martin Joos. 1953. "The Spectrographic Analysis of Sound Signals of the Domestic Fowl." *Behaviour* 5:175–88.

Collini, Stefan. 1979. *Liberalism and Sociology: L. T. Hobhouse and Political Argument in England, 1880–1914*. Cambridge: Cambridge University Press.

Conn, H. W. 1892. Review of Garner, *The Speech of Monkeys. Science* 20:221–22.

Conrad, Joseph. 1902. *Heart of Darkness*. In *Joseph Conrad: Selected Works*. London: Leopard Books, 1994.

Conway, Flo, and Jim Siegelman. 2005. *Dark Hero of the Information Age: In Search of Norbert Wiener, the Father of Cybernetics*. New York: Basic Books.

Conway, M. D. 1900. "Memories of Max Müller." *North American Review* 171:884–93.

Cornish, C. J. 1895. *Life at the Zoo: Notes and Traditions of the Regent's Park Zoo*. London: Seeley.

Cosans, Christopher. 1994. "Anatomy, Metaphysics, and Values: The Ape Brain Debate Reconsidered." *Biology and Philosophy* 9:129–65.

Costall, A. 1993. "How Lloyd Morgan's Canon Backfired." *Journal of the History of the Behavioral Sciences* 29:113–22.

———. 1998. "Lloyd Morgan and the Rise and Fall of 'Animal Psychology.'" *Society and Animals* 6:13–29.

Costall, A., J. F. M. Clark, and R. H. Wozniak. 1997. "Conwy Lloyd Morgan (1852–1936): An Introduction to His Work and a Bibliography of His Writings." *Teorie & Modelli*, n.s., 2:65–92.

Craig, Wallace E. 1908. "The Voices of Pigeons Regarded as a Means of Social Control." *American Journal of Sociology* 14:86–100.

Craik, Kenneth J. W. 1943. *The Nature of Explanation*. Cambridge: Cambridge University Press.

Crail, Ted. 1981. *Apetalk and Whalespeak: The Quest for Interspecies Communication*. Los Angeles: J. P. Tarcher.

Cravens, Hamilton. 1978. *The Triumph of Evolution: American Scientists and the Heredity-Environment Controversy, 1900–1941*. Philadelphia: University of Pennsylvania Press.

Critchley, Macdonald. 1958. "Animal Communication." *Transactions of the Hunterian Society* 16:89–112.

———. 1960. "The Evolution of Man's Capacity for Language." In *Evolution after Darwin*, vol. 2, *The Evolution of Life*, ed. Sol Tax, 289–308. Chicago: University of Chicago Press.

———. 1970. "The Nature of Animal Communication and Its Relation to Language in Man." In *Aphasiology and Other Aspects of Language*, 126–43. London: Edward Arnold.

Critchley, Macdonald, and Eileen A. Critchley. 1998. *John Hughlings Jackson: Father of English Neurology*. Oxford: Oxford University Press.

Crystal, David. 1987. *The Cambridge Encyclopedia of Language*. Cambridge: Cambridge University Press.

Cunningham, Andrew, and Nicholas Jardine, eds. 1990. *Romanticism and the Sciences*. Cambridge: Cambridge University Press.

Cunningham, D. J., and Alfred Cort Haddon. 1892. "The Anthropometric Laboratory of Ireland." *Journal of the Anthropological Institute of Great Britain and Ireland* 21:35–39.

D'Agostino, F. 1983. "Darwinism and Language." In *The Wider Domain of Evolutionary Thought*, ed. D. Oldroyd and I. Langham, 159–73. London: D. Reidel.

Darley, Frederic L., ed. 1967. *Brain Mechanisms Underlying Speech and Language*. New York: Grune and Stratton.

Darnell, Regna. 1976. "Daniel Brinton and the Professionalization of American Anthropology." In *American Anthropology: The Early Years*, ed. J. V. Murra, 69–98. New York: West Publishing.

———. 1988. *Daniel Garrison Brinton: The "Fearless Critic" of Philadelphia*. University of Pennsylvania Publications in Anthropology, no. 3. Philadelphia: Department of Anthropology, University of Pennsylvania.

———. 1990. *Edward Sapir: Linguist, Anthropologist, Humanist*. London: University of California Press.

———. 1998. *And Along Came Boas: Continuity and Revolution in American Anthropology*. Amsterdam: John Benjamins.

Darwin, Charles. 1859. *On the Origin of Species by Means of Natural Selection, or the Preservation of Favoured Races in the Struggle for Life*. London: John Murray.

———. 1868. *The Variation of Animals and Plants under Domestication*. 2 vols. London: John Murray.

———. 1871. *The Descent of Man, and Selection in Relation to Sex*. 2 vols. London: John Murray.

———. 1872. *The Expression of the Emotions in Man and Animals*. London: John Murray.

———. 1874. *The Descent of Man and Selection in Relation to Sex*. 2nd ed. London: John Murray.

———. 1877. "A Biographical Sketch of an Infant." In Gruber 1974, 465–74. First published in *Mind* 2:285–94.

———. 1881. "A Letter to Mrs. Emily Talbot on the Mental and Bodily Development of Infants." In *The Collected Papers of Charles Darwin*, ed. P. H. Barrett,

2:232–33. Chicago: University of Chicago Press, 1977. First published in *Nature* 24:459.

Darwin, Francis, and A. C. Seward, eds. 1903. *More Letters of Charles Darwin.* Vol. 2. London: John Murray.

Darwin, George. 1874. "Professor Whitney on the Origin of Language." In Harris 1996, 277–90. First published in *Contemporary Review* 24:894–904.

Daston, Lorraine, and Gregg Mitman, eds. 2005. *Thinking with Animals: New Perspectives on Anthropomorphism.* New York: Columbia University Press.

Dawkins, Richard. 2003. "The 'Information Challenge.'" In *A Devil's Chaplain: Selected Essays,* ed. Latha Menon, 107–22. London: Weidenfeld and Nicolson. First published in *The Skeptic,* December 1998.

Deacon, Terence. 1997. *The Symbolic Species: The Co-evolution of Language and the Human Brain.* London: Allen Lane / Penguin Press.

Degler, Carl N. 1991. *In Search of Human Nature: The Decline and Revival of Darwinism in American Social Thought.* Oxford: Oxford University Press.

De Laguna, Grace Andrus. 1927. *Speech: Its Function and Development.* New Haven: Yale University Press. Reprint. Bloomington: Indiana University Press, 1963.

De Mortillet, G. 1883. *Le préhistorique: Antiquité de l'homme.* Paris: C. Reinwald.

Dennett, Daniel C. 1983. "Intentional Systems in Cognitive Ethology: The 'Panglossian Paradigm' Defended." *Behavioral and Brain Sciences* 6:343–90. Including extensive peer commentary and Dennett's response. Main article reprinted with further reflections in *The Intentional Stance,* 237–86. Cambridge, MA: Bradford Books / MIT Press, 1987.

———. 1987. "Reflections: Interpreting Monkeys, Theorists, and Genes." In *The Intentional Stance,* 269–86. Cambridge, MA: Bradford Books / MIT Press.

———. 1988. "Out of the Armchair and into the Field." *Poetics Today* 9:205–21. Reprinted in *Brainchildren: Essays on Designing Minds,* 289–306. London: Penguin, 1998.

———. 1989. "Cognitive Ethology: Hunting for Bargains or a Wild Goose Chase." In *Goals, No-Goals, and Own Goals: A Debate on Goal-Directed and Intentional Behaviour,* ed. A. Montefiore and D. Noble, 101–16. London: Unwin Hyman. Reprinted in *Brainchildren: Essays on Designing Minds,* 307–22. London: Penguin, 1998.

Depew, David J., and Bruce H. Weber. 1995. *Darwinism Evolving: Systems Dynamics and the Geneaology of Natural Selection.* Cambridge, MA: Bradford Books / MIT Press.

Desmond, Adrian J. 1979. *The Ape's Reflexion.* New York: Dial Press / James Wade.

———. 1994. *Huxley: The Devil's Disciple.* London: Michael Joseph.

———. 1997. *Huxley: Evolution's High Priest.* London: Michael Joseph.

Desmond, Adrian J., and James Moore. 1991. *Darwin.* London: Penguin.

Dessalles, J.-L., and L. Ghadakpour, eds. 2000. *Proceedings of the Third International Conference on the Evolution of Language, 3–6 April 2000.* Paris: École Nationale Supérieure des Télécommunications.

DeVore, Irven, ed. 1965. *Primate Behavior: Field Studies of Monkeys and Apes*. New York: Holt, Rinehart and Winston.

De Waal, Frans. 2001. *The Ape and the Sushi Master: Cultural Reflections of a Primatologist*. New York: Basic Books.

Dewsbury, Donald A. 1989a. "A Brief History of the Study of Animal Behavior in North America." In *Perspectives in Ethology*, vol. 8, *Whither Ethology?* ed. P. P. G. Bateson and Peter H. Klopfer, 85–122. New York: Plenum Press.

———, ed. 1989b. *Studying Animal Behavior: Autobiographies of the Founders*. Chicago: University of Chicago Press. First published in 1985 as *Leaders in the Study of Animal Behavior* by Associated University Presses.

———. 1992a. "Comparative Psychology and Ethology: A Reassessment." *American Psychologist* 47:208–15.

———. 1992b. "Triumph and Tribulation in the History of American Comparative Psychology." *Journal of Comparative Psychology* 106:3–19.

———. 1997. "Rhetorical Strategies in the Presentation of Ethology and Comparative Psychology in the Magazines after World War II." *Science in Context* 10:367–86.

———. 1998. "Celebrating E. L. Thorndike a Century after *Animal Intelligence*." *American Psychologist* 53 (October): 1121–24.

———. 2000a. "Issues in Comparative Psychology at the Dawn of the 20th Century." *American Psychologist* 55 (7): 750–53.

———. 2000b. "Yerkes, Robert Mearns." In Kazdin 2000, 8:295–96.

———. 2006. *Monkey Farm: A History of the Yerkes Laboratories of Primate Biology, Orange Park, Florida, 1930–1965*. Lewisburg, PA: Bucknell University Press.

Diamond, Jared. 1991. *The Rise and Fall of the Third Chimpanzee*. London: Radius.

Dickson, W. K. L., and A. Dickson. 1895. *History of the Kinetograph, Kinetoscope, and Kinetophonograph*. New York: A. Bunn.

Di Gregorio, Mario A. 1990. *Charles Darwin's Marginalia*. Vol. 1. London: Garland Publishing.

———. 2002. "Reflections of a Nonpolitical Naturalist: Ernst Haeckel, Wilhelm Bleek, Friedrich Müller, and the Meaning of Language." *Journal of the History of Biology* 35:79–109.

Dixon, Thomas. 2003. *From Passions to Emotions: The Creation of a Secular Psychological Category*. Cambridge: Cambridge University Press.

Dixson, A. F. 1981. *The Natural History of the Gorilla*. London: Weidenfeld and Nicolson.

Dowling, Linda. 1982. "Victorian Oxford and the Science of Language." *Publication of the Modern Language Association* 97:160–78.

Dubois, Eugène. 1888. "Over de Wenschelijkheid van een onderzoek naar de diluviale fauna van Ned. Indië, in het bijzander van Sumatra." *Natuurkundig tijdschrift voor Nederlandsch-Indië* 48:148–65.

Dukelow, W. Richard. 1995. *The Alpha Males: An Early History of the Regional Primate Research Centers*. Lanham, MD: University Press of America.

Dunbar, Robin. 1996. *Grooming, Gossip, and the Evolution of Language*. London: Faber and Faber.

Durant, John R. 1981. "Innate Character in Animals and Man: A Perspective on the Origins of Ethology." In *Biology, Medicine, and Society, 1840–1940*, ed. Charles Webster, 157–92. Cambridge: Cambridge University Press.

———. 1985. "The Science of Sentiment: The Problem of the Cerebral Localization of Emotion." In *Perspectives in Ethology*, vol. 6, *Mechanisms*, ed. P. P. G. Bateson and Peter H. Klopfer, 1–31. New York: Plenum Press.

———. 1986. "The Making of Ethology: The Association for the Study of Animal Behaviour, 1936–1986." *Animal Behaviour* 34:1601–16.

Dybowski, J. 1893. *La route du Tchad: Du Loango au Chari*. Paris: Librairie de Firmin-Didot.

Edwards, A. W. F. 2003. "Human Genetic Diversity: Lewontin's Fallacy." *BioEssays* 25:798–801.

E. H. T. 1893. "Re-volution." *Blackwood's Edinburgh Magazine* 153:253–55.

Eldredge, Niles, and Stephen Jay Gould. 1972. "Punctuated Equilibria: An Alternative to Phyletic Gradualism." In *Models in Paleobiology*, ed. T. J. M. Schopf, 82–115. San Francisco: Freeman, Cooper.

Eling, Paul, ed. 1994. *Reader in the History of Aphasia: From Gall to Geschwind*. Amsterdam: John Benjamins.

Ellegård, Alvar. 1990. *Darwin and the General Reader: The Reception of Darwin's Theory of Evolution in the British Periodical Press, 1859–1872*. Chicago: University of Chicago Press. First published in 1958.

Engels, Friedrich. 1876. "The Part Played by Labour in the Transition from Ape to Man." In *Dialectics of Nature*, translated by Clemens Dutt, 170–83. Moscow: Progress, 1966.

Evans, B. Ifor. 1955. Introduction to Ayer et al. 1955, 1–9.

Evans, E. P. 1891. "Speech as a Barrier between Man and Beast." *Atlantic Monthly* 68:299–312.

———. 1893. "Studies of Animal Speech." *Popular Science Monthly* 43:433–49.

———. 1897. *Evolutional Ethics and Animal Psychology*. New York: D. Appleton.

Falls, J. Bruce. 1963. "Properties of Bird Song Eliciting Responses from Territorial Males." *Proceedings of the XIII International Ornithological Congress*, 13:259–71.

———. 1992. "Playback: A Historical Perspective." In *Playback and Studies of Animal Communication*, ed. P. K. McGregor, 11–33. New York: Plenum Press.

Fancher, Raymond E. 1996. *Pioneers of Psychology*. 3rd ed. New York: W. W. Norton.

Farrar, F. W. 1865. *Chapters on Language*. London: Longmans, Green.

Ferguson, Christine. 2006. *Language, Science, and Popular Fiction in the Victorian Fin-de-Siècle: The Brutal Tongue*. Aldershot: Ashgate.

Fewkes, Jesse W. 1890. "On the Use of the Phonograph in the Study of the Languages of American Indians." *Science* 15:267–69.

Fichman, Martin. 2004. *An Elusive Victorian: The Evolution of Alfred Russel Wallace*. Chicago: University of Chicago Press.

Finger, Stanley. 1994. *Origins of Neuroscience: A History of Explorations into Brain Function*. Oxford: Oxford University Press.

Fitzpatrick, Simon. Forthcoming. "Doing Away with Morgan's Canon." *Mind and Language*.

FitzRoy, Robert. 1839. *Proceedings of the Second Expedition, 1831–1836*. Vol. 2 of *Narrative of the Surveying Voyages of His Majesty's Ships Adventure and Beagle*. London: Henry Colburn.

Fleure, H. J. 2004. "Haddon, Alfred Cort." Revised by Sandra Rouse. In *Oxford Dictionary of National Biography*, online version, Oxford University Press (accessed 15 March 2006).

Flourens, M.-J.-P. 1842. *Recherches expérimentales sur les propriétés et les fonctions du système nerveux, dans les animaux vertébrés*. 2nd ed. Paris: Baillière. First edition published in 1824.

———. 1846. *Phrenology Examined*. Translated by C. de Lucena Meigs from French 2nd ed., 1845. Philadelphia: Hogan and Thompson. First edition published in 1842.

Fodor, Jerry. 1999. "Not So Clever Hans." *London Review of Books*, 4 February, 12–13.

Fodor, Jerry, T. G. Bever, and M. F. Garrett. 1974. *The Psychology of Language: An Introduction to Psycholinguistics and Generative Grammar*. New York: McGraw-Hill.

Foucault, Michel. 1970. *The Order of Things: An Archaeology of the Human Sciences*. New York: Vintage Books. First published in French, 1966.

Fouts, Roger. 2003. *Next of Kin: My Conversations with Chimpanzees*. With Stephen Tukel Mills. Introduction by Jane Goodall. New York: Quill / HarperCollins. First published in 1997.

Freeman, Derek. 1983. *Margaret Mead and Samoa: The Making and Unmaking of an Anthropological Myth*. London: Harvard University Press.

———. 1999. *The Fateful Hoaxing of Margaret Mead: A Historical Analysis of Her Samoan Research*. Boulder, CO: Westview Press.

Freeman, R. B. 1977. *The Works of Charles Darwin: An Annotated Bibliographical Handlist*. Folkestone, UK: Dawson and Sons.

Freud, Sigmund. 1953. *On Aphasia: A Critical Study*. London: Imago. First published in German, 1891.

Fries, Charles C. 1963. "The Bloomfield 'School.'" In *Trends in European and American Linguistics, 1930–1960*, ed. C. Mohrmann, A. Sommerfelt, and J. Whatmough, 196–224. Utrecht: Spectrum.

Frisch, Karl von. 1950. *Bees: Their Vision, Chemical Senses, and Language*. Ithaca: Cornell University Press.

Frisch, Karl von, and Martin Lindauer. 1956. "The 'Language' and Orientation of the

Honey Bee." *Annual Review of Entomology* 1:45–58. Reprinted in Houck and Drickamer 1996, 539–52.

Furness, William H. III. 1916. "Observations on the Mentality of Chimpanzees and Orang-Utans." *Proceedings of the American Philosophical Society* 55:281–90.

Galaburda, Albert M. 1997. "Norman Geschwind (1926–1984)." In Schachter and Devinsky 1997, 295–97.

Gale, R. L. 1999. "McClure, Samuel Sidney." In Garraty and Carnes 1999, 14:887–89.

Galef, Bennett G. Jr. 1996. "Historical Origins: The Making of a Science." In Houck and Drickamer 1996, 5–12.

———. 1998. "Edward Thorndike: Revolutionary Psychologist, Ambiguous Biologist." *American Psychologist* 53 (October): 1128–34.

Galison, Peter. 1994. "The Ontology of the Enemy: Norbert Wiener and the Cybernetic Vision." *Critical Inquiry* 21:228–66.

Gall, Franz J. 1835. *On the Functions of the Brain and of Each of Its Parts: With Observations on the Possibility of Determining the Instincts, Propensities, and Talents, or the Moral and Intellectual Dispositions of Man and Animals, by the Configuration of the Brain and Head.* 6 vols. Translated by W. Lewis Jr. Boston: Marsh, Capen and Lyon. First published in French, 1822–25.

Galton, Francis. 1887. "Thought without Words." *Nature* 36:28–29.

Gardinier, D. E. 1994. *Historical Dictionary of Gabon.* 2nd ed. London: Scarecrow Press.

Gardner, Howard. 1987. *The Mind's New Science: A History of the Cognitive Revolution.* New York: Basic Books. First published in 1985.

———. 1995. "Green Ideas Sleeping Furiously." *New York Review of Books,* 23 March, 32–38.

Gardner, Martin. 1980. "Monkey Business." *New York Review of Books,* 20 March, 3–6.

Gardner, R. Allen, and Beatrice T. Gardner. 1969. "Teaching Sign Language to a Chimpanzee." *Science,* 15 August, 664–72.

Garland, J. R. 1987. "An Economic Survey of Southwest Virginia during the Ante-Bellum Period." *Historical Society of Washington County, Va., Bulletin,* ser. 2, no. 24:9–15.

Garner, Richard Lynch. 1889. *Nancy Bet: The Story of Sloomy Perkins and His Transaction in Real Estate.* Norfolk, VA: Norfolk Landmark Co.

———. 1891a. *The Psychoscope.* Warrenton, VA: True Index Co.

———. 1891b. "The Simian Tongue [I]." In Harris 1996, 314–21. First published in *New Review* 4:555–62 and *Littell's Living Age* 190:218–21. Abridged in *Review of Reviews* 3:574 and *Phonogram* 1:200–202.

———. 1891c. "The Simian Tongue II." In Harris 1996, 321–27. First published in *New Review* 5:424–30.

———. 1892a. "A Mission to the Monkeys." *New York Times,* 25 September, 17. Also published, with minor variations, as Garner 1892b.

———. 1892b. "A Monkey's Academy in Africa." *New Review* 7:282–92.

———. 1892c. "Phonographic Studies of Speech." *Forum* 13:778–87.

———. 1892d. "Simian Speech and Simian Thought." *Cosmopolitan* 13:72–79.

———. 1892e. "The Simian Tongue III." In Harris 1996, 327–32. First published in *New Review* 6:181–86.

———. 1892f. *The Speech of Monkeys*. London: William Heinemann.

———. 1892g. "The Speech of Monkeys." *Forum* 13:246–56.

———. 1892h. "What I Expect to Do in Africa." *North American Review* 154:713–18. Abridged in *Review of Reviews* 6 (1892): 49.

———. 1893. "Among the Gorillas: A Voice from the Wilderness." *McClure's Magazine* 1:364–72.

———. 1894. "Gorillas and Chimpanzees." *Harper's Weekly* 38:302–3, illustration on 297. Also published (with more illustrations) in *Pall Mall Magazine* 2:919–32.

———. 1896. *Gorillas and Chimpanzees*. London: Osgood, McIlvaine.

———. 1899. *Yazyk obez'yan*. Russian translation of Garner 1892f. Translated by S. L. Khalyutina. Ed. V. V. Bitnera. St. Petersburg: Press of P. P. Soikina.

———. 1900. *Apes and Monkeys: Their Life and Language*. Introduction by E. E. Hale. London: Ginn / Athenaeum Press.

———. 1902. "Native Institutions of the Ogowe Tribes of West Central Africa: An Interpretation of Their Meaning as Viewed from the Standpoint of the Native Philosopher." *Journal of the African Society* 1 (3): 369–80.

———. 1905a. *Die Sprache der Affen*. German translation of Garner 1892f. Translated by W. Marshall. 2nd ed. Dresden: Hans Schultze. First edition published in 1900.

———. 1905b. "Psychological Studies of the Chimpanzee." *North American Review* 181 (August): 272–80.

———. 1910. "My Recent Work, and Susie." *Independent*, 8 September, 518–23.

———. 1920. "Adventures in Central Africa." *Century* 99:595–604, 842–52; 100:125–35.

———. 1930. *Autobiography of a Boy: From the Letters of Richard Lynch Garner*. Introduction by Aleš Hrdlička. Washington, DC: Huff Duplicating Co.

"Garner, Richard Lynch." 1906. In *The National Cyclopaedia of American Biography*, 13:314. New York: James T. White.

"Garner, Richard Lynch." 1968. In *Who Was Who in America*, 4:346. Chicago: Marquis Who's Who.

Garraty, J. A., and M. C. Carnes, eds. 1999. *American National Biography*. New York: Oxford University Press.

Gaulme, F. 1974. "Un problème d'histoire du Gabon: Le sacre du P. Bichet par les Nkomi en 1897." *Revue Française d'Histoire d'Outre-Mer* 61:395–416.

Gelatt, Roland. 1977. *The Fabulous Phonograph, 1877–1977*. 2nd rev. ed. London: Cassell.

Geschwind, Norman. 1964. "The Development of the Brain and the Evolution of Language." In *Monograph Series on Languages and Linguistics*, no. 17, ed. C. I. J. M.

Stuart, 155–69. Washington, DC: Georgetown University Press. Reprinted in Geschwind 1974, 86–104.

———. 1965. "Disconnexion Syndromes in Animals and Man: Parts I and II." *Brain* 88:17–294, 585–644. Reprinted in Geschwind 1974, 105–236.

———. 1974. *Selected Papers on Language and the Brain.* Ed. Robert S. Cohen and Marx W. Wartofksy. Boston Studies in the Philosophy of Science, vol. 16. Boston: D. Reidel.

Gessinger, J. von, and W. von Rahden, eds. 1989. *Theorien vom Ursprung der Sprache.* 2 vols. New York: Walter de Gruyter.

Giles, Eugene. 1997. "Hooton, E(arnest) A(lbert) (1887–1954)." In Spencer 1997, 1:499–501.

Gillham, Nicholas Wright. 2001. *A Life of Sir Francis Galton: From African Exploration to the Birth of Eugenics.* Oxford: Oxford University Press.

Gillispie, C. C., ed. 1970–80. *Dictionary of Scientific Biography.* New York: Charles Scribner's Sons.

Gladstone, D. 1996. "The Changing Dynamics of Institutional Care: The Western Counties Idiot Asylum, 1864–1914." In *From Idiocy to Mental Deficiency: Historical Perspectives on People with Learning Disabilities,* ed. D. Wright and A. Digby, 134–60. London: Routledge.

Goldstein, D. 1994. "'Yours for Science': The Smithsonian Institution's Correspondents and the Shape of Scientific Community in Nineteenth Century America." *Isis* 85:573–99.

Goodall, Jane. 2001. *Beyond Innocence: An Autobiography in Letters: The Later Years.* Ed. Dale Peterson. New York: Houghton Mifflin.

Gould, James L. 1975. "Honey Bee Recruitment: The Dance-Language Controversy." *Science* 189:685–93.

Gould, James L., and Carol Grant Gould. 1994. *The Animal Mind.* New York: Scientific American Library.

Gould, Stephen Jay. 1977. *Ontogeny and Phylogeny.* Cambridge, MA: Harvard University Press.

———. 1985. "The Hardening of the Modern Synthesis." In *Dimensions of Darwinism,* ed. M. Grene, 71–93. Cambridge: Cambridge University Press

———. 1991. "Knight Takes Bishop?" In *Bully for Brontosaurus: Reflections in Natural History,* 385–401. New York: W. W. Norton.

———. 1996. "Speaking of Snails and Scales." In *Dinosaur in a Haystack: Reflections in Natural History,* 344–55. London: Jonathan Cape.

———. 1997. *The Mismeasure of Man.* Revised and expanded edition. London: Penguin.

———. 1998. "A Seahorse for All Races." In *Leonardo's Mountain of Clams and the Diet of Worms,* 119–40. London: Vintage.

———. 2002. *The Structure of Evolutionary Theory.* Cambridge, MA: Belknap Press / Harvard University Press.

Gould, Stephen Jay, and Niles Eldredge. 1977. "Punctuated Equilibria: The Tempo and Mode of Evolution Reconsidered." *Paleobiology* 3:115–51.

Green, Steven. 1975a. "Dialects in Japanese Monkeys: Vocal Learning and Cultural Transmission of Locale-Specific Vocal Behavior?" *Zeitschrift für Tierpsychologie* 38:304–14.

———. 1975b. "Variation of Vocal Pattern with Social Situation in the Japanese Monkey (*Macaca fuscata*): A Field Study." In *Primate Behavior: Developments in Field and Laboratory Research*, ed. Leonard A. Rosenblum, 4:1–102. New York: Academic Press.

———. 1977. "Comparative Aspects of Vocal Signals Including Speech: Group Report." In *Recognition of Complex Acoustic Signals*, ed. Theodore H. Bullock, 209–37. Berlin: Dahlem Konferenzen.

Green, Steven, and Peter Marler. 1979. "The Analysis of Animal Communication." In *Handbook of Behavioral Neurobiology*, vol. 3, *Social Behavior and Communication*, ed. Peter Marler and J. Vandenbergh, 73–158. New York: Plenum Press.

Greenblatt, Samuel H. 1965. "The Major Influences on the Early Life and Work of John Hughlings Jackson." *Bulletin of the History of Medicine* 39:346–76.

———. 1970. "Hughlings Jackson's First Encounter with the Work of Paul Broca: The Physiological and Philosophical Background." *Bulletin of the History of Medicine* 44:555–70.

———. 1977. "The Development of Hughlings Jackson's Approach to Diseases of the Nervous System, 1863–1866: Unilateral Seizures, Hemiplegia, and Aphasia." *Bulletin of the History of Medicine* 51:412–30.

Griffin, Donald R. 1976. *The Question of Animal Awareness: Evolutionary Continuity of Mental Experience*. New York: Rockefeller University Press.

———, ed. 1982. *Animal Mind — Human Mind*. Berlin: Springer-Verlag.

———. 1989. "Recollections of an Experimental Naturalist." In Dewsbury 1989b, 121–42.

Griffiths, Paul E. 2004. "Instinct in the '50s: The British Reception of Konrad Lorenz's Theory of Instinctive Behavior." *Biology and Philosophy* 19:609–31.

Gross, Charles G. 1998. "The Hippocampus Minor and Man's Place in Nature: A Case Study in the Social Construction of Neuroanatomy." In *Brain, Vision, Memory: Tales in the History of Neuroscience*, 136–78. Cambridge, MA: Bradford Books / MIT Press.

Gruber, Howard E. 1974. *Darwin on Man: A Psychological Study of Scientific Creativity*. London: Wildwood House.

Gruber, J. W. 1967. "Horatio Hale and the Development of American Anthropology." *Proceedings of the American Philosophical Society* 111:5–37.

Gunson, Niel. 1994. "British Missionaries and Their Contribution to Science in the Pacific Islands." In MacLeod and Rehbock 1994, 283–316.

Hacking, Ian. 1983. *Representing and Intervening: Introductory Topics in the Philosophy of Natural Science*. Cambridge: Cambridge University Press.

———. 1990. "Artificial Phenomena." *British Journal for the History of Science* 24:235–41.

———. 1992a. "How, Why, When, and Where Did Language Go Public?" *Common Knowledge* 1:74–91.

———. 1992b. "The Self-Vindication of the Laboratory Sciences." In *Science as Practice and Culture*, ed. Andrew Pickering, 29–64. Chicago: University of Chicago Press.

———. 1999. *The Social Construction of What?* Cambridge, MA: Harvard University Press.

Haddon, Alfred Cort. 1890. "Manners and Customs of the Torres Straits Islanders." *Nature*, 30 October, 637–42.

———. 1895. *Evolution in Art: As Illustrated by the Life-Histories of Designs*. London: Walter Scott.

———. 1898a. "The Cambridge Expedition to Torres Straits and Borneo." *Nature*, 20 January, 276.

———. 1898b. *The Study of Man*. Progressive Science Series. London: Bliss, Sands and Co.; New York: G. P. Putnam's Sons.

———. 1899. "The Cambridge Anthropological Expedition to Torres Straits and Sarawak." *Nature*, 31 August, 413–16.

———. 1900. "Studies in the Anthropogeography of British New Guinea." *Geographical Journal* 16:265–91, 414–41.

———, ed. 1901–35. *Reports of the Cambridge Anthropological Expedition to Torres Straits*. Cambridge: Cambridge University Press.

Haeckel, Ernst. 1868. *Natürliche Schöpfungsgeschichte*. Berlin: Georg Reimer.

———. 1876. *The History of Creation*. Translated by E. Ray Lankester from the German edition of 1868. 2 vols. London: Henry King.

Hagen, Joel B. 1999. "Retelling Experiments: H. B. D. Kettlewell's Studies of Industrial Melanism in Peppered Moths." *Biology and Philosophy* 14:39–54.

Haggerty, Melvin E. 1913. "Plumbing the Minds of Apes." *McClure's Magazine* 41:151–54.

Hahn, Emily. 1971. "Washoese." *New Yorker*, 11 December, 54–98.

Haldane, J. B. S. 1936. "Some Principles of Causal Analysis in Genetics." *Erkenntnis* 6:346–57.

———. 1938. *Heredity and Politics*. New York: W. W. Norton.

———. 1941. *New Paths in Genetics*. London: George Allen and Unwin.

———. 1946. "The Interaction of Nature and Nurture." *Annals of Eugenics* 13:197–205.

———. 1952. "The Origin of Language." In *The Rationalist Annual*, ed. Frederick Watts, 38–45. London: Watts and Co.

———. 1953. "Animal Ritual and Human Language." *Diogenes* 4:61–73.

———. 1954. "La signalisation animale." *L'Année Biologique* 30:89–98.

———. 1955a. "Animal Communication and the Origin of Human Language." *Science Progress* 43:385–401. Reprinted in *Current Science* 63 (1992): 604–11.

———. 1955b. "Communication in Biology." In Ayer et al. 1955, 29–43.

Haldane, J. B. S., and Helen Spurway. 1954. "A Statistical Analysis of Communication in 'Apis mellifera' and a Comparison with Communication in Other Animals." *Insectes Sociaux* 1:247–83.

Hale, Horatio. 1884. "On Some Doubtful or Intermediate Articulations." *Journal of the Royal Anthropological Institute* 14:233–43.

———. 1886. "The Origin of Languages, and the Antiquity of Speaking Man." *Proceedings of the American Association for the Advancement of Science* 35:279–323. An abbreviated version of the address appeared in *Science*, 27 August 1886, 191–96.

———. 1888. *The Development of Language.* Toronto: Copp, Clark.

———. 1891. "Language as a Test of Mental Capacity." *Transactions of the Royal Society of Canada* 9:77–112.

———. 1892. Review of Garner, *The Speech of Monkeys. American Anthropologist* 5:381–82.

Hall, G. Stanley. 1909. "Evolution and Psychology." In *Fifty Years of Darwinism: Modern Aspects of Evolution*, ed. T. C. Chamberlin, 251–67. New York: Holt.

Hamilton, Gilbert Van Tassel. 1907. "An Experimental Study of an Unusual Type of Reaction in a Dog." *Journal of Comparative Neurology and Psychiatry* 17 (4): 329–41.

———. 1911. "A Study of Trial and Error Reactions in Mammals." *Journal of Animal Behavior* 1:33–66.

Hammond, W. A. 1877. Review of Bateman, *Darwinism Tested by Language. Journal of Nervous and Mental Diseases* 5:362–70, 744–52.

———. 1878. "New York College of Veterinary Surgeons." *Medical Records* 13:197–98.

Haraway, Donna J. 1981–82. "The High Cost of Information in Post–World War II Evolutionary Biology: Ergonomics, Semiotics, and the Sociobiology of Communication Systems." *Philosophical Forum* 13:244–78.

———. 1983. "Signs of Dominance: From a Physiology to a Cybernetics of Primate Society, C. R. Carpenter, 1930–1970." *Studies in the History of Biology* 6:129–219.

———. 1989. *Primate Visions: Gender, Race, and Nature in the World of Modern Science.* London: Routledge, Chapman and Hall.

Hardy, Alister. 1973. Memoir of David Lack. *Ibis* 115:434–36.

Harnad, Stevan R., Horst D. Steklis, and Jane Lancaster, eds. 1976. *Origins and Evolution of Language and Speech.* New York: New York Academy of Sciences.

Harrington, Anne. 1987. *Medicine, Mind, and the Double Brain: A Study in Nineteenth-Century Thought.* Princeton: Princeton University Press.

———. 1996. *Reenchanted Science: Holism in German Culture from Wilhelm II to Hitler.* Princeton: Princeton University Press.

Harrington, John P. 1912. "Notes on Learning a New Language." In *Notes and Que-*

ries on Anthropology, 4th ed., ed. Barbara Freire-Marreo and John Linton Myres, 192–97. London: Royal Anthropological Institute.

Harris, Randy Allen. 1993. *The Linguistics Wars*. Oxford: Oxford University Press.

Harris, Roy, ed. 1996. *The Origin of Language*. Bristol: Thoemmes Press.

Hartmann, R. 1885. *Anthropoid Apes*. London: Kegan Paul, Trench.

Hauser, Marc D. 1996. *The Evolution of Communication*. Cambridge, MA: MIT Press.

———. 2000. *Wild Minds: What Animals Really Think*. London: Allen Lane / Penguin Press.

———. 2004. "Intellectual Promiscuity." In *When We Were Kids: How a Child Becomes a Scientist*, ed. John Brockman, 153–61. London: Jonathan Cape.

Hauser, Marc D., Noam Chomsky, and W. Tecumseh Fitch. 2002. "The Faculty of Language: What Is It, Who Has It, and How Did It Evolve?" *Science*, 22 November, 1569–79.

Hayes, Cathy. 1952. *The Ape in Our House*. London: Victor Gollancz. First published in 1951.

Hayes, Keith J. 1950. "Vocalization and Speech in Chimpanzees." *American Psychologist* 5:275–76.

Hayes, Keith J., and Hayes, Catherine. 1951. "The Intellectual Development of a Home-Raised Chimpanzee." *Proceedings of the American Philosophical Society* 95:105–9.

———. 1954. "The Cultural Capacity of Chimpanzee." *Human Biology* 26:288–303.

Head, Henry. 1926. *Aphasia and Kindred Disorders of Speech*. 2 vols. Cambridge: Cambridge University Press.

Hearnshaw, L. S. 1964. *A Short History of British Psychology, 1840–1940*. London: Methuen.

Hecht, Jennifer Michael. 2003. *The End of the Soul: Scientific Modernity, Atheism, and Anthropology in France*. New York: Columbia University Press.

Heims, Steve Joshua. 1991. *The Cybernetics Group*. Cambridge, MA: MIT Press.

Henson, Hilary. 1974. *British Social Anthropologists and Language: A History of Separate Development*. Oxford: Clarendon Press.

Henson, Louise, Geoffrey Cantor, Gowan Dawson, Richard Noakes, Sally Shuttleworth, and Jonathan R. Topham, eds. 2004. *Culture and Science in the Nineteenth-Century Media*. Aldershot: Ashgate.

Herder, Johann Gottfried. 1772. *Essay on the Origin of Language*. In *On the Origin of Language: Jean Jacques Rousseau, Essay on the Origin of Languages; Johann Gottfried Herder, Essay on the Origin of Language*. Ed. and translated by J. H. Moran and A. Gode, 85–166. Chicago: University of Chicago Press, 1986.

Herle, Anita. 1998. "The Life-Histories of Objects: Collections of the Cambridge Anthropological Expedition to the Torres Strait." In *Herle and Rouse 1998*, 77–105.

Herle, Anita, and Sandra Rouse, eds. 1998. *Cambridge and the Torres Strait: Cente-*

nary Essays on the 1898 Anthropological Expedition. Cambridge: Cambridge University Press.

Herrnstein, Richard J., and Donald H. Loveland. 1964. "Complex Visual Concept in the Pigeon." *Science* 146:549–51.

Herrnstein, Richard J., Donald H. Loveland, and Cynthia Cable. 1976. "Natural Concept in Pigeons." *Journal of Experimental Psychology: Animal Behavior Processes* 2:285–302.

Herzig, Rebecca. 2005. *Suffering for Science: Reason and Sacrifice in Modern America.* New Brunswick, NJ: Rutgers University Press.

Hewes, Gordon. 1975. *Language Origins: A Bibliography.* 2 vols. 2nd ed. The Hague: Mouton.

———. 1996. "A History of the Study of Language Origins and the Gestural Primacy Hypothesis." In *Handbook of Human Symbolic Evolution,* ed. A. Lock and C. R. Peters, 571–95. Oxford: Clarendon Press.

Higham, John. 1955. *Strangers in the Land: Patterns of American Nativism, 1860–1925.* New Brunswick, NJ: Rutgers University Press.

Hilgard, Ernest R. 1965. "Robert Mearns Yerkes, 1876–1956." In *Biographical Memoirs of the National Academy of Sciences of the United States of America,* 38:385–425. New York: Columbia University Press.

Hill, Jane H. 1980. "Apes and Language." In Sebeok and Umiker-Sebeok 1980, 331–51. First published in *Annual Review of Anthropology* 7 (1978): 89–112.

Hinde, Robert A. 1958. "Alternative Motor Patterns in Chaffinch Song." *Animal Behaviour* 6:211–18.

———. 1970. *Animal Behaviour: A Synthesis of Ethology and Comparative Psychology.* 2nd ed. New York: McGraw-Hill.

———. 1974. *Biological Bases of Human Social Behaviour.* New York: McGraw-Hill.

———. 1976. "Interactions, Relationships, and Social Structure." *Man,* n.s., 11:1–17.

———. 1987. "William Homan Thorpe, 1902–1986." *Biographical Memoirs of Fellows of the Royal Society* 33:620–39.

———. 1989. "Ethology in Relation to Other Disciplines." In Dewsbury 1989b, 193–203.

———. 2000. "Some Reflections on Primatology at Cambridge and the Science Studies Debate." In Strum and Fedigan 2000, 104–15.

Hinsley, Curtis M. Jr. 1981. *Savages and Scientists: The Smithsonian Institution and the Development of American Anthropology, 1846–1910.* Washington, DC: Smithsonian Institution Press.

Hirshfield, C. 1984. "Truth." In *British Literary Magazines: The Victorian and Edwardian Age, 1837–1913,* ed. A. Sullivan, 3:423–32. London: Greenwood Press.

Hobhouse, Leonard T. 1901. *Mind in Evolution.* London: Macmillan.

———. 1902. "The Diversions of a Psychologist." *Pilot,* 4 January, 12–13; 11 January, 36–37; 1 February, 126–27; 1 March, 232–33; 29 March, 344–45; 26 April, 449–51.

———. 1913. *Development and Purpose: An Essay towards a Philosophy of Evolution.* London: Macmillan.

Hockett, Charles Francis. 1948. "Biophysics, Linguistics, and the Unity of Science." In Hockett 1977, 1–18. First published in *American Scientist* 36:558–72.

———. 1953. Review of Shannon and Weaver, *The Mathematical Theory of Communication.* In Hockett 1977, 19–52. First published in *Language* 29:69–93.

———. 1955a. "How to Learn Martian." In Hockett 1977, 97–106. First published in *Astounding Science Fiction* 55:97–106.

———. 1955b. *A Manual of Phonology.* Indiana University Publications in Anthropology and Linguistics, memoir 11. Baltimore: Waverly Press.

———. 1956. Review of W. La Barre, *The Human Animal* (1954); C. S. Coon, *The Story of Man* (1954); V. G. Childe, *Man Makes Himself* (1951); G. G. Simpson, *The Meaning of Evolution* (1950). *Language* 32:460–69.

———. 1958. *A Course in Modern Linguistics.* New York: Macmillan.

———. 1959. "Animal 'Languages' and Human Language." In *The Evolution of Man's Capacity for Culture,* arranged by J. N. Spuhler, 32–39. Detroit: Wayne State University Press, 1965. First published in *Human Biology* 31 (February): 32–39.

———. 1960a. "Logical Considerations in the Study of Animal Communication." In Lanyon and Tavolga 1960, 392–430. Reprinted in Hockett 1977, 124–62.

———. 1960b. "The Origin of Speech." *Scientific American,* September, 89–96, with author's note on 48 and bibliography on 276.

———, ed. 1970. *A Leonard Bloomfield Anthology.* Bloomington: Indiana University Press.

———. 1973. *Man's Place in Nature.* New York: McGraw-Hill.

———. 1977. *The View from Language: Selected Essays, 1948–1974.* Athens: University of Georgia Press.

———. 1980. "Preserving the Heritage." In *First Person Singular,* ed. Boyd H. Davis and Raymond K. O'Cain, 99–107. Amsterdam: John Benjamins.

Hockett, Charles Francis, and Stuart A. Altmann. 1968. "A Note on Design Features." In Sebeok 1968, 61–72.

Hockett, Charles Francis, and Robert Ascher. 1964. "The Human Revolution." With commentaries from G. G. Simpson, Theodosius Dobzhansky, Margaret Mead, and many others. *Current Anthropology* 5:135–68.

Hodge, Jonathan, and Gregory Radick, eds. 2003. *The Cambridge Companion to Darwin.* Cambridge: Cambridge University Press.

Hodge, M. J. [Jonathan] S. 1977. "The Structure and Strategy of Darwin's 'Long Argument.'" *British Journal for the History of Science* 10:237–46.

———. 2003. "The Notebook Programmes and Projects of Darwin's London Years." In Hodge and Radick 2003, 40–68.

Hofstadter, Richard. 1992. *Social Darwinism in American Thought*. Rev. ed. Boston: Beacon Press. First edition published in 1944.

Hook, Ernest B., ed. 2002. *Prematurity in Scientific Discovery: On Resistance and Neglect*. Berkeley: University of California Press.

Hooper, Judith. 2002. *Of Moths and Men: Intrigue, Tragedy, and the Peppered Moth*. London: Fourth Estate.

Hooton, Earnest A. 1947. *Up from the Ape*. Rev. ed. New York: Macmillan.

Hopkins, E. W. 1900. "Max Müller." In *Portraits of Linguists: A Biographical Source Book for the History of Western Linguistics, 1746–1963*, ed. Thomas A. Sebeok, 1:395–99. Bloomington: Indiana University Press, 1967.

Houck, Lynne D., and Lee C. Drickamer, eds. 1996. *Foundations of Animal Behavior: Classic Papers with Commentary*. Chicago: University of Chicago Press.

Houghton, W., and E. R. Houghton, eds. 1979. *The Wellesley Index to Victorian Periodicals, 1824–1900*. Vol. 3. London: Routledge / Kegan Paul.

Hovelacque, A. 1877. *The Science of Language: Linguistics, Philology, Etymology*. Translated by A. H. Keane. London: Chapman and Hall.

Hrdlička, Aleš. 1908. "Physical Anthropology and Its Aims." *Science*, 10 July, 33–43.

———. 1914. "The Most Ancient Skeletal Remains of Man." In *Annual Report of the Smithsonian Institution, 1913*, 491–522. Washington, DC: Government Printing Office.

———. 1918. "Physical Anthropology: Its Scope and Aims; Its History and Present Status in America." *American Journal of Physical Anthropology* 1:3–23.

———. 1920. *Anthropometry*. Philadelphia: Wistar Institute.

———. 1940. "Locomotion in Gibbon." *American Journal of Physical Anthropology* 27:481.

———. 1947. *Hrdlička's Practical Anthropometry*. 3rd ed. Ed. T. D. Stewart. Philadelphia: Wistar Institute of Anatomy and Biology. First edition published in 1920.

Hubrecht, A. A. W. 1892. "Spreken De Apen?" *De Gids*, ser. 4, 10:508–22. Abridged in translation in *Review of Reviews* 7 (1893): 61.

Hull, David L. 2003. "Darwin's Science and Victorian Philosophy of Science." In Hodge and Radick 2003, 168–91.

Humphrey, Nicholas K. 1976. "The Social Function of Intellect." In *Growing Points in Ethology*, ed. P. P. G. Bateson and Robert Hinde, 303–17. Cambridge: Cambridge University Press.

———. 1977. Review of Griffin, *The Question of Animal Awareness*. *Animal Behaviour* 25:521–22.

Hunt, James. 1868. "On the Localisation of the Functions of the Brain with Special Reference to the Faculty of Language [Part 1]." *Anthropological Review* 6:329–45.

Hurford, James R., Michael Studdert-Kennedy, and Chris Knight, eds. 1998. *Approaches to the Evolution of Language*. Cambridge: Cambridge University Press.

Huxley, Julian S. 1914. "The Courtship-Habits of the Great Crested Grebe (*Podiceps cristatus*), with an Addition to the Theory of Sexual Selection." *Proceedings of the Zoological Society of London* 35:491–562.

———. 1942. *Evolution: The Modern Synthesis*. London: George Allen and Unwin.

Huxley, Julian S., and Ludwig Koch. 1938. *Animal Language*. London: Country Life.

Huxley, Thomas H. 1863a. *Evidence as to Man's Place in Nature*. London: Williams and Norgate.

———. 1863b. *On Our Knowledge of the Causes of the Phenomena of Organic Nature*. London: Robert Hardwicke. Reprinted as *On the Origin of Species, or the Causes of the Phenomena of Organic Nature*. Ann Arbor: University of Michigan, 1968.

———. 1869. "On the Physical Basis of Life." *Fortnightly Review* 11:129–45.

———. 1874. "On the Hypothesis that Animals Are Automata, and Its History." *Fortnightly Review* 16:555–80.

———. 1881. *Hume*. London: Macmillan.

Israel, Paul. 1998. *Edison: A Life of Invention*. New York: John Wiley and Sons.

Itani, Jun'ichiro. 1963. "Vocal Communication of the Wild Japanese Monkey." *Primates* 4:11–66.

Jackson, John Hughlings. 1868. *Abstract of a Paper in the Physiology of Language*. Norwich: Jarrold and Sons.

———. 1958. *Selected Writings of John Hughlings Jackson*. Ed. J. Taylor. 2 vols. London: Staples Press.

Jacyna, L. Stephen. 2000. *Lost Words: Narratives of Language and the Brain, 1825–1926*. Princeton: Princeton University Press.

James, William. 1878. "Brute and Human Intellect." *Journal of Speculative Philosophy* 12:236–76.

———. 1890a. *The Principles of Psychology*. 2 vols. New York: Henry Holt. Reprinted in facsimile by Dover, 1950.

———. 1890b. *The Principles of Psychology*. 3 vols. Including volume of James's annotations etc. in *The Works of William James*, vols. 8–10. Cambridge, MA: Harvard University Press, 1981.

———. 1892. *Psychology: Briefer Course*. In *The Works of William James*, vol. 14. Cambridge, MA: Harvard University Press, 1984.

Jankowsky, Kurt R. 1979. "F. Max Müller and the Development of Linguistic Science." *Historiographia Linguistica* 6:339–59.

Jardine, Nicholas. 1991. *The Scenes of Inquiry: On the Reality of Questions in the Sciences*. Oxford: Clarendon Press.

Jastrow, Joseph. 1892. "Conversations with the Simians." *Dial* 13:215–16.

Jay, Phyllis C., ed. 1968. *Primates: Studies in Adaptation and Variability*. New York: Holt, Rinehart and Winston.

Johansen, Donald, and Blake Edgar. 1996. *From Lucy to Language*. New York: Simon and Schuster.

Johnson, Francis E., and Alan Mann. 1997. "United States of America." In Spencer 1997, 2:1069–81.

Johnson-Laird, P. N., and P. C. Wason, eds. 1977. *Thinking: Readings in Cognitive Science*. Cambridge: Cambridge University Press.

Jolly, Alison. 2000. "The Bad Old Days of Primatology?" In Strum and Fedigan 2000, 71–84.

Jonçich, Geraldine. 1968. *The Sane Positivist: A Biography of Edward L. Thorndike*. Middletown, CT: Wesleyan University Press.

Jones, H. F., ed. 1912. *The Note-books of Samuel Butler*. London: A. C. Fifield.

Joos, Martin. 1948. *Acoustic Phonetics*. Supplement to *Language*, monograph 23.

Joseph, John E. 2002. "How Behaviourist Was *Verbal Behavior*?" In *From Whitney to Chomsky: Essays in the History of American Linguistics*, 169–80. Amsterdam Studies in the Theory and History of Linguistic Science, vol. 103. Amsterdam: John Benjamins.

Kalmus, Hans. 1950. "A Cybernetical Aspect of Genetics." *Journal of Heredity* 41:19–22.

Kavanagh, James F., and James E. Cutting, eds. 1975. *The Role of Speech in Language*. Cambridge, MA: MIT Press.

Kay, Lily E. 2000. *Who Wrote the Book of Life? A History of the Genetic Code*. Stanford: Stanford University Press.

Kazdin, Alan E., ed. 2000. *Encyclopedia of Psychology*. Oxford: Oxford University Press.

Kellogg, Winthrop N. 1968. "Communication and Language in the Home-Raised Chimpanzee." *Science*, 25 October, 423–27.

Kellogg, Winthrop N., and Luella A. Kellogg. 1933. *The Ape and the Child: A Study of Environmental Influence upon Early Behavior*. New York: Whittlesey House.

Kelly, Mervin J. 1950. "The Bell Telephone Laboratories: An Example of an Institute of Creative Technology." *Proceedings of the Royal Society of London* A 203:287–301.

Kevles, Daniel J. 1985. *In the Name of Eugenics: Genetics and the Uses of Human Heredity*. London: Penguin.

Kingsland, Sharon E. 1995. *Modeling Nature: Episodes in the History of Population Ecology*. Chicago: University of Chicago Press.

Kingsley, Charles. 1863. *The Water Babies*. Illustrated edition of 1916. Reprint. London: Penguin, 1995.

Kingsley, Mary. 1897. *Travels in West Africa*. Abridged edition of 1976. Reprint. London: J. M. Dent, 1993.

Knight, Chris, Michael Studdert-Kennedy, and James R. Hurford, eds. 2000. *The Evolutionary Emergence of Language: Social Function and the Origins of Linguistic Form*. Cambridge: Cambridge University Press.

Knoll, Elizabeth G. 1986. "The Science of Language and the Evolution of Mind: Max Müller's Quarrel with Darwinism." *Journal of the History of the Behavioral Sciences* 22:3–22.

———. 1987. "Mental Evolution and the Science of Language: Darwin, Müller, and

Romanes on the Development of the Human Mind." PhD dissertation, University of Chicago.

Koch, Ludwig. 1955. *Memoirs of a Birdman*. London: Phoenix House.

Kohler, R. E. 2002. *Landscapes and Labscapes: Exploring the Lab-Field Border in Biology*. Chicago: University of Chicago Press.

Köhler, Wolfgang. 1925. *The Mentality of Apes*. Translated by Ella Winter from 1917 German edition. Harmondsworth: Penguin, 1957.

Kohn, Marek. 1995. *The Race Gallery: The Return of Racial Science*. London: Jonathan Cape.

Kohts, Nadia. 1935. "Infant Ape and Human Child (Instincts, Emotions, Play, and Habits)." *Scientific Memoirs of the Darwin Museum, Moscow*, 3, 1–596. Reprinted as *Infant Chimpanzee and Human Child: A Classic 1935 Study of Ape Emotions and Intelligence*. Oxford: Oxford University Press, 2002.

Koren, H. J. 1983. *To the Ends of the Earth: A General History of the Congregation of the Holy Ghost*. Pittsburgh: Duquesne University Press.

Krebs, John. 1977. "Mental Imagery." *Nature* 266:792.

Kroeber, Alfred L. 1917. "The Superorganic." *American Anthropologist*, n.s., 19:163–213.

———. 1928. "Sub-human Cultural Beginnings." *Quarterly Review of Biology* 3:325–42.

———. 1948. *Anthropology*. London: George G. Harrap.

———. 1952. "Sign and Symbol in Bee Communications." *Proceedings of the National Academy of Sciences* 38:753–57.

———. 1955. "On Human Nature." *Southwestern Journal of Anthropology* 11:195–204.

Kruuk, Hans. 2003. *Niko's Nature: The Life of Niko Tinbergen and His Science of Animal Behavior*. Oxford: Oxford University Press.

Kuklick, Bruce. 1977. *The Rise of American Philosophy: Cambridge, Massachusetts, 1860–1930*. New Haven: Yale University Press.

Kuklick, Henrika. 1991. *The Savage Within: The Social History of British Anthropology, 1885–1945*. Cambridge: Cambridge University Press.

———. 1994. "The Color Blue: From Research in the Torres Strait to an Ecology of Human Behavior." In MacLeod and Rehbock 1994, 339–67.

———. 1998. "Fieldworkers and Physiologists." In Herle and Rouse 1998, 158–80.

Kuklick, Henrika, and Robert E. Kohler, eds. 1996. *Science in the Field. Osiris* 11.

[Labouchere, Henry.] 1894a. "A Few Questions for Professor Garner." *Truth* 35:480–81.

[———]. 1894b. "Garner's Unpaid Bill." *Truth* 36:1513–14.

[———]. 1894c. "I Am Much Disappointed . . ." *Truth* 35:583.

[———]. 1894d. "Professor Garner at Home." *Truth* 35:887–88.

[———]. 1894e. "'Professor' Garner on the War Path." *Truth* 36:83.

[———]. 1895. "Garner Still Booming." *Truth* 38:370–71.

[———]. 1896. "The Munchausen of Monkey-land." *Truth* 40:1232–33.

Lack, David. 1943. *The Life of the Robin*. London: H. F. and G. Witherby.

———. 1946. *The Life of the Robin*. 2nd ed. London: H. F. and G. Witherby.

———. 1947. *Darwin's Finches*. Cambridge: Cambridge University Press.

———. 1953. "Darwin's Finches." *Scientific American*, April, 67–72.

———. 1973. "My Life as an Amateur Ornithologist." *Ibis* 115:421–31.

Lack, David, and H. N. Southern. 1949. "Birds on Tenerife." *Ibis* 91:607–26.

Laird, Carobeth. 1975. *Encounter with an Angry God: Recollections of My Life with John Peabody Harrington*. Foreword by H. Lawton. Banning, CA: Malki Museum Press, Morongo Indian Reservation.

Lakoff, George, and Mark Johnson. 1980. *Metaphors We Live By*. Chicago: University of Chicago Press.

Lancaster, Jane B. 1968. "Primate Communication Systems and the Emergence of Human Language." In Jay 1968, 439–57.

Lancaster, Jane B., and Sherwood L. Washburn. 1971. "On the Evolution and Origin of Language." *Current Anthropology* 12:384–86.

Langer, Suzanne K. 1957. *Philosophy in a New Key: A Study in the Symbolism of Reason, Rite, and Art*. 3rd ed. Cambridge, MA: Harvard University Press. First edition published in 1942.

Lanyon, W. E., and W. N. Tavolga, eds. 1960. *Animal Sounds and Communication*. Washington, DC: American Institute of Biological Sciences.

Larson, Edward J. 2001. *Evolution's Workshop: God and Science on the Galápagos Islands*. New York: Basic Books.

Lasky, Marvin. 1977. "Review of Undersea Acoustics to 1950." *Journal of the Acoustical Society of America* 61:283–97.

Leach, M. 1996. *The Great Apes: Our Face in Nature's Mirror*. London: Blandford.

Lenneberg, Eric H. 1967. *Biological Foundations of Language*. With appendices by Noam Chomsky and Otto Marx. New York: John Wiley and Sons.

Leopold, Joan. 1990. "Darwin on Expression and the Origin of Language." In *History and Historiography of Linguistics*, ed. H.-J. Niederehe and K. Koerner, 2:633–45. Amsterdam: John Benjamins.

Le Roy, A. 1895. "La langue des singes." *Cosmos* 31:144–47, 173–76.

Lestel, Dominique. 1995. *Paroles des singes: L'impossible dialogue homme/primate*. Paris: La Découverte, 1995.

Lewis, Herbert S. 2001. "Boas, Darwin, Science, and Anthropology." *Current Anthropology* 42 (3): 381–94, with comments from others and a reply on 394–406.

Lewontin, Richard C. 1972. "The Apportionment of Human Diversity." In *Evolutionary Biology*, ed. Theodosius Dobzhansky et al., 6:381–98. New York: Appleton-Century-Crofts.

Lieberman, Philip. 1968. "Primate Vocalizations and Human Linguistic Ability." *Journal of the Acoustical Society of America* 44:1574–84. Reprinted in Lieberman 1972, 11–40.

———. 1972. *The Speech of Primates*. The Hague: Mouton.

———. 1998. *Eve Spoke: Human Language and Human Evolution.* London: Picador.

Lightman, Bernard, ed. 2004. *Dictionary of Nineteenth-Century British Scientists.* 4 vols. Bristol: Thoemmes Continuum.

Linden, Eugene. 1974. *Apes, Men, and Language: How Teaching Chimpanzees to "Talk" Alters Man's Notions of His Place in Nature.* New York: Saturday Review Press / E. P. Dutton.

Locke, John. 1689. *An Essay Concerning Human Understanding.* Ed. P. H. Nidditch. Oxford: Clarendon Press, 1975.

Lorenz, Konrad. 1935. "Companions as Factors in the Bird's Environment: The Conspecific as the Eliciting Factor of Social Behaviour Patterns." In Lorenz 1970–71, 1:101–258. First published as "Der Kumpan in der Umwelt des Vogels," *Journal für Ornithologie* 83:137–213, 289–413.

———. 1941. "Comparative Studies of the Motor Patterns of Anatinae." In Lorenz 1970–71, 2:14–114. Digested in Houck and Drickamer 1996, 683–96. First published as "Vergleichende Bewegungsstudien an Anatinen," *Journal für Ornithologie* 89:194–293.

———. 1950. "The Comparative Method in Studying Innate Behaviour Patterns." In J. F. Danielli and R. Brown, eds., *Physiological Mechanisms in Animal Behaviour,* Symposia of the Society for Experimental Biology, no. 4, 221–68. Cambridge: Cambridge University Press.

———. 1952a. *King Solomon's Ring: New Light on Animal Ways.* Translated by Marjorie Kerr Wilson from the German. Introduction by W. H. Thorpe. London: Methuen and Co., 1982.

———. 1952b. "The Past Twelve Years in the Comparative Study of Behavior." In C. H. Schiller 1957, 288–310.

———. 1956. "The Objectivistic Theory of Instinct." In Autuori et al. 1956, 51–76.

———. 1970–71. *Studies in Animal and Human Behaviour.* 2 vols. Translated by R. Martin. London: Methuen.

———. 1989. "My Family and Other Animals." In Dewsbury 1989b, 259–87.

Lorenz, Konrad, [and Niko Tinbergen]. 1938. "Taxis and Instinctive Behaviour in Egg-Rolling by the Greylag Goose." In Lorenz 1970–71, 1:316–50.

Lyell, Charles. 1863. *The Geological Evidences of the Antiquity of Man, with Remarks on the Theories of the Origin of Species by Variation.* London: John Murray.

Lyons, John, ed. 1970a. *New Horizons in Linguistics.* Harmondsworth: Penguin.

———. 1970b. *Noam Chomsky.* Fontana Modern Masters. Glasgow: Fontana / Collins.

———. 1977. *Chomsky.* Rev. ed. Glasgow: Fontana / Collins. First edition published in 1970.

Macdonald, Helen. 2002. "'What Makes You a Scientist Is the Way You Look at Things: Ornithology and the Observer, 1930–1955." *Studies in History and Philosophy of Biological and Biomedical Sciences* 33C:53–77.

MacLeod, Roy. 1994. "Embryology and Empire: The Balfour Students and the Quest

for Intermediate Forms in the Laboratory of the Pacific." In MacLeod and Rehbock 1994, 140–65.

MacLeod, Roy, and Philip F. Rehbock, eds. 1994. *Darwin's Laboratory: Evolutionary Theory and Natural History in the Pacific*. Honolulu: University of Hawai'i
Press.

MacNamara, N. C. 1908. *Human Speech: A Study in the Purposive Action of Living
Matter*. London: Kegan Paul, Trench, Trübner.

Manning, Aubrey. 1998. "The Potential of Parrots: Do Animals Try to Deceive Each
Other?" *Times Literary Supplement*, 28 August, 4–5.

Marks, Jonathan. 2000. "Ashley Montagu, 1905–1999." *Evolutionary Anthropology*
9:111–12.

Marler, Peter. 1952. "Variation in the Song of the Chaffinch *Fringilla coelebs*." *Ibis*
94:458–72.

———. 1954. "Vocal Communication in the Chaffinch, *Fringilla coelebs*." *Proceedings of the Association for the Study of Animal Behaviour*. Abstracts of meeting
held on 6 July.

———. 1955a. "Characteristics of Some Animal Calls." *Nature*, 2 July, 6–8.

———. 1955b. "Evolution of Finch Colours." *Middle-Thames Naturalist* 8:4–6.

———. 1955c. "Studies of Fighting in Chaffinches: 1, Behaviour in Relation to the
Social Hierarchy." *British Journal of Animal Behaviour* 3:111–17.

———. 1955d. "Studies of Fighting in Chaffinches: 2, The Effect on Dominance Relations of Disguising Females as Males." *British Journal of Animal Behaviour*
3:137–46.

———. 1956a. "Behaviour of the Chaffinch *Fringilla coelebs*." *Behaviour* suppl. 5:1–
184.

———. 1956b. "Studies of Fighting in Chaffinches: 3, Proximity as a Cause of Aggression." *British Journal of Animal Behaviour* 4:23–30.

———. 1956c. "Territory and Individual Distance in the Chaffinch *Fringilla coelebs*."
Ibis 98:496–501.

———. 1956d. "Über die Eigenschaften einiger tierlicher Rufe." *Journal für Ornithologie* 97:220–27.

———. 1956e. "The Voice of the Chaffinch." *New Biology* 20:70–87.

———. 1956f. "The Voice of the Chaffinch and Its Function as a Language." *Ibis*
98:231–61.

———. 1957a. "Specific Distinctiveness in the Communication Signals of Birds." *Behaviour* 11:13–39.

———. 1957b. "Studies of Fighting in Chaffinches: 4, Appetitive and Consummatory
Behaviour." *British Journal of Animal Behaviour* 5:29–37.

———. 1959. "Developments in the Study of Animal Communication." In *Darwin's
Biological Work: Some Aspects Reconsidered*, ed. P. R. Bell, 150–206. Cambridge: Cambridge University Press.

———. 1960. "Bird Songs and Mate Selection." In Lanyon and Tavolga 1960, 348–67.

———. 1961a. "The Filtering of External Stimuli during Instinctive Behaviour." In

Current Problems in Animal Behaviour, ed. W. H. Thorpe and O. L. Zangwill, 150–66. Cambridge: Cambridge University Press.

———. 1961b. "The Logical Analysis of Animal Communication." *Journal of Theoretical Biology* 1:295–317. Reprinted in facsimile in Houck and Drickamer 1996, 649–71.

———. 1963a. "Inheritance and Learning in the Development of Animal Vocalizations." In Busnel 1963, 228–43, 794–97.

———. 1963b. Review of G. Schaller, *The Mountain Gorilla: Ecology and Behavior. Science*, 7 June, 1081–82.

———. 1965. "Communication in Monkeys and Apes." In DeVore 1965, 544–84.

———. 1967a. "Acoustical Influences in Birdsong Development." *Rockefeller University Review*, September–October, 8–13.

———. 1967b. "Animal Communication Signals." *Science*, 18 August, 769–74.

———. 1968. "Aggregation and Dispersal: Two Functions in Primate Communication." In Jay 1968, 420–38.

———. 1969a. "Animals and Man: Communication and Its Development." In *Communication*, ed. John D. Roslansky, 25–62. Amsterdam: North-Holland.

———. 1969b. "*Colobus guereza*: Territoriality and Group Composition." *Science*, 3 January, 93–95.

———. 1969c. "Vocalizations of Wild Chimpanzees: An Introduction." *Recent Advances in Primatology* 1:94–100.

———. 1970a. "Birdsong and Speech Development: Could There Be Parallels?" *American Scientist* 58 (November–December): 669–73.

———. 1970b. "A Comparative Approach to Vocal Learning: Song Development in White-crowned Sparrows." In *Biological Boundaries of Learning*, ed. M. E. P. Seligman and J. L. Hager, 336–76. New York: Appleton-Century-Crofts, 1972. First published in *Journal of Comparative and Physiological Psychology* 71:1–25.

———. 1970c. "Vocalizations of East African Monkeys: 1, Red Colobus." *Folia Primatologica* 13:81–91.

———. 1972. "Vocalizations of East African Monkeys: 2, Black and White Colobus." *Behaviour* 42:175–97.

———. 1973a. "A Comparison of Vocalizations of Red-tailed Monkeys and Blue Monkeys, *Cercopithecus ascanius* and *C. mitis*, in Uganda." *Zeitschrift für Tierpsychologie* 33:223–47.

———. 1973b. "Les communications animale." *La Recherche* 36 (July–August): 644–60.

———. 1974. "Animal Communication." In *Nonverbal Communication: Advances in the Study of Communication and Affect*, ed. L. Krames, P. Pliner, and T. Alloway, 25–50. New York: Plenum Press.

———. 1977a. "The Evolution of Communication." In *How Animals Communicate*, ed. Thomas A. Sebeok, 45–70. Bloomington: Indiana University Press.

———. 1977b. "Primate Vocalization: Affective or Symbolic?" In Sebeok and Umiker-Sebeok 1980, 221–29. First published in *Progress in Ape Research*, ed. Geoffrey H. Bourne, 85–96. New York: Academic Press.

———. 1977c. "The Structure of Animal Communication Sounds." In *Recognition of Complex Acoustic Signals*, ed. Theodore H. Bullock, 17–35. Berlin: Dahlem Konferenzen.

———. 1978. "Affective and Symbolic Meaning: Some Zoosemiotic Speculations." In *Sight, Sound, and Sense*, ed. Thomas A. Sebeok, 113–23. Bloomington: Indiana University Press.

———. 1982. "Avian and Primate Communication: The Problem of Natural Categories." *Neuroscience and Biobehavioral Reviews* 6:87–94.

———. 1985a. "The Predator Alarm Calls of Free-Ranging Vervet Monkeys." *National Geographic Society Research Reports* 18:505–16.

———. 1985b. "Representational Vocal Signals of Primates." *Fortschritte der Zoologie* 31:211–21.

———. 1986. Commentary on Marler 1957a as a "citation classic." *Current Contents*, no. 11, 17 March, 16.

———. 1989. "Hark Ye to the Birds: Autobiographical Marginalia." In Dewsbury 1989b, 315–45.

———. 1992. "Functions of Arousal and Emotion in Primate Communication: A Semiotic Approach." In *Topics of Primatology*, vol. 1, *Human Origins*, ed. T. Nishida, W. C. McGrew, Peter Marler, Martin Pickford, and Frans de Waal, 225–33. Tokyo: University of Tokyo Press.

———. 1995. "Rating the Thorndike Inheritance: By Worth or by Weight?" *Contemporary Psychology* 40 (1): 11–14.

———. 1999. "How Much Does a Human Environment Humanize a Chimp?" *American Anthropologist* 101:432–36.

———. 2003. "Innateness and the Instinct to Learn." Opening address, International Bioacoustics Conference, Brazil, August. Manuscript.

———. 2004. "Science and Birdsong: The Good Old Days." In *Nature's Music: The Science of Birdsong*, ed. Peter Marler and Hans Slabbekoorn, 1–38. London: Elsevier.

Marler, Peter, and Derrick J. Boatman. 1951. "Observations on the Birds of Pico, Azores." *Ibis* 93:90–99.

———. 1952. "An Analysis of the Vegetation of the Northern Slopes of Pico: The Azores." *Journal of Ecology* 40:143–55.

Marler, Peter, and William J. Hamilton III. 1966. *Mechanisms of Animal Behavior*. New York: John Wiley and Sons.

Marler, Peter, and Linda Hobbett. 1975. "Individuality in a Long-Range Vocalization of Wild Chimpanzees." *Zeitschrift für Tierpsychologie* 38:97–109.

Marler, Peter, Marcia Kreith, and Miwako Tamura. 1962. "Song Development in Hand-Raised Oregon Juncos." *Auk* 79:12–30.

Marler, Peter, and Miwako Tamura. 1962. "Song 'Dialects' in Three Populations of White-crowned Sparrows." *Condor* 64:368–77.

———. 1964. "Culturally Transmitted Patterns of Vocal Behavior in Sparrows." *Sci-*

ence 146:1483–86. Reprinted in facsimile in Houck and Drickamer 1996, 293–96.

Marler, Peter, and Richard Tenaza. 1977. "Signaling Behavior of Apes with Special Reference to Vocalization." In *How Animals Communicate*, ed. Thomas A. Sebeok, 965–1033. Bloomington: Indiana University Press.

Marshall, John C. 1970. "The Biology of Communication in Man and Animals." In Lyons 1970a, 229–41.

Martland, Peter. 1998. "Thomas Edison's British Phonograph Business, 1903–1914." Faculty of History, University of Cambridge. Manuscript.

Mason, William A. 1976. "Windows on Other Minds." *Science* 194:930–31.

Massin, Benoit. 1996. "From Virchow to Fischer: Physical Anthropology and 'Modern Race Theories' in Wilhelmine Germany, 1890–1914." In Stocking 1996, 79–154.

Mathews, D. G. 1979. "Religion." In *The Encyclopedia of Southern History*, ed. D. C. Roller and R. W. Twyman, 1046–47. Baton Rouge: Louisiana State University Press.

Maynard Smith, John. 1965. "The Evolution of Alarm Calls." *American Naturalist* 99:59–63.

———. 1989. "In Haldane's Footsteps." In Dewsbury 1989b, 347–54.

Mayr, Ernst. 1942. *Systematics and the Origin of Species*. New York: Columbia University Press.

———. 1973. Memoir of David Lack. *Ibis* 115:432–34.

———. 1982. *The Growth of Biological Thought*. Cambridge, MA: Belknap Press / Harvard University Press.

"M. Broca at Norwich." 1868. *Lancet*, 15 August, 226.

McClure, Samuel Sidney. 1914. *My Autobiography*. London: John Murray.

McCook, Stuart. 1996. "'It May Be Truth, but It Is Not Evidence': Paul Du Chaillu and the Legitimation of Evidence in the Field Sciences." *Osiris*, n.s., 11:177–97.

McDougall, William. 1928. *An Outline of Psychology*. 4th rev. ed. London: Methuen. First edition published in 1923.

McKendrick, John G. 1897. *Waves of Sound and Speech — as Revealed by the Phonograph*. London: Macmillan.

McLean, Steven. 2002. "Animals, Language, and Degeneration in H. G. Wells's *The Island of Doctor Moreau*." *Undying Fire* 1:43–50.

McReynolds, Paul. 1997. *Lightner Witmer: His Life and Times*. Washington, DC: American Psychological Association.

———. 2000. "Witmer, Lightner." In Kazdin 2000, 8:253–55.

Menand, Louis. 2001. "The Socrates of Cambridge." *New York Review of Books*, 26 April, 52–55.

———. 2002. *The Metaphysical Club*. London: Flamingo.

Menzel, Emil W. Jr. 1971. "Communication about the Environment in a Group of Young Chimpanzees." *Folia Primatologica* 15:220–32.

———. 1974. "A Group of Young Chimpanzees in a One-Acre Field." In *Behavior of Nonhuman Primates: Modern Research Trends*, ed. Allan M. Schrier and Fred Stollnitz, 84–153. New York: Academic Press.

———. 1975. "Purposive Behavior as a Basis for Objective Communication between Chimpanzees." *Science* 189:652–54.

Mergoupi-Savaidou, Eirini. 2004. "The Function of Correspondence in the Victorian Scientific Periodicals: A Comparison between *Nature* and the *English Mechanic*." Master's dissertation, University of Leeds.

Middleton, Harvey. 1912. "Teaching Animals and Birds to Talk." *New England Magazine*, February, 565–73.

Millard, A. 1990. *Edison and the Business of Innovation*. Baltimore: Johns Hopkins University Press.

Miller, Jonathan. 1983. *States of Mind*. New York: Pantheon Books.

Mills, T. Wesley. 1899. "The Nature of Intelligence and the Methods of Investigating It." *Psychological Review* 6:262–74.

Mindell, David A. 2002. *Between Human and Machine: Feedback, Control, and Computing before Cybernetics*. Baltimore: Johns Hopkins University Press.

Mitchell, R. W. 1999. "Scientific and Popular Conceptions of the Psychology of Great Apes from the 1790s to the 1970s: Déjà Vu All Over Again." *Primate Report* 53:3–118.

Mitchell, R. W., N. S. Thompson, and H. L. Miles, eds. 1997. *Anthropomorphism, Anecdotes, and Animals*. Albany: State University of New York Press.

Mithen, Steven R. 2005. *The Singing Neanderthals: The Origins of Music, Language, Mind, and Body*. Cambridge, MA: Harvard University Press.

Mitman, Gregg. 1999. *Reel Nature: America's Romance with Wildlife on Film*. Cambridge, MA: Harvard University Press.

[Mivart, St. G. J.] 1871. "Darwin's Descent of Man." In *Darwin and His Critics: The Reception of Darwin's Theory of Evolution by the Scientific Community*, ed. David Hull, 354–84. Cambridge, MA: Harvard University Press, 1973. First published in *Quarterly Review* 131:47–90.

———. 1873. *Man and Apes: An Exposition of Structural Resemblances and Differences Bearing upon Questions of Affinity and Origin*. London: Robert Hardwicke.

[———]. 1874. "Primitive Man: Tylor and Lubbock." *Quarterly Review* 137:40–77.

———. 1875. "Instinct and Reason." *Contemporary Review* 25:763–88.

[———]. 1889. *Origin of Human Reason*. London: Kegan Paul, Trench.

Möbius, Karl. 1873. "Die Bewegungen der Thiere und ihr psychischer Horizont." Separatabdruck (Kiel), from *Schriften des Naturwissenschaftenlichen Vereins für Schleswig-Holstein* 1:111–30.

Moles, Abraham. 1963. "Animal Language and Information Theory." In Busnel 1963, 112–31.

Montagu, Ashley. 1944. "Aleš Hrdlička, 1869–1943." *American Anthropologist* 46:113–17.

————, ed. 1974. *Frontiers of Anthropology*. New York: Capricorn Books / G. P. Putnam's Sons.

Montgomery, Georgina M. 2005. "Place, Practice, and Primatology: Clarence Ray Carpenter, Primate Communication, and the Development of Field Methodology, 1931–1945." *Journal of the History of Biology* 38:495–533.

Montgomery, W. 1985. "Charles Darwin's Thought on Expressive Mechanisms in Evolution." In *The Development of Expressive Behavior: Biology-Environment Interactions*, 27–50. Orlando, FL: Academic Press.

Moore, James R. 1979. *The Post-Darwinian Controversies: A Study of the Protestant Struggle to Come to Terms with Darwin in Great Britain and America, 1870–1900*. Cambridge: Cambridge University Press.

Morgan, Conwy Lloyd. 1882. "Animal Intelligence." *Nature* 26:523–24.

————. 1884. "Instinct." *Nature* 29:370–75.

————. 1885. *The Springs of Conduct: An Essay in Evolution*. London: Kegan Paul, Trench.

————. 1886. "On the Study of Animal Intelligence." *Mind* 11:174–85.

————. 1890–91. *Animal Life and Intelligence*. London: E. Arnold.

————. 1891. "Psychology in America." *Nature* 43:506–9.

————. 1892a. "The Limits of Animal Intelligence." In *International Congress of Experimental Psychology*, 2nd session, 44–48. London: Williams and Norgate.

————. 1892b. "The Limits of Animal Intelligence." *Nature* 46:417.

————. 1892c. "The Speech of Monkeys." *Nature* 46:509–10.

————. 1892d. "Text-books of Psychology." *Nature* 46:1–4.

————. 1893a. "The Limits of Animal Intelligence." *Fortnightly Review* 54:223–39.

————. 1893b. "The Method of Comparative Psychology." In *Report of the British Association for the Advancement of Science*, 754–55. London: John Murray. Abstract of paper delivered in Edinburgh, August 1892.

————. 1894. *An Introduction to Comparative Psychology*. London: Walter Scott.

————. 1898a. "Animal Intelligence as an Experimental Study." In *Report of the 68th Meeting of the British Association for the Advancement of Science, Bristol, September 1898*, 909. London: John Murray.

————. 1898b. "Animal Intelligence as an Experimental Study." *Natural Science* 13 (October): 265–72.

————. 1898c. Review of Thorndike, *Animal Intelligence*. Nature, 14 July, 249–50.

————. 1900. *Animal Behaviour*. London: Edward Arnold.

————. 1903a. "The Ascent of Mind." *Nature*, 1 January, 199.

————. 1903b. *An Introduction to Comparative Psychology*. 2nd rev. ed. London: Walter Scott.

————. 1923. *Emergent Evolution*. London: Williams and Norgate.

————. 1932. "C. Lloyd Morgan." In *A History of Psychology in Autobiography*, ed. C. Murchison, 2:237–64. London: Clark University Press.

Morgan, L. H. 1868. *The American Beaver and His Works*. Philadelphia: Lippincott.

Morrell, Jack. 1997. *Science at Oxford, 1914–1939: Transforming an Arts University.* Oxford: Clarendon Press.

Morris, Charles W. 1938. *Foundations of the Theory of Signs.* International Encyclopedia of Unified Science Series, vol. 1, no. 2. Chicago: University of Chicago Press.

———. 1946. *Signs, Language, and Behavior.* New York: Prentice Hall.

Morton, David. 2000. *Off the Record: The Technology and Culture of Sound Recording in America.* New Brunswick, NJ: Rutgers University Press.

Moser, Don. 1974. "Teaching Animals to Talk: From Chimps and Sparrows Come Clues to Language." *Nature/Science Annual, 1975 Edition,* ed. Jane D. Alexander, 157–65. New York: Time-Life Books.

Mott, F. L. 1957. *A History of American Magazines.* Vol. 4, 1885–1905. Cambridge, MA: Belknap Press / Harvard University Press.

Müller, Friedrich Max. 1861. *Lectures on the Science of Language.* Ser. 1. London: Longman, Green, Longman, and Roberts.

———. 1864. *Lectures on the Science of Language.* Ser. 2. London: Longman, Green, Longman, Roberts, and Green.

———. 1873. "Lectures on Mr. Darwin's Philosophy of Language." In Harris 1996, 146–233. First published in *Fraser's Magazine* 7:525–41, 659–78; 8:1–24.

———. 1875a. "In Self-Defence: Present State of Scientific Studies." In *Chips from a German Workshop,* 4:473–549. London: Longmans, Green.

———. 1875b. "My Reply to Mr. Darwin." *Contemporary Review* 25:305–26. Reprinted in *Chips from a German Workshop,* 4:433–72. London: Longmans, Green.

———. 1876. "'Light, Delight, Alight.'" *Academy* 9:34.

———. 1878. "On the Origin of Reason." *Contemporary Review* 31:465–93.

———. 1881. "Kant's Critique of Pure Reason." In *Last Essays: Essays on Language, Folklore, and Other Subjects,* 218–50. London: Longmans, Green, 1901.

———. 1887a. "No Language without Reason—No Reason without Language." *Nature* 36 (14 July): 249–51.

———. 1887b. *The Science of Thought.* London: Longmans, Green.

———. 1888a. "Language–Reason." *Nature* 37 (2 February): 323–25.

———. 1888b. "Language = Reason." *Nature* 37 (1 March): 412–14.

———. 1891a. Section H. Anthropology. Opening address by Prof. F. Max Müller, president of the section. *Nature* 44 (3 September): 428–34. Subsequently published as "On the Classification of Mankind by Language or by Blood," in *Chips from a German Workshop,* 1894 ed., vol 1.

———. 1891b. "On Thought and Language." *Monist* 1:572–89.

———. 1898. *Auld Lang Syne.* Vol. 1. London: Longmans, Green.

———. 1901. *My Autobiography.* London: Longmans, Green.

Müller, Georgina G. 1902. *The Life and Letters of the Rt. Honourable Friedrich Max Müller.* 2 vols. London: Longmans, Green.

Munz, Tania. 2005. "The Bee Battles: Karl von Frisch, Adrian Wenner, and the

Honey Bee Dance Language Controversy." *Journal of the History of Biology* 38:535–70.

Musser, C., and C. Nelson. 1991. *High-Class Moving Pictures: Lyman H. Howe and the Forgotten Era of Traveling Exhibition, 1880–1920.* Oxford: Oxford University Press.

Nagel, Thomas. 1974. "What Is It Like to Be a Bat?" *Philosophical Review* 83:435–50.

National Defense Research Committee. 1946. *Recognition of Underwater Sounds.* Summary Technical Report of Division 6, NDRC. Washington, DC: National Defense Research Committee with the assistance of Columbia University Press.

"National Zoological Park." 1996. In *Guide to the Smithsonian Archives 1996,* 187–88. Washington, DC: Smithsonian Institution Press.

Newbury, E. 1954. "Current Interpretation and Significance of Lloyd Morgan's Canon." *Psychological Bulletin* 51:70–74.

Newmeyer, Frederic J. 1986. *The Politics of Linguistics.* Chicago: University of Chicago Press.

———. 1998. "On the Supposed 'Counterfunctionality' of Universal Grammar: Some Evolutionary Implications." In Hurford, Studdert-Kennedy, and Knight 1998, 305–19.

Nice, Margaret Morse. 1943. *Studies in the Life History of the Song Sparrow: 2, The Behavior of the Song Sparrow and Other Passerines.* Transactions of the Linnaean Society of New York, vol. 6. New York: Dover.

Nottebohm, Fernando. 1971. "Neural Lateralization of Vocal Control in a Passerine Bird: 1, Song." *Journal of Experimental Zoology* 177:229–61.

———. 1972. "Neural Lateralization of Vocal Control in a Passerine Bird: 2, Subsong, Calls, and a Theory of Vocal Learning." *Journal of Experimental Zoology* 179:35–49.

Numbers, Ronald L., and L. D. Stephens. 1998. "Darwinism and the American South: From the Early 1860s to the Late 1920s." In *Darwinism Comes to America,* R. L. Numbers, 58–75. Cambridge, MA: Harvard University Press.

Nye, Bill. 1894. "Personal Experiences in Monkey Language." *Pall Mall Magazine,* May–August, 648–55.

O'Donnell, John. 1985. *The Origins of Behaviorism: American Psychology, 1870–1920.* New York: New York University Press.

Oldroyd, David R. 1983. *Darwinian Impacts: An Introduction to the Darwinian Revolution.* 2nd rev. ed. Milton Keynes, UK: Open University Press.

Olender, M. 1992. *The Languages of Paradise: Race, Religion, and Philology in the Nineteenth Century.* Translated by A. Goldhammer. Cambridge, MA: Harvard University Press.

"The Origin of Language." 1866. In Harris 1996, 100–139. First published in *Westminster Review,* n.s., 30:88–122.

"The Origin of Language." 1874. In Harris 1996, 234–76. First published in *Westminster Review,* n.s., 46:381–418.

Ottavi, Dominique. 2002. *De Darwin à Piaget: Pour une histoire de la psychologie de l'enfant*. Paris: CNRS Editions.

Owen, John E. 1974. *L. T. Hobhouse, Sociologist*. London: Nelson.

Owings, Donald H., and Eugene S. Morton. 1998. *Animal Vocal Communication: A New Approach*. Cambridge: Cambridge University Press.

Patterson, Francine. 1978. "Conversations with a Gorilla." *National Geographic* 154 (4): 438–65.

———. 1981. "More on Ape Talk." *New York Review of Books*, 2 April, 43.

Paul, Diane. 1983. "'A War on Two Fronts': J. B. S. Haldane and the Response to Lysenkoism in Britain." *Journal of the History of Biology* 16:1–37.

Pauly, Daniel. 2004. *Darwin's Fishes: An Encyclopedia of Ichthyology, Ecology, and Evolution*. Cambridge: Cambridge University Press.

Pearson, Hesketh. 1936. *Labby: The Life and Character of Henry Labouchere*. London: Hamish Hamilton.

Pedersen, Holger. 1962. *The Discovery of Language: Linguistic Science in the Nineteenth Century*. Translated by J. W. Spargo. Bloomington: Indiana University Press.

Penfield, Wilder, and Theodore Rasmussen. 1950. *The Cerebral Cortex of Man: A Clinical Study of Localization of Function*. New York: Macmillan.

Petersen, Michael R., Michael D. Beecher, Stephen R. Zoloth, David B. Moody, and William C. Stebbins. 1978. "Neural Lateralization of Species-Specific Vocalizations by Japanese Macaques." *Science* 202:324–26.

Pfungst, Oskar. 1911. *Clever Hans (the Horse of Mr. Van Osten)*. Translated by Carl L. Rahn from the 1907 German original. Reprinted in facsimile with a new introduction by Robert Rosenthal. New York: Holt, Rinehart and Winston, 1965.

Phillips, Barnet. 1891. "A Record of Monkey Talk." *Harper's Weekly* 35:1050, illustration on 1036.

Phillips, V. N. 1997. *Between the States: Bristol Tennessee/Virginia during the Civil War*. Johnson City, TN: Overmountain Press.

Pierce, George W. 1948. *The Songs of Insects*. Cambridge, MA: Harvard University Press.

Pierce, John R. 1972. "Communication." *Scientific American*, September, 30–41.

Pierpont, Claudia Roth. 2004. "The Measure of America." *New Yorker*, 8 March, 48–63.

Pilet, P. E. 1976. "Carl Vogt." In Gillispie 1970–80, 14:57–58.

Pinker, Steven. 1992. Review of Bickerton, *Language and Species*. *Language* 68 (2): 375–82.

———. 1994. *The Language Instinct: How the Mind Creates Language*. New York: HarperPerennial.

Pinker, Steven, and Paul Bloom. 1990. "Natural Language and Natural Selection." *Behavioral and Brain Sciences* 13:707–84. Includes peer commentary and authors' response.

Ploog, Detlev W. 1967. "The Behavior of Squirrel Monkeys (*Saimiri sciureus*) as Revealed by Sociometry, Bioacoustics, and Brain Stimulation." In Altmann 1967a, 149–84.

Poquette, Ryan. 2000. "Earthly Passions." *Caltech News* 34 (1). Online at http://pr .caltech.edu/periodicals/CaltechNews/articles/v34/n1.green.html (accessed 30 November 2004).

Porter, Roy. 1997. *The Greatest Benefit to Mankind: A Medical History of Humanity from Antiquity to the Present*. London: HarperCollins.

Potter, Ralph K. 1930. "Transmission Characteristics of a Short-Wave Telephone Circuit." *Proceedings of the Institute of Radio Engineers* 18:581–648.

———. 1945. "Visible Patterns of Sound." *Science*, 9 November, 463–70.

———. 1946. "Visible Speech." *Bell Laboratories Record*, January, 7–11.

Potter, Ralph K., George A. Kopp, and Harriet C. Green. 1947. *Visible Speech*. New York: D. Van Nostrand.

Poulsen, Holger. 1951. "Inheritance and Learning in the Song of the Chaffinch (*Fringilla coelebs* L.)." *Behaviour* 3:216–27.

Powell, J. W. 1880. *Introduction to the Study of Indian Languages*. Washington, DC: Government Printing Office.

Prelli, Lawrence J. 1989. "The Rhetorical Construction of Scientific Ethos." In *Rhetoric in the Human Sciences*, ed. Herbert W. Simons, 48–68. London: Sage.

Premack, Ann James. 1976. *Why Chimps Can Read*. New York: Harper and Row.

Premack, Ann James, and David Premack. 1972. "Teaching Language to an Ape." *Scientific American*, October, 92–99.

Premack, David. 1971. "Language in Chimpanzee?" *Science* 172:808–22.

———. 1973. Brief autobiography in *American Men and Women of Science*, 12th ed., 2:1974. New York: Jacques Cattell Press / R. R. Bowker.

———. 1974. "Concordant Preferences as a Precondition for Affective but Not for Symbolic Communication (or How to Do Experimental Anthropology)." In *Experimental Behaviour: A Basis for the Study of Mental Disturbance*, ed. John H. Cullen, 346–61. Dublin: Irish University Press.

———. 1975. "On the Origins of Language." In *Handbook of Psychobiology*, ed. Michael S. Gazzaniga and Colin Blakemore, 591–605. New York: Academic Press.

Premack, David, and Arthur Schwartz. 1966. "Preparations for Discussing Behaviorism with Chimpanzee." In *The Genesis of Language: A Psycholinguistic Approach*, ed. Frank Smith and George A. Miller, 295–346, including a response by P. B. Denes and a summary of the general discussion. Cambridge, MA: MIT Press.

Pringle, J. W. S. 1951. "On the Parallel between Learning and Evolution." *Behaviour* 3:174–215.

Promptov, A. N. 1930. "Die geographische Variabilität des Buchfinkenschlags (*Fringilla coelebs* L.) in Zusammenhang mit etlichen allgemeinen Fragen der Saisonvögelzüge." *Biologisches Zentralblatt* 50:478–503.

Promptov, A. N., and E. V. Lukina. 1945. "Conditioned-Reflectory Differentiation of Calls in *Passeres* and Its Biological Value." *Comptes Rendus (Doklady) de l'Académie des Sciences de l'URSS* 46:382–84.

Pullum, G. K. 1991. "The Great Eskimo Vocabulary Hoax." In *The Great Eskimo Vocabulary Hoax and Other Irreverent Essays in the Study of Language*, 159–71. Chicago: University of Chicago Press.

Quastler, Henry. 1958. "A Primer on Information Theory." In *Symposium on Information Theory in Biology*, ed. Hubert P. Yockey, 3–49. London: Pergamon Press.

Radick, Gregory. 2000a. "Morgan's Canon, Garner's Phonograph, and the Evolutionary Origins of Language and Reason." *British Journal for the History of Science* 33:3–23.

———. 2000b. Review of S. Alter, *Darwinism and the Linguistic Image*. *British Journal for the History of Science* 33:122–24.

———. 2002. "Darwin on Language and Selection." *Selection* 3:7–16.

———. 2003. "Is the Theory of Natural Selection Independent of Its History?" In Hodge and Radick 2003, 143–67.

———. 2004a. "Morgan, Conwy Lloyd." In Lightman 2004, 3:1425–26.

———. 2004b. "Müller, Friedrich Max." In Lightman 2004, 3:1435–39.

———. 2005a. "The Case for Virtual History." *New Scientist*, 20 August, 34–35.

———. 2005b "Other Histories, Other Biologies." In *Philosophy, Biology, and Life*, ed. A. O'Hear, 21–47. Royal Institute of Philosophy Lectures 2003–4. Cambridge: Cambridge University Press.

———. 2006. "What's in a Name? The Vervet Predator Calls and the Limits of the Washburnian Synthesis." *Studies in History and Philosophy of Biological and Biomedical Sciences* 37:334–62.

———. In press a. "The Ethologist's World." *Journal of the History of Biology*.

———. In press b. "Thorpe, William Homan." In *New Dictionary of Scientific Biography*, ed. Noretta Koertge.

———. In preparation. "The Reactionary Origins of the Chomskyan Revolution: The Making and Unmaking of a Modern Synthesis." Manuscript.

Radick, Gregory, and Graeme J. N. Gooday. 2004. "Geddes, Patrick." In Lightman 2004, 2:764–68.

Ramsay, Alexandra. 1966. "Animal Communication: Report of a Conference." *Current Anthropology* 7:251–53.

Read, O., and W. L. Welch. 1976. *From Tin Foil to Stereo: Evolution of the Phonograph*. 2nd ed. Indianapolis: Howard W. Sams.

Reade, W. Winwood. 1863. *Savage Africa*. London: Smith, Elder.

Reader, John. 1990. *Missing Links: The Hunt for Earliest Man*. 2nd ed. London: Penguin.

Rée, Jonathan. 1999. *I See a Voice: A Philosophical History of Language, Deafness, and the Senses*. London: HarperCollins.

Reed, James W. 1987. "Robert M. Yerkes and the Mental Testing Movement." In

Psychological Testing and American Society, 1890–1930, 75–94. New Brunswick, NJ: Rutgers University Press.

———. 1999. "Yerkes, Robert Mearns." In Garraty and Carnes 1999, 24:132–34.

Rees, Amanda. 2006. "A Place That Answers Questions: Primatological Field Sites and the Making of Authentic Observations." *Studies in History and Philosophy of Biological and Biomedical Sciences* 37:311–33.

Reeve, Arthur B. 1909. "Men and Monkeys: Primates." *Hampton's Broadway Magazine*, January, 95–103.

Reisch, George A. 2005. *How the Cold War Transformed Philosophy of Science: To the Icy Slopes of Logic*. Cambridge: Cambridge University Press.

Renwick, Chris. 2005. *Sex and the City: The Evolution of Patrick Geddes, 1874–1889*. Master's dissertation, University of Leeds.

Ribnick, Rosalind. 1982. "A Short History of Primate Field Studies: Old World Monkeys and Apes." In *A History of American Physical Anthropology, 1930–1980*, ed. Frank Spencer, 49–73. New York: Academic Press.

Richards, Graham. 1998. "Getting a Result: The Expedition's Psychological Research, 1898–1913." In Herle and Rouse 1998, 136–57.

Richards, Robert J. 1987. *Darwin and the Emergence of Evolutionary Theories of Mind and Behavior*. Chicago: University of Chicago Press.

———. 1992. *The Meaning of Evolution: The Morphological Construction and Ideological Reconstruction of Darwin's Theory*. Chicago: University of Chicago Press.

———. 2002. "The Linguistic Creation of Man: Charles Darwin, August Schleicher, Ernst Haeckel, and the Missing Link of Nineteenth-Century Evolutionary Theory." In *Experimenting in Tongues: Studies in Science and Language*, ed. Matthias Dörries, 21–48. Stanford: Stanford University Press.

———. 2003. "Darwin on Mind, Morals, and Emotions." In Hodge and Radick 2003, 92–115.

Richmond, Marsha L. 2006. "The 1909 Darwin Celebration." *Isis* 97:447–84.

Riordan, Michael, and Lillian Hoddeson. 1997. *Crystal Fire: The Invention of the Transistor and the Birth of the Information Age*. New York: W. W. Norton.

Ristau, Carolyn A. 1996. "Animal Language and Cognition Projects." In *Handbook of Human Symbolic Evolution*, ed. Andrew Lock and Charles R. Peters, 644–85. Oxford: Clarendon Press.

Rivers, W. H. R. 1901. Part I (excluding appendix). *Reports of the Cambridge Anthropological Expedition to Torres Straits* 2 (1903): 1–132. In Haddon 1901–35.

Robinson, Bryan W. 1967a. "Neurological Aspects of Evoked Vocalizations." In Altmann 1967a, 135–47.

———. 1967b. "Vocalization Evoked from Forebrain in *Macaca mulatta*." *Physiology and Behavior* 2:345–54.

Rocher, R. 1995. "Discovery of Sanskrit by Europeans." In *Concise History of the Language Sciences: From the Sumerians to the Cognitivists*, ed. E. F. K. Koerner and R. E. Asher, 188–91. Oxford: Pergamon.

Roe, Anne, and George Gaylord Simpson. 1958. *Behavior and Evolution*. New Haven: Yale University Press.

Romanes, George John. 1878. "Animal Intelligence." *Nineteenth Century* 4:653–72.

———. 1879. "Intellect in Brutes." *Nature* 20:122–25.

———. 1882. *Animal Intelligence*. London: Kegan Paul, Trench.

———. 1883. *Mental Evolution in Animals*. London: Kegan Paul, Trench.

———. 1888. *Mental Evolution in Man: Origin of Human Faculty*. London: Kegan Paul, Trench.

———. 1889. "On the Mental Faculties of *Anthropopithecus calvus*." *Nature* 40:160–62.

———. 1892. "Thought and Language." *Monist* 2:56–69.

Rooy, Piet de. 1995. "In Search of Perfection: The Creation of a Missing Link." In *Ape, Man, Ape-Man: Changing Views since 1600*, ed. Raymond Corbey and Bert Theunissen, 195–207. Leiden: Leiden University.

Rosenfield, Israel. 1985. "A Hero of the Brain." *New York Review of Books*, 21 November, 49–55.

———. 1988. *The Invention of Memory: A New View of the Brain*. New York: Basic Books.

Rothenberg, David. 2005. *Why Birds Sing: One Man's Quest to Solve an Everyday Mystery*. London: Allen Lane.

Rowell, Thelma E. 1962. "Agonistic Noises of the Rhesus Monkey (*Macaca mulatto*)." *Symposia of the Zoological Society of London* 8:91–96.

Royce, Josiah. 1903. *Outlines of Psychology: An Elementary Treatise with Some Practical Applications*. New York: Macmillan.

Rumbaugh, Duane M., ed. 1977. *Language Learning by a Chimpanzee: The LANA Project*. New York: Academic Press.

Rupke, Nicolaas A. 1994. *Richard Owen: Victorian Naturalist*. London: Yale University Press.

Ruse, Michael. 1996. *Monad to Man: The Concept of Progress in Evolutionary Biology*. Cambridge, MA: Harvard University Press.

———. 1999a. *The Darwinian Revolution: Science Red in Tooth and Claw*. 2nd ed. Chicago: University of Chicago Press.

———. 1999b. *Mystery of Mysteries: Is Evolution a Social Construction?* Cambridge, MA: Harvard University Press.

———. 2004. "Adaptive Landscapes and Dynamic Equilibrium: The Spencerian Contribution to Twentieth-Century American Evolutionary Biology." In *Darwinian Heresies*, ed. Abigail Lustig, Michael Ruse, and Robert J. Richards, 131–50. Cambridge: Cambridge University Press.

Rushing, Homer D. 1990. *The Gorilla Comes to Darwin's England: A History of the Impact of the Largest Anthropoid Ape on British Thinking from Its Rediscovery to the End of the Gorilla War, 1846–1863*. Master's thesis, University of Texas, Austin.

Ryan, Frank X., ed. 2000. *The Evolutionary Philosophy of Chauncey Wright*. 3 vols. Bristol: Thoemmes Press.

Sacks, Oliver. 1996. *The Island of the Colour-Blind and Cycad Island*. London: Picador.

———. 2001. *Uncle Tungsten: Memories of a Chemical Boyhood*. London: Picador.

Sanford, M. B. 1892. "Bridget O'Flanagan on the Language of the Monkeys." *Puck* 31:259.

Sapir, Edward. 1921. *Language: An Introduction to the Study of Speech*. New York: Harcourt, Brace & World.

———. 1933. "Language." In *Encyclopaedia of the Social Sciences*, ed. E. R. A. Seligman and A. Johnson, 9:155–69. New York: Macmillan.

Sarkar, Sahotra. 1992. "Science, Philosophy, and Politics in the Work of J. B. S. Haldane, 1922–1937." *Biology and Philosophy* 7:385–409.

Saussure, Ferdinand de. 1974. *Course in General Linguistics*. Translated by Wade Baskin from the 1916 French original. Introduction by Jonathan Culler. Glasgow: Fontana / Collins.

Savage-Rumbaugh, Sue, and Roger Lewin. 1994. *Kanzi: The Ape at the Brink of the Human Mind*. New York: Doubleday.

Savage-Rumbaugh, Sue, Stuart G. Shanker, and Talbot J. Taylor. 1998. *Apes, Language, and the Human Mind*. Oxford: Oxford University Press.

Scarborough, Elizabeth. 2000. "Washburn, Margaret Floy." In Kazdin 2000, 8:230–32.

Schachter, Steven C., and Orrin Devinksy, eds. 1997. *Behavioral Neurology and the Legacy of Norman Geschwind*. Philadelphia: Lippincott-Raven.

Schaffer, Simon. 1994. *From Physics to Anthropology—and Back Again*. Prickly Pear Pamphlet no. 3. Cambridge: Prickly Pear Press.

———. 2000–2001. "Rivers Lecture: The Disappearance of Useful Sciences." *Cambridge Anthropology* 22 (1): 1–23.

Schiller, Claire H, ed. 1957. *Instinctive Behavior: The Development of a Modern Concept*. Translated by Claire H. Schiller. Introduction by Karl S. Lashley. London: Methuen and Co.

Schiller, F. 1979. *Paul Broca: Founder of French Anthropology, Explorer of the Brain*. Berkeley: University of California Press.

Schleicher, August. 1983. "On the Significance of Language for the Natural History of Man." Translated by J. P. Maher from German original published in 1865. In *Linguistics and Evolutionary Theory: Three Essays by August Schleicher, Ernst Haeckel, and Wilhelm Bleek*, ed. K. Koerner. Amsterdam: John Benjamins.

Schmeck, Harold M. Jr. 1980. "Studies in Africa Find Monkeys Using Rudimentary 'Language.'" *New York Times*, 28 November, A1, A22.

Schneirla, T. C. 1950. "The Relationship between Observation and Experimentation in the Field Study of Behavior." *Annals of the New York Academy of Sciences* 51:1022–44.

Schrempp, Gregory. 1983. "The Re-education of Friedrich Max Müller: Intellectual

Appropriation and Epistemological Antinomy in Mid-Victorian Evolutionary Thought." *Man*, n.s., 18:90–110.

Schwartz, Joel S. 2002. "Out from Darwin's Shadow: George John Romanes's Efforts to Popularize Science in *Nineteenth Century* and Other Victorian Periodicals." *Victorian Periodicals Review* 35:133–59.

Schwidetzky, Georg. 1932. *Do You Speak Chimpanzee? An Introduction to the Study of the Speech of Animals and of Primitive Men*. Translated by Margaret Gardiner. London: George Routledge and Sons. First published as *Sprechen Sie schimpansisch? Einführung in die Tier- und Ursprachenlehre* in 1931.

Schwidetzky, Ilse, ed. 1973. *Über die Evolution der Sprache: Anatomie, Verhaltensforschung, Sprachwissenschaft, Anthropologie*. Frankfurt am Main: S. Fischer Verlag.

"Scientia Scientiarum." 1867. *Journal of the Transactions of the Victoria Institute* 1:5–29.

Scripture, Edward Wheeler. 1902. *The Elements of Experimental Phonetics*. London: Edward Arnold; New York: Charles Scribner's Sons.

Sebeok, Thomas A., ed. 1968. *Animal Communication: Techniques of Study and Results of Research*. Bloomington: Indiana University Press.

———, ed. 1978. *Sight, Sound, and Sense*. Bloomington: Indiana University Press.

———. 1980. "Looking in the Destination for What Should Have Been Sought in the Source." In Sebeok and Umiker-Sebeok 1980, 407–27.

Sebeok, Thomas A., and Alexandra Ramsay, eds. 1969. *Approaches to Animal Communication*. The Hague: Mouton.

Sebeok, Thomas A., and Robert Rosenthal, eds. 1981. *The Clever Hans Phenomenon: Communication with Horses, Whales, Apes, and People*. New York: New York Academy of Sciences.

Sebeok, Thomas A., and Jean Umiker-Sebeok, eds. 1980. *Speaking of Apes: A Critical Anthology of Two-Way Communication with Man*. New York: Plenum Press.

Secord, James A. 2000. *Victorian Sensation: The Extraordinary Publication, Reception, and Secret Authorship of* Vestiges of the Natural History of Creation. Chicago: University of Chicago Press.

Segerstråle, Ullica. 2000. *Defenders of the Truth: The Sociobiology Debate*. Oxford: Oxford University Press.

Seyfarth, Robert M. 1975. "The Social Relationships among Adults in a Troop of Free-Ranging Baboons *(Papio cynocephalus ursinus)*." PhD thesis, University of Cambridge.

———. 1976. "Social Relationships among Adult Female Baboons." *Animal Behaviour* 24:917–38.

———. 1977. "A Model of Social Grooming among Adult Female Monkeys." *Journal of Theoretical Biology* 65:671–98.

———. 1978a. "Social Relationships among Adult Male and Female Baboons: 1, Behaviour during Sexual Consortship." *Behaviour* 64:204–26.

———. 1978b. "Social Relationships among Adult Male and Female Baboons: 2, Behaviour throughout the Female Reproductive Cycle." *Behaviour* 64:227–47.

———. 1980. "The Distribution of Grooming and Related Behaviours among Adult Female Vervet Monkeys." *Animal Behaviour* 28:798–813.

———. 1981. "Do Monkeys Rank Each Other?" *Behavioral and Brain Sciences* 4:447–48.

———. 1982. "Talking with Monkeys and Great Apes." *International Wildlife*, March–April, 13–18.

Seyfarth, Robert M., and Dorothy L. Cheney. 1980. "The Ontogeny of Vervet Monkey Alarm Calling Behavior: A Preliminary Report." *Zeitschrift für Tierpsychologie* 54:37–56.

———. 1992. "Meaning and Mind in Monkeys." *Scientific American*, December, 122–28.

———. 2002. "Dennett's Contribution to Research on the Animal Mind." In *Daniel Dennett*, ed. A. Brook and D. Ross, 117–39. Cambridge: Cambridge University Press.

Seyfarth, Robert M., Dorothy L. Cheney, and Robert A. Hinde. 1978. "Some Principles Relating Social Interactions and Social Structure among Primates." In *Recent Advances in Primatology*, ed. D. J. Chivers and J. Herbert, 1:39–51. New York: Academic Press.

Seyfarth, Robert M., Dorothy L. Cheney, and Peter Marler. 1980a. "Monkey Responses to Three Different Alarm Calls: Evidence of Predator Classification and Semantic Communication." *Science*, 14 November, 801–3.

———. 1980b. "Vervet Monkey Alarm Calls: Semantic Communication in a Free-Ranging Primate." *Animal Behaviour* 28:1070–94.

Shannon, Claude E., and Warren Weaver. 1949. *The Mathematical Theory of Communication*. Urbana: University of Illinois Press.

Shelemay, Kay Kaufman. 1991. "Recording Technology, the Record Industry, and Ethnomusicological Scholarship." In *Comparative Musicology and Anthropology of Music: Essays on the History of Ethnomusicology*, ed. Bruno Nettl and Phillip V. Bohlman, 277–92. Chicago: University of Chicago Press.

Shipman, Pat. 2001. *The Man Who Found the Missing Link: Eugène Dubois and His Lifelong Quest to Prove Darwin Right*. New York: Simon and Schuster.

Silcock, Bryan. 1980. "Words That Monkeys Invented." *Sunday Times* (London), 23 November, 14.

Simms, Eric, and G. F. Wade. 1953. "Recent Advances in the Recording of Bird-Songs." *British Birds* 46:200–210.

Simpson, George Gaylord. 1950. *The Meaning of Evolution: A Study of the History of Life and of Its Significance for Man*. Oxford: Oxford University Press.

Singer, B. 1981. "History of the Study of Animal Behaviour." In *The Oxford Companion to Animal Behaviour*, ed. D. McFarland, 255–72. Oxford: Oxford University Press.

Singer, Peter. 1975. *Animal Liberation*. New York: New York Review / Random House.

Skinner, B. F. 1957. *Verbal Behavior*. Englewood Cliffs, NJ: Prentice Hall.

Sleigh, Charlotte. 2007. *Six Legs Better: A Cultural History of Myrmecology*. Baltimore: Johns Hopkins University Press.

Smith, Merritt Roe, and Leo Marx, eds. 1994. *Does Technology Drive History? The Dilemma of Technological Determinism*. Cambridge, MA: MIT Press.

Smith, Roger. 1992. *Inhibition: History and Meaning in the Sciences of Mind and Brain*. London: Free Association Books.

———. 1997. *The Fontana History of the Human Sciences*. London: Fontana Press.

Smith, W. John. 1963. "Vocal Communication of Information in Birds." *American Naturalist* 97:117–25.

———. 1965. "Message, Meaning, and Context in Ethology." *American Naturalist* 99:405–9.

———. 1968. "Message-Meaning Analyses." In Sebeok 1968, 44–60.

———. 1977. *The Behavior of Communicating: An Ethological Approach*. Cambridge, MA: Harvard University Press.

Smocovitis, Vassiliki Betty. 1996. *Unifying Biology: The Evolutionary Synthesis and Evolutionary Biology*. Princeton: Princeton University Press.

Sober, Elliott. 1998. "Morgan's Canon." In *The Evolution of Mind*, ed. D. D. Cummins and C. Allen, 224–42. Oxford: Oxford University Press

Sparks, John. 1982. *The Discovery of Animal Behaviour*. London: William Collins Song / British Broadcasting Corporation.

Specter, Michael. 2001. "Rethinking the Brain." In *The Best American Science Writing 2002*, ed. Matt Ridley, 151–70. New York: Ecco, 2002. First published in *New Yorker*.

The Speech of Man and Holy Writ. 1894. London: William R. Gray.

Spencer, Frank. 1981. "The Rise of Academic Physical Anthropology in the United States (1880–1980): A Historical Review." *American Journal of Physical Anthropology* 56:353–64.

———. 1990. *Piltdown: A Scientific Forgery*. London: Natural History Museum Publications / Oxford University Press.

———, ed. 1997. *History of Physical Anthropology*. 2 vols. London: Garland.

———. 1999. "Hrdlička, Aleš." In Garraty and Carnes 1999, 11:376–78.

Spencer, Herbert. 1870–72. *The Principles of Psychology*. 2 vols. London: Williams and Norgate.

Stam, Henderikus J., and Tanya Kalmanovitch. 1998. "E. L. Thorndike and the Origins of Animal Psychology: On the Nature of the Animal in Psychology." *American Psychologist* 53 (October): 1135–44.

Stam, J. H. 1976. *Inquiries into the Origin of Language: The Fate of a Question*. London: Harper and Row.

Star, S. L. 1989. *Regions of the Mind: Brain Research and the Quest for Scientific Certainty*. Stanford: Stanford University Press.

Starr, S. Z. 1979. "Forrest's Raids." In *The Encyclopedia of Southern History*, ed. D. C. Roller and R. W. Twyman, 468–69. Baton Rouge: Louisiana State University Press.

[Stead, W. T.] 1892. "Off to Monkey Land: A Visit from Mr. Garner." *Review of Reviews* 6:256.

Steinthal, H. 1888. *Der Ursprung der Sprache im Zusammenhange mit den letzten Fragen alles Wissens*. 4th ed. Berlin: F. Dümmler.

Steklis, H. Dieter. 1999. "The Primate Brain and the Origins of Intelligence." In Strum, Lindburg, and Hamburg 1999, 49–63.

Stephen, L. 1873. "Darwinism and Divinity." In *Essays in Freethinking and Plainspeaking*, 72–109. London: Longmans, Green. First published (in slightly different form) in *Fraser's Magazine* 5 (1872): 409–21.

Stewart, T. D. 1940. "The Life and Writings of Dr. Aleš Hrdlička (1869–1939)." *American Journal of Physical Anthropology* 26:3–40.

Stocking, George W. Jr. 1968a. "The Critique of Racial Formalism." In Stocking 1968d, 161–94.

———. 1968b. "The Dark-Skinned Savage: The Image of Primitive Man in Evolutionary Anthropology." In Stocking 1968d, 110–32.

———. 1968c. "From Physics to Ethnology." In Stocking 1968d, 133–60.

———. 1968d. *Race, Culture, and Evolution: Essays in the History of Anthropology*. New York: Free Press; London: Collier-Macmillan.

———. 1974a. "The Basic Assumptions of Boasian Anthropology." In Stocking 1974b, 1–20.

———, ed. 1974b. *The Shaping of American Anthropology, 1883–1911: A Franz Boas Reader*. New York: Basic Books. Reprinted (with title and subtitle reversed) in 1982 and 1989.

———. 1987. *Victorian Anthropology*. New York: Free Press; Oxford: Maxwell Macmillan International.

———. 1992a. "The Boas Plan for the Study of American Indian Languages." In Stocking 1992b, 60–91.

———. 1992b. *The Ethnographer's Magic and Other Essays in the History of Anthropology*. Madison: University of Wisconsin Press.

———. 1992c. "The Ethnographer's Magic: Fieldwork in British Anthropology from Tylor to Malinowski." In Stocking 1992b, 12–59.

———. 1992d. "Ideas and Institutions in American Anthropology: Thoughts Toward a History of the Interwar Years." In Stocking 1992b, 114–77.

———. 1992e. "Paradigmatic Traditions in the History of Anthropology." In Stocking 1992b, 342–61.

———. 1995. *After Tylor: British Social Anthropology, 1888–1951*. London: Athlone Press.

———, ed. 1996. *"Volksgeist" as Method and Ethic: Essays on Boasian Ethnography and the German Anthropological Tradition*. History of Anthropology, vol. 8. Madison: University of Wisconsin Press.

Stone, Jon R., ed. 2002. *The Essential Max Müller: On Language, Mythology, and Religion*. New York: Palgrave Macmillan.

Strick, James E. 2000. *Sparks of Life: Darwinism and the Victorian Debates over Spontaneous Generation*. Cambridge, MA: Harvard University Press.

Struhsaker, Thomas T. 1967a. "Auditory Communication among Vervet Monkeys (*Cercopithecus aethiops*)." In Altmann 1967a, 281–324.

———. 1967b. "Behavior of Vervet Monkeys and Other Cercopithecines." *Science* 156:1197–1203.

———. 1967c. "Ecology of Vervet Monkeys (*Cercopithecus aethiops*) in the Masai-Amboseli Game Reserve, Kenya." *Ecology* 48:891–904.

———. 1967d. "Social Structure among Vervet Monkeys (*Cercopithecus aethiops*)." *Behaviour* 29:83–121.

———. 1975. *The Red Colobus Monkey*. Chicago: University of Chicago Press.

Strum, Shirley C., and Linda Marie Fedigan, eds. 2000. *Primate Encounters: Models of Science, Gender, and Society*. Chicago: University of Chicago Press.

Strum, Shirley C., Donald G. Lindburg, and David Hamburg, eds. 1999. *The New Physical Anthropology: Science, Humanism, and Critical Reflection*. Upper Saddle River, NJ: Prentice Hall.

Sussman, Robert W. 1997. "Primate Field Studies." In Spencer 1997, 2:842–48.

Takasaki, Hiroyuki. 2000. "Traditions of the Kyoto School of Field Primatology in Japan." In Strum and Fedigan 2000, 151–64.

Tallerman, Maggie, ed. 2005. *Language Origins: Perspectives on Evolution*. Oxford: Oxford University Press.

Taschwer, Klaus, and Benedikt Föger. 2003. *Konrad Lorenz*. Vienna: Zsolnay.

Taub, Liba. 1993. "Evolutionary Ideas and 'Empirical' Methods: The Analogy between Language and Species in Works by Lyell and Schleicher." *British Journal for the History of Science* 26:171–93.

Taylor, S. 1878. "Sound Colour-Figures." *Nature* 17:426–27.

Terborgh, John. 1996. "Cracking the Bird Code." *New York Review of Books*, 11 January, 40–44.

Terrace, Herbert S. 1979. *Nim*. New York: Knopf.

———. 1983. "Nonhuman Intentional Systems." *Behavioral and Brain Sciences* 6:378–79.

Thayer, J. B., ed. 1878. *Letters of Chauncey Wright, with Some Account of His Life*. Cambridge, MA: John Wilson and Son.

Theunissen, B. 1989. *Eugène Dubois and the Ape-Man from Java: The History of the First 'Missing Link' and Its Discoverer*. Dordrecht: Kluwer.

Thomas, Marion. 2003. "Rethinking the History of Ethology: French Animal Behaviour Studies in the Third Republic (1870–1940)." PhD thesis, University of Manchester.

———. 2005. "Are Animals Just Noisy Machines? Louis Boutan and the Co-invention of Animal and Child Psychology in the French Third Republic." *Journal of the History of Biology* 38:425–60.

————. 2006. "Yerkes, Hamilton, and the Experimental Study of the Ape Mind: From Evolutionary Psychiatry to Eugenic Politics." *Studies in History and Philosophy of Biological and Biomedical Sciences* 37:273–94.

Thompson, Emily Ann. 2002. *The Soundscape of Modernity: Architectural Acoustics and the Culture of Listening in America, 1900–1933*. Cambridge, MA: MIT Press.

Thorndike, Edward Lee. 1898. *Animal Intelligence: An Experimental Study of the Associative Processes in Animals*. In Thorndike 1911, 20–155. First published in *Psychological Review*, monograph suppl. 2:1–109.

————. 1899a. "The Instinctive Reactions of Young Chicks." In Thorndike 1911, 156–68. First published in *Psychological Review* 6 (3): 282–91.

————. 1899b. "A Note on the Psychology of Fishes." In Thorndike 1911, 169–71. First published in *American Naturalist* 33 (396): 923–95

————. 1899c. "A Reply to 'The Nature of Animal Intelligence and the Methods of Investigating It.'" *Psychological Review* 6:412–20.

————. 1901a. "The Evolution of the Human Intellect." In Thorndike 1911, 282–94. First published in *Popular Science Monthly*, November.

————. 1901b. *The Mental Life of the Monkeys: An Experimental Study*. In Thorndike 1911, 172–240. First published in *Psychological Review*, monograph suppl. 3:1–57.

————. 1909. "Darwin's Contribution to Psychology." *University of California Chronicle* 12:65–80.

————. 1911. *Animal Intelligence: Experimental Studies*. New York: Macmillan. Reprinted in facsimile with an introduction by Darryl Bruce. New Brunswick, NJ: Transaction Publishers, 2000. Pages given in notes refer to reprint edition.

————. 1913. *Educational Psychology*. Vol. 1. *The Original Nature of Man*. New York: Teachers College, Columbia University.

————. 1936. Autobiography. In *A History of Psychology in Autobiography*, ed. Carl Murchison, 3:263–70. Worcester, MA: Clark University Press.

Thorpe, William Homan. 1940. "Ecology and the Future of Systematics." In *The New Systematics*, ed. Julian Huxley, 341–64. London: Oxford University Press.

————. 1943–44. "Types of Learning in Insects and Other Arthropods: Part 3." *British Journal of Psychology* 34:66–76.

————. 1951. "The Learning Abilities of Birds, Parts 1 and 2." *Ibis* 93:1–52, 252–96.

————. 1954. "The Process of Song-Learning in the Chaffinch as Studied by Means of the Sound Spectrograph." *Nature* 173:465–69.

————. 1956. *Learning and Instinct in Animals: A Study of the Integration of Acquired and Innate Behaviour*. London: Methuen.

————. 1958. "The Learning of Song Patterns by Birds, with Especial Reference to the Song of the Chaffinch *Fringilla coelebs*." *Ibis* 100:535–70.

————. 1961. *Bird-Song: The Biology of Vocal Communication and Expression in Birds*. Cambridge: Cambridge University Press.

———. 1972. "The Comparison of Vocal Communication in Animals and Man." In *Non-Verbal Communication*, ed. Robert A. Hinde, 27–47. Cambridge: Cambridge University Press.

———. 1974. "David Lambert Lack, 1910–1973." *Biographical Memoirs of Fellows of the Royal Society* 20:271–93.

———. 1979. *The Origins and Rise of Ethology: The Science of the Natural Behaviour of Animals.* London: Heinemann Educational Books.

Tinbergen, Nikolaas. 1942. "An Objectivistic Study of the Innate Behaviour of Animals." *Bibliotheca Biotheoretica* 1:39–98.

———. 1951. *The Study of Instinct.* Oxford: Clarendon Press.

———. 1953. *The Herring Gull's World: A Study of the Social Behaviour of Birds.* London: Collins.

———. 1957. Preface to C. H. Schiller 1957, xv–xix.

———. 1960. "Behaviour, Systematics, and Natural Selection." In *Evolution after Darwin*, vol. 1, *The Evolution of Life*, ed. Sol Tax, 595–613. Chicago: University of Chicago Press. First published in *Ibis* 101 (1959): 318–30.

———. 1972–73. *The Animal in Its World.* 2 vols. London: George Allen and Unwin.

———. 1989. "Watching and Wondering." In Dewsbury 1989b, 431–63.

Tort, Patrick. 1996a. "Möbius, Karl August." In *Dictionnaire du darwinisme et de l'evolution*, ed. P. Tort, 2:2992–93. Paris: Presses Universitaires de France.

———. 1996b. "Vogt, Karl." In *Dictionnaire du darwinisme et de l'evolution*, ed. P. Tort, 3:4485–88. Paris: Presses Universitaires de France.

Tracy, Julia Lowndes. n.d. *Terrier VC: A Black and Tan Hero Who Won the Victoria Cross.* London: Aldine.

Trautmann, T. R. 1997. *Aryans and British India.* Berkeley: University of California Press.

Triplett, N. 1901. "The Educability of the Perch." *American Journal of Psychology* 12:354–60.

Trivers, Robert L. 1971. "The Evolution of Reciprocal Altruism." *Quarterly Review of Biology* 46:35–57.

———. 1972. "Parental Investment and Sexual Selection." In *Sexual Selection and the Descent of Man, 1871–1971*, ed. B. Campbell, 136–79. Chicago: Aldine. Reprinted in Houck and Drickamer 1996, 795–838.

———. 1974. "Parent-Offspring Conflict." *American Zoologist* 14:249–64.

Turner, Frank M. 1978. "The Victorian Conflict between Science and Religion: A Professional Dimension." *Isis* 69:356–76. Reprinted in revised form in *Contesting Cultural Authority: Essays in Victorian Intellectual Life*, 171–200. Cambridge: Cambridge University Press, 1993.

Tylor, E. B. 1865. *Researches into the Early History of Mankind and the Development of Civilization.* London: John Murray.

———. 1866a. "On the Origin of Language." In Harris 1996, 81–99. First published in *Fortnightly Review* 4:544–59.

———. 1866b. "The Science of Language." *Quarterly Review* 119:394–435.

Uexküll, Jakob von. 1934. "A Stroll through the Worlds of Animals and Men." In C. H. Schiller 1957, 5–80. First published (with Georg Kriszat) as *Streifzüge durch die Umwelten von Tieren und Menschen*. Berlin: J. Springer.

Urry, James. 1984. "Englishmen, Celts, and Iberians: The Ethnographic Survey of the United Kingdom." In *Functionalism Historicized: Essays on British Social Anthropology*, ed. G. W. Stocking Jr., 83–105. History of Anthropology vol. 2. Madison: University of Wisconsin Press.

———. 1998. "Making Sense of Diversity and Complexity: The Ethnological Context and Consequences of the Torres Strait Expedition and the Oceanic Phase in British Anthropology, 1890–1935." In Herle and Rouse 1998, 201–33.

Valone, David A. 1996. "Language, Race, and History: The Origin of the Whitney-Müller Debate and the Transformation of the Human Sciences." *Journal of the History of the Behavioral Sciences* 32:119–34.

Van Lawick-Goodall, Jane. 1968. "A Preliminary Report on Expressive Movements and Communication in the Gombe Stream Chimpanzees." In Jay 1968, 313–74.

Van Riper, A. Bowdoin. 1993. *Men among the Mammoths: Victorian Science and the Discovery of Prehistory*. Chicago: University of Chicago Press.

Van Wyhe, John. 2004. *Phrenology and the Origins of Victorian Scientific Naturalism*. Aldershot, UK: Ashgate.

———. 2005. "The Descent of Words: Evolutionary Thinking, 1780–1880." *Endeavour* 29:94–100.

Verne, Jules. 1901. *Le village aérien*. Paris: J. Hetzel.

———. 1964. *The Village in the Treetops*. Translated from *Le village aérien* by I. O. Evans. London: Arco Publications.

Voget, F. W. 1970. "Boas, Franz." In Gillispie 1970–80, 2:207–13.

Vogt, Charles. 1867. *Mémoire sur les microcéphales ou hommes-singes*. Geneva: L'institute National Genevois.

Von Mayrhauser, Richard T. 2000. "Edward Lee Thorndike." In Kazdin 2000, 8:80–83.

Vygotsky, Lev Semenovich. 1934. *Thought and Language*. Ed. and translated by Eugenia Hanfmann and Gertrude Vakar from the Russian original. Preface by Jerome Bruner. Cambridge, MA: MIT Press, 1962.

Wade, Nicholas. 1980. "Does Man Alone Have Language? Apes Reply in Riddles, and a Horse Says Neigh." *Science* 208:1349–51.

Wake, Charles Staniland. 1868. *Chapters on Man, with the Outlines of a Science of Comparative Psychology*. London: Trübner.

Wallace, Alfred Russel. 1870. "The Limits of Natural Selection as Applied to Man." In *Contributions to the Theory of Natural Selection*, 332–71. London: Macmillan.

Wallman, Joel. 1992. *Aping Language*. Cambridge: Cambridge University Press.

Walsh, J. 1999. "Harrington, John Peabody." In Garraty and Carnes 1999, 10:149–51.

Warden, C. J., and L. H. Warner. 1927. "The Development of Animal Psychology in the United States during the Past Three Decades." *Psychological Review* 34:196–205.

Waser, Peter M. 1975. "Experimental Playbacks Show Vocal Mediation of Intergroup Avoidance in a Forest Monkey." *Nature* 255:56–58.

———. 1976. "Individual Recognition, Intragroup Cohesion, and Intergroup Spacing: Evidence from Sound Playback to Forest Monkeys." *Behaviour* 60:28–74.

Washburn, Margaret Floy. 1908. *The Animal Mind: A Textbook of Comparative Psychology*. New York: Macmillan.

Washburn, Sherwood L. 1951. "The Analysis of Primate Evolution with Particular Reference to the Origin of Man." In *Ideas on Human Evolution: Selected Essays, 1949–1961*, ed. William Howells, 154–71. New York: Atheneum, 1967. First published in *Cold Spring Harbor Symposia on Quantitative Biology* 15:67–78.

———. 1952. "The Strategy of Physical Anthropology." In *Anthropology Today: An Encyclopedic Inventory*, ed. A. L. Kroeber, 714–27. Chicago: University of Chicago Press.

———. 1959. "Speculations on the Interrelations of the History of Tools and Biological Evolution." In *The Evolution of Man's Capacity for Culture*, arranged by J. N. Spuhler, 21–31. Detroit: Wayne State University Press, 1965. First published in *Human Biology* 31:21–31.

———. 1973. "The Promise of Primatology." In Strum, Lindburg, and Hamburg 1999, 43–48. First published in *American Journal of Physical Anthropology* 38:177–82.

———. 1983. "Evolution of a Teacher." *Annual Review of Anthropology* 12:1–24.

Washburn, Sherwood L., and Irven DeVore. 1961. "Social Behavior of Baboons and Early Man." In *Social Life of Early Man*, ed. Sherwood L. Washburn, 91–105. London: Methuen.

Washburn, Sherwood L., and Ruth Moore. 1980. *Ape into Human: A Study of Human Evolution*. 2nd ed. Boston: Little, Brown.

Watson, James D. 1970. *The Double Helix: A Personal Account of the Discovery of the Structure of DNA*. Harmondsworth: Penguin. First published in 1968.

Watson, John Broadus. 1907. *Kinesthetic and Organic Sensations: Their Rôle in the Reactions of the White Rat to the Maze*. Psychological Review, monograph suppl. 8:1–100.

———. 1908. Review of Pfungst on Clever Hans. *Journal of Comparative Neurology and Psychology* 18:329–31.

———. 1913. "Psychology as the Behaviorist Views It." Reprinted in *Modern Philosophy of Mind*, ed. William Lyons, 24–42. London: J. M. Dent, 1995. First published in *Psychological Review* 20:158–77.

———. 1914. *Behavior: An Introduction to Comparative Psychology*. New York: Henry Holt.

———. 1936. Autobiography. In *A History of Psychology in Autobiography*, ed. Carl Murchison, 3:271–81. Worcester, MA: Clark University Press.

Weber, G. 1993. "Henry Labouchere, *Truth*, and the New Journalism of Late Victorian Britain." *Victorian Periodicals Review* 26:36–43.

Webster, M. M., and Peter Marler. 1952. "A Contribution to the Flora of West Sutherland." *Watsonia* 2:163–79.

[Wedgwood, F. J.] 1862. "The Origin of Language: The Imitative Theory and Mr. Max Müller's Theory of Phonetic Types." *Macmillan's Magazine* 7:54–60.

Wedgwood, Hensleigh. 1859. *A Dictionary of English Etymology*. Vol. 1. London: Trübner.

———. 1866. *On the Origin of Language*. London: N. Trübner.

Weeden, Judith Stenger, and J. Bruce Falls. 1959. "Differential Responses of Male Ovenbirds to Recorded Songs of Neighboring and More Distant Individuals." *Auk* 76:343–51.

Weidman, Nadine. 1996. Review of Dukelow, *The Alpha Males*. *Isis* 87 (4): 756–57.

Weightman, John. 1997. "Cosmic Adventurer." *New York Review of Books*, 17 July, 53–56.

Wells, G. A. 1987. *The Origin of Language: Aspects of the Discussion from Condillac to Wundt*. La Salle, IL: Open Court.

West, R. 1972. *Brazza of the Congo: European Exploration and Exploitation in French Equatorial Africa*. London: Jonathan Cape.

White, Paul S. 2005. "The Experimental Animal in Victorian Britain." In Daston and Mitman 2005, 59–81.

Whitney, William Dwight. 1867. *Language and the Study of Language*. London: N. Trübner.

———. 1871. "Strictures on the Views of August Schleicher Respecting the Nature of Language and Kindred Subjects." *Transactions of the American Philological Association* 2:35–64.

———. 1873. *Oriental and Linguistic Studies: The Veda; The Avesta; The Science of Language*. New York: Scribner, Armstrong.

———. 1874. "Darwinism and Language." *North American Review* 119:61–88.

———. 1875a. "Are Languages Institutions?" *Contemporary Review* 25:713–32.

———. 1875b. "Nature and Origin of Language." In Harris 1996, 291–313. First published in *The Life and Growth of Language*. New York: D. Appleton.

———. 1876. "A Rejoinder." *Academy* 9:11–12.

———. 1892. *Max Müller and the Science of Language*. New York: Appleton.

Wiener, J. H. 1988. "How New Was the New Journalism?" In *Papers for the Millions: The New Journalism in Britain, 1850s to 1914*, ed. J. H. Wiener, 47–71. London: Greenwood Press.

Wiener, Jonathan. 1994. *The Beak of the Finch: A Story of Evolution in Our Time*. London: Jonathan Cape.

Wiener, Norbert. 1948. *Cybernetics, or Control and Communication in the Animal and the Machine*. Cambridge, MA: MIT Press. Reprinted, with additional material, in 1961.

———. 1950. *The Human Use of Human Beings: Cybernetics and Society.* Boston: Houghton Mifflin.

Wilson, David A. H. 2001. "A 'Precipitous *Dégringolade*'? The Uncertain Progress of British Comparative Psychology in the Twentieth Century." In *Psychology in Britain: Historical Essays and Personal Reflections,* ed. G. C. Bunn, A. D. Lovie, and G. D. Richards, 243–66. Leicester: BPS Books / Science Museum.

———. 2002a. "Animal Psychology and Ethology in Britain and the Emergence of Professional Concern for the Concept of Ethical Cost." *Studies in History and Philosophy of Biological and Biomedical Sciences* 33:235–61.

———. 2002b. "Experimental Animal Behaviour Studies: The Loss of Initiative in Britain 100 Years Ago." *History of Science* 40:291–320.

Wilson, Edward O. 1972. "Animal Communication." *Scientific American,* September, 53–60.

———. 1975. *Sociobiology: The New Synthesis.* Cambridge, MA: Belknap Press / Harvard University Press.

———. 1994. *Naturalist.* Washington, DC: Island Press.

Wilson, H. S. 1970. *McClure's Magazine and the Muckrakers.* Princeton: Princeton University Press.

Wilson, Leonard G. 1996. "The Gorilla and the Question of Human Origins: The Brain Controversy." *Journal of the History of Medicine and Allied Sciences* 51:184–207.

Witmer, Lightner. 1909. "A Monkey with a Mind." *Psychological Clinic* 3 (7): 179–205.

Wray, Alison, ed. 2002. *The Transition to Language.* Oxford: Oxford University Press.

Wright, Chauncey. 1870. "Limits of Natural Selection." *North American Review* 111:282–311.

———. 1873. "Evolution of Self-Consciousness." *North American Review* 116:245–310.

Wundt, Wilhelm. 1894. *Lectures on Human and Animal Psychology.* Translated by J. E. Creighton and E. B. Titchener, from the 1892 German 2nd ed. London: Swan Sonnenschein; New York: Macmillan.

Yerkes, Robert Mearns. 1901. "The Formation of Habits in the Turtle." Reprinted in *A Source Book in the History of Psychology,* ed. Richard J. Herrnstein and Edwin G. Boring, 544–51. Cambridge, MA: Harvard University Press, 1965. First published in *Popular Science Monthly* 58:519–25.

———. 1905. "Animal Psychology and Criteria of the Psychic." *Journal of Philosophy, Psychology, and Scientific Methods* 2 (6): 141–49.

———. 1906. "Objective Nomenclature, Comparative Psychology, and Animal Behavior." *Journal of Comparative Neurology and Psychology* 16:380–89.

———. 1907a. *The Dancing Mouse: A Study in Animal Behavior.* New York: Macmillan.

———. 1907b. Review of Watson, *Kinesthetic and Organic Sensations*. *Journal of Philosophy, Psychology, and Scientific Methods* 4 (21): 584–86.

———. 1910. "Psychology in Its Relations to Biology." *Journal of Philosophy, Psychology, and Scientific Methods* 7 (5): 113–24.

———. 1911. *Introduction to Psychology*. New York: Henry Holt.

———. 1913. "Comparative Psychology: A Question of Definitions." *Journal of Philosophy, Psychology, and Scientific Methods* 10 (21): 580–82.

———. 1914. "The Study of Human Behavior." *Science*, 1 May, 625–33.

———. 1916a. *The Mental Life of Monkeys and Apes: A Study of Ideational Behavior*. Behavior Monographs, vol. 3. New York: Henry Holt. Reprinted in facsimile, with an introduction by George M. Haslerud. Delmar, NY: Scholars Facsimiles and Reprints, 1979.

———. 1916b. "A New Method of Studying Ideational and Allied Forms of Behavior in Man and Other Animals." *Proceedings of the National Academy of Sciences of the United States of America*, 15 November, 631–33.

———. 1921. "A New Method of Studying the Ideational Behavior of Mentally Defective and Deranged as Compared with Normal Individuals." *Journal of Comparative Psychology* 1 (5): 369–94.

———. 1925. *Almost Human*. London: Jonathan Cape.

———. 1932. "Psychobiologist." In *A History of Psychology in Autobiography*, ed. Carl Murchison, 2:381–407. Worcester, MA: Clark University Press.

———. 1936. Review of Kohts, *Infant Ape and Human Child*. *Science* 83:466–67.

———. 1943a. *Chimpanzees: A Laboratory Colony*. New Haven: Yale University Press.

———. 1943b. "Early Days of Comparative Psychology." *Psychological Review* 50:74–76.

Yerkes, Robert Mearns, and Charles A. Coburn. 1915. "A Study of the Behavior of the Pig *Sus scrofa* by the Multiple Choice Method." *Journal of Animal Behavior* 5:185–225.

Yerkes, Robert Mearns, and Blanche W. Learned. 1925. *Chimpanzee Intelligence and Its Vocal Expressions*. Baltimore: Williams and Williams.

Yerkes, Robert Mearns, and S. Morgulis. 1909. "The Method of Pavlov in Animal Psychology." *Psychological Bulletin* 6:257–73.

Yerkes, Robert Mearns, and Henry W. Nissen. 1939. "Pre-linguistic Sign Behavior in Chimpanzee." Reprinted in *Language Intervention from Ape to Child*, ed. Richard L. Schiefelbusch and John H. Hollis, 145–51. Baltimore: University Park Press, 1979. First published in *Science* 89:585–87.

Yerkes, Robert Mearns, and John B. Watson. 1911. *Methods of Studying Vision in Animals*. Behavior Monographs 1 (2).

Yerkes, Robert Mearns, and Ada W. Yerkes. 1929. *The Great Apes: A Study of Anthropoid Life*. New Haven: Yale University Press.

Young, Emma. 2004. "The Beast with No Name." *New Scientist*, 9 October, 33–35.

Young, John Z. 1951. *Doubt and Certainty in Science: A Biologist's Reflections on the Brain.* BBC Reith Lectures for 1950. Oxford: Clarendon Press.

———. 1953. "Discrimination and Learning in Octopus." In *Cybernetics: Circular Causal and Feedback Mechanisms in Biological and Social Systems,* ed. Heinz von Foerster, 109–19. New York: Josiah Macy Jr. Foundation.

———. 1954. "Memory, Heredity, and Information." In *Evolution as a Process,* ed. Julian Huxley, A. C. Hardy, and E. B. Ford, 281–99. London: George Allen and Unwin.

———. 1955. "The Influence of Language on Medicine." In Ayer et al. 1955, 91–107.

Young, Robert M. 1990. *Mind, Brain, and Adaptation in the Nineteenth Century: Cerebral Localization and Its Biological Context from Gall to Ferrier.* Oxford: Oxford University Press. First published in 1970.

Zhinkin, N. I. 1963. "An Application of the Theory of Algorithms to the Study of Animal Speech: Methods of Vocal Intercommunication between Monkeys." In Busnel 1963, 132–80.

Zimmerman, Andrew. 2001. *Anthropology and Antihumanism in Imperial Germany.* Chicago: University of Chicago Press.

Zoloth, Stephen R., Michael R. Peterson, Michael D. Beecher, Steven Green, Peter Marler, David B. Moody, and William Stebbins. 1979. "Species-Specific Perceptual Processing of Vocal Sounds by Monkeys." *Science* 204:870–73.

Zuckerman, S. 1932. *The Social Life of Monkeys and Apes.* London: Kegan Paul.

Index

Page numbers in italics refer to figures. Unless otherwise indicated, "Darwin" refers to Charles Darwin, "Garner" refers to Richard Lynch Garner, "Marler" refers to Peter Marler, "Morgan" refers to Conwy Lloyd Morgan, and "Yerkes" refers to Robert Mearns Yerkes.

AAAS. See American Association for the Advancement of Science

Aaron (chimpanzee), 135, 144, 416n87

Abercrombie, John, 60, 394n41

Abingdon (Va.), 87–88, 188

Abreu, Rosalie, 188, 233, 235

Academy, 48–49

Adamic origin of language, 119

Admiralty Research Laboratory, Teddington, 265–66

affective communication: as grading into symbolic communication, 329 (Marler's theory); as informationally sufficient when preferences are concordant (Premack's theory), 327. *See also* emotional expression; symbolic communication

Africa: as origin of *Homo sapiens*, 374–75. *See also* Amboseli; Congo; French Congo; Gombe Stream Chimpanzee Reserve; South Africa; Uganda

Africa Fund of Chicago, 417n97

African Society, 160, 168–69, 423n35

alarm calls: of chimpanzee according to Garner, 415n78; as depicted by the sound spectrograph, 268; distinguished from vocalizations mediating social interactions, 340; as examples of cooperative animal behavior, 331, 340; as first words according to Darwin, 116, 389n99;

Marler's analysis of, 267–70, 278, 295, 307–10, 328–29. *See also* vervet alarm calls

Alexander, Caroline, 411n28–29

Allen, A. A., 453n61

Allen, Grant, 95

Allen, Woody, 324

Altenberg. *See* Austria

Alter, Stephen, 387–88n82

Altmann, Jeanne: advises the Seyfarths, 333; "Observational Study of Behavior," 333, 482n61; with Struhsaker in Amboseli, 305–6, 342

Altmann, Stuart: as pioneer in applying information theory to animal communication, 461n126; primate vocalization research, 300–301, 305–6, 342; publishes with Hockett, 468n60, 472–73n107; with Struhsaker in Amboseli, 305–6, 342

Amboseli, 10–11, 305–6, 309, 336, 342–52, 378

American Anthropologist: *American Journal of Physical Anthropology* founded as reaction to, 430n140; as a Boasian journal, 430n140; Hale's review of *The Speech of Monkeys* in, 115–16; Kroeber's "The Superorganic" in, 195–96

American Association for the Advancement of Science, 176–81

American Indian languages: Bloomfield surveys, 195; Boas's *Handbook of American Indian Languages*, 194–95, 429n130; Garner's interest in, 172–75, 195; Harrington and, 160; recordings at the Smithsonian 166. *See also* Amerindians

American Journal of Physical Anthropology: founded in reaction to *American Anthropologist*, 430n140; Garner's letter published in, 186; Garner's obituary in, 188; Hrdlička as founder and editor, 184–85, 186, 195, 430n140

American Museum of Natural History, 183–84, 191, 452n49

American Philosophical Society, 113, 227

American Sign Language. *See* sign language

Amerindians, 8, 257, 287–88, 402n21. *See also* American Indian languages

Amsterdam Zoo, 254

Amtsberg's pike: Darwin's response to, 65–66, 395–96n74, 396n80; Möbius's description of, 64–65; Müller's discussion of, 64–68, 395–96n74, 396n87; other responses in the nineteenth century, 66–68; responses in the twentieth century, 396n87

anatomy, study of human and nonhuman, 15, 32–33, 50–64, 78–81, 143, 162–66, 176–93, 199–200, 296–98, 310–13, 369–76. *See also* Boas, Franz; Dubois, Eugène; Gall, Franz Joseph; Geschwind, Norman; Hooten, Earnest; Hrdlička, Aleš; Huxley, Thomas Henry; Mivart, St. George Jackson; Montagu, Ashley; Owen, Richard; Schultz, A. H.; Virchow, Rudolph; Vogt, Carl; Washburn, Sherwood; Young, J. Z.

Anderson, Stephen, 496n36

Ancon sheep, 369–70

Andrew, Richard, 301

Anglican Church, 18, 20

Anglin, Jeremy, 359

angular gyrus, 311–12, 477n8

animal ecology, 245, 258, 453n56

animal mind. *See* Amtsberg's pike; comparative psychology; ethology; Morgan's canon

anthropocentrism and its opponents, 9–10, 105, 232, 255–57, 259, 293, 307, 313–22, 362–64, 366–68

anthropology, xii, 7–8, 20, 24, 27–29, 37, 49, 50–58, 112–16, 159–98, 287–93, 296–301, 310–12, 361, 368–72, 379, 388n91, 393n24. *See also American Anthropologist*; *American Journal of Physical Anthropology*; anatomy, study of human and nonhuman; anthropometry; Ascher, Robert; Boas, Franz; Brinton, Daniel; Broca, Pierre Paul; De Laguna, Grace; DeVore, Irven; École d'anthropologie, Paris; ethnology; Fewkes, Jesse Walter; Galton, Francis; Haddon, Alfred Cort; Hale, Horatio; Harrington, John Peabody; Hill, Andrew; Hockett, Charles; Hrdlička, Aleš; Hunt, James; Kroeber, Alfred; Lancaster, Jane; Lombroso, Cesare; Manouvrier, Léon; Mead, Margaret; Montagu, Ashley; Müller, Friedrich Max; new physical anthropology; Rivers, W. H. R.; Schwidetsky, Ilse; Société d'anthropologie, Paris; Torres Straits expedition; Tylor, Edward Burnett; Virchow, Ruldoph; Washburn, Sherwood; Wenner-Gren Foundation

anthropometry: Boas and, 371–72, 430n140, 431n4, 495n21; Galton and, 114, 372; Hooten and, 426n90; Hrdlička and, 183–84, 430n140; *Hrdlička's Practical Anthropometry*, 184, 443n126; in medicine, 425n83; in physical anthropology, 162–63, 298, 470n77; Seligmann and, 163

anthropomorphism and its opponents, 5–6, 50–53, 64–78, 81–83, 117–18, 199–215, 219, 225, 210–11, 221, 225, 359–60, 382n13. *See also* Amtsberg's pike; Griffin, Donald; Morgan's canon

ants: in the French Congo, 145; Garner offers specimens to Smithsonian, 130; and information theory, 461n126; Lubbock researches, 77, 433n23; Morgan and Müller discuss, 77; and Schneirla, 465n18; and Wilson, 300

ape language projects, 10–11, 313–21, 325–29, 358, 362–64, 366–68, 376, 378,

491n148. *See also* Lana; Nim Chimpsky; Sarah; Washoe

apes. *See* baboons; capuchin monkeys; *Cebus* monkeys; chimpanzees; diana monkeys; gibbons; gorillas; Japanese macaques; langurs; macaques; mangabeys; orangutans; spider monkeys; vervet alarm calls

aphasia: Bateman on, 56–63; Broca, on, 54–56, 79–80; *Darwinism Tested by Language*, 63, 399n130; Darwin on, 60–62; in debates about cerebral localization, 54–63, 79–80; Evans on, 80–81; Geschwind on, 310–11, 475n126; Jackson on, 50–51, 56, 62, 78–79; motor aphasia, 79–81; *On Aphasia*, 51, 56–58, 60, 62, 80, 392n1, 399n130; Pinker on, 496n35; sensory aphasia, 79–80; Viki compared to sufferers of, 314

Applied Psychology Research Unit, Cambridge, 270

arbitrariness. *See* properties of language

archaeology: anthropology and, 184, 385n41; Brinton and 113, 160; ethnology and, 184; and fossils, 182; as furnishing evidence of early man's lowly state, 27–28; Haddon and 160, 162–63

Armstrong, Edward, 460n119

Arnold, Matthew, 139

articulate ideas (Hobhouse's term), 214, 440n103

Aschemeier, Charles, 185–87

Ascher, Robert, 468n60

Asiatic Primate Expedition, 298, 446n457, 447n160

association areas (in the brain), 311

Association for the Study of Animal Behaviour, 275–6. *See also* Institute for the Study of Animal Behaviour

associative learning: Amtsberg's pike and, 65–66; and comparative psychology, 201, 204, 206–7, 209–10, 217, 376, 431–32n11; Darwin on, 31, 65–66; evolutionism and, 75–76; Garner on, 103; James on, 81; Morgan and, 75–76, 202, 207; and sensory association area in the brain, 311–13; Thorndike and, 201, 204, 206–7, 209–10, 217; and trial-and-error learning,

200; Wundt on, 433n28; Yerkes and 217, 221–23, 230. *See also* Geschwind, Norman; trial-and-error learning

astronomy, 26, 199

Athenaeum, 15, 116, 127

Atlantic Monthly, 73, 80

Australia, 6, 8, 28, 162, 175, 370

Austria: B. Gardner from, 316; Lorenz's home laboratory in Altenberg, 254–55, 258; 1965 Wenner-Gren symposium in, 307, 310, 313

Ayer, A. J., 271

Azores, 243, 245–48, 261, 451n38, 458n99

BAAS. *See* British Association for the Advancement of Science

baboons: the Altmanns' social behaviour project with, 305; DeVore observes, 298–300, 496n39; Lee and, 486n94; the Seyfarths' research on, 11, 332–36, 341–42; vervet alarm call for, as identified by Struhsaker, 306, 485n89; Washburn observes, 298–99

Bailey, James, 108

Baker, Frank, 93–94

Baldwin effect, 451n35

Balfour, F. M., 421n19

Balloon Society of Great Britain, 410n14

Barnett, S. A., 453n51

Bastian, Adolf, 372

Bateman, Frederic, 55; on aphasia, 51, 56–63, 80; on cerebral localization, 5, 51, 52–53, 56–61, 78–80; on evolution, 5, 51, 58–63, 78–79; influences on the work of, 55–59; materialism of, 61–63

Baume, Robert, 179

BBC, 166, 267, 274

Beach, Frank, 201, 299

Beer, Thomas, 218, 439n92

bees, dance language of: Birch on, 461n126; as exception to the rule that symbolic communication is humans-only, 1; Frisch on, 1, 273–74, 288–89, 359; J. Gould and, 360; Griffin on, 359; Haldane on, 273–74, 285; Hockett on, 289–92; Kroeber and, 288–89; Lindauer on, 273–74; Marler on, 285–86 behavioral sciences. *See* anthropology; comparative psychology; cybernetics; ethology; primatology

behaviorism, 8–9, 52, 195, 200, 210, 215, 219–26, 236–38, 256–58, 275, 288–89, 313–19, 323–24, 376–77, 435–56n57, 491n152

Behaviour, 258, 283–84

Bell, Alexander Graham, 93, 111

Bell, Alexander Melville, 88–90, 93, 107–11, 143–44, 264

Bell, Peter, 284–85

Belle Vue Zoological Gardens, 198, 211

Bell Telephone Laboratories (Bell Labs), 263–65, 270–71, 282

Berkeley, University of California at, 10, 172, 195, 282–84, 295–96, 299, 303, 310, 361, 452n44, 454n67, 469n72, 492n159

Berlin: Academy of Science, 24; Clever Hans in, 211; Congress of, 130; Deutsche Gesellschaft für Tierpsychologie in, 451n41; Harrington in, 172, 174–75; Heinroth in, 254; Müller in, 19; Stresemann in, 457n97; Von Holst in, 254; Whitney in, 42

Bethe, Albrecht, 218, 439n92

Betjemen, John, 244

Bible, the, 20, 68, 158, 384n29

biblical anthropology, 20, 24, 37

Bibliothèque national, 154

Bickerton, Derek, 374–78

bioacoustics, 265, 284, 336–37, 455n81

biology: as furnishing useful training to the psychologist according to Yerkes, 217; as irrelevant to understanding human language and culture on the Boasian view, 161, 189–98; as unwelcoming of racial prejudice after Second World War, 368, 373

biophysics, 288

Birch, Herbert, 461n126

birds. *See* chaffinches; robins

birdsong, 9–10, 36, 166, 243–78, 282–86, 293–96, 303, 305, 329, 360, 366, 478–79n19

Blackwood's Edinburgh Magazine, 123–24

Blanshard, Brand, 379–80

Bloom, Paul, 377

Bloomfield, Leonard, 195, 287–90, 323

Boas, Franz, *192*; case against primitive language, 190–95; cultural anthropology as focus for many of his students, 166, 379; and Hale, 190–91, 372; as influence on Bloomfield's linguistics, 287–88; as Kroeber's teacher, 172; as Montagu's teacher, 495n24; opposed to evolutionist race ranking, 8, 161, 189–90; saltationism of Darwin centenary lecture (1909), 371–73

Boasianism, 8, 189–90, 195, 379. *See also* Bloomfield, Leonard; Boas, Franz; Hockett, Charles; Kroeber, Alfred; Mead, Margaret; Montagu, Ashley; Sapir, Edward

Boatman, Derrick, 243, 247–48

Bopp, Franz, 19–20, 383n13, 384n23

Borges, Jorge Luis, 380

Borneo, 164, 228

Borror, Donald, 265, 457n89

Boston Psychiatric Hospital, 218, 440–41n108

botany, 244–45, 253, 342, 449n14. *See also* Bell, Peter; Boatman, Derrick; Marler, Peter; Pearsall, W. H.

Bouillaud, Jean Baptiste, 54

Boule, Marcellin, 185

Boutan, Louis: gibbon language research, 166, 226–27, 230, 232, 285; influences on, 166, 281; and Pépée 226–27; use of the phonograph, 166, 226–27, 230

Bowlby, John, 300

bow-wow theory (of origin of language). *See* onomatopoeic theory

brains, animal and human, 5, 15–16, 53–54, 59, 62, 78, 80, 113, 228, 310–14, 323, 478n18. *See also* angular gyrus; Broca's convolution; hippocampus minor; limbic system

Brave New World (A. Huxley), 319

Brehm, Alfred Edmund, 118

Brinton, Daniel, 113, 115, 162, 190

British Association for the Advancement of Science: animal intelligence discussed at, 397n93; cerebral localization discussed at, 6, 56; funds Boas's fieldwork, 190–91; Garner and, 83, 117, 126–28, 139, 167; Harrington writes for, 172; Morgan and, 81–83, 117, 127, 433–34n32; Müller and,

20, 49, 126; Prichard and, 20; publishes *Notes and Queries*, 172; sponsors Ethnographic Survey of the United Kingdom, 163

British Museum, 154

Broadbent, Donald, 267–70, 275, 282

broadcast transmission and directional reception. *See* design features of language

Broca, Pierre Paul: on aphasia, 54–56, 79–80; influence of, 55–59, 183. *See also* Broca's convolution

Broca's convolution, 51, 54–58, 59, 79–81, 115–16, 179–80, 398–99n134

Bronx Zoo: Ditmars, R. as curator of, 166, 230; Garner collects specimens for, 167; and New York City Zoological Society, 304; primate research at, 221, 301, 443n125; Yerkes aquires chimps through, 230

Brooklyn Daily Eagle, 116, 414n64, 415n63

Brooks, W. K., 493–94n8

Buckland, William, 18

Buckman, S. S., 413n51

Budongo Forest. *See* Uganda

Buléon, Joachim, *150*; betrayal of Garner, 147–58; and Garner, 133, 141, 142, 148–53

Bureau of American Ethnology, 92, 166, 172, 193, 402n21

Butler, Samuel, 84

Cabanis, Pierre Jean George, 69

Cambridge University, 159–60, 162–66, 244, 251, 253, 256–58, 260–63, 266–67, 269–70, 275, 283–84, 301, 340–41, 420–21n13, 484n77. *See also* Madingley Ornithological Field Station

Canary Islands, 198, 222, 246–48. *See also* Köhler, Wolfgang

Candland, Douglas Keith, 442n116

cannibals, 130, 135

Cannon, Walter B., 227, 461n127

capuchin monkeys: Garner's research on, 99, 101, 104, 117, 381n6; "Spud," 147; Watson and, 232

Carpenter, Clarence Ray: De Laguna and, 236–38; impact of, 300, 305–6; playbacks among gibbons of Siam, 237, 237–39, 284–85, 298; researches howler monkeys

in Panama, 235–36, 238–39, 301; training of, 235–36

Carr, Harvey, 435n55

Carus, Paul, 72

catastrophism, 17–20, 31–33, 39–40, 114, 496n34–35. *See also* punctuated equilibria, theory of; saltationism

Cayo Santiago, 300, 471n89

Cebus monkeys, 209, 444n136

Center for Advanced Studies in the Behavioral Sciences, Stanford, 288, 299, 301

Central Park, New York City, 96–99, 112, 405n91

cerebral localization of language, 5, 50–64, 78–81, 310–13. *See also* brains, animal and human

Cervantes, Miguel de, 380

chaffinches, 246–52, 259, 262, 267–70, 275–79, 282, 284–85, 294–99

Chambers, Robert, 29–30, 33, 383n3

Chantek (orangutan), 320

Chapman, Frank M., 446n150, 453n61

Cheney, Dorothy. *See* Seyfarth, Robert and Dorothy

Cherfas, Jeremy, 361

Cherry, Colin, 271, 275, 286–87, 294–95

Chim (Yerkes's chimpanzee), 227, 229–31, 235, 461n127

chimpanzees: Garner and, 85, 106–7, 109, 115–16, 118, 134–35, 140, 142–44, 152, 154, 167, 186, 188, 405n94, 415n78, 415n83–84, 416n89, 417n96; Goodall observes at Gombe Stream, 300, 303–4, 325, 331–32; Hobhouse's experiments with, 211–12; Köhler's experimental psychology with, 214–15; Kohts on facial expressions of, 285, 443n123; kulu kambas (most gregarious chimpanzees according to Garner), 144; in Marler's research, 285, 300–304, 325–26; in Menzel's research, 479–80n28; Premack and Schwartz's joystick-based research with, 315; the Reynolds observe in Uganda, 300, 471n90; Romanes's number experiments with, 117. *See also* Aaron; Chim; Elishiba; Gua; Lana; Moses; Nim Chimpsky; Panzee; Peter; Sally; Sarah; Susie; Viki; Washoe

choice chamber, 9. *See also* multiple choice apparatus
Chomsky, Noam, 320–25, 374, 377–80
Church, F. S., 98, 109
Churchill, Winston, 276
Clark University, 200, 431n4
Cleveland, Grover, 108–9
Clever Hans, 211, 229, 320, 435n54, 477n3
Clockwork Orange, A (Burgess), 319
Coetzee, J. M., 437n71, 442n119
cognitive ethology, 2, 322, 359–60, 490n137
cognitive science, 287, 358, 478n11, 490n142. *See also* communication theory; cybernetics; information theory
Colburn, Zerah, 369–70
Collias, Nicholas, 265, 284, 458n98, 472n99
Collins, Alfred. M., 185–86
Collins-Garner expedition, 185–87
colobus monkeys, 303, 337–40, 343, 361
Columbia Graphophone Company, 93
Columbia Phonograph Company, 402n22
Columbia University, 11, 113, 191, 195, 198, 200, 206–8, 221, 237, 296, 320, 371–72, 470n77
communication. *See* communication theory; descriptive communication; emotional expression; evocative communication
Communication Research Centre, London, 271–75
communication theory, 270–79, 285–95, 309–10. *See also* cognitive science; cybernetics; information theory
comparative anatomy. *See* anatomy, study of human and nonhuman
comparative musicology, 166, 239
comparative philology, 19–20, 28, 40, 52, 58, 70–73, 78, 383n12, 388n86, 390n112, 401n16. *See* Bopp, Franz; Müller, Friedrich Max; Schleicher, August; Whitney, William Dwight
comparative psychology 5–6, 8–10, 50–53, 64–83, 103–5, 118, 127, 197–239, 245, 255–57, 259, 276, 283, 293, 313–22, 325–29, 358, 362–64, 367–68, 378, 410n18. *See also* anthropocentrism and its opponents; anthropomorphism and its opponents; Beach, Frank; Carpenter,

Clarence Ray; Carr, Harvey; Craig, Wallace; Darwin, Charles; Gardner, Allen and Beatrice; Haggerty, Melvin; Hamilton, Gilbert; Hayes, Keith and Catherine; Herrnstein, Richard; Hobhouse, Leonard; *Journal of Animal Behavior; Journal of Comparative Neurology and Psychology; Journal of Comparative Psychology*; Kinnaman, Andrew; Köhler, Wolfgang; Kohts, Nadia; Lehrman, Daniel; Menzel, Emil; Mills, Wesley; Morgan, Conway Lloyd; Morgan's canon; Premack, David; puzzle box; Romanes, George John; Schneirla, T. C.; Scott, John Paul; Skinner, B. F.; Terrace, Herbert; Thorndike, Edward Lee; Tolman, Edward; Washburn, Margaret Floy; Watson, John B.; Yerkes, Robert Mearns
complexity theory, 493n3
Comte, Auguste, 26–27
Congo, 300, 422–23n34. *See also* French Congo
Contemporary Review, 45, 47, 49, 69
Coolidge, Harold, 238
Cornell University, 218, 288
Cornish, C. J., 408n143, 417n98
Cosmopolitan, 96, 402n19
Cosmos, 155
Craig, Wallace, 236–37, 254
craniofacial angle (Garner's measurement), 99
creole, 374–76
Critic, 116–17
Cross, Dr. (animal dealer), 144
cryptography, 271
Cuba. *See* Abreu, Rosalie
cultural anthropology, 7, 161, 189–90, 197–98, 371, 369, 379. *See also* anthropology; ethnography; ethnology
cultural relativism, 8, 168–70. *See also* Boas, Franz; Boasianism
cultural transmission. *See* properties of language
Cuvier, Georges, 299
cybernetics: S. Altmann and, 300; and communication, 274; etymology of, 458n103; Haldane and, 271–74; Kalmus on, 274; Lorenz on, 463n146; Macy Foundation

conferences on, 274–75; Pringle and, 462n135; Wiener and, 270–72, 274, 294, 469n67; Young on, 274–75. *See also* cognitive science; communication theory; information theory

cylinder phonograph. *See* phonograph

Daily News, 417n96

Darwin, Charles, *30*; accused by Müller of dogmatism, 41; on aphasia, 60–62; and Bateman, 51, 58, 60–61, 63; and Boas, 194, 371–72; and Brinton, 113; on cerebral localization of speech, 58, 78; and debate over anthropomorphism in science (1870s), 65–69; debate over the *Origin* as context for Müller's 1861 lecture, 15–16; defends linguistic hierarchy among human races against creationist criticism, 37–38; and Garner, 86, 89, 96, 104; and T. H. Huxley, 32–33; linked to British empiricist tradition by Müller, 40; linked to British saltationist tradition by Hale, 115; Marler on, 285–86, 322; meets Müller, 45–46; pre-*Descent* reflections on language origins, 29, 31–32; prophesy in *Origin*, 380; provokes Müller's "Lectures on Mr. Darwin's Philosophy of Language," 39; relation to the Wedgwoods, 25; response to Müller, 4, 31–32, 37, 41, 43, 45–48, 65–66; and Romanes, 70–72, 258; theoretical writings on animal communication and the origin of language, 35–39, 60–61, 389n99; and uniformitarianism, 18–19, 376; and the *vera causa* doctrine, 27; views on embryology linked to origin-of-language question, 46; views on link between arousal and vocal pitch recalled, 388; and Vogt, 57–58; and Whitney, 43–48; and Wright, 34–35, 38, 43–44, 81. *See also* Darwin centenary; *Descent of Man*; *Expression of the Emotions in Man and Animals*; *Origin* centenary; *Origin of Species*

Darwin, George, 47–48, 50

Darwin centenary (1909), 199, 228–29, 371–72. See also *Origin* centenary

Darwinians, on language origins (1860s), 32–35

Davenport, Charles, 438n81

De Gids, 116

Degler, Carl, 195

De Laguna, Grace, 236–38

De Mortillet, Gabriel, 178–79

Dennett, Daniel, 2, 491n152, 493n166

Densmore, Frances, 166

Descent of Man (Darwin), 16, 29, 32, 35, 39, 43, 45, 47, 51, 57, 60–61, 65–66, 68–69, 96, 104, 322, 383n6, 387n73, 387n75, 389n99, 392n1, 393–94n27, 397n92

descriptive communication: Haldane on, 272–74; Marler on, 279, 285; Morgan on, 203

designators (semiotic category), 294, 447n166

design features of language (according to Hockett), 291–93. *See also* properties of language

Deutsche Gesellschaft für Tierpsychologie, 451n41

DeVore, Irven, 296, 298–300, 330, 471n90–91

Dewey, John, 371

Dewsbury, Donald, 431–32n11, 452n43

Dial, 116, 118

dialectical regeneration (Müller's theory of), 24, 385n32

diana monkeys, 378

ding-dong theory (of origin of language). *See* vanished-instinct theory

discreteness. *See* design features of language

disk recorder, 237

displacement. *See* properties of language

Ditmars, Raymond, 166–67, 230

Dixson, A. F., 416n85

Dobzhansky, Theodosius, 296, 450–51n33, 495n24

Donaldson, J. W., 20

Don Quixote, 380

Draper, John W., 126, 153

duality. *See* properties of language

Dublin Anthropometric Laboratory, 162, 420n10

Dubois, Eugène, 181–82

Du Chaillu, Paul, 15–16, 107, 130, 387n67

Duke of Argyll, 49
Dutch East Indies, 181
Dybowski, Jean, 141–42, 157, 414n74

Eastern Daily Press, 62
East India Company, 19–20
École d'anthropologie, Paris, 178, 183
ecologists. *See* Boatman, Derrick; Gause,
    G. F.; Lack, David; Marler; Pearsall,
    W. H.; Southern, Mick
Edison, Thomas: the Edison Phonograph,
    83, 90–92, 105, 109, 128–29; Edison
    Phonograph Works, 111; Edison United
    Phonograph Company, 109, 112; Garner
    and, 109–12, 128–29, 133, 188; predic-
    tions for the phonograph, 111; research at
    Menlo Park, 90. *See also* phonograph
Edward Grey Institute for Field Ornithology.
    See Oxford University
Eldredge, Niles, 374
Elishiba (Garner's chimpanzee), 135, 144
embryology, 46, 162, 272, 421n19. *See also*
    Balfour, F. M.; Haddon, Alfred Cort
emotional expression: Darwin's explana-
    tory principles of, 285; Geschwind on,
    310–12; as governed by the limbic system,
    312–13; Green on, 340; Itani on, 326;
    Jackson and, 79, 312; Kroeber on, 196;
    Lancaster on, 310–12; Lorenz on, 278;
    Magoun on, 313; Marler on, 279, 313,
    325–26, 328–29, 375; J. Marshall on,
    322–23; Morgan on, 117, 203; Premack
    on, 10, 327–28; Robinson on, 313; Sapir
    on, 196–97; Tinbergen on, 277. *See also*
    affective communication; evocative com-
    munication; *Expression of the Emotions
    in Man and Animals*; vervet alarm calls
empiricism, 20, 40
Encyclopaedia of the Social Sciences, 197
Engels, Friedrich, 391–92n144, 460n121
Eshira tribe, 141, 144, 151–52, 155, 157,
    416n86
Eskimos, 193–94, 429n132
Esthwaite Water, 245, 247–48
Ethnographic Survey of the United Kingdom,
    163
ethnography: Brinton and, 113, 162; defini-

tion of, 162; Ethnographic Survey of the
    United Kingdom, 163–64; Garner and, 8,
    135, 160, 167–70; Japanese primatology
    and, 337
ethnology: anthropology as, 184; Brinton on,
    162; Bureau of American Ethnology, 92,
    166, 172, 193; Harrington and, 172; Mül-
    ler and 20, 28; and the origin of humanity,
    369; Prichardian ethnology, 24; use of the
    phonograph in, 93, 166, 239. *See also* cul-
    tural anthropology; Haddon, Alfred Cort;
    Wake, Charles Staniland
ethology, xii, 2, 9–11, 234, 236–37, 245,
    252–86, 289, 293–96, 303–8, 316, 322,
    325–64, 366–67, 452n43, 490n137. *See
    also* Association for the Study of Animal
    Behavior; *Behaviour*; cognitive ethology;
    Frisch, Karl von; Gardner, Beatrice; Hal-
    dane, J. B. S; Hinde, Robert; Lack, David;
    Lorenz, Konrad; Manning, Aubrey; Mar-
    ler, Peter; Morris, Desmond; Schwidetzky,
    Georg; Seyfarth, Robert and Dorothy;
    Smith, W. John; Thorpe, William Ho-
    man; Tinbergen, Nikolaas; *Zeitschrift für
    Tierpsychologie*
eugenics: Haldane and, 250–51, 272;
    Hrdlička on, 184–85; opposition to, 195,
    250–51, 258, 372; Thorndike supports,
    435n47
Evans, E. P.: on aphasia and language, 80–81;
    on comparative philology, 73; on com-
    parative psychology, 73; discusses work
    of Brehm, 118; on Garner, 3, 118, 146,
    156, 417n98; on the language barrier, 80;
    on the phonograph, 3, 118; on the roots
    of language, 73; "Speech as a Barrier Be-
    tween Man and Beast," 73, 80
Evans, Ifor, 271
Evidence as to Man's Place in Nature (T. H.
    Huxley), 33, 178, 289, 380
evocative communication, 273–74, 279, 285
evolutionary theory. *See* catastrophism;
    Darwin, Charles; Dobzhansky, Theodo-
    sius; Galton, Francis; Gould, Stephen Jay;
    Gray, Asa; Hamilton, W. D.; Hobhouse,
    Leonard; Huxley, Julian; Huxley, Thomas
    Henry; Mayr, Ernst; Morgan, Conway

Lloyd; saltationism; Simpson, George
Gaylord; Smith, John Maynard; Spencer,
Herbert; Wallace, Alfred Russell; uniformitarianism
experimental psychology. *See* comparative
psychology; Torres Straits expedition
*Expression of the Emotions in Man and
Animals* (Darwin), 38–39, 60, 72, 277,
387n75, 388n91, 388n92

Falls, J. Bruce, 284, 303, 307
Fan, the. *See* cannibals
Faraday, Michael, 91–92, *92*
Farrar, Frederic, 28, 36, 95
Fernan Vaz, 130, *131*, 133–58, 167–70,
185–89, 232
Fernan Vaz testament (Garner), 7, 126, 133,
147–58
Ferrier, David, 62, 78, 312
Fewkes, Jesse Walter, 92–93
field experiments: as characteristic techniques
of Tinbergian ethology, 254; as combining
moral and epistemic authority, 364, 366,
368; as unusual within primatology, 336
Fitch, W. Tecumseh, 378
Fitzpatrick, Simon, 382n13
Fletcher, Alice Cunningham, 166
Flourens, Marie-Jean-Pierre, 53–56
focal animal sampling, 333
Fodor, Jerry, 382n13
Fort Gorilla (Garner's cage), 106–7, 111,
123–24, 134, 136–37, 139–42, 144–46,
145, 150–51, 156–57, 366, 409n6,
411n19, 411n22, 412–13n46, 413n53,
416n93, 419n131, 419n128
Forum, 96, 112–13, 129
Fossey, Dian, 282, 481n44
fossils: Cro-Magnon, 7, 176–82; found in
Germany, 176, 182; found in La Quina,
182; Garner on, 187; Java man, 7, 161,
176, 425n74, 425n78; "missing link,"
181, 200, 395n58; Neanderthal fossil
found in Schipka cave, 179, 424n69; Neanderthal found in La Naulette, 178–79
Fouts, Roger, 321
Fraser's Magazine, 41, 396n83
Frederic, Harold, 127–28, 146–47, 156

French Congo: administrative formalities of,
131–32; Congress of Berlin as establishing governance of, 130; different groups
in, 130–33, 411n29; Dybowski, Jean in,
141–42, 157, 414n74; Garner's arrival
in, 129–30; Garner's research in, 6–8,
123–26, 134–60, 167–70, 185–86; as
represented before Garner's first expedition, 85, 98, 105–8; trade in, 132,
412n33, 412n42. *See also* Du Chaillu,
Paul; Fernan Vaz; St. Anne's Mission
Frisch, Karl von: awarded Nobel Prize, 273;
on dance language of bees, 1, 273–74,
288–89, 359; and ethology, 273; influence of, 273–74, 288–89; at Paris Instinct
Conference, 276
Fritsch, Gustav, 62, 78
Fry, D. B., 271
Fuegians, 6, 62, 114, 370, 395n58
Furlong, Charles, 186
Furness, William. H. (referred to as "Hor-
ace"), 227–29, 233

Gabon. *See* French Congo
Gaboon. *See* French Congo
Galapagos Islands, 246, 448nn8–9, 449n16
Gall, Franz Joseph, 53–54, 78. *See also* Phre-
nology
Gallen, Judith, 276, 283, 342, 484–85n84
Galoi, the, 135
Galton, Francis, 49, 114–15, 180, 369, 372
Gandhi, Mahatma, 174
Gardner, Allen and Beatrice, 10, 316–21, 325
Gardner, Beatrice ("Trixie"), 316
Garner, Harry, 174, 233–34
Garner, Richard Lynch, *87*, *98*; and Carpen-
ter, 239; difficulties with British phono-
graph firm, 128–29; early life, education,
and career, 87–90; and Edison, 109–12,
128–29, 133, 188; as ethnographer,
167–70; fails to appear at BAAS meeting
(1892), 126–27; first expedition to Africa
and controversies following, 129–58;
Hale on, 115–16; and Harrington, 8, 161,
171–76; and Hrdlička, 8, 161, 185–89;
ideas about language and reason, 99–100,
104; as inventor of the primate playback

Garner, Richard Lynch (*cont.*)
  experiment, 2–4, 364, 366–68, 374,
  380; on linguistic hierarchy in evolved
  nature, 6–8, 99, 104–5, 366; linked to
  new comparative psychologists (1909),
  199–202; and McClure, 95–96, 108, 129,
  131; Morgan's response to, 82–83, 117,
  202–4; and Müller, 89, 100, 112; as outsider to elite debates, 86, 104, 366; plans
  first expedition to Africa, 105–9; psychological experiments with monkeys and
  apes, 103; religious consternation over his
  claims, 118–19, 159–60; responses to his
  work in the popular press, 84–85, 94–97,
  112, 116–18, 123–26; G. Schwidetzky
  on, 280; and the shift toward the fossil
  record for clues to human origins, 161,
  186–87; and the Smithsonian, 2, 90,
  93–94, 129–30; use of the phoneidoscope,
  102–3, 263; use of the phonograph, 2–4,
  90, 93–94, 97–103; Verne's Dr. Johausen
  as fictional counterpart, 124–25; Whitney's praise for, 86, 112–13; and Yerkes,
  8, 202, 230–34, 379–80. *See also* Fernan
  Vaz testament; *Gorillas and Chimpanzees*; "Simian Tongue, The"; *Speech of
  Monkeys, The*
Gause, G. F., 448n8
Geddes, Patrick, 163
genetics. *See* Dobzhansky, Theodosius;
  Haldane, J. B. S; Kalmus, Hans; Watson,
  James D.
genial tubercles, 179, 180, 189
genio-glossal muscles, 181
Geographical Journal, 164–65
Germany: Deutsche Gesellschaft für Tierpsychologie, Berlin, 451n41; fossils found in,
  176, 182; German Society for the Study
  of Animal and Primeval Languages, 281;
  Harrington studies under Wundt in, 3–4,
  173, 174–75; Max Planck Institute of Psychiatry, Munich, 313; Max Planck Station,
  Buldern, 260; Müller's philological studies
  in, 19; romantic science in, 19; University of Jena, 58. *See also* Amtsberg's pike;
  Bastian, Adolph; Beer, Thomas; Bethe,

Albrecht; Boas, Franz; Haeckel, Ernst;
  Heinroth, Oskar; Heyse, Lorenz; Herder,
  Johann Gotfried; Koch, Ludwig; Köhler,
  Wolfgang; Lorenz, Konrad; Müller,
  Friedrich Max; Nazism; Noiré, Ludwig;
  Schwidetsky, Georg; Schwidetsky, Ilse;
  Stresemann, Erwin; Virchow, Rudolph;
  Von Uexküll, Jakob
Geschwind, Norman, 310–13, 477n8,
  479n22
Gestalt psychology, 214
gibbons: Boutan's studies of, 166, 226–27,
  230, 232, 285; Carpenter's studies of,
  237–39, 284–85, 298; Darwin on, 39;
  Hockett on, 289–91; Marler on, 285,
  308; Washburn dissects, 298. *See also*
  Pépée
Glave, E. J., 107, 130
Globe, 417n96
gnathic index (Garner's measurement), 99,
  404n58
Gombe Stream Chimpanzee Reserve, 300,
  303–4, 325, 331–32
Goodall, Jane, 282, 300, 303–4, 325,
  331–32, 481n44
Goode, Adolphus Clemens, 412n37
gorillas: Du Chaillu and, 15–16; as featured in the hippocampus minor debate,
  15–16, 51; Garner and, 82–83, 85, 106,
  109, 116, 118, 133, 134, 140–47, 152,
  154–57, 167, 185–86; Müller and, 16;
  Schaller's studies of, 299–300. *See also*
  Koko; Othello
Gorillas and Chimpanzees (Garner), 140,
  142–47, 155, 156, 405n94, 414–15n74,
  415n78
Gould, James, 360
Gould, Stephen Jay, 374
Gouraud, George Edward, 128
gramophone, 90, 263, 455n78
graphophone. *See* phonograph
Gray, Asa, 31–32, 386n60
Greek, 19, 279, 430n138, 458n103
Green, Steve, 337–40, 343, 361
Greite, Walter, 281
Griffin, Donald, 359–60, 379

Grimm's law, 16, 21, 23–24, 384n21
Grove, Archibald, 95, 403n45, 411n30
Gua (chimpanzee), 235

habituation/dishabituation experiment, 376–77
Hacking, Ian, 365, 404n64
Haddon, Alfred Cort, 8, 160–66, 191, 261, 493n4
Haeckel, Ernst: and Darwin, 58, 78, 388n86, 394n33; and *Pithecanthropus*, 71, 181, 425n73, 425n77
Haggerty, Melvin, 200, 221–22, 431n6, 441n109
Haldane, J. B. S., 245, 250–51, 270–79, 283–86. *See also* Spurway, Helen
Hale, Horatio, 113–16, 176–81, *177*, 190–91, 369–73
Hall, Ron, 300
Hamilton, Gilbert ("Dick"), 221–22
Hamilton, W. D., 331, 333
Hampton's Broadway Magazine, 199–201
Haraway, Donna, 447n166, 461n126
Harper and Brothers, 188
Harper's Weekly, 96, 98, 109, 140. *See also* Church, F. S.; Phillips, Barnet
Harrington, John Peabody, 3–4, 8, 160–61, 166, *171*, 172–76, 195
Hartmann, Robert, 178, 370
Harvard University: animal behavior research at, 205, 216–18; faculty of, 81, 200, 231, 300, 323, 426n90, 438n81, 469–70n75; physical anthropology at, 296; Seyfarth works at, 331; students of, 113, 204, 296, 300, 308, 330; Wundt's influence on psychology there, 204–5
Haskins Laboratories, 324
Hatton & Cookson, 132, 412n42
Hauser, Marc, 377–78
Hay, John, 108
Hayes, Keith and Catherine, 313–16, 328, 445n146, 467n47, 479n22
Heart of Darkness (Conrad), 425n83
Heidelberg jawbone (fossil), 182
Heinemann, William, 96
Heinroth, Oskar, 254, 281, 451n40

Herder, Johann Gottfried, 24–25, 384n24, 424n69
Herring Gull's World, The (Tinbergen), 277–78
Herrnstein, Richard, 358
Herschel, John, 27
Herzig, Rebecca, 406n105
heterogenesis, 369–70
Heyse, Lorenz, 23, 424n69
Hill, Andrew, 352
Hinde, Robert, 260–63, 282, 284, 300, 331–34
hippocampus minor, 15–16, 32, 51, 312
Hitler, Adolf, 257
Hitzig, Eduard, 62, 78
Hobhouse, Emily, 436n66
Hobhouse, Leonard, 200–201, 211–15, 221, 223, 258, 432n20
Hockett, Charles, 287–93, 301, 322–24, 430n138
Holmes, W. H., 402n21, 426n86
holophrasis, 193–94
Holy Ghost Fathers. *See* St. Anne's Mission
Homo alalus, 71, 370, 425n73
Homo erectus, 374–75. *See also* Java man
Hooker, Joseph, 31, 32, 57
Hooten, Earnest, 426n90, 469–70n75
howler monkeys, 235–36, 238–39, 285, 301
How Monkeys See the World (Cheney and Seyfarth), 2, 377
Hrdlička, Aleš, 8, 161, 172, 176, 182–89, 230, 430n140
Hubrecht, A. A. W., 408–9n147
Hughlings Jackson, John. *See* Jackson, John Hughlings
Hunt, James, 393n24
Huxley, Julian, 167, 251–52, 254
Huxley, Thomas Henry: debate with Owen, 15–16; and Garner, 130, 415n83; as Morgan's teacher, 6, 73, 76; on Neanderthal fossils, 178–79; at the Norwich BAAS meeting (1868), 57; on the origin and importance of the animal-human language barrier, 19, 32–33, 176, 180, 208–9, 380, 394n33; on protoplasm, 383n6; as public

Huxley, Thomas Henry (*cont.*)
man of science, 95, 383n7; saltationism of, 33, 76, 369–70, 372. See also *Evidence as to Man's Place in Nature*

Ibis, 247, 277
"ideational behavior" (Yerkes's term), 214–15, 225
Illustrated London News, 136, 382n2
Imanishi, Kinji, 479n23
imitation: Bateman on aphasics and, 61; Carpenter on gibbons and, 238; Darwin on 18–19, 29, 36, 60–61, 389n99; as explanatorily preferable under Morgan's canon to reason, 52; Garner as uninvolved with the psychological debates over, 86; Garner's monkey subjects as responding to, 404n70; Haggerty on, 200, 221; Haldane on, 273; Kinnaman on, 200; Kohts on absence in apes, 443n123; Müller's repudiation of significance of, 25–26; Thorndike on, 200, 208–9; Vygotsky on absence in apes, 443n123; Watson on, 200; the Wedgwoods' backing of the importance of, 25–26, 36; Yerkes on, 200, 443n123. See also birdsong; onomatopoeic theory
imitation theory (of origin of language). See onomatopoeic theory
Imperial College London, 271
imprint learning, as ethological concept, 252, 254
India: Jay observes langurs in, 301; Kellogg on wolf-raised children of, 235
information theory, 263, 270–79, 285–87, 294–95. See also cognitive science; communication theory; cybernetics
instinct, 115, 204, 236, 276
Institute for Research in Animal Behavior, 304. See also Millbrook laboratory, Rockefeller University
Institute for the Study of Animal Behaviour, 451n41, 452n50, 453n52, 462n136. See also Association for the Study of Animal Behaviour
interjectional theory (of origin of language), 22–23, 25, 39
interchangability. See properties of language

International Wildlife, 363, 379
Introduction to Comparative Psychology (Morgan), 202–3, 206, 432n15, 433n28, 434n40
introspective psychology, 217–20. See also behaviorism
Itani, Jun'ichiro, 326

Jackson, John Hughlings: on aphasia and the cerebral localization of language, 50–51, 56, 62, 78–80; on emotional expression versus rational language, 79, 312; influence of Spencer on, 79, 165; as teacher of Rivers, 165
James, Henry, 95
James, William, 3, 81, 84, 94, 216, 231, 273
Japanese macaques, 300–301, 326, 337–40, 349–50, 360, 482n62
Jardine, Nicholas, 365
Jastrow, Joseph, 3, 118
Java man: discovery of, 7, 161, 176, 181–83; Dubois and, 181–82; Hrdlička and, 161, 176, 182, 426n93; as stimulating search of fossil record for evidence of human origins, 7, 161, 176, 425n74, 425n77–78. See also *Homo erectus*
Jay, Phyllis, 301
Jesus College. See Cambridge University
Johausen, Dr., 124–25
John Holt station, 132, 148–49
Johns Hopkins University, 216, 219, 225, 493–94n8
Jolly, Alison, 473n108, 486n102, 492n152
Jonçich, Geraldine, 432n22
Joos, Martin, 265, 284, 458n98, 472n99
Josiah Macy Foundation, 271, 274–75
Journal of Animal Behavior, 216, 220, 438n75
Journal of Comparative Neurology and Psychology, 218–19
Journal of Comparative Psychology, 256
Journal of the African Society. See African Society
Journal of Theoretical Biology, 293, 334
Julius (orangutan), 222–24

Kafka, Franz, 442n119
Kalmus, Hans, 274

Kant, Immanuel, 20, 40, 66–67, 191
Kay Electric Company, 264
Keith, Arthur, 185
Kellogg, Winthrop, 235
Kempster, Walter, 410n18
Kenya. *See* Amboseli
Kettlewell, Bernard, 246
King's College London, 262
Kingsley, Charles, 131, 387n67
Kingsley, Mary, 131–34, 143, 423n35
King Solomon's Ring (Lorenz), 278
Kinnaman, Andrew, 200, 432n20
Kipling, Rudyard, 123, 445–46n146
Koch, Ludwig, 166–67, 237
Kohler, Robert, 493n7
Köhler, Wolfgang, 214–15, 222, 224,
   437n71, 442n119
Kohts, Nadia, 285, 443n123, 437n74
Koko (gorilla), in ape language project, 320,
   477n3
Konishi, Masakazu (Mark), 295
Kroeber, Alfred, 172, 195–96, 288–89,
   445n142
kulu-kambas. *See* chimpanzees
kymograph, 230, 423n46

Labouchere, Henry, 6–7, 125–27, 137–39,
   140–42, 146–49, 154–57
Lack, David, 198, 243–48, 258, 260–63, 269,
   282, 303
Laird, Carobeth. *See* Tucker, Carobeth
Lamarckism, 250–51, 259, 388n90, 451n35
Lana (chimpanzee), 320, 328
La Naulette jawbone (fossil), 178–79
Lancaster, Jane, 302, 310–14
Lancet, 55–57
Lang, Andrew, 95
Langer, Suzanne, 428n116
language. *See* aphasia; comparative philology;
   interjectional theory; natural selection, as
   a factor in the evolutionary origin of language; onomatopoeic theory; phonetics;
   reason as counterpart to language; roots of
   language, in Müller's philology; semiotic,
   three levels of; sign language; sympathic
   theory; syntax; transformational generative grammar; vanished-instinct theory

langurs, 301
La Quina skull (fossil), 182
Lashley, Karl, 314
law of effect, 435n47
law of exercise, 435n47
Leakey, Louis, 300, 303
Learned, Blanche, 215, 226, 229–31, 234–35,
   379. *See also* Yerkes, Robert Mearns
learning curves, 201, 211–12, 215, 224,
   436–37n67. *See also* time curve
Lederberg, Joshua, 330
Lee, Phyllis, 351–52, 485n85, 486n94
Lehrman, Daniel, 276, 452n49, 465n18
Lenneberg, Eric, 478n12
Le Roy, Alexandre, 155, 419n125
Lieberman, Philip, 324
Life, 264
limbic system (brain), 311–13, 323
Lindauer, Martin, 273–74
linguistics. *See* Anderson, Stephen; Bickerton,
   Derek; Bloomfield, Leonard; Boas, Franz;
   Chomsky, Noam; Hockett, Charles; Joos,
   Martin; Lyons, John; Pinker, Steven; Sapir,
   Edward; Saussure, Ferdinand de; Sebeok,
   Thomas; Whorf, Benjamin Lee
Linnaeus, Carolus, 281
Linton, Eliza Lynn, 95
Littell's Living Age, 403n40
Liverpool Courier, 159–60
Liverpool Post, 417n96
Lloyd Morgan, Conwy. *See* Morgan, Conwy
   Lloyd
Lloyd Morgan's canon. *See* Morgan's canon
Locke, John, 22, 40, 42, 74
Loeb, Jacques, 435n56
Lombroso, Cesare, 182–83
London School of Economics, 331, 436n65
London Zoological Garden, 117
Lorenz, Konrad: as founder of ethology, 9,
   254–61, 273, 293, 457n97; and Haldane,
   276; Marler's divergence from the views
   of, 283–84; on the mechanical nature of
   animal interaction, 278, 289; and Nazism,
   257–58, 263, 281
Lubbock, John, 45–46, 77, 433n23
Lukina, E. V., 277
Lyell, Charles, 18, 27, 33–34

Lyons, John, 322–23
Lysenko, T. D., 250–51

macaques, 166–67, 326, 339–40, 350, 378.
   See also Japanese macaques
Macmillan, 216
MacMillan's Magazine, 25–26, 386n60
Madingley Ornithological Field Station,
   258–62, 283–84, 300–301, 331–34, 351,
   358, 361
Madison Square Garden, 96, 111, 113
Magoun, Horace, 313, 479n27
Mallery, Garrick, 402n21
Manchester, 198, 211, 412n33, 436n66
mangabeys, 167, 339
Manning, Aubrey, 52
Manouvrier, Léon, 183
Marler, Chistopher, 342, 344–45, 484–85n84
Marler, Peter, 244, 304, 337, 344; on alarm
   calls as on a continuum between affective and symbolic extremes, 329, 375;
   and the ape language projects, 325–29;
   at Cambridge, 253, 261–63, 266–70,
   275, 282–84; early life and education,
   244–45; and ethology, 9–10, 245, 256,
   259, 261–63, 283–84, 293; experimental studies of how birds learn their songs,
   252–53, 261–62, 295–96; and Haldane,
   245, 250–51, 274–79, 285; and Hockett,
   287, 292–93, 303; on "information" as
   a key concept for understanding animal
   communication, 274–79, 285–87, 294;
   and Lack, 245, 247–48, 261; non-vervet
   primatological studies, 284–85, 299,
   301–4, 313; as reinventor of the primate
   playback experiment, 1–2, 4, 10–11, 243,
   321–22, 324, 367–68, 379–80; and the
   Seyfarths, 1–2, 309, 321–22, 329–30,
   334–47, 349, 352–53, 355, 358, 360,
   362–63; spectrographically aided studies
   of structure and function (localization)
   in animal calls, 267–70; studies of the
   geographical ecology of birdsong, 243,
   245–53; on vervet alarm calls, 307–10,
   328–29; and Washburnian anthropology,
   296, 299, 310–12
Marshall, William, 116

Marxism, 250, 272
Masai-Amboseli Game Reserve, Kenya. *See*
   Amboseli
Massachusetts Fish and Game Protective As-
   sociation, 231, 443n129
materialism, 40, 57, 61–63, 215
mathematicians. *See* Altmann, Jeanne; Col-
   burn, Zerah; Wiener, Norbert
Max Müller, Friedrich. See Müller, Friedrich
   Max
Maya glyphs, 90, 93, 113
Mayr, Ernst, 251–52, 296, 371
McClure, Samuel Sidney, 95–96, 108, 129,
   131–35, 221, 403n45
McDougall, William, 163, 236, 254
McLaughlin, John, 148–49, 151
Mead, Margaret, 430n137
Menard, Pierre, 380
Menlo Park, NJ, 90
mental tubercles. *See* genial tubercles
Menzel, Emil, 479–80n28
Mer. *See* Torres Straits expedition
microcephali, 57–58, 80, 398–99n124
Millbrook Laboratory, Rockefeller University,
   336–41, 357–61
Miller, George, 459n109, 478–79n19
Mills, Wesley, 434n34
mimesis theory (of origin of language). *See*
   onomatopoeic theory
minds, animals vs. human. *See* anthropomorphism and its opponents
missing link: in the fossil record, 200, 395n58;
   Garner "finds," 187; Peter described as,
   229, 373; *Pithecanthropus erectus* as,
   181; speech of gorillas and chimpanzees as
   language's, 83
missionaries, 130–33, 159–60. *See also* St.
   Anne's Mission
MIT, 270, 323, 462n134
Mivart, St. George Jackson, 68–70, 76
Mizuhara, Hiroki, 300–301
Möbius, Karl, 64–65, 67, 395–96n74,
   396n87
modern synthesis, of genetics and natural
   selection theory, 251, 289–90
monogenesis, 20
Montagu, Ashley, 495n24

Montecito, Yerkes's laboratory at, 222–26
Morgan, Conwy Lloyd, 74; and the canon, 5, 51–53, 73–78, 81–83, 202–4, 210–11; and Garner, 6, 82–83, 86, 104, 117–18, 127, 203–4, 367; on Hobhouse, 213–14; and T. H. Huxley, 73, 76; and Müller, 5, 52, 75–77, 204; and Thorndike, 204–7; on Yerkes, 230. See also *Introduction to Comparative Psychology*; Morgan's canon
Morgan, Lewis Henry, 52
Morgan's canon, 5–6, 9, 11, 51–53, 73, 81, 201–4, 207, 209–10, 367, 372, 382n13, 432nn14–15, 433n23
Moriarty, Stephen, 112, 128
Morris, Charles, 294, 447n166
Morris, Desmond, 282
Moses (chimpanzee), 135, 136, 143–44, 156, 416n89–90
Müller, Friedrich Max, 17, 323; and Bateman, 61; and catastrophism, 18–19, 25–27; critical reponses in the 1860s, 25–29; and Darwin, 3–4, 31–32, 37, 39–41, 43, 45–48, 65–66; and degenerationism, 24, 27–28; dispute with Whitney, 4, 19, 41–49, 67, 89, 401n16; education and early career, 19–20; and Garner, 89, 100, 112; influence on Morgan, 5, 52, 75–77, 204; and Jackson, 62; and James, 81; on the origin of language, 21–25, 39, 49; role in provoking debate about anthropomorphism, 64–68, 73, 218; and Romanes, 70–72; as setting the agenda in the nineteenth-century debate on animal language, 4–6, 15–17, 51. See also reason as counterpart to language; roots of language, in Müller's philology
multiple choice apparatus, 221–25. See also choice chamber
Münsterberg, Hugo, 231, 438n81, 443n129
Murray Island. See Torres Straits expedition
Museum d'histoire naturelle, 154–55, 414n71
Myers, Charles, 163–64

Nagel, Thomas, 491n147
Nassau, Robert, 133
Nation, 49, 116, 141, 414n67

National Coal Board, 271
National Museum of Natural History. *See* Smithsonian Institution
National Research Council, Washington, DC, 226
National Zoological Park, 93. *See also* Smithsonian Institution
Natural Science, 127, 433–34n32
natural selection, as a factor in the evolutionary origin of language: Darwin on, 36; Hockett on, 290–91; Marler on, 286, 329; Pinker and Bloom on, 377; Wallace on, 34; in *The Water Babies* (1863), 387n67; Wright on, 34–35
Nature: Darwin's letter to, 388n93; Haddon's expedition plan in, 163; Marler in, 267–69, 270, 283–84; Morgan's reviews in, 81–83, 116–17, 207, 213–14; reaction to Müller in, 49; review of *The Speech of Monkeys* in, 116–17; Romanes's report on Sally in, 71–72; *The Science of Thought* provokes response in, 49; Thorpe in, 266; Waser's playback report in, 339
Nature Conservancy, 248, 252–53
Nazism, 257–58, 281–82, 373
Neanderthal fossils, 7, 176–82, 185, 189, 324
neocortic system (brain), 312–13
neurology. *See* Geschwind, Norman; Jackson, John Hughlings
neuroscience. *See* brains, animal and human; Ferrier, David; Fritsch, Gustav; Geschwind, Norman; Hitzig, Eduard; Jackson, John Hughlings; Lenneberg, Eric; Magoun, Horace; Ploog, Detlev; Robinson, Bryan; Washburn, Sherwood
new journalism (of the late nineteenth century), 139
new physical anthropology (of Washburn), 296–98, 312
New Review, 84, 94–96, 106, 108, 410–11n18, 411n30
Newsday, 361
Newton, Issac, 27, 96
New York: Josiah Macy conference in, 274–75; New York City Zoological Society, 304; Pathological Institute at, 182–83. *See also* Bronx Zoo; Columbia University

New York Daily Tribune, 116, 414n64,
414–15n74, 417n97
New York Herald, 406n99
New York Times: H. Frederic as reporter for,
127–28, 146–47; Garner in, 116, 127–29,
135, 140, 161, 381n6, 411n22; Rocke-
feller team in, 1, 321; vervet playback
research in, 361
New York World, 96, 142, 400n7
New York Zoological Park. *See* Bronx Zoo
Nice, Margaret Morse, 450–51n33, 453n64
Nicholson, E. M., 252–53
Nim Chimpsky (chimpanzee), 320–21, 362
Nineteenth Century, 95, 397n93
Nineteenth Century Club, 96
Nissen, Henry, 314, 445n145, 446n149
Nkami (language), 144, 413n47
Nobel Prize, awarded to Lorenz, Tinbergen
and Frisch (1973), 273
Noiré, Ludwig, 49, 424n69, 460n121
North American Review, 35, 44, 106–7,
444n135
Nottebohm, Fernando 295, 305, 360

objectivism, 218–19
object naming, as demarcating humans from
nonhumans, 307–8, 310–13
octopus, as cybernetical system, 274–75
Ogowé River: Garner travels around the,
130, 133–34, 148–49, 150–51, 169–70;
Kingsley's travels around the, 131–32;
Moses (Garner's chimpanzee) found in,
135; trade around the, 132; tribes of the,
169–70
Ohio State University, 216, 265, 287
onomatopoeic theory (of origin of language),
22, 23, 25, 112
orangutans: Garner's reseach with, 128–29,
410–11n18, 416n89; Haggerty's reseach
with, 221–22; Hrdlička's studies of, 185;
Witmer's research with, 228–29; Yerkes's
reseach with, 222–24. *See also* Chantek;
Julius
organic selection, 252. *See also* Baldwin effect
Origin centenary, 281, 284–85, 457n97
Origin of Species (Darwin), 4, 31–32, 199,
281, 285, 369, 380, 384n20

ornithology, 244–45, 253, 254, 466n23. *See
also* Allen, A. A.; Armstrong, Edward;
bioacoustics; Borror, Donald; Chapman,
Frank M.; Collias, Nicholas; Heinroth,
Oskar; Hinde, Robert; *Ibis;* Joos, Martin;
Lack, David; Lehrman, Daniel; Mading-
ley Ornithological Field Station; Marler,
Peter; Nice, Margaret Morse; Nicolson,
E. M.; Promptov, A. N.; Selous, Edmond;
sound spectrograph; Stresemann, Erwin;
Thorpe, William Homan
oscilloscope, 263–64
Othello (gorilla), 144
Ott, John T., 112
Owen, Richard, 15–16, 32, 51
Oxford University, 211, 244–46, 258,
260–61, 316

Page, W. H., 112–13, 129
paleoanthropology, 7–8, 161, 176–89,
296–99, 324, 375. *See also* Dubois, Eugène; Hrdlička, Aleš; Leakey, Louis
paleontology, 173, 182–85, 199–200, 289.
See also Buckman, S. S.; Eldredge, Niles;
Gould, Stephen Jay
palethnology, 162
Pall Mall Magazine, 413n53
Panama, 235
Pangwe. *See* cannibals
Panzee (Yerkes's chimpanzee), 227, 229–30,
443n125, 461n127
Papua New Guinea. *See* Torres Straits expedi-
tion
Paris Linguistic Society, ban on
language-origin papers, xii
parsimony considerations, in explainations of
animal behavior, 203, 207, 354, 432n15,
433n29. *See also* Morgan's canon
particle physics, 263
Passamaquoddy Indians, 93
Pavlovian conditioned reflex, 215, 256, 275,
376
Payn, James, 136
Pearsall, W. H., 245
Peirce, Charles Sauders, 294
Penfield, Wilder, 474n124
Pépée (gibbon), 226–27

Perry, E. D., 113

Peter (chimpanzee), 228–29, 235, 373

Pfungst, Oskar. *See* Clever Hans

Philadelphia Evening Bulletin, 232–33, 401n16

Philadelphia Natural History Society, 232

Phillips, Barnet, 96, 106–7, 112, 401n13, 402n19

philology. *See* comparative philology

philosophers. *See* Ayer, A. J.; Dennett, Daniel; Fitzpatrick, Simon; Fodor, Jerry; Herder, Johann Gottfried; James, William; Langer, Suzanne; Morris, Charles; Nagel, Thomas; Noiré, Ludwig; Peirce, Charles Sauders; Renan, Ernest; Royce, Josiah; Schelling, Friedrich; Singer, Peter; Sober, Elliot; Wright, Chauncey

phoneidoscope, 102–3, 263, 405n76, 454n75

phonetic decay (Müller's theory of), 24, 385n32

phonetics: Boas as critical of evolutionist version, 193; growth of popular interest in during late nineteenth century, 88–89; in paleolithic speech as reconstructed by Brinton, 113; studied by Harrington, 172

Phonogram, 91–4, 111–12, 402n22

phonograph, 2–4, 11, 83–86, 90–94, 92, 96–102, 98, 105–8, 109–12, 115–16, 118–19, 128–29, 135–37, 143–44, 160, 163, 164, 166–67, 170, 171, 172, 203, 226, 230, 239, 365–66

phrenology, 53–54

physical anthropology, 45, 162, 176–88, 296–301, 310–12. *See also* anthropology

Pico. *See* Azores

pidgin languages, 116, 376, 496n35

Pinker, Steven, 377

Pithecanthropus erectus. *See* Java man

playback experiments, xi, 1–4, 6–11, 101, 160–61, 166–67, 197–98, 201–2, 210, 215, 230, 234, 237–39, 243, 245, 282, 284–85, 295, 298, 303–4, 307, 309, 321–22, 329–30, 338–39, 361–74, 376, 379, 380, 378–80, 465n20

Ploog, Detlev, 313

poetry: about Garner, 84, 85, 123–24,

136–37; by Garner, 153, 403–4n52; on evolution, 173

polydactylism, 181, 369–70

pooh-pooh theory (of origin of language). *See* interjectional theory

Portielje, A. F. J., 254–55

Potawatami Indians, 288

Potter, Ralph, K., 263–65

Poulsen, Holger, 252, 259

Powell, John Wesley, 193

pragmatics. *See* semiotic, three levels of

Premack, Ann, 480n29

Premack, David: doubts natural symbolic communication in nonhuman species, 10, 326–27, 358; influence of, 346–47, 358, 363–64; and Marler, 325–28, 346–47, 358, 367; primate vocalizations as expressions of emotion, 326–27, 346–47; Sarah (chimpanzee), 10, 315–16, 317, 320, 324, 326–28, 346

Prichard, James Cowles, 20, 24, 383n16

Priestley, J. B., 258

Primate Project, 299–302, 308–10, 326, 338

primates, nonhuman. *See* baboons; capuchin monkeys; *Cebus* monkeys; chimpanzees; diana monkeys; gibbons; gorillas; Japanese macaques; langurs; macaques; mangabeys; orangutans; spider monkeys; vervet alarm calls

primatology, 300–301, 337–38, 367. *See also* Altmann, Jeanne; Altmann, Stuart; De-Vore, Irven; Dixson, A. F.; Fossey, Dian; Goodall, Jane; Hall, Ron; Hinde, Robert; Jay, Phyllis; Jolly, Alison; Lancaster, Jane; Leakey, Louis; Saymann, Graham; Schaller, George; Seyfarth, Robert and Dorothy; Struhsaker, Thomas; Washburn, Sherwood

Prince's Hall, London, 135–37, 139, 140, 156

Pringle, John, 269, 462n135, 463n146

productivity. *See* properties of language

progress in science, nineteenth-century conceptions of, 21, 26–27

Promptov, A. N., 251–52, 259–60, 277

properties of language (according to Hockett), 289–90, 292

protozoa, 70, 99
psychiatry, 313
Psychological Review, 206, 219
psychology. *See* behaviorism; Bloom, Paul;
    Bowlby, John; comparative psychol-
    ogy; James, William; Jastrow, Joseph;
    McDougall, William; Miller, George; Riv-
    ers, W. H. R.; Witmer, Lightner; Wundt,
    William
Psychoscope, The, 153, 170, 403–4n52
Puck, 85
Punch, 15–16, 84–85, 91–92, 94, 136–37
punctuated equilibria, theory of, 374. *See also*
    catastrophism; saltationism
Putnam, Frederic Ward, 109
puzzle box, 9, 200–211, 214, 219, 256, 259.
    See also comparative psychology

Quarterly Review, 28, 45, 68–69
Quatrefages, M. de, 369–70
Quirk, Randolph, 271

racial hierarchy, and nineteenth-century
    evolutionist thinking about language: in
    Brinton's work, 113; in Darwin's work,
    37; in Garner's work, 99, 107; pulled-back
    from by Torres Strats expeditioneers, 165;
    repudiated by Boasians, 161, 195; repudi-
    ated by new physical anthropologists, 296;
    role of saltationism in catalyzing shift to
    evolutionary theory predicting no racial
    hierarchy, 373–74; sometimes com-
    bined with respect for "inferior" culture,
    159–60, 189
racial prejudice: Boas's firsthand experience
    of, 191; decline in evolutionary biology in
    the twentieth century, 373–74; Garner's
    as representative of evolutionist thinking
    of his era, 6; Haldane as vigilant against,
    250, 273; as officially unwelcome in
    recent science, 8, 368; as resisted in Ger-
    man anthropology in the age of Virchow,
    494–95n20
rapid fading. *See* design features of language
Ray, Sidney, 163
Reade, W. Winwood, 412n33
reason as counterpart to language: in Garner's

work, 99–100; in Morgan's work, 77–78,
    203; in Müller's work, 22
reason as mental climax of increased powers
    of association: in Darwin's work, 65–66;
    in Thorndike's work, 207, 209
Red Peter. *See* Kafka, Franz
Reeve, Arthur, 199–201, 216, 228–29
Regent's Park, London: in *Animal Language,*
    167; Cornish on, 417n98; Garner's orang-
    utan subject Jim in, 128–29, 410–11n18,
    416n89; Romanes observes "Sally" in,
    71–72
releasers, 254, 267, 273, 283, 453n62,
    463n147–49
Religious Tract Society, 409n152
Renan, Ernest, 384n31
Review of Reviews, 403n42, 406n99,
    408–9n147
Reynolds, Vernon and Frances, 300, 471n90
rhesus monkeys: S. Altmann observes on Cayo
    Santiago, 300, 471n89; Garner's research
    with 99, 100, 107; Hobhouse's experimen-
    tal psychology with, 211; at Madingley,
    studied by Hinde and Rowell, 300–301
Richards, I. A., 470n77
Richards, Robert J., 388n86, 392n8, 432n15
Rig Veda, 19–20, 43, 383n15
Rivers, W. H. R., 163, 165, 420n5, 422n24,
    493n4
robins, 245–46, 260–61, 264, 303
Robinson, Bryan, 312–13, 323
Rockefeller Foundation: funds Thorpe's
    purchase of the sound spectrograph, 266;
    funds Yerkes's primatological research sta-
    tion, 234; Natural Sciences Division run
    by Warren Weaver, 270
Rockefeller University, 1–2, 4, 10–11, 282,
    304–5, 329–30, 336–42, 357–62
Romanes, George John: anthropomorphism
    of, 52, 70–76, 117, 399–400n138,
    416n89; on cerebral localization, 78–80;
    and comparative psychology, 70–73, 76,
    258; criticism of, 72, 117, 206; on the
    langage barrier, 86
romanticism, German, 19, 384n24
roots of language, in Müller's philology, 4,
    16–17, 39–41, 49

Rowell, Thelma, 300–301, 332–33, 481n44
Royce, Josiah, 216
Rumbaugh, Duane, 328, 493n166
Rutherford, Ernest, 163

Sally (chimpanzee), 71–72
saltationism, 115, 368–78, 387n69, 428n122, 493n8. *See also* catastrophism; punctuated equillibria, theory of
Sanskrit: Bopp studies, 383n13; Harrington, J. P. on, 175; Hockett on, 430n138; Müller's study of, 4, 19–20, 24, 43, 112, 383n15; Romanes on, 70, 72; Whitney studies, 42
Sapir, Edward, 195–98, 287
Sarah (chimpanzee), 316–17, *317*, 320, 324, 326–28, 346
Saussure, Ferdinand de, 468n62
Saymann, Graham, 332
Schaller, George, 299–300
Schelling, Friedrich, 19, 424n69
Schipka cave jawbone (fossil), 179
Science: achievements at Millbrook in, 360; ape symbolic communication in, 321; Fewkes on the phonograph in, 92–93; the Gardners in, 316; Hale and Boas in, 190; Hrdlička in, 184; Marler in, 299, 307; review of *The Speech of Monkeys* in, 116; sound spectrograph in, 264
science-religion conflict: Catholicism especially vilified, 153; Garner and *The Speech of Monkeys and Holy Writ* (1894), 118–19, 158; Garner portrays his difficulties with Fernan Vaz missionaries as an instance of, 152–58; role of J. W.'s *History of the Conflict between Religion and Science (1876)* in promoting idea of ageless conflict, 126, 153; F. Turner's sociological analysis, 410n8; White, A. D., *History of the Warfare of Science with Theology in Christendom* (1896), 126
Schaller, George, 299–300
Schelling, Friedrich, 19
Schleicher, August, 58, 78, 388n86, 390n112
Schneirla, T. C., 452n49, 465n18
Schreiner, Olive, 95
Schultz, A. H., 298

Schweitzer, Albert, 130
Schwidetsky, Ilse, 464n5
Schwidetzky, Georg, 280–81
Scientific American, 96, 293, 317
Scott, John Paul, 467n49
S. C. Tinsley and Co., 405n76
Sebeok, Thomas, 321, 473n113, 492n160
Sefarth, Robert. *See* Seyfarth, Robert and Dorothy
Seligmann, Charles, 163–64
Selous, Edmond, 254
semanticity: as design feature of language (according to Hockett), 292; discussed as property of vervet alarm calls, 1–2, 307–8, 310–11, 322–23, 328–30, 339–64
semantics. *See* semiotic, three levels of
semiotic, three levels of, 294–95, 468–69n62
Seyfarth, Robert and Dorothy, *332, 344, 348, 352*; pre-Rockefeller careers, 329–35; as reinventors of the primate playback experiment, 1–2, 4, 10–11, 282, 309, 321–22, 329–30, 366–68, 376–80; the vervet alarm-call project, 336–64; work with Marler before deciding on the vervet project, 335–36. See also *How Monkeys See the World*
Shannon, Claude, 270–71, 274, 294
Siam. *See* Asiatic Primate Expedition
sign language, 10, 60, 316–19, 325, 362–63, 496n33
"Simian Tongue, The" (Garner), 84, 94–96, 99, 105–8, 231, 401n18, 403n46, 410n18, 411n30
Simpson, George Gaylord, 289, 296
Singer, Peter, 324
Skinner, B. F., 318, 323, 431n9, 491n152
Slough, 244–45, 248, 282
Smith, Aloyisius, 132
Smith, John Maynard, 457–58n97
Smith, W. John, 308–10, 354
Smithsonian Institution: Amerindian language recordings stored at, 166; Bureau of American Ethnology at, 92; Division of Physical Anthropology at, 184; Garner offers specimens to, 129–30; Garner's research at, 2–3, 90–94, 175, 185; Harrington at,

Smithsonian Institution (*cont.*)
172, 175; Hrdlička at, 172, 184–85; National Museum of Natural History at, 167, 184–86

Sober, Elliott, 382n13

Société d'anthropologie, Paris, 54

Society for Psychical Research, 135

Sociobiology: and the Ecological Society of America, 467n49; Hockett and, 288–89, 467n49; *Sociobiology* (1975), 324; Wilson and, 324, 330–31, 466n31, 471n89

sociology. *See* Hobhouse, Leonard

somatology. *See* physical anthropology

sona-graph. *See* sound spectrograph

song learning. *See* birdsong; vocal learning

Song Tutor, 284

sound spectrograph, 9–10, 263–79, 266, 268, 277, 282, 284, 287, 293, 295, 303, 305, 338–39, 343

South Africa, 28, 73, 332–34

Southern, Mick, 246–48

Southwest Museum, California, 172

Specialization. *See* properties of language

Spectator, 94–95, 116, 417n96

Spectrograms. *See* sound spectrograph

Spectrograph. *See* sound spectrograph

Speech of Monkeys, The (Garner), 3, 96–97, 102, 112, 115–19, 127, 232, 401n16, 406n107, 408n143–44, 409n151, 415n83

Spencer, Herbert: coins "the superorganic," 196; Haddon's and Rivers's debt to, 165; influence on Hobhouse, 213; influence on Jackson, 79–80; influence on Powell, 193; Müller acknowledges the work of, 40; as "old lion" of nineteenth-century public science, 95; "the vogue of Spencer," 89

spider monkeys, 99

Spiritans. *See* St. Anne's Mission

spiritualism, 135

Spurway, Helen, 274, 276. *See also* Haldane, J. B. S.

squirrel monkey, 313

Stanford, 172, 235, 313. *See also* Center for Advanced Studies in the Behavioral Sciences, Stanford

Stanley, Henry, 93, 107, 130

St. Anne's Mission: Garner and, 130–42, 144, 147–58, 160, 167; testimony of missionaries against Garner, 126, 141–42, 147–58, 160. *See also* Buléon, Joachim; Fernan Vaz

Stead, W. T., 401n12

Stephen, Leslie, 66–67, 396n83

sticklebacks, 253–55, 289

stimulus situation, 236, 305–7

Stresemann, Erwin, 457n97

Struhsaker, Thomas, 10, 305–8, 306, 312, 322, 324–25, 335–36, 340–41, 361, 367

Sunday Times, 361–62

Susie (chimpanzee), 232

symbolic communication: as humans-only except with tutoring, 10, 322–23, 327; as originating with the disordering of preferences, 327. *See also* affective communication; ape language projects; vervet alarm calls

sympathic theory, 49, 460n121

syntactics. *See* semiotic, three levels of

syntax: as the acquisition that changed protolanguage into language according to Bickerton, 374, 376; as the distinguishing feature of human language according to Chomsky, 320–21, 325

Taine, Hippolyte, 388–89n93

Tamura, Mivako, 295

Tanganyika. *See* Gombe Stream Chimpanzee Reserve

Tanzania. *See* Gombe Stream Chimpanzee Reserve

Tate, Alfred Ord, 109

Teddington. *See* Admiralty Research Laboratory, Teddington

telestimulation, 312–13

Tenerife. *See* Canary Islands

Terrace, Herbert, 11, 320–21, 362

Thames, the, 243, 249, 282

third frontal convolution. *See* Broca's convolution

Thompson, D'Arcy, 288

Thorndike, Edward Lee, 205; and behaviorism, 221, 435n47; and Hobhouse, 211–14; and Morgan, 204–7; and puzzle-box psychology, 8–9, 200–201, 204–19, 221, 224, 372; and Yerkes, 202, 214–18. *See also* associative learning; learning curves; Morgan's canon; puzzle box

Thorpe, William Homan, 9–10, 251–52, 256, 258–62, 265–67, 282–84, 313, 463n146, 479n22

Time, 264, 320–21

time curve, 206, 434n42. *See also* learning curves

Times, The, 116, 127, 410n9, 427n111

Tinbergen, Nikolaas: animal behavior experiments of, 253–56, 277–78, 293; awarded Nobel Prize, 273; on behaviorism, 256–57; founds *Behaviour,* 258; and Lorenz, 254–55, 258, 272–73, 277–78, 282–83; programmatic evangelism of, 273; students of, 282–83, 316; study of herring gulls, 255, 277–78; "triggers," 289

Titchener, Edward, 218

Today, 361–63

Tolman, Edward, 431n10, 452n44

Toronto Globe and Mail, 361–62

Torres Straits expedition, 162–68, 191, 261, 493n4

Tortugas, Florida, 219, 232

total feedback. *See* design features of language

Tracy, Julia Lowndes, 381n6

Trader Horn. *See* Smith, Aloyisius

transformational generative grammar, 323

trial-and-error learning: "Clever Hans" and, 211; Haggerty on, 221; Hamilton on, 221–22; Hobhouse on, 211–13, 220–21; Morgan and, 52, 104, 201–6; in psychology, 5, 6, 9, 255; Thorndikian, 200–201, 204–8, 221; Yerkes's research with Julius, 222–24. *See also* associative learning

Trivers, Robert, 330–31, 333–34

Trousseau, Armand, 56, 80–81

True, F. W., 129–30

Truth, 6–7, 127, 137–39, 141–42, 148–49, 167

Tucker, Carobeth, 173–74

Turner, Frank, 410n8

Tylor, Edward Burnett, 28, 165

Tyndall, John, 57, 89, 383n7

UCL. *See* University College London

Uganda: Marler's colobus research in, 303;

307, 335–38; the Reynolds observe chimpanzees in, 300, 471n90; Waser's playbacks with grey cheeked mangabeys in Kibale Forest, 338–39

umwelt, 254

uniformitarianism, 17–19, 33, 39

Unity of Science movement: Bloomfield and, 288; Carpenter and, 447n166; Hockett and, 288; Morris and, 447n166. *See also* semiotic, three levels of

University College London, 245–47, 250, 271–75

University of Bristol, United Kingdom, 73, 331

University of California at Santa Barbara, 315

University of Chicago, 200, 216, 219, 294, 298

University of Jena. *See* Germany

University of Michigan, 314, 483n68

University of Minnesota, 315, 441–42n114

University of Wisconsin, 265

vanished-instinct theory: Heyse and, 23; Müller and, 18–19, 23, 26–27, 39, 43

Van Lawick, Baron Hugo, 303–4. *See also* Goodall, Jane

variation: Boas on, 371–72; the congruence of Lorenz's attitude with ethological practice and National Socialist themes, 257–58; Galton on, 114–15; Lack's and Marler's adaptationist interpretation of, 268–69; Lorenz's dismissive attitude toward, as contrasted with Marler's attitude, 283

venesection, 80–81

vera causa doctrine, 31, 34, 48, 385n39

Verne, Jules: *The Aerial Village,* 124–25, 155, 409n4; and Garner, 106, 409n5; on Garner's chimpanzees, 135; influences Edison, 109

vervet alarm calls, 1–2, 10–11, 305–12, 306, 321–25, 328–30, 336, 339–64, 374–78, 480n29

vervet monkeys. *See* vervet alarm calls

Verwey, Jan, 254

Vickers, Geoffrey, 271

Vienna. *See* Austria

Vietnam War, 195, 325, 330

Viki (chimpanzee). *See* Hayes, Keith and Catherine
Virchow, Rudolph, 89, 191, 372, 494–95n20
vitalism, 288
vocal-auditory channel. *See* design features of language
vocal learning, 296, 302, 358. *See also* birdsong
Vogt, Carl: and Bateman, 57, 78–79; cerebral localization of language, 78–79; and Darwin, 57–58, 78; defense of Broca, 57; *Mémoire sur les microcéphales ou hommes-singes*, 57–58
volapuk. *See* pidgin languages
Von Bunsen, Baron Christian: at the BAAS, 20; encourages Müller, 384n31; on the "language barrier," 384n18; on transmutation theories, 384n18
Von Frisch, Karl. *See* Frisch, Karl von
Von Holst, Erich, 254
Von Humboldt, Alexander, 21, 26, 384n24
Von Humboldt, Wilhelm, 424n69
Von Uexküll, Jakob, 218, 254, 439n92
Vygotsky, Lev, 215, 441n110–11, 443n123

Wake, Charles Staniland, 37, 385n35, 398n111
Wallace, Alfred Russell, 33–35, 57, 81, 95, 394n33
Wall Street Journal, 361
Waser, Peter, 339, 484n80
Washburn, Margaret Floy, 210–11, 214, 216
Washburn, Sherwood, 10, 161, 295–300, 297, 310, 312, 361
Washington Post, 95, 427n111
Washoe (chimpanzee), 11, 316–21, 325, 328, 492n159
Watson, James D., 262
Watson, John B., 200, 219–21, 225–26, 231–32, 373
Weaver, Warren, 270, 274, 294, 477n10
Webster, Charles L., 96
Webster, T. B. L., 271
Wedgwood, Frances Julia, 25–26, 41, 67, 386n60

Wedgwood, Hensleigh: and Darwin, 25, 36, 386n60, 388n85, 388n92; views on language, 28, 45–46
Wells, H. G., 409n2
Wenner-Gren Foundation: funds Cheney, 340, 348, 354; funds Washburn, 298; symposia in Austria in summer 1965, 307, 310, 313
Wernicke's region. *See* Broca's convolution
West Africa. *See* French Congo
Western Reserve University, Cleveland, 207–8
Westminster Gazette, 413n51
Westminster Review: anonymous reviews of Müller in, 26–29, 46, 62; criticism of Müller, in, 67–68; physiological experiments in, 78; review of Garner in, 417n96
Whewell, William, 20–21, 26
White, Andrew Dickson, 126
Whitney, William Dwight: affinities between his position on language and Boas's, 190; "Darwinism and Language," 44–45, 47, 67, 385n42; dispute with Müller, 4, 19, 41–49, 67, 89, 401n16; on the "language barrier," 42–45, 47, 58, 67, 113; *The Life and Growth of Language*, 47–48; support of Garner, 86, 112–13
Whorf, Benjamin Lee, 429n132
Wiener, Norbert, 270–72, 274, 294, 469n67
Wiener-Shannon theory, 270–71
Wilkes expedition, 113. *See also* Hale, Horatio
Wilkin, Anthony, 163–64
Williams, Monier, 20
Wilson, Edward O., 300, 324, 331, 466n31, 460–61n125. *See also* sociobiology
witchcraft, 169–70
Witmer, Lightner, 227–29, 233, 235, 373
Wittkower, R., 271
Wolfgang, Alfred, 149, 151
Woods Hole, Cape Cod, 208, 216–17
Wright, Chauncey, 19, 34–35, 38, 44, 81, 294
Wundt, William 3–4, 81, 94, 173–75, 204–6, 218, 231

Yale: faculty of, 42, 201, 234, 287, 301, 379; Laboratories of Primate Biology and Comparative Psychobiology at, 234; students of, 287

Yerkes, Ada, 234, 284

Yerkes, Robert Mearns: and Abreu's colony, 188; and Carpenter, 235–36; as comparative psychologist, 200, 202, 214–36, 379–80; Garner, 8, 202, 230–34, 379–80; investigations into ape language, 226, 229–31, 379; and the Kelloggs, 235; and Thorndike, 202, 214–18; and Watson, 200, 219–21, 225–26. See also Yerkes Laboratory of Primate Biology, Orange Park, Florida; Yerkes Regional Primate Research Center, Atlanta, Georgia

Yerkes Laboratory of Primate Biology, Orange Park, Florida, 234–35, 313–16, 326

Yerkes Regional Primate Research Center, Atlanta, Georgia, 312–13, 329, 445n146

yo-heave-ho theory (of origin of language). See sympathic theory

yo-he-yo theory (of origin of language). See sympathic theory

Yorkshire Post, 409n1

Young, J. Z., 245, 271, 274–75

Zangwill, Oliver, 262, 282

Zeitschrift für Tierpsychologie, 255–56

Zola, Émile, 95

Zoloth, Steve, 360

Zoöglottology, 118

Zoological Gardens, London, 71, 147–48

Zuckerman, Solly, 284, 452n50, 470n78